First and Second Order Circuits and Equations

First and Second Order Circuits and Equations

Technical Background and Insights

Robert O'Rourke
USA

IEEE Press
WILEY

Published by John Wiley & Sons, Inc., Hoboken, New Jersey.
Published simultaneously in Canada.

Library of Congress Cataloging-in-Publication Data

Names: O'Rourke, Robert (Electronic engineer), author.
Title: First and second order circuits and equations : technical background and insights / Robert O'Rourke.
Description: Hoboken, New Jersey : Wiley, [2024] | Includes index.
Identifiers: LCCN 2024008985 (print) | LCCN 2024008986 (ebook) | ISBN 9781119913535 (hardback) | ISBN 9781119913542 (adobe pdf) | ISBN 9781119913559 (epub)
Subjects: LCSH: Electric circuits. | Electronic circuits. | Differential equations.
Classification: LCC TK454 .O75 2024 (print) | LCC TK454 (ebook) | DDC 621.381501/5153--dc23/eng/20240314
LC record available at https://lccn.loc.gov/2024008985
LC ebook record available at https://lccn.loc.gov/2024008986

Cover Design: Wiley
Cover Image: Courtesy of Robert O'Rourke

Set in 9.5/12.5pt STIXTwoText by Straive, Chennai, India

To my wife Evelyn, whose loving support made this book possible.

Contents

About the Author

Robert O'Rourke has an electrical engineering degree from UC Berkeley, including coursework in circuit theory and electromagnetics. Robert has taught electronics and calculus at college level. Robert worked with circuit theory and circuit and system simulation at Analogy, now a part of Synopsys, Inc. During several positions in his career, Robert hired and managed electronic engineers in applications engineering roles that involved circuit theory. Robert has taught circuit and electromagnetic simulation training courses for multiple vendors. As part of Ansys, Robert also created circuit and electromagnetic simulation product training courses for the Ansys Learning Hub.

Acknowledgments

The simulations made for this book and the simulation schematics and graphs appearing in this book were made with the QUCS simulator, available from SourceForge.net.

The author would also like to thank Professor Alberto Sangiovanni-Vincentelli, who first taught him circuit theory at UC Berkeley.

Part 1

Circuit Elements and Resistive Circuits

1

Ohm's Law, Branch Relationships, and Sources

1.1 Chapter Summary and Polarity Reference

1.1.1 Chapter Summary

Ohm's law describes the relationship between voltage, current, and resistance for resistive circuits. This chapter describes Ohm's law and the related circuit elements, resistors, current sources, and voltage sources. This chapter also covers the more general idea of branch relationships, the relation between the voltage across a circuit element and the current through a circuit element, for resistors, voltage sources, and current sources.

Current source and resistor

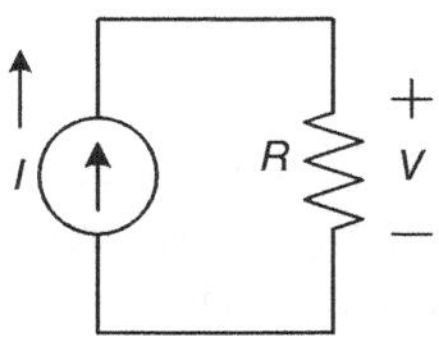

In Figure 1.1a, the current source drives current through the resistor causing a voltage drop across the resistor.

Figure 1.1a Current source driving a resistor.

The ideal independent ***current source*** supplies a fixed amount of current I through the resistor R regardless of the amount of voltage across the source.

The resistor R resists the flow of current through it. The amount of voltage V, that develops across the resistor R, as a result of the current I flowing through the resistor R, is determined by Ohm's law, V equals IR, shown in Equation 1.1a

Voltage source and resistor

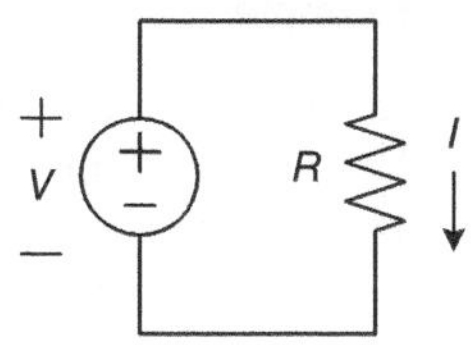

In Figure 1.1b, the voltage source across the resistor causes current to flow through the resistor.

Figure 1.1b Voltage source across a resistor.

The ideal independent ***voltage source*** supplies a fixed amount of voltage V across the resistor R regardless of the amount of current through the source.

The resistor R resists the flow of current through it. The amount of current, that flows through the resistor, as a result of the voltage V across the resistor R, is determined by Ohm's law, I equals V over R, shown in Equation 1.1b.

1.1.1.1 Ohm's Law

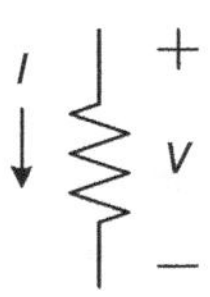

Figure 1.2 Resistor schematic symbol with passive sign convention.

The voltage across a resistor, shown in Figure 1.2, equals the current through the resistor times the resistance of the resistor.

$$V = IR \tag{1.1a}$$

The current through a resistor equals the voltage across the resistor divided by the resistance of the resistor.

$$I = V/R \tag{1.1b}$$

First and Second Order Circuits and Equations: Technical Background and Insights, First Edition. Robert O'Rourke.

1.1.1.2 Branch Relationships

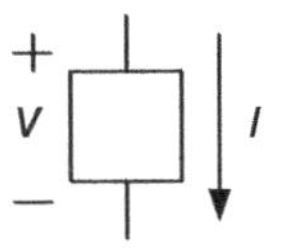

Figure 1.3 Passive sign convention on a generic fictitious schematic symbol.

The branch relationship is the equation describing the relationship between current through a circuit element (in a branch of a circuit) and the voltage across the circuit element. For example, Ohm's law $V = IR$ or $I = V/R$ is the branch relationship of a resistor. Figure 1.3, with a simple square shape, is a generic, non-specific circuit element.

1.1.2 Polarity Reference

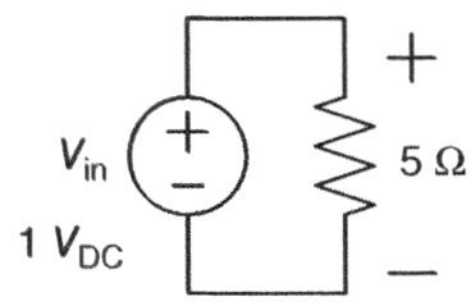

Figure 1.4a 1 volt DC voltage source across a 5 ohm resistor.

In Figure 1.4a, 1 V divided by 5 Ω equals 200 mA (0.2 A).

In each of these two direct current (DC) examples, Figures 1.4a and 1.4b, the current in the loop is 200 mA DC, going down through the resistor from + to −, and the voltage across the 5 Ω resistor is 1 V DC.

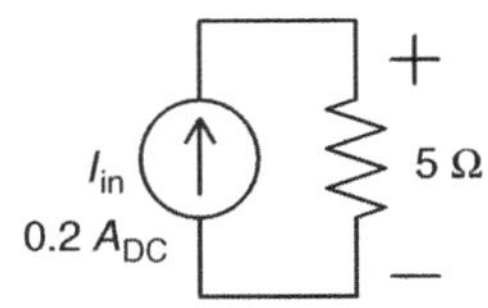

Figure 1.4b 200 mA DC current source driving a 5 ohm resistor.

In Figure 1.4b, 0.2 A multiplied by 5 Ω equals 1 V.

1.1.2.1 DC Voltage Source Polarity Example

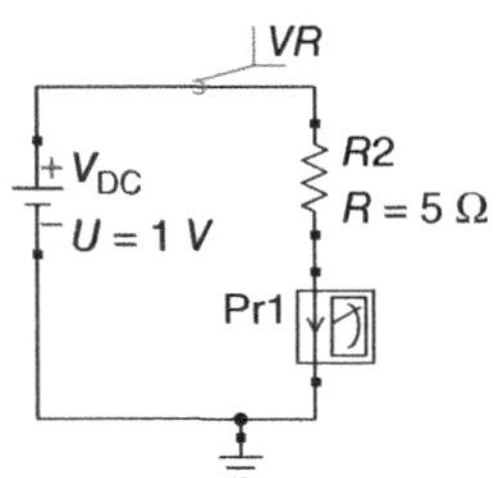

Number	Pr1.I	VR.V
1	0.2	1

Figure 1.5 Circuit simulation schematic with avoltage source and a resistor.

The voltage source in Figure 1.4a applies voltage across the resistor, and the + (plus) and − (minus) signs on the resistor indicate the polarity of the voltage.

Figure 1.5 shows a circuit simulation schematic corresponding to the circuit in Figure 1.4a. The voltage source symbol in Figure 1.5 is specific to DC voltage sources.

There is a current meter in the right-hand leg of the circuit, just below the resistor. The downward arrow in the current meter symbol indicates a reference direction pointing down; the current meter considers clockwise flow of current in the circuit to be positive.

For the DC circuit simulation, there is a table of results in Figure 1.5. VR, measured at the top of the circuit, is positive 1 V, and the current Pr1.I, measured by the current meter, is 200 mA, verifying that current flows downward through the resistor.

1.1.2.2 DC Current Source Polarity Example

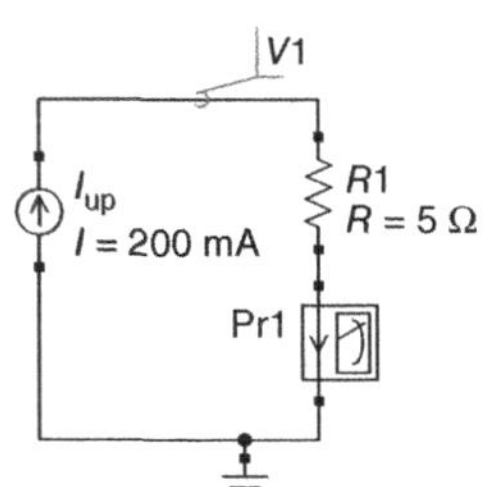

Number	Pr1.I	V1.V
1	0.2	1

Figure 1.6 Circuit simulation schematic with a current source and a resistor.

The arrow, pointing up in the current source in Figure 1.4b, indicates the direction of a positive current from the source. Applying Ohm's law, multiplying the current times the resistance, tells us the voltage across the resistor, both the amount and the polarity of the voltage. The + (plus) and − (minus) signs on the resistor indicate the polarity of the voltage.

Figure 1.6 shows a circuit simulation schematic corresponding to the circuit in Figure 1.4b. The downward arrow in the current meter symbol indicates a reference direction pointing down; the current meter considers clockwise flow of current in the circuit to be positive.

In Figure 1.6, the current source arrow points up, telling us that current flows up and out of the current source, and then down through the resistor. As expected from Ohm's law, 200 mA multiplied by 5 Ω yields 1 mV across the resistor. This 1 V result appears in the table under V1.V. The current direction of the current source and the current meter are the same, and the measured current, under Pr1.I, is positive 200 mA.

At the top of the schematic in Figure 1.6, there is a V1 marker, indicating a voltage measurement. This voltage is referenced to ground, so it is equivalent to the way schematics show a + (plus) sign above the resistor and a – (minus) sign below the resistor.

In Figure 1.7 the current source arrow points down, telling us that current flows down and out of the current source, and then up through the resistor, the opposite direction from the current source in Figure 1.6.

This makes the measured voltage V2.V –1 V, as it is referenced from the top V2 voltage reference.

As expected from Ohm's law, –0.2 A multiplied by 5 Ω yields – 1 V across the resistor. This result appears under V2.V in units of volts.

In Figure 1.8, the current source arrow indicates counterclockwise circulation of current around the loop. The meter, pointing the same direction in Figure 1.8 as it does in Figure 1.7, indicates opposite circulation direction from the current source and correspondingly the current measured under Pr2.1 is negative 0.2 A.

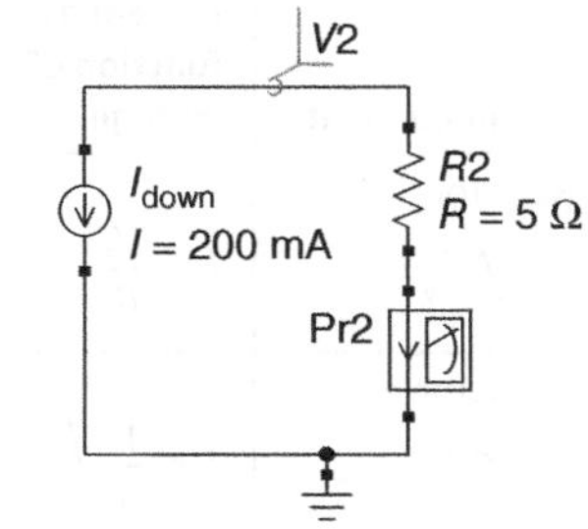

Number	Pr2.1	V2.V
1	-0.2	-1

Figure 1.7 Circuit simulation schematic with a DC current source, pointed down, and a resistor.

1.1.2.3 Reference Polarity versus Physical Current Flow Direction

It is important to distinguish between actual current flow direction and reference polarity. If we reversed the direction of the current meter in Figure 1.7, the simulation would indicate a positive current value for Pr2.1, but the current still flows counterclockwise in the circuit.

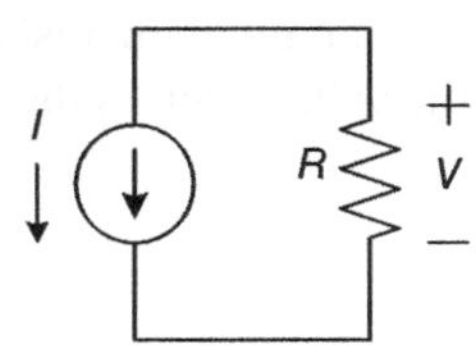

Figure 1.8 Schematic with a current source, pointed up, and a resistor.

The arrow pointing down, in the current source in Figure 1.8, indicates positive current going down from the current source and then up through the resistor. Using the same + (plus) and – (minus) signs on the resistor as a reference polarity for the resistor, we would get a negative voltage.

In Figure 1.7 the reference direction for the current is down, indicated by an arrow, in both schematics (Figures 1.7 and 1.8). The current polarity will be measured relative to this counter clockwise reference direction.

1.2 Branch Relationships and *I*–*V* Characteristics

1.2.1 Circuit Element Branch Relationships

1.2.1.1 Ohm's Law is a Resistor's Branch Relationship

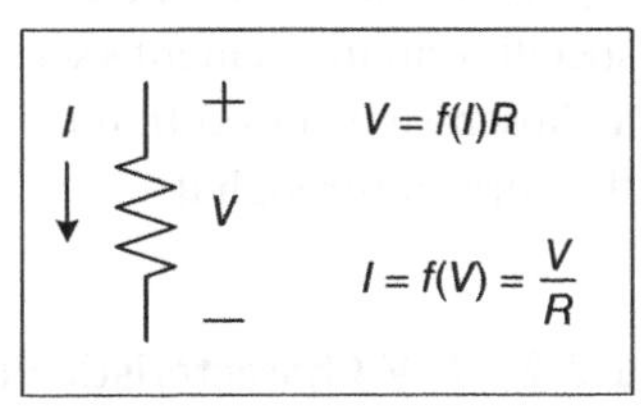

Figure 1.9 Resistor schematic diagram and ohm's law expressions where current or voltage are dependent.

Ohm's law $V = IR$, describing the behavior of a resistor, is the branch relationship for a resistor. Figure 1.9 shows how voltage V can be expressed as a function of I (voltage as a dependent variable) or current can be expressed as a function of V (current as the dependent variable).

1.2.1.2 Capacitor and Inductor Branch Relationships

The branch relationships for capacitors and inductors involve first derivatives and integrals, as shown in Figure 1.10. The branch relationships for capacitors and inductors will be studied in more detail in subsequent chapters on reactive elements.

Branch Relationships for R, L, and C

Circuit element	Current as a function of voltage	Voltage as a function of current
Resistor	$I = \frac{V}{R}$	$V = IR$
Inductor	$I = \frac{1}{L}\int_0^T V\,dt + I_0$	$V = L\frac{di}{dt}$
Capacitor	$I = C\frac{dV}{dt}$	$V = \frac{1}{C}\int_0^T I dt + V_0$

Figure 1.10 Table showing schematic symbols and branch relationships for resistor, inductor, and capacitor.

A circuit elements' defining equation (model) typically describes the current through the element in terms of the voltage across the element.

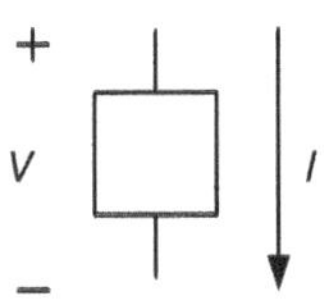

Figure 1.11a Fictitious generic circuit schematic symbol.

Figure 1.11a shows an example of a generic (non-specific) circuit element's schematic symbol, including a chosen sign convention for the voltage across it and the current through it.

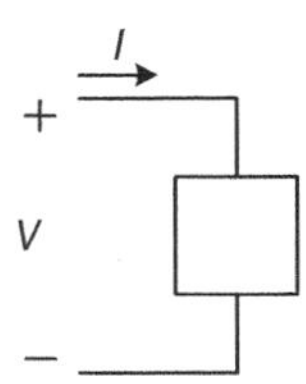

Figure 1.11b Fictitious generic circuit schematic symbol.

Drawn differently than Figure 1.11a, Figure 1.11b shows the same sign convention as Figure 1.11a for the voltage across a circuit element and the current through it that make up an I–V characteristic.

1.2.2 *I*–*V* Characteristic Plots

The branch relationship is sometimes called the I–V (current-voltage) characteristic of a circuit element. The I–V characteristic can be plotted with current on the vertical (dependent) axis and voltage on the horizontal (independent) axis. Some publications put voltage the vertical axis and current on the horizontal axis.

Resistors obeying Ohm's law are linear devices and this linearity appears as a straight line in the I–V characteristic.

1.2.2.1 *I–V* Characteristics for Circuit Elements

An *I–V* characteristic of a circuit element is the equation and graph of the current through the circuit element as a function of the voltage across the circuit element. (The shape of the curve in Figure 1.12 is a fictitious generic example.)

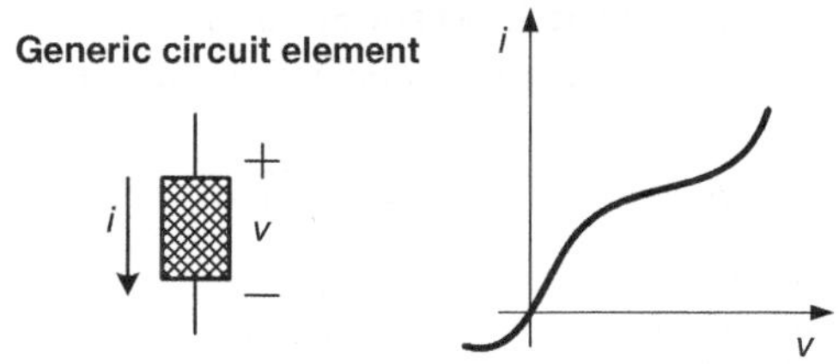

Figure 1.12 Fictitious generic circuit schematic symbol.

1.2.2.2 Resistor *I–V* Characteristic

A linear resistor's *I–V* characteristic is Ohm's law *v* equals *iR*. The *I–V* characteristic line, shown in Figure 1.13, is straight because there is a linear relationship between current and voltage. The slope of the line is 1 over *R* (1 over *R* equals *G* equals conductance).

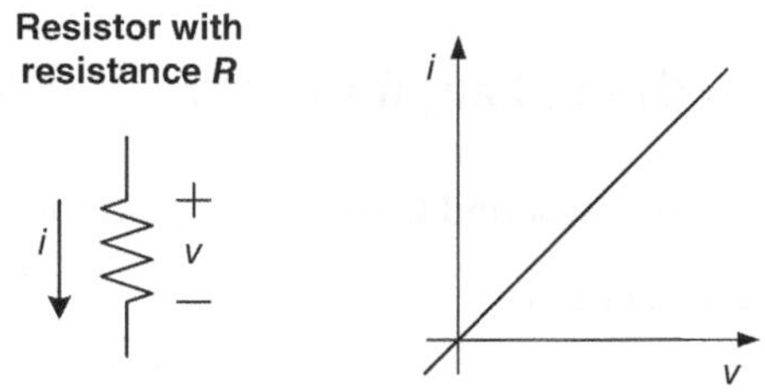

Figure 1.13 Resistor schematic symbol, marked with passive sign convention, adjacent to its linear I-V characteristic graph.

1.2.2.3 Non-resistor *I–V* Example

Diodes have a nonlinear exponential relationship between the voltage across them and the current through them, shown in Figure 1.14 and given by Equation 1.2.

$$i(v) = i_s e^{(qv/kT-1)} \tag{1.2}$$

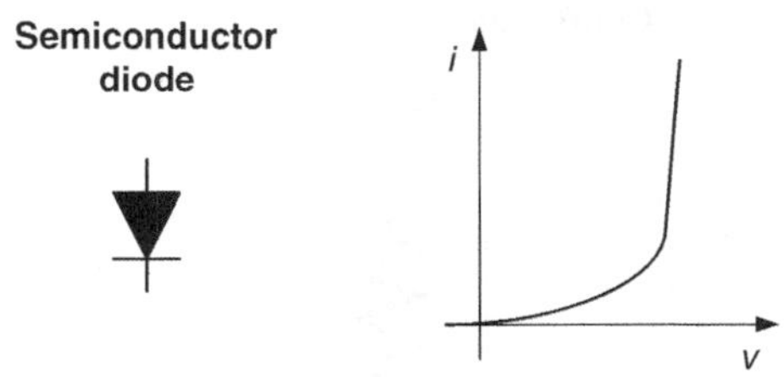

Figure 1.14 Diode schematic symbol and nonlinear *IV* characteristic.

1.2.3 Circuit Elements Models and Schematic Symbols

1.2.3.1 Circuit Elements are Model Descriptions and Schematic Symbols

In electronics, a circuit element is a description (model) and a schematic symbol of a physical device in an electric or electronic circuit, as shown in Figure 1.15 for a resistor. Resistors, capacitors, inductors, opamps (operational amplifiers), transistors, and current sources are all examples of circuit elements. Branch relationships represent the defining model of a lumped circuit element such as a resistor, inductor, or diode.

Figure 1.15 Resistor circuit element visual summary including the resistor schematic symbol, a drawing of a chip resistor, and Ohm's law, the branch relationship, and ideal lumped element modeling equation for a resistor.

1.2.3.2 Circuit Schematic Symbols

A circuit schematic symbol represents the circuit element in a schematic of the circuit, as shown in Figure 1.16.

Figure 1.16 Examples of common circuit element schematic symbols.

1.3 Ohm's Law, Resistance, and Resistors

1.3.1 Resistor and Conductor Equations

1.3.1.1 Resistors

Figure 1.17 Resistor schematic symbol.

A resistor, shown in Figure 1.17, is a circuit element which resists the flow of electrical current. An applied voltage is required in order to pass current through a resistor. The resistance R is the ratio of the voltage applied to the current that flows through the resistor. This ratio relationship is called Ohm's law. The unit of resistance is the ohm (Ω).

1.3.1.2 Ohm's Law

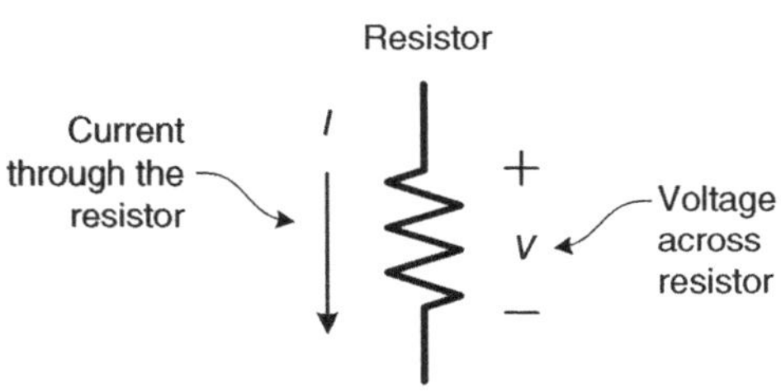

Figure 1.18 Resistor schematic symbol labeled with passive sign convention for visualizing Ohm's law.

Ohm's law states that the voltage across a resistor, shown in Figure 1.18, equals the current through the resistor times the resistance R of the resistor.

Equivalently, Ohm's law states that the current through the resistor equals the voltage across the resistor divided by the resistance R.

$$V = IR \tag{1.3}$$

$$IG = \frac{V}{R} = GV \tag{1.4}$$

$$R = \frac{V}{I} \tag{1.5}$$

Equations 1.3, 1.4, and 1.5 are all statements of Ohm's law, which describes the behavior of a resistor. V is the voltage across the resistor. I is the current through the resistor. R is the resistance and G is conductance, the inverse of resistance, shown in Equation 1.6. The unit of conductance is the siemen. It used to be called the mho.

Ohm's law assumes that the resistor is linear, that is, that the resistance of the resistor does not vary as a function of time, current, or voltage.

$$G = \frac{1}{R} = \frac{I}{V} \tag{1.6}$$

1.3.1.3 Ohm's Law Notation with Time-Dependent Functions

Ohm's law is also valid for currents and voltages that vary as a function of time. (But the resistance is still constant with respect to time, voltage, and current.)

$$v = iR = f(t) = Ri(t) \quad (1.7)$$

$$i = \frac{v}{R} = f(t) = \frac{1}{R}v(t) \quad (1.8)$$

The notation $v = iR$ makes no indication of time; it applies to DC. The notations $f(t)$, $i(t)$, and $v(t)$ in Equations 1.7 and 1.8, indicate functions of time, as also shown in Figure 1.19.

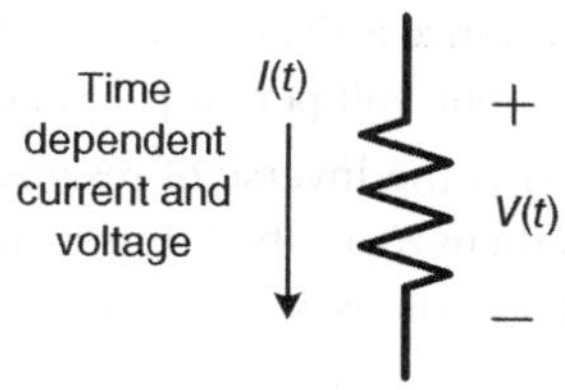

Figure 1.19 Resistor schematic symbol labeled with passive sign convention for visualizing Ohm's law with time-dependent functions explicitly annotated.

1.3.2 Impedance and Ohm's Law Beyond Resistive Circuits

Ohm's law applies beyond resistive circuits, where the ratio of the voltage function to the current function is more generally called impedance. For reactive circuits, which can include capacitors and inductors, we speak of the impedance. Resistance is one type of impedance, a subset of impedance.

Impedance is a generalization of resistance. Often a resistance is referred to in terms of impedance, such as "high impedance" for a large value of resistance.

$$V(j\omega) = I(j\omega)Z(j\omega) \quad (1.9)$$

$$I(j\omega) = \frac{V(j\omega)}{Z(j\omega)} \quad (1.10)$$

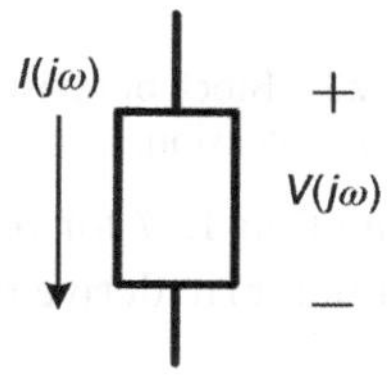

Equations 1.9 and 1.10 are both statements of Ohm's law in frequency domain.

$$Z(j\omega) = \frac{v(j\omega)}{i(j\omega)} \quad (1.11)$$

Equation 1.11 is the impedance in Fourier $j\omega$ frequency domain.

Figure 1.20 Generic circuit element schematic symbol labeled in terms of sinusoidal steady state frequency domain.

The rectangular symbol, shown in Figures 1.20 and 1.21 denotes a generic circuit element. It could be a capacitor, inductor, or a resistor. However, in European notation this rectangular symbol usually denotes a resistor.

$$V(s) = I(s)Z(s) \quad (1.12)$$

$$I = \frac{V(s)}{Z(s)} \quad (1.13)$$

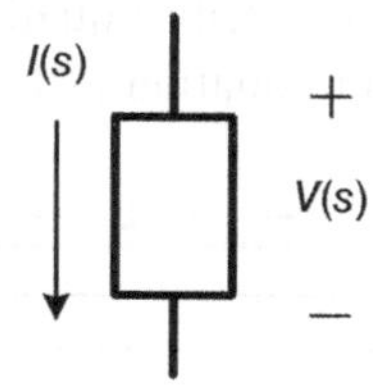

Equations 1.12 and 1.13 are both statements of Ohm's law in Laplace s-domain.

$$Z(s) = \frac{v(s)}{i(s)} \quad (1.14)$$

Equation 1.14 is the impedance in Laplace s frequency domain.

Figure 1.21 Generic circuit element schematic symbol labeled in terms of Laplace s-domain.

1.3.3 Resistance and Bulk Resistivity

Resistance is specific to an individual resistor or conductor, as shown in Figure 1.22. The resistance depends upon the size and shape and resistivity of the material. Resistance is **defined** by Ohm's law: R equals V over I.

Apply a voltage V across a conductor and measure the resulting current I. The resistance of the conductor is defined by Ohm's law to be V/I, the voltage across the conductor divided by the current through the conductor.

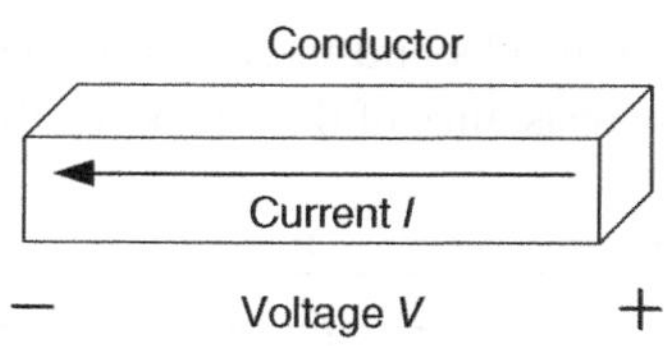

Figure 1.22 Block of resistive material labeled with current for visualizing resistance and Ohm's law.

The unit of resistance is the ohm (Ω). Equation 1.15 says that one ohm equals one volt per ampere of current.

$$1\ \Omega = \frac{1\ \text{V}}{1\ \text{A}} \tag{1.15}$$

Conductance G is the inverse of resistance. The unit of conductance is siemens S or mho (ohm spelled backward). The SI units symbol for siemens is an upside-down capital omega (℧).

$$G = \frac{1}{R} \tag{1.16}$$

$G =$ conductance

1.3.3.1 Bulk Resistivity and Resistance

Resistivity and conductivity are the resistive or conductive property of materials. Resistivity is often denoted by the Greek variable ρ (rho) which has the units of ohm-meter in SI units. Sometimes the term ***bulk*** resistivity is used. Conductivity is often denoted with the variable σ and the units of siemens per meter S/m.

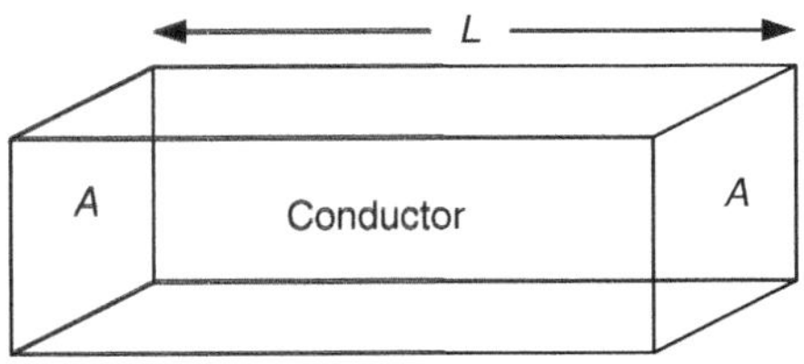

Figure 1.23 Block of resistive material for resistivity discussion.

In terms of the rectangular conductor, of uniform cross sectional area A, length L, and resistance R, shown in Figure 1.23, the resistivity, ρ (rho) is given in Equation 1.17.

$$\rho = \frac{R \cdot A}{L} \tag{1.17}$$

From this physical situation, the resistivity of a material could be calculated from the dimensions of the block and a measurement of the resistance of the block.

In Equation 1.17 for resistivity ρ, we can cancel one unit of distance (d) on the top (numerator) and bottom (denominator) to derive the units of resistivity as ohm-meter in SI units shown in Equation 1.18.

$$\text{Resistivity}: \rho = \frac{RA}{L} = \frac{Rd^2}{d} = \frac{\text{ohm} \cdot d^2}{d} = \Omega\text{m} \tag{1.18}$$

The conductivity of a material sigma σ, shown in Equation 1.19, is the inverse of the resistivity.

$$\text{Conductivity}: \sigma = \frac{1}{\rho} = \frac{L}{RA} = \frac{L}{RL^2} = \frac{d}{\text{ohm} \cdot d^2} = \text{S} - \text{m}^{-1} \tag{1.19}$$

The conductance G, in Equation 1.20, is the inverse of resistance.

$$G = \frac{1}{R} = \frac{A}{\rho\ L} \tag{1.20}$$

In the resistivity Equation 1.18, solve for resistance R, as shown in Equation 1.21.

$$R = \frac{\rho \cdot L}{A} \tag{1.21}$$

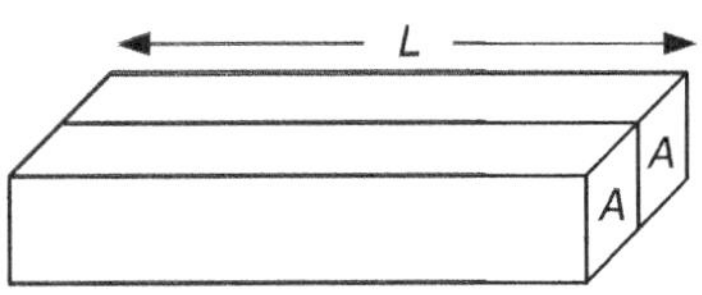

Figure 1.24 Two adjacent blocks of resistive material.

The resistance of the three-dimensional rectangular block is inversely proportional to the area A, and proportional to the length L. This makes intuitive sense because a larger area provides more parallel path for current, reducing overall resistance. Conversely, more length means every current path is longer, increasing the total resistance.

Joining two blocks side-by-side, shown in Figure 1.24, would double the area but maintain the same length, cutting the resistance of this block in half relative to the original rectangular block.

Joining two blocks end-to-end, in series, Figure 1.25, would double the length but maintain the same area, doubling the resistance of this block relative to the original rectangular block.

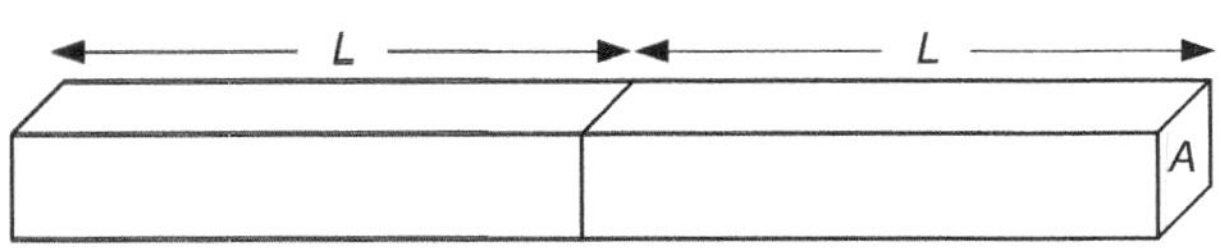

Figure 1.25 Two blocks of resistive material connected end-to-end.

Combining conductive blocks side-by-side or end-to-end and examining the combined resistance is analogous to the study of parallel and series combinations of lumped element resistors. Like side-by-side resistive blocks, two identical lumped element resistors in parallel have more path for current and therefore less resistance than one resistor. Like two resistive blocks end-to-end, two identical series resistor circuit elements have more resistance than one resistor alone.

1.4 Current, Voltage, and Sources Overview

1.4.1 Description and Units of Current and Voltage

Current is the amount of charge flowing as a function of time, as shown in Equation 1.22a. By convention, positive current is the movement of positive charge.

$$I = \frac{dq}{dt} \tag{1.22a}$$

One amp is defined as one coulomb per second.

Voltage is the electric potential difference and has the units of work (force times distance – the same units as energy) divided by charge, as shown in Equation 1.22b.

$$V = \frac{W}{q} \tag{1.22b}$$

One volt is defined as one joule per coulomb.

1.4.2 Current and Voltage Source Definitions

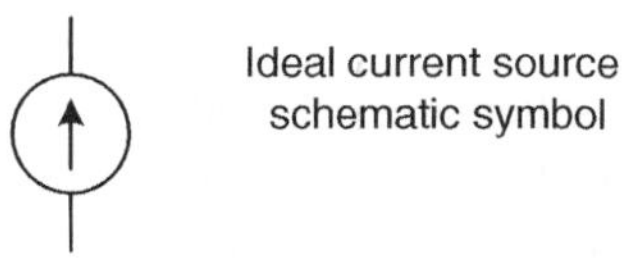

Figure 1.26a Ideal current source schematic symbol.

An ideal independent current source, Figure 1.26a, supplies the specified current regardless of the amount of voltage across the source.

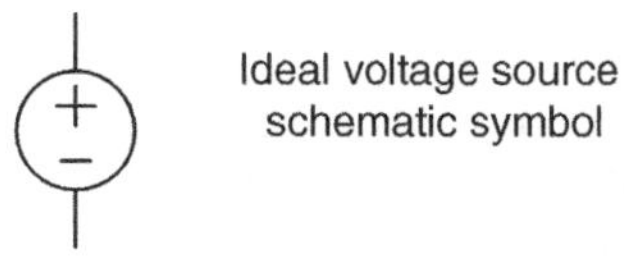

Figure 1.26b Ideal voltage source schematic symbol.

An ideal independent voltage source, Figure 1.26b, supplies the specified voltage regardless of the amount of current that goes through the source.

1.4.3 Current and Voltage Source Branch Relationships

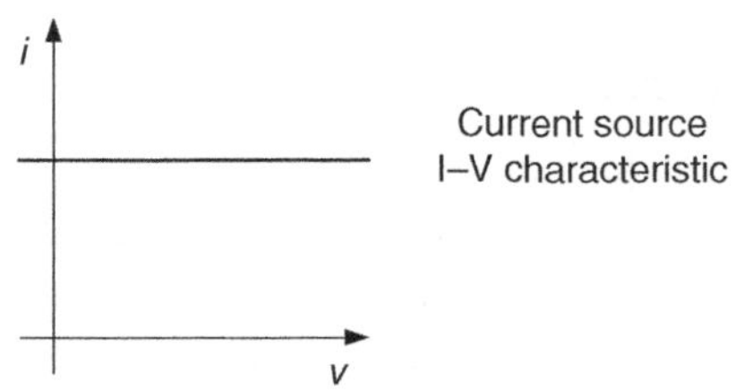

Figure 1.27a *IV* characteristic showing the branch relationship of an ideal current source.

The I–V characteristic of an ideal current source is a horizontal line, as shown in Figure 1.27a, indicating that it provides a constant current for any value of voltage across it.

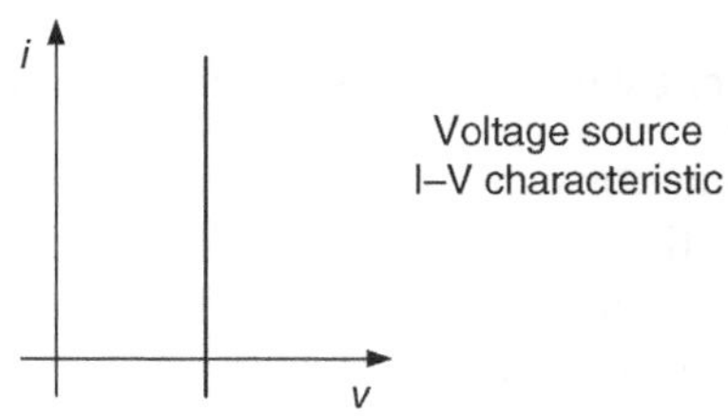

Figure 1.27b *IV* characteristic showing the branch relationship of an ideal voltage source.

The I–V characteristic of an ideal voltage source is a vertical line, as shown in Figure 1.27b, indicating that it provides a constant voltage for any value of current through it.

1.5 Voltage Sources

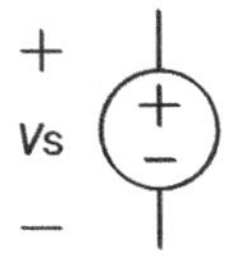

Figure 1.28 Independent ideal voltage source schematic symbol.

An ideal independent voltage source, Figure 1.28, supplies the specified voltage Vs regardless of the amount of current that goes through the source.

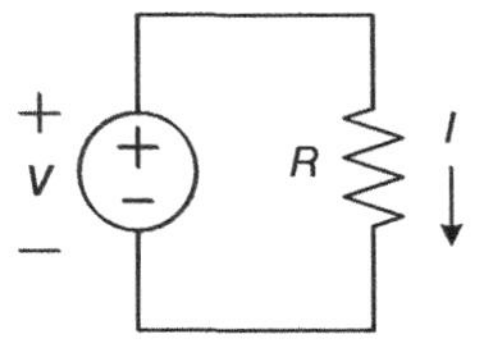

Figure 1.29 Schematic with voltage source across a resistor.

The current in Figure 1.29 could be any value in order to maintain the specified voltage across the ideal voltage source.

The voltage supplied by the source is *independent* from the amount of current flowing through the source and other circuit quantities. (There are also dependent or controlled voltage sources where the voltage depends upon another voltage or current in the circuit.)

1.5.1 Independent Voltage Source Schematic symbols

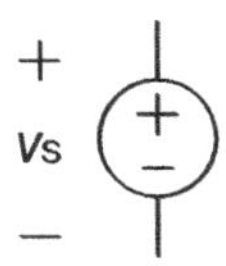

Figure 1.30 Independent ideal voltage source schematic symbol.

This commonly used schematic symbol, in Figure 1.30 can apply to any type of source, DC, AC, or even Laplace domain.

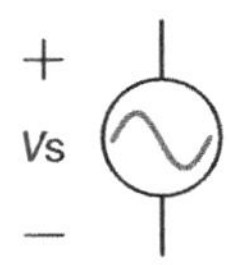

Figure 1.31 Independent ideal AC voltage source schematic symbol.

This schematic symbol in Figure 1.31 is often used for a sinusoidal steady-state (AC – alternating current) voltage source.

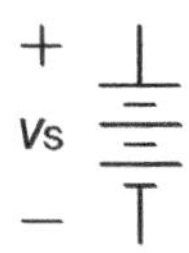

Figure 1.32 Independent ideal DC voltage source schematic symbol.

This schematic symbol, in Figure 1.32, is used for DC (direct current – zero frequency) independent voltage sources.

The S in these examples, in V_S, is common to indicate source voltage.

1.5.2 Independent Voltage Source Branch Relationship (*I*—*V* Characteristic)

For an independent constant (DC) voltage source, shown in Figure 1.33, the I–V characteristic, shown in Figure 1.34, is a vertical line, indicating that the voltage source provides a constant voltage for any value current through it. The $I-V$ characteristic of an ideal independent constant (DC) voltage source is a vertical line, indicating that it provides a constant voltage for any value of current through it. A zero value of voltage for the source would be the same as an short circuit. (Note that a short circuit is any amount of current and zero voltage.)

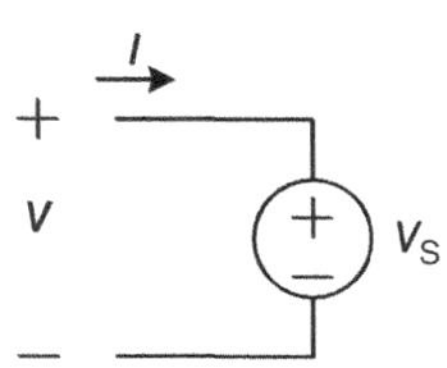

Figure 1.33 Independent ideal voltage source schematic symbol, labeled with reference polarity.

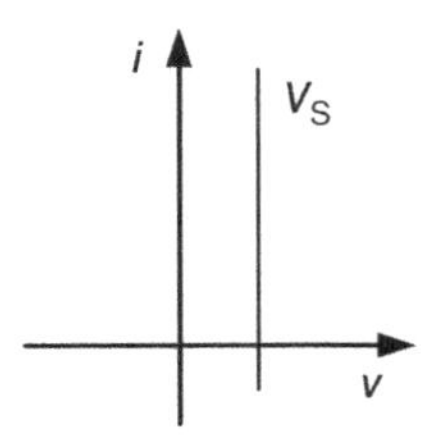

Figure 1.34 $I-V$ characteristic curve for independent voltage source with voltage value Vs.

Ideal voltage sources cannot be short circuited, as shown in Figure 1.35. Short-circuit means no voltage, which conflicts with the defined voltage value of an ideal source.

In real-world examples, short circuits occur and resulting large current flows can cause damage or injury.

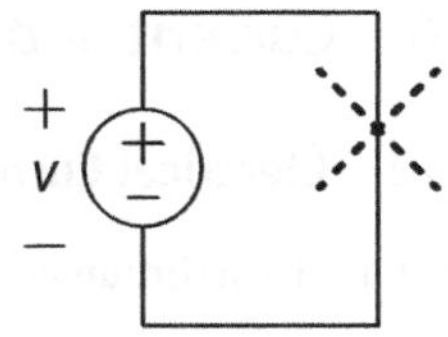

Figure 1.35 Short-circuited voltage source with dashed X to indicate that it is not allowed.

1.5.3 Intuitive Description of an Independent Voltage Source

An informal and intuitive description is that the voltage source "pushes apart" positive and negative charge, as shown in Figure 1.36. Opposite charges attract, so by pushing the opposite charges apart, the voltage source is supplying energy to the circuit.

The positive charge flows out of the positive terminal of the voltage source, and then through the resistor, creating the voltage across the resistor equal to the voltage source. The current flows from + voltage to – voltage through the resistor.

Think of the source as pushing the positive charges out of the top of the source and/or the negative charges out of the bottom of the source. The charges seek equilibrium by moving through the circuit, as shown in Figure 1.36. Electrical potential energy is lost going through the resistive circuit element (resistor). Figure 1.37 shows charges pushed out of a voltage source to capacitor plates.

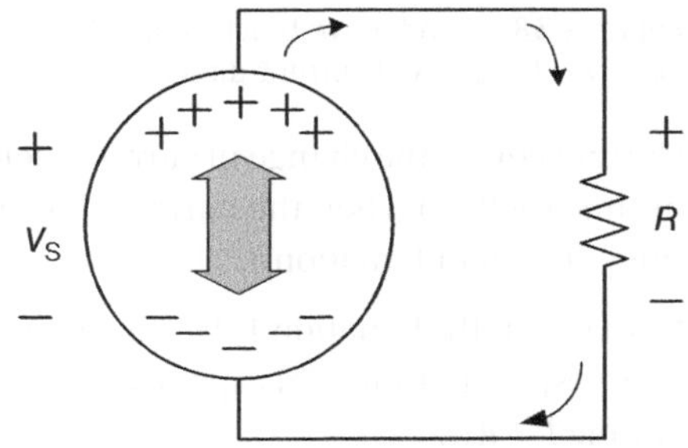

Figure 1.36 Voltage source embellished with separated charge to visualize voltage.

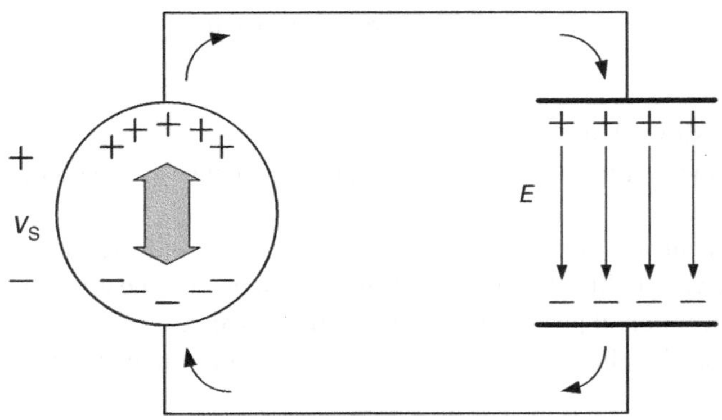

Figure 1.37 Drawing of a voltage source in parallel with a capacitor showing EMF inside the voltage source and the electric field that the voltage sets up on the capacitor.

$$\text{voltage } V = \Delta V \text{ potential} \tag{1.23}$$

Think of a source pushing apart the electrons as creating potential ΔV in Equation 1.23. Charge going through the circuit can dissipate that potential energy. A voltage source continuously creates or maintains that potential and current can flow continuously.

1.6 Current and Current Sources

1.6.1 Electrical Current

1.6.1.1 Description and Definition of Electrical Current

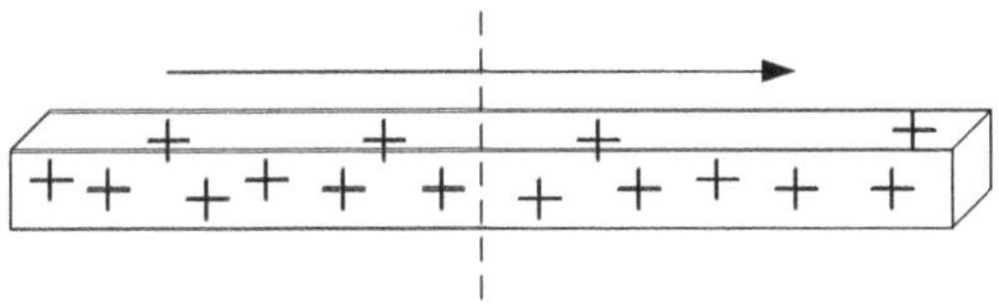

Figure 1.38 Conductive block, labeled with positive charges and current direction arrow.

Electrical current is net electrical charge in motion. Current is the amount of charge flowing as a function of time. An applied voltage or electrical potential difference makes charges flow from the higher potential towards the lower potential.

Figure 1.38 shows positive charges in a conductor flowing to the right across a dotted line indicating a plane through the conductor.

Current equals the change in current (delta q) divided by the change in time (delta t) when the current flow is constant as a function of time, shown in Equation 1.24.

$$I = \frac{\Delta q}{\Delta t} \tag{1.24}$$

Most generally, Equation 1.25, current is the derivative of the charge with respect to time. This applies when current varies as a function of time $I = i(t)$.

$$I = \frac{dq}{d} \tag{1.25}$$

1.6.1.2 Units of Current and Charge

The unit of current is the ampere, or the amp for short. One ampere is 1 coulomb of charge per second, shown in Equation 1.26.

$$1\ \text{A} = \frac{1\ \text{C}}{\text{s}} \tag{1.26}$$

One coulomb C of negative charge equals 6.25 times 10 to the 18 electrons, shown in Equation 1.27.

$$1\ \text{C} = 6.25 \cdot 10^{18} \tag{1.27}$$

The reciprocal of one coulomb gives us the amount of charge on one electron in terms of coulombs, as shown in Equation 1.28.

$$1\ \text{electron} = 0.16 \times 10^{-18}\ \text{C} \tag{1.28}$$

1.6.1.3 Positive Current Convention

By convention, positive current is the movement of positive charge.

This means that we assign positive mathematical value to current from positive valued charge flow as a reference convention.

This does not mean that electrical current consists of only positive charges. In common metals it is the negatively charged electrons that actually move and produce current. In this case of electrons moving in metals, the direction of electron current is negative with respect to our convention of positive value for positive current.

A reference doesn't imply a description of the physical reality; a reference tells you when the value is plus or minus. It might also tell you where zero is.

1.6.2 Current Sources

1.6.2.1 Independent Current Source Definition

Figure 1.39 Independent current source schematic symbol.

An ideal independent current source, Figure 1.39, supplies the specified current regardless of the amount of voltage across the source.

The direction of the current arrow in the current source symbol specifies the direction of positive current flow.

Note that depending upon the circuit, the voltage across the current source may end up positive or negative.

The current supplied by the source is *independent* from the amount of voltage across the source and other circuit quantities. (By comparison, there are also dependent or controlled voltage sources where the voltage depends upon another voltage or current in the circuit.)

This voltage across the resistor and the current source, in Figure 1.40, could be any value in order to maintain the specified current through the ideal current source.

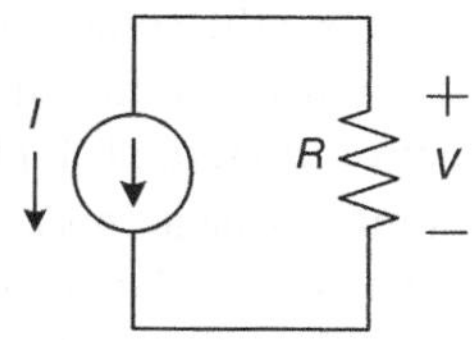

Figure 1.40 Circuit schematic with current source and resistor.

1.6.2.2 Alternative Independent Current Source Schematic Symbol

This is another schematic symbol, in Figure 1.41, sometimes used to designate an independent current source.

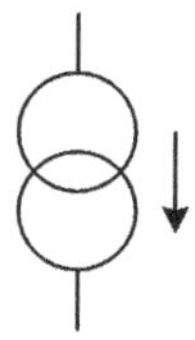

Figure 1.41 An alternative circuit schematic symbol sometimes used for current sources.

1.6.2.3 Current Source Water Analogy

Flowing water is analogous to electrical current and water pressure is analogous to electrical voltage. You can think of a current source as a pump that continuously sends a constant amount of water flow out the end of a pipe or hose. In the analogy to a current source, regardless of where or how you might connect that water pump to various plumbing, the same amount of water will always flow out of the water source per second. Regardless of how you connect an ideal independent current source in a schematic, the specified amount of current flows from the source.

1.6.2.4 Current Sources Can't Be Open Circuit

This open-circuited source is disconnected from the rest of the circuit at one end.

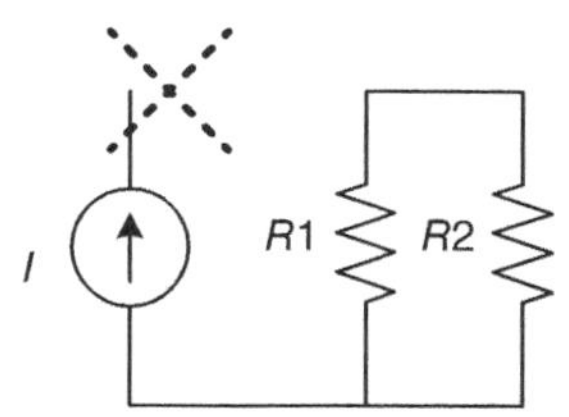

Figure 1.42 Resistive circuit schematic with a current source disconnected at one end.

Since the current source always generates the specified amount of current, you can't have an open-circuited source. The current has no place to go as indicated by the X in Figure 1.42.

1.6.3 Current Source Branch Relationship (I–V Characteristic)

For an ideal constant current source, shown in Figure 1.43, the I–V characteristic, shown in Figure 1.44 is a horizontal line, which indicates that the ideal current source provides a constant current for any value of voltage across the current source. A zero value of current for the source, putting the horizontal current line on the voltage axis, would be the same as an open circuit; no amount of voltage would cause current to flow.

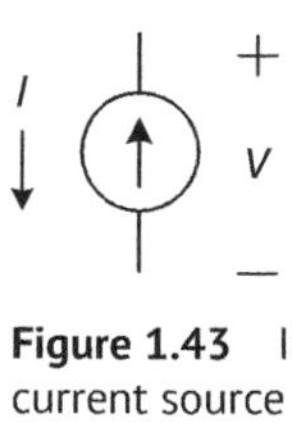

Figure 1.43 Independent current source labeled with reference polarity.

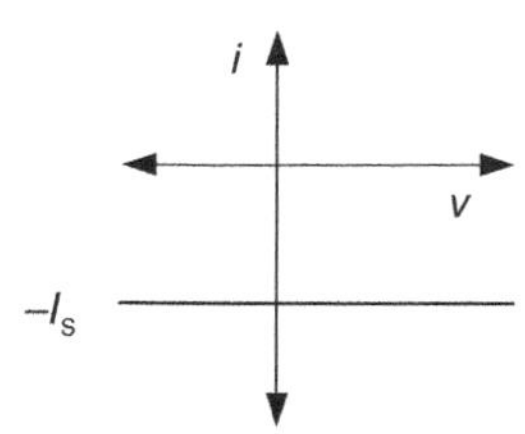

Figure 1.44 *IV* characteristic and branch relationship of an ideal constant current source.

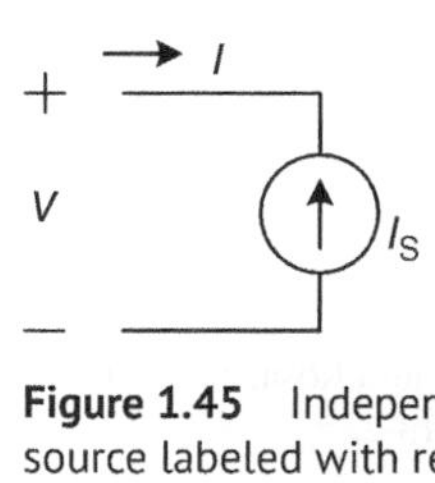

Figure 1.45 Independent current source labeled with reference polarity.

Notice the voltage and current sign conventions for the current source in Figures 1.43 and 1.45, with I flowing down through the current source, opposite the direction of the arrow in the current source symbol. By this sign convention, the I–V characteristic is a negative value of current I_S, as shown in Figure 1.44. Also note that on the summary page this I–V characteristic was shown generically as positive; no polarity was given.

Some other books, such as Thomas and Rosa,[1] use a reversed sign convention, shown in Figure 1.46, on the I–V characteristic, in order to get a positive value of I on the I–V graph, shown in Figure 1.47.

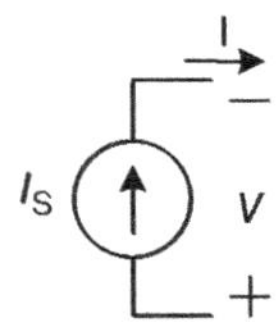

Figure 1.46 Independent current source labeled with reference polarity.

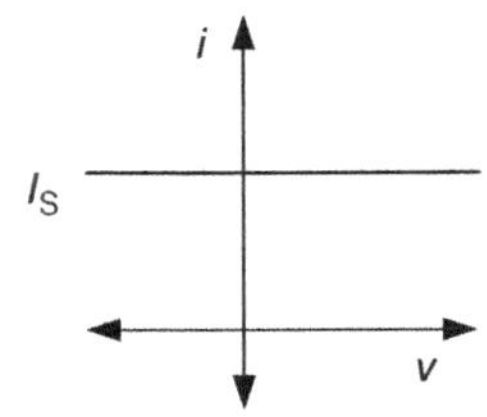

Figure 1.47 *I*–*V* characteristic for an independent current source of value I_s.

1 Thomas, R.E. and Rosa, c., A.J. (ed.) (2004). *The Analysis and Design of Linear Circuits*, Fourthe, by, 19. John Wiley & Sons, Inc., ISBN 0-471-27213-2.

2

Kirchhoff's Laws and Resistive Dividers

2.1 Kirchhoff's Laws and Dividers Comparison Summary

Kirchhoff's voltage law (KVL)

Kirchhoff's voltage law states that the algebraic sum of the voltage drops around a loop is zero (Figure 2.1a and Equation 2.1a).

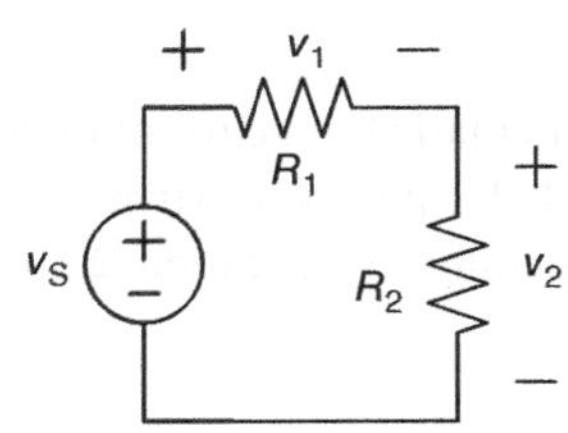

Figure 2.1a Voltage divider circuit marked with voltage polarities corresponding to the equation $v_s = v_1 + v_2$.

$$-v_S + v_1 + v_2 = 0 \tag{2.1a}$$

Kirchhoff's current law (KCL)

Kirchhoff's current law states that the algebraic sum of all the currents entering or leaving a node is zero (Figure 2.1b and Equation 2.1b).

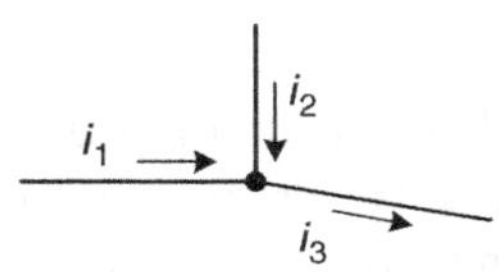

Figure 2.1b Junction of three wires with two currents marked as entering the node.

$$i_1 + i_2 - i_3 = 0 \tag{2.1b}$$

Resistors in series

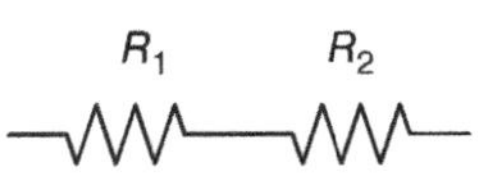

Figure 2.2a Two resistors in series.

$$R_S = R_1 + R_2 \tag{2.2a}$$

R_S in Equation 2.2a is the resistance of two resistors in series (Figure 2.2a).

Resistors in parallel

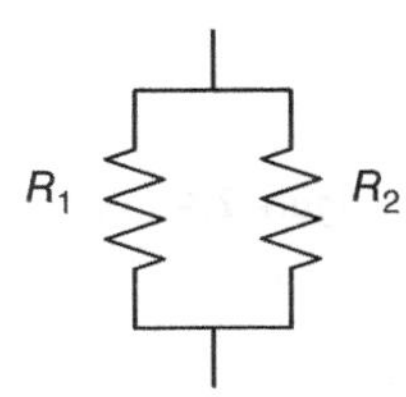

Figure 2.2b Two resistors in parallel.

$$R_P = \frac{R_1 R_2}{R_1 + R_2} \tag{2.2b}$$

R_P in Equation 2.2b is the resistance of two resistors in parallel (Figure 2.2b).

First and Second Order Circuits and Equations: Technical Background and Insights, First Edition. Robert O'Rourke.

These series and parallel resistance formulae, in Equations 2.2a and 2.2b, generalize to impedances Z in Equations 2.3a and 2.3b.

$$Z_S = Z_1 + Z_2 \tag{2.3a}$$

Z_S in Equation 2.3a is the impedance of two circuit elements in series.

$$Z_P = \frac{Z_1 Z_2}{Z_1 + Z_2} \tag{2.3b}$$

Z_P in Equation 2.3b is the impedance of two circuit elements in parallel.

Voltage divider

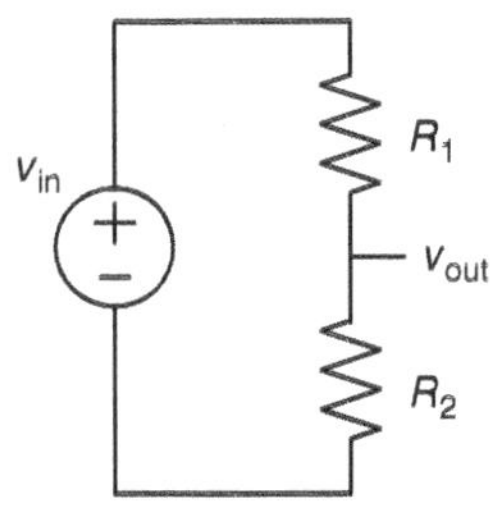

Figure 2.3a Resistive voltage divider circuit.

$$\frac{v_{out}}{v_{in}} = \frac{R_2}{R_1 + R_2} \tag{2.4a}$$

$$\frac{v_{out}}{v_{in}} = \frac{Z_2}{Z_1 + Z_2} \tag{2.5a}$$

Figure 2.3a shows the schematic for a voltage divider, described by Equation 2.4a for resistive circuits and described more generally by Equation 2.5a for impedances.

Current divider

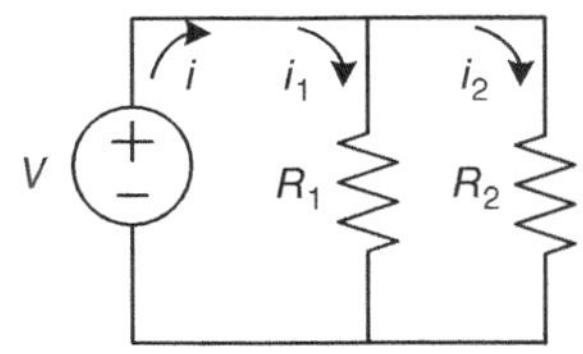

Figure 2.3b Resistive current divider circuit.

$$i_1 = i\frac{R_2}{(R_1 + R_2)} \tag{2.4b}$$

$$i_2 = i\frac{Z_1}{(Z_1 + Z_2)} \tag{2.5b}$$

Figure 2.3b shows the schematic for a current divider, described by Equation 2.4b for purely resistive circuits and described more generally by Equation 2.5b for impedances.

2.2 Kirchhoff's Laws Physical Analogies

2.2.1 Kirchhoff's Voltage Law – Gravitational (Elevation) Analogy

Visualize voltage like elevation on the surface of the earth. At each elevation there is a certain gravitational potential energy. A path can go up or down as it moves around the surface of the earth, as shown by the dashed line in Figure 2.4. If the path comes back to the original location, then the end of the path is at the same elevation, and therefore the same gravitational potential energy, as the beginning of the path as indicated by the X in Figure 2.4.

Figure 2.4 Got skipped in numbering.

2.2.1.1 Voltage Doesn't Accumulate Around a Loop

You can't traverse the circuit loop, starting at the ground symbol and come back around to ground, and have accumulated voltage (Figure 2.5). Instead, the voltages must sum to zero. Voltage works like in the gravitational analogy where you can't walk around in a loop and come back to the same point and have it be at a different elevation (Figure 2.4).

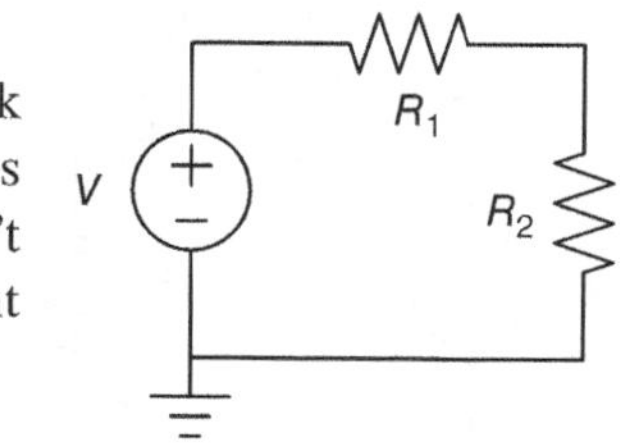

Figure 2.5 Two-resistor circuit for visualizing voltage loops.

2.2.2 Kirchhoff's Current Law – Water Analogy

Visualize current flowing in circuit wires like water flowing in pipes. At a pipe junction, water doesn't accumulate. Water can only accumulate in a tank or against a flexible membrane, something that is not an ideal pipe.

2.2.2.1 Current Doesn't Accumulate in a Node

A circuit node is a junction of ideal wires. Current can't accumulate charge in an ideal wire or any junction of ideal wires. If charge did accumulate, then the wire would be a capacitor, not an ideal wire (Figure 2.6).

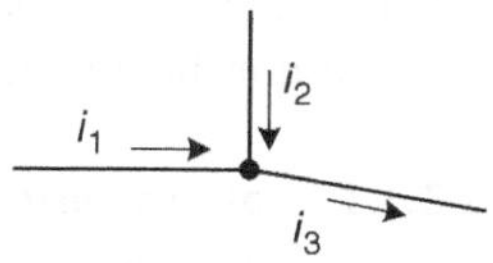

Figure 2.6 Junction of three wires with currents marked as entering the node.

2.2.2.2 Water and Current Flow in a Loop

You can have a pump moving water through a loop of piping as long as it continues to flow. The water flow can't just stop at the end of the pipe (unless there is a tank there for accumulating the water). Similarly, a current source can move current through a conductor loop as long as the charge keeps moving. Current can't flow out to the end of a wire and stop (unless there is a capacitor there for accumulating current) (Figures 2.7 and 2.8).

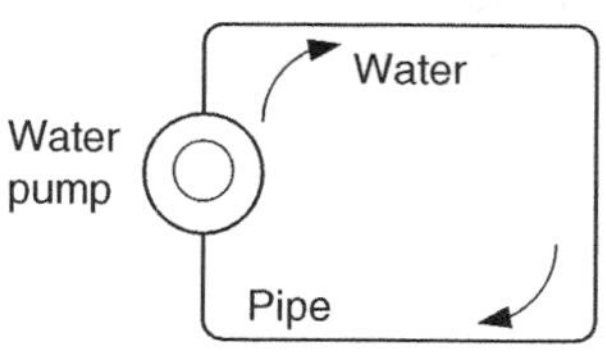

Figure 2.7 Water pump and pipe loop diagram for comparison to current flow.

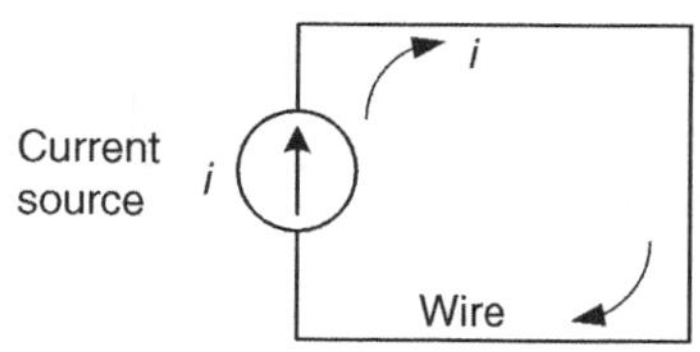

Figure 2.8 Current source and ideal wire for demonstrating current flow only in loops.

2.3 Source Polarity in KVL – Time and Frequency Domains

2.3.1 DC Source Polarity and Kirchhoff's Voltage Law

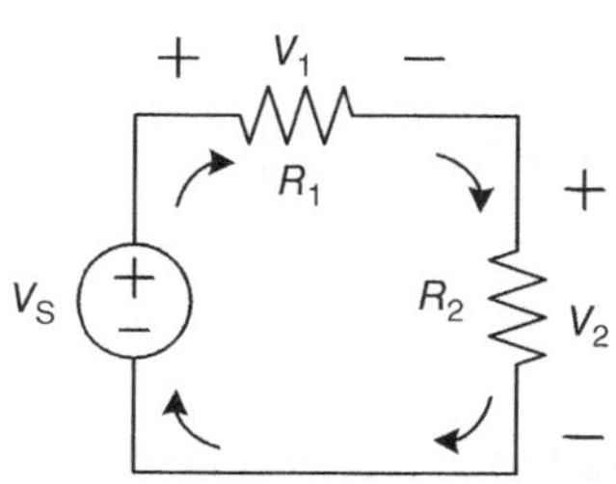

Figure 2.9 Voltage divider circuit marked with voltage polarities corresponding to the equation $v_s = v_1 + v_2$.

In Figure 2.9, the arrows indicate a clockwise direction around the loop to apply Kirchhoff's voltage law (KVL). The path goes up through the voltage source, against its reference polarity, so the source voltage is negative in the KVL Equation 2.6.

$$-v_S + v_1 + v_2 = 0 \tag{2.6}$$

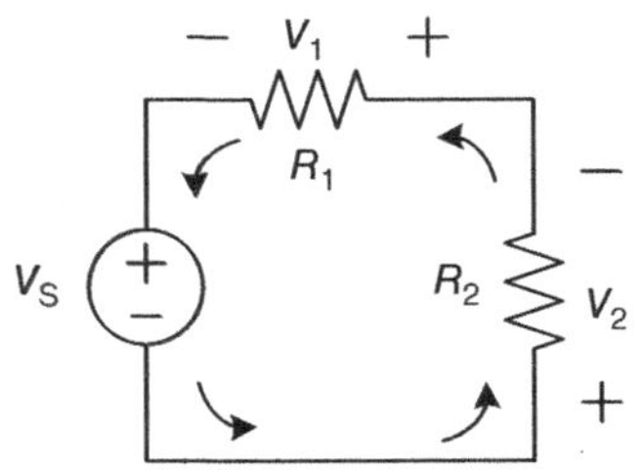

Figure 2.10 Voltage divider circuit marked with voltage polarities corresponding to the equation $v_s = v_1 + v_2$.

In Figure 2.10, the arrows indicate the counterclockwise direction around the loop to apply KVL. The path goes down through the voltage source, with its reference polarity, so the source voltage and the two resistor polarities are positive in the KVL Equation 2.7.

$$v_S + v_1 + v_2 = 0 \tag{2.7}$$

It is important to keep track of the polarities being used, but the choice of reference polarity won't change the amount nor the direction of current flow.

2.3.1.1 Positive Current Convention Distinction

In electrical engineering, current is considered to be the flow of positive charge by convention. This may differ from a physical reality, such as electrons flowing in a wire.

This positive current convention is also distinct from the source polarity modeling discussed here.

2.3.2 Time Domain Current Source Reference Polarity Example

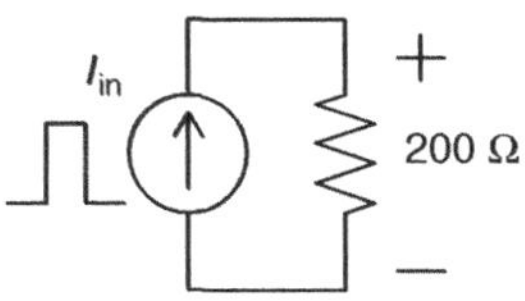

Figure 2.11 Current source and resistor schematic labeled with voltage reference polarity.

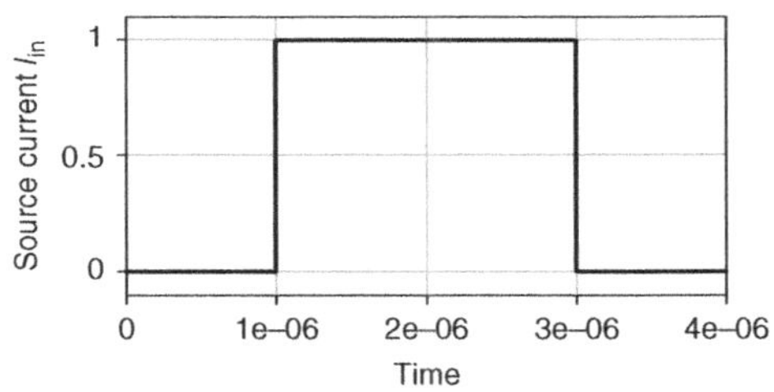

Figure 2.12 Circuit simulation result from current square wave circuit showing positive voltage.

The 1 A current pulse, I_{in} in Figures 2.11 and 2.12, starts at 1 μs and lasts for 2 μs, returning to zero at 3 μs.

Assume that the arrow on the current source in the schematic in Figure 2.11 points in the direction of positive current supplied by the source. This reference polarity corresponds to current flowing vertically down through the resistor, creating a voltage drop across the resistor. We can check this reference polarity assumption with circuit simulation.

Because there are only two circuit elements in Figure 2.14, the current source and the resistor, they are in parallel and have the same single voltage is across both the current source and the resistor.

2.3.2.1 Circuit Simulation Results

The "V" at the top of the simulation schematic in Figure 2.13 indicates a voltage label; V.Vt in this simulation corresponds to the + and − across the resistor. Pr1 is a current meter, with the arrow pointed down, in the same clockwise circulation direction as the current source.

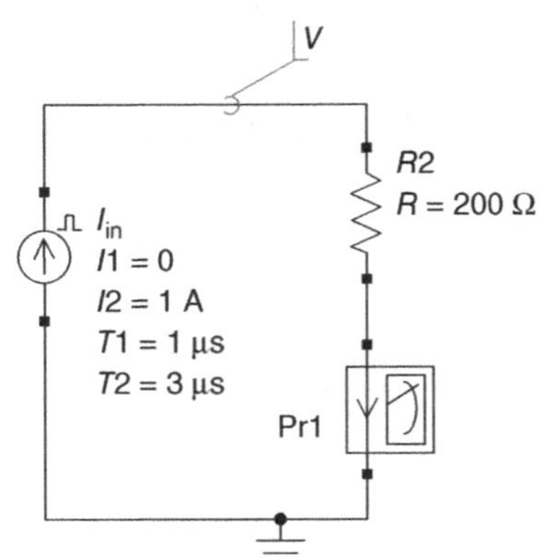

Figure 2.13 Circuit simulation schematic with current source and current meter for reference polarity.

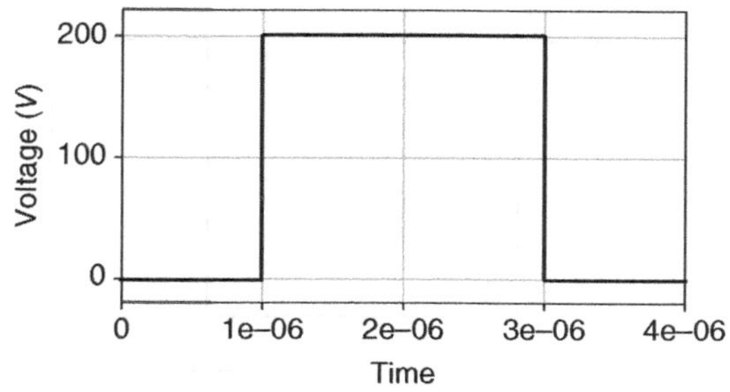

Figure 2.14 Circuit simulation result voltage plot showing voltage polarity.

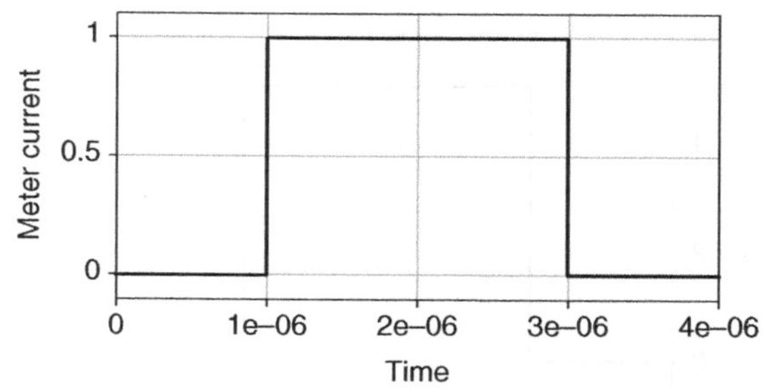

Figure 2.15 Circuit simulation result current plot showing current polarity.

In Figure 2.14, the voltage V, across the current source and the resistor, is V.Vt and it is a positive 200 V as we expect from Ohm's law $V = IR$, corresponding to current flowing down through the resistor.

In Figure 2.15 the current through the meter is positive 1 A. This confirms the assumed reference polarity of the current source, as positive current flowing out of the top of the current source.

2.3.3 Time Domain Voltage Source Reference Polarity Example

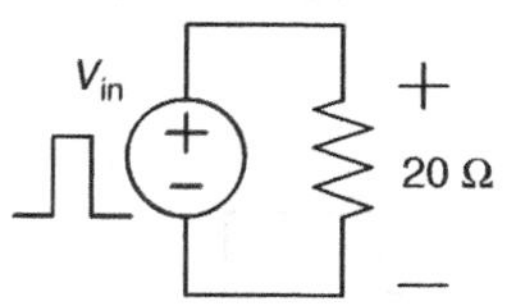

Figure 2.16 Voltage and resistor schematic labeled with voltage reference polarity.

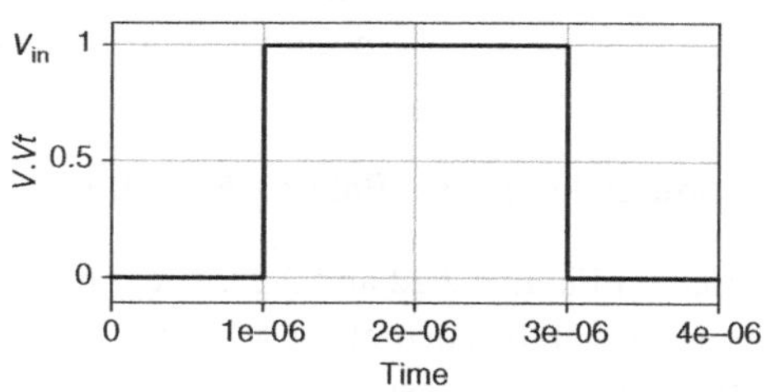

Figure 2.17 Circuit simulation result voltage showing voltage polarity.

The 1 V voltage pulse from V_{in}, in Figures 2.16 and 2.17, starts at 1 μs and lasts for 2 μs, returning to zero at 3 μs.

For a voltage source, what is the reference polarity of current going through the source? Does the voltage source consider current, going up through the voltage source from minus to plus, to be positive? (No, it doesn't) This can be checked using circuit simulation.

2.3.3.1 Circuit Simulation Results – Time Domain Voltage Source

V at the top of the simulation schematic, shown in Figure 2.18, indicates a voltage label. V.Vt in this simulation, shown in Figure 2.19, corresponds to the + and – across the 20 Ω resistor in Figures 2.16 and 2.18. Pr1 is a current meter with the arrow pointed down.

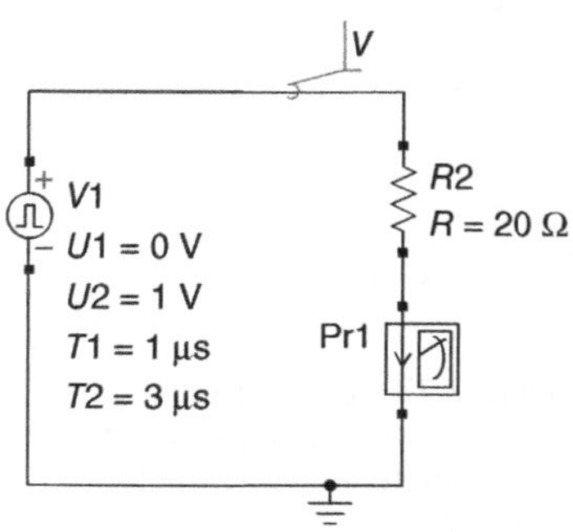

Figure 2.18 Circuit simulation schematic with voltage source, resistor, and current meter.

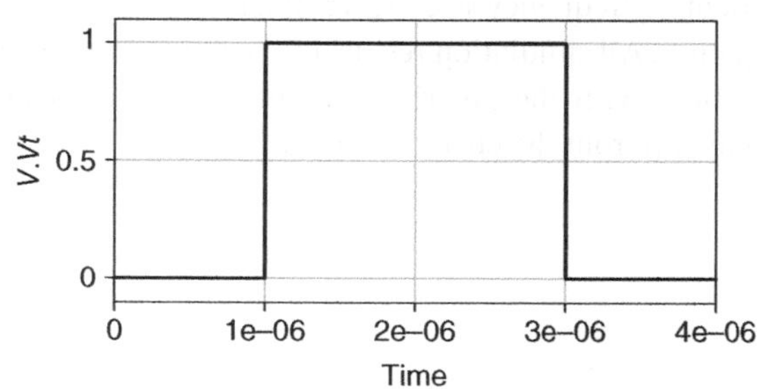

Figure 2.19 Circuit simulation result showing voltage source polarity.

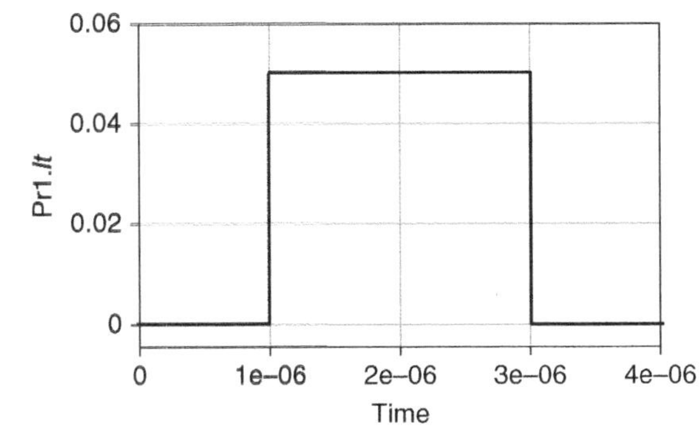

Figure 2.20 Circuit simulation result showing current polarity according to the current meter.

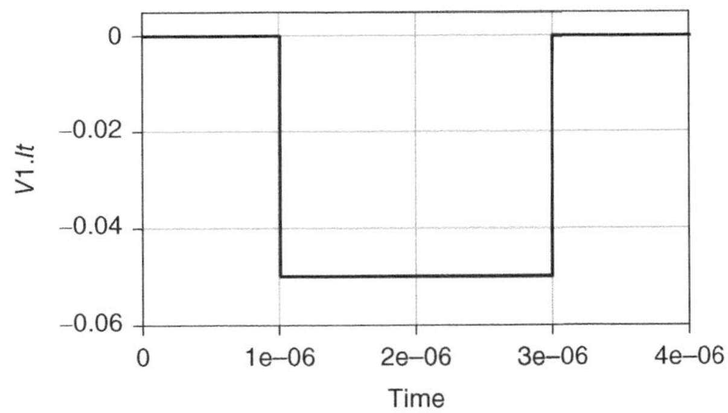

Figure 2.21 Circuit simulation result showing current polarity through the voltage source.

The current through the meter Pr1, shown in Figure 2.20, is positive 50 mA, corresponding to current flowing down through the resistor, in the same direction as the current meter reference arrow.

The current through voltage source, shown in Figure 2.21, is negative 50 mA. This tells us that the reference polarity of the voltage source is positive current moving down through the voltage source. Current flowing up through the voltage source is negative to the voltage source.

2.3.4 Frequency Domain Current Source Reference Polarity Example

Given a 1 mA sinusoidal current source, in Figures 2.22 and 2.23, and a reference polarity that positive current comes out of the current source in the direction of the arrow, we expect to see 30 mV across the resistor R_1 in Figure 2.23. Because there are no reactive circuit elements, we expect to see no phase shifts for either circuit element.

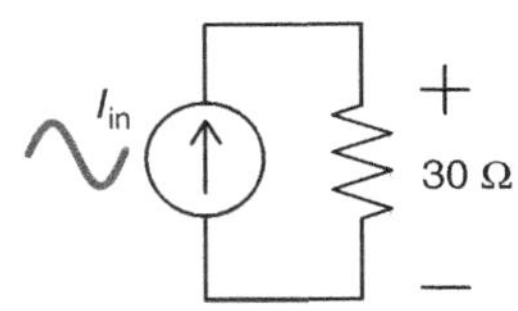

Figure 2.22 Current source and resistor schematic showing voltage polarity.

2.3.4.1 Circuit Simulation Results – Frequency Domain Current Source

A current meter is placed in the circuit loop as a reference for the source model polarity as shown in Figure 2.23. The direction arrow on the current meter is the same as the direction on the current source. Because there are no frequency-dependent reactive circuit elements, a frequency sweep gives flat graphs, constant values over frequency. All simulation results in Figures 2.24–2.26, and 2.27 are consistent with the polarity reference model of positive current flowing up from the current source.

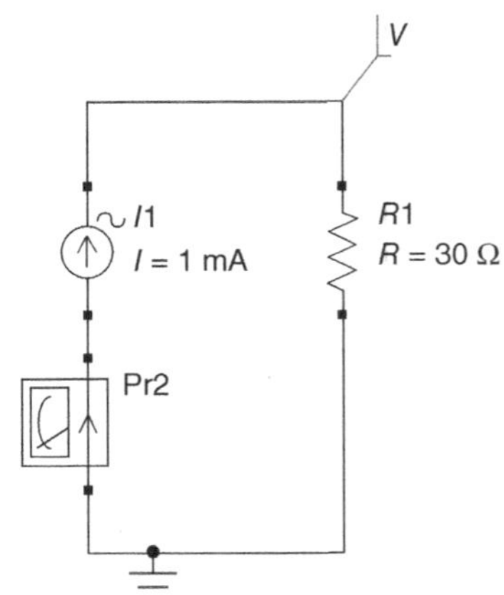

Figure 2.23 Circuit simulation schematic with current source and current meter pointing the same direction.

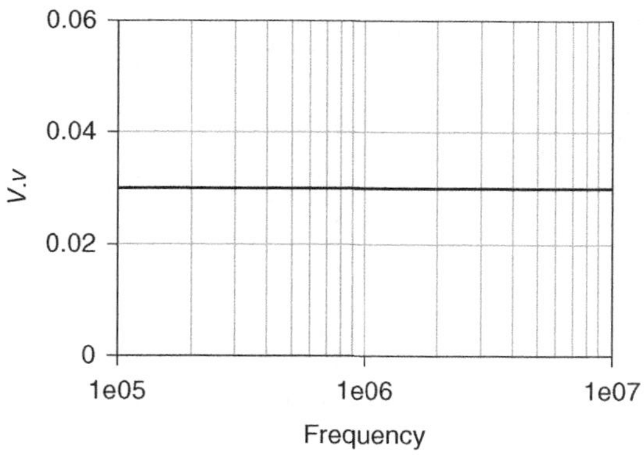

Figure 2.24 Circuit simulation result showing positive-node voltage.

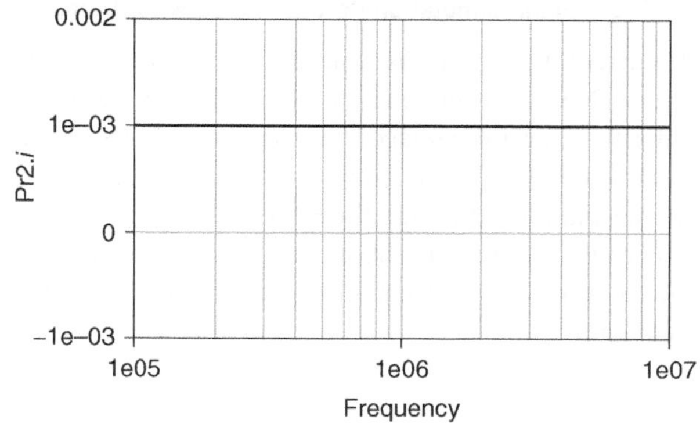

Figure 2.25 Circuit simulation result showing positive current in clockwise direction.

Across the resistor in Figure 2.23, the positive 30 mV voltage V, seen in Figure 2.24, is consistent with positive 1 mA current flowing up through the voltage source and then down through the 30 Ω resistor.

The current, through the meter Pr2 shown in Figure 2.23, is also positive 1 mA (0.001 A) current flowing up and consistent with the polarity of the source, as shown in Figure 2.25.

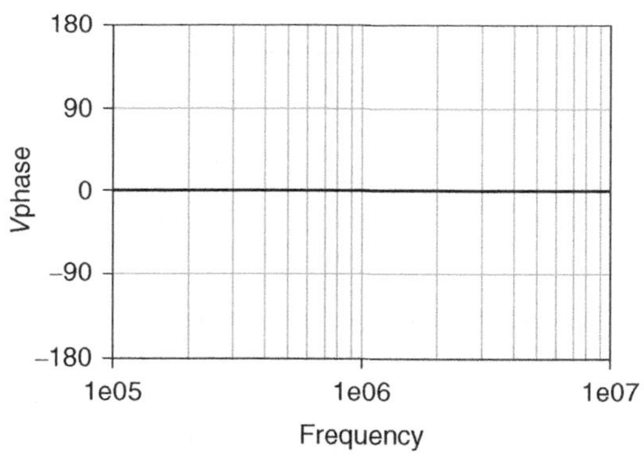

Figure 2.26 Circuit simulation result showing zero-phase change.

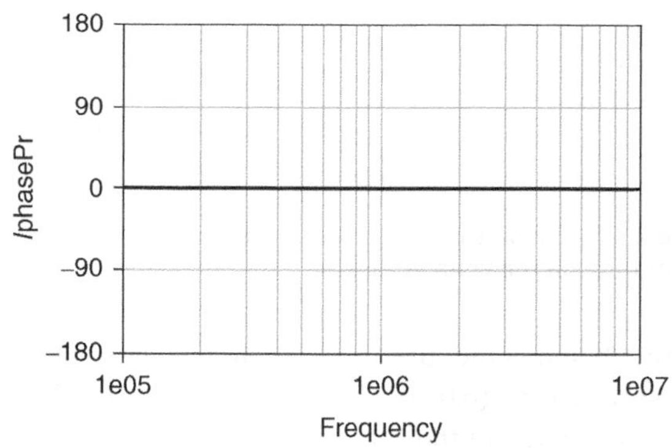

Figure 2.27 Circuit simulation result showing no phase shift.

The voltage phase V phase, shown in Figure 2.26, has no phase shift, as expected.

Figure 2.27 shows that the current source has no phase shift, as expected.

2.3.5 Frequency Domain Voltage Source Reference Polarity Example

Given a 1 V sinusoidal voltage source across the 50 Ω resistor, shown in Figures 2.28 and 2.29, what voltage magnitudes and phases do we expect to see throughout the circuit?

Is the voltage source reference polarity a positive current flowing down through the voltage source from top to bottom? (Yes, it is.)

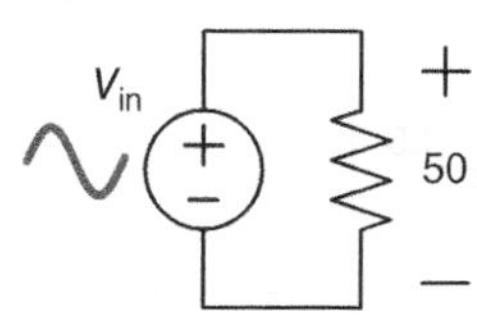

Figure 2.28 Schematic showing voltage polarity.

2.3.5.1 Circuit Simulation Results – Frequency Domain Voltage Source

A current meter Pr2 is placed in the circuit loop, shown in Figure 2.29, as a reference to check the source model reference polarity. A positive voltage V, across the 50 Ω resistor, corresponds to current flowing up and out of the positive terminal of the voltage source and down through the 50 Ω resistor.

Because there are no frequency-dependent reactive circuit elements, a frequency sweep gives flat graphs, constant values over frequency.

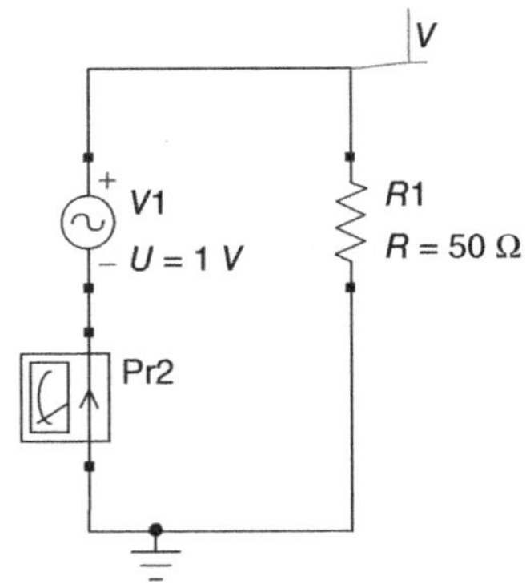

Figure 2.29 Circuit simulation schematic with alternating current voltage source.

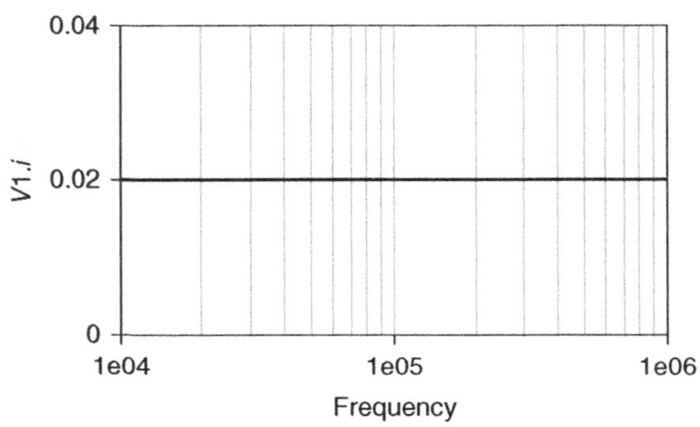

Figure 2.30 Circuit simulation result showing positive current through the voltage source.

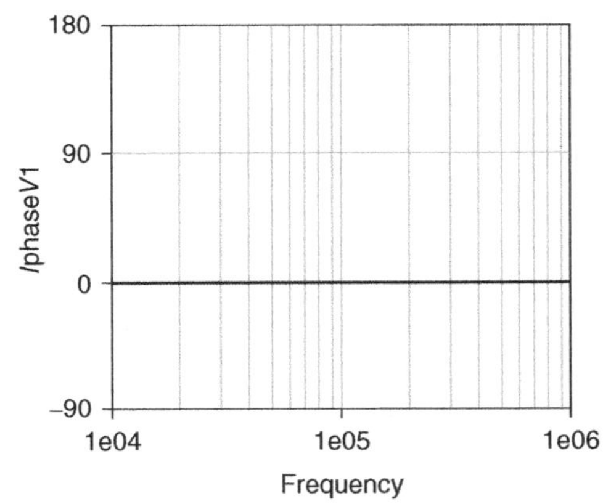

Figure 2.31 Circuit simulation result showing zero-phase shift across the voltage source.

Figure 2.30 shows the current V1.i in the voltage source as positive 20 mA (0.02 A). But we need to verify the reference polarity of the voltage source. Is source current positive coming up through the voltage source and out the top? (No, it isn't.)

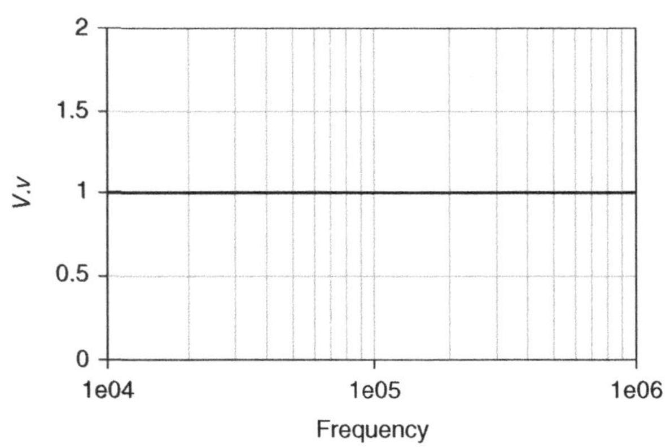

Figure 2.32 Circuit simulation result showing positive voltage across the resistor.

In Figure 2.31 the phase of the voltage source is zero, which we expect in a resistive circuit with no reactive elements.

The positive value of V.v across the resistor, shown in Figure 2.32, matches a reference polarity of current coming up through the voltage source, going out of the top of the source, and going down through the resistor.

It can be important to persevere and check more information for consistency.

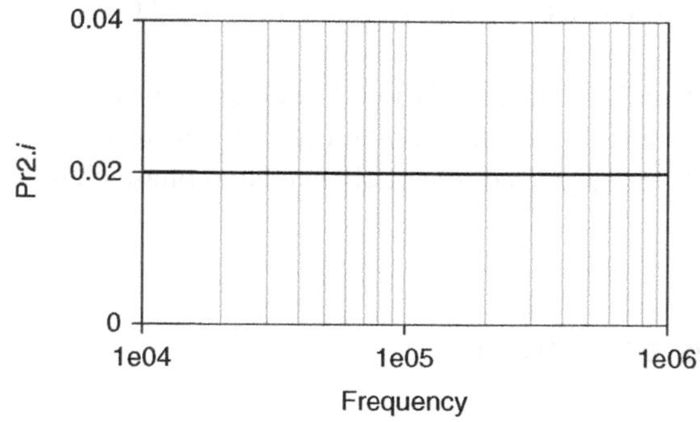

Figure 2.33 Circuit simulation result showing positive value of current in the current meter.

The current Pr2.1 through the current meter, shown in Figure 2.33, is 20 mA, corresponding to a reference polarity of current going up through the voltage source.

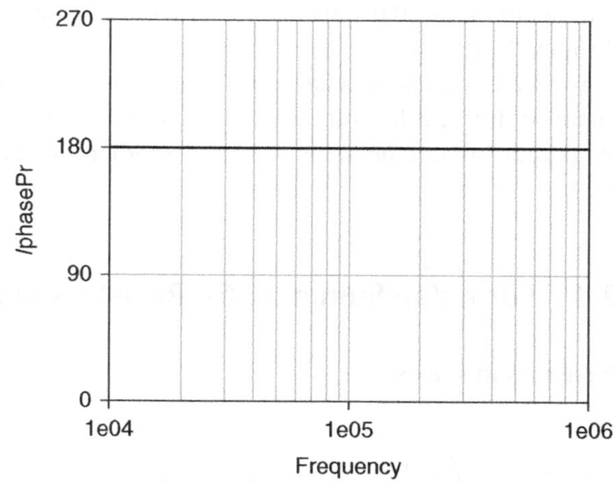

Figure 2.34a Circuit simulation result showing 180° phase shift in current.

The phase *I*phasePr, shown in Figure 2.34a, of the current meter in Figure 2.29, is 180.

In Figure 2.34a, the 180° phase *I*phasePr, and the 0° phase *I*phaseV_1 in Figure 2.31, appear to disagree with one another. They also disagree with a voltage source reference polarity that has positive current traveling up through the voltage source. However, the current meter current in Figure 2.33 appears to match the voltage source current in Figure 2.30, both oriented upward.

Where is the inconsistency? Is the assumption, that the voltage source current reference is positive going up and out the positive terminal, incorrect? (Hint: yes)

Test a different voltage source reference polarity; assume positive current traveling ***down*** through the voltage source and out the bottom minus terminal of the source.

Positive current flowing down through the voltage source would move up through the resistor resulting in ***negative*** 1 V.

The simulation results V.v, in Figure 2.32, indicate ***positive*** 1 voltage across the resistor, but in Figure 2.34b we see a 180° phase shift in the voltage across the resistor. For sinusoidal steady-state circuit analysis, a 180° phase shift is equivalent to a negative value.

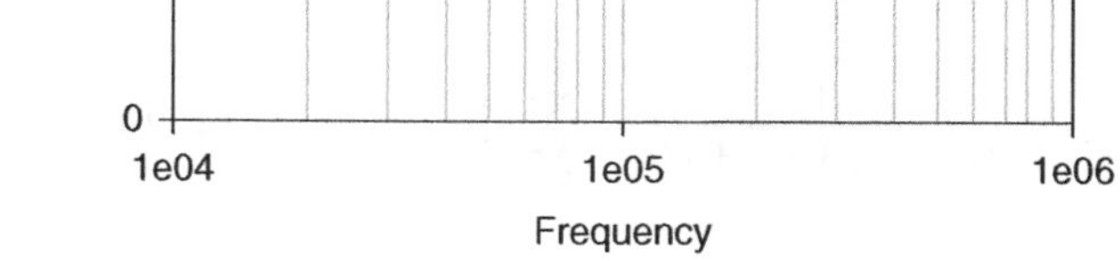

Figure 2.34b Circuit simulation result showing 180° phase shift in voltage.

The 180° phase shift across the resistor, effectively making the AC voltage negative, validates the new reference polarity assumption that positive current flows down through the voltage source from plus to minus.

In addition, the 180° phase shift across the current meter is also consistent, indicating a negative AC current through the meter.

Adequate consistency checks are crucial to successful circuit analysis and simulation. The polarity conventions might be different from these in a different simulator. Two currents, for example, might have the same current magnitude but one phase at +180 and another phase at −180. 360 degrees of phase change is no phase change at all.

2.4 Formulae Summary for Resistors in Series and Parallel

Resistors in series

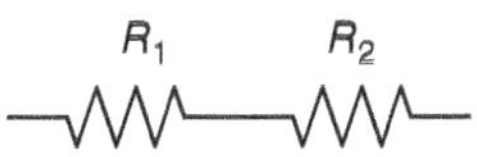

Figure 2.35a Two resistors in series.

R_S is the resistance of two resistors in series (Figure 2.35a).

$$R_S = R_1 + R_2 \quad (2.8a)$$

Resistors in parallel

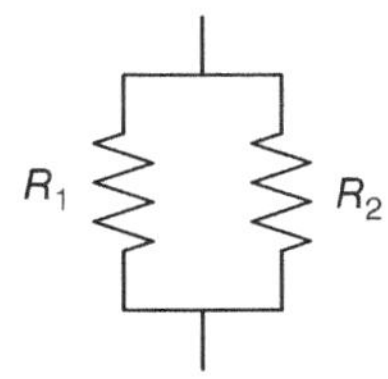

Figure 2.35b Two resistors in parallel.

R_P is the resistance of two resistors in parallel (Figure 2.35b).

$$R_P = \frac{R_1 R_2}{R_1 + R_2} \quad (2.8b)$$

Figure 2.36a *n* resistors in series.

R_S is the resistance of n resistors in series (Figure 2.36a).

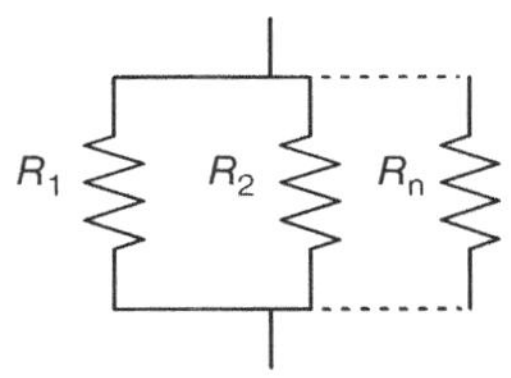

Figure 2.36b *n* resistors in parallel.

R_P is the resistance of n resistors in parallel (Figure 2.36b).

$$R_S = R_1 + R_2 + \cdots + R_n = \sum_{m=1}^{n} R_m \quad (2.9a)$$

$$\frac{1}{R_P} = \frac{1}{R_1} + \frac{1}{R_2} + \cdots + \frac{1}{R_n} = \sum_{m=1}^{n} \frac{1}{R_m} \quad (2.9b)$$

$$G_S = \frac{1}{R_S} \quad (2.10a)$$

G_S is the conductance of n resistors in series.

$$G_P = \frac{1}{R_P} \quad (2.10b)$$

G_P is the conductance of n resistors in parallel.

$$\frac{1}{G_S} = \frac{1}{G_1 + G_2 + \cdots + G_n} = \sum_{m=1}^{n} \frac{1}{G_m} \quad (2.11a)$$

$$G_P = G_1 + G_2 + \cdots + G_n = \sum_{m=1}^{n} G_m \quad (2.11b)$$

Impedances in series

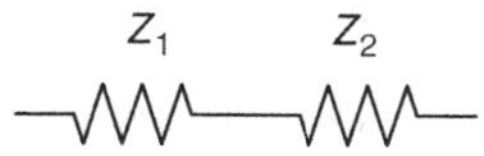

$$Z_S = Z_1 + Z_2 \quad (2.12a)$$

Figure 2.37a Two purely resistive impedances in series.

Z_S is the impedance of two circuit elements in series (Figure 2.37a).

Impedances in parallel

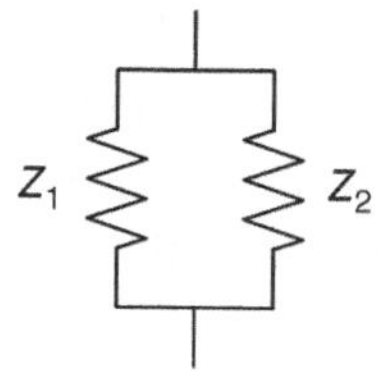

Figure 2.37b Two purely resistive impedances in parallel.

$$Z_P = \frac{Z_1 Z_2}{Z_1 + Z_2} \quad (2.12b)$$

Z_P is the impedance of two circuit elements in parallel (Figure 2.37b).

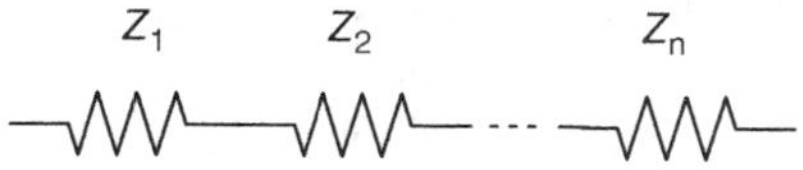

Figure 2.38a *n* purely resistive impedances in series.

Z_S = the impedance of *n* circuit elements in series (Figure 2.38a)

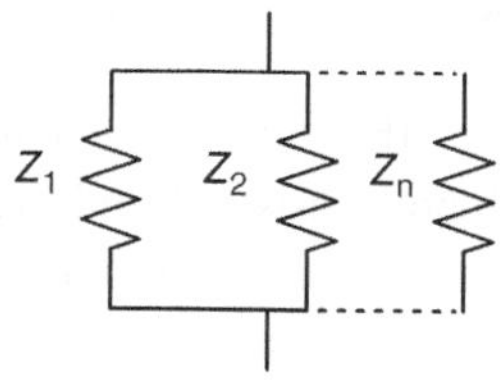

Figure 2.38b *n* purely resistive impedances in parallel.

Z_P = the impedance of *n* circuit elements in parallel (Figure 2.38b)

$$Z_S = Z_1 + Z_2 + \cdots + Z_n \quad (2.13a)$$

$$\frac{1}{Z_P} = \frac{1}{Z_1} + \frac{1}{Z_2} + \cdots \frac{1}{Z_n} \quad (2.13b)$$

2.5 Resistors in Series

When resistances are connected in series, the combined resistance is the sum of the individual resistances. The combined resistance must be higher than either of the separate resistances because the current must go through all resistors (Figure 2.39a).

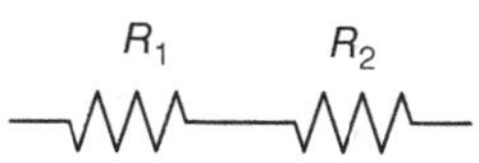

Figure 2.39a Two resistors in series.

$$R_S = R_1 + R_2 \quad (2.14)$$

R_S in Equation 2.14 is the resistance of **two** resistors in series (Figure 2.39b).

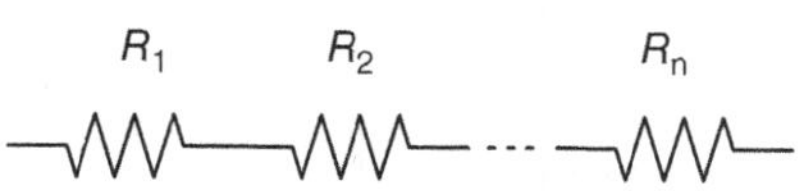

Figure 2.39b *n* resistors in series.

R_S in Equation 2.15 is the resistance of *n* resistors in series.

$$R_S = R_1 + R_2 + \cdots + R_n = \sum_{m=1}^{n} R_m \quad (2.15)$$

2.5.1 Derivation of the Formulae for Two Resistors in Series

We can use Kirchhoff's laws and Ohm's law to derive the formulae for resistors in series.

1. Put a voltage source v across two resistors, R_1 and R_2, connected in series, as shown in Figure 2.40. The resistors R_1 and R_2 might not be equal values of resistance.
2. We want to solve for R_S, the resistance of both resistors taken together, expressed in terms of R_1 and R_2.
3. According to Kirchhoff's current law (KCL), the sum of the currents at node 1, in between R_1 and R_2, must be zero. Since there are only two elements connected to node 1, the current must be the same through both resistors. (There might be a different voltage across each resistor.)
4. Since the current is the same through both resistors, we can apply Ohm's law to both of the resistors R_1 and R_2 grouped together, the equivalent series resistance R_S. $I = v/R_S$.
   (We don't yet "know" that $R_S = R_1 + R_2$; that is what we're deriving here!)
5. Since the same current flows through both resistors, and the total voltage drops across both of the resistors, we can group those two resistors together in Ohm's law, as shown in Equation 2.16.
   (This is the crux of the derivation.)
6. Solve Equation 2.16 for R_S as follows.
   Multiply both sides of the Equation 2.16 by R_S, as shown in Equation 2.17.

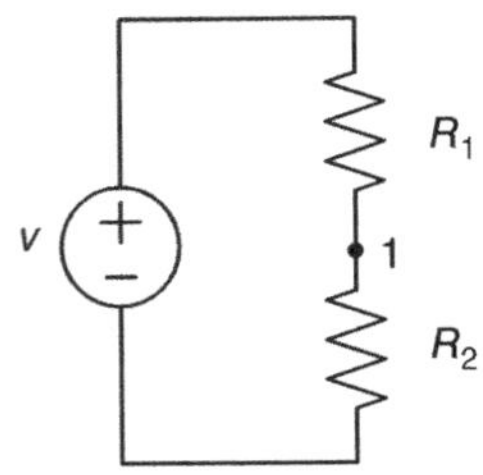

Figure 2.40 Voltage divider circuit with two resistors.

$$i = \frac{v}{R_S} = \frac{v}{R_1 + R_2} \tag{2.16}$$

$$\frac{R_S v}{R_S} = \frac{R_S v}{R_1 + R_2} \tag{2.17}$$

Multiply both sides of the Equation 2.17 by R_1 plus R_2, and cancel R_S terms on the left-hand side, as shown in Equation 2.18.

$$v(R_1 + R_2) = R_S v \tag{2.18}$$

Divide both sides of Equation 2.18 by v to get the final result Equation 2.19.

$$R_S = R_1 + R_2 \tag{2.19}$$

Equation 2.19 is the formula for the equivalent resistance of two resistors in series.

2.5.2 *n* Resistors in Series

The same approach, for deriving the equivalent series resistance of two resistors, can be applied to any number n of resistors in series, since the loop current is the same through all n of the resistors and the voltage is applied across all n of the resistors, as shown in Figure 2.41 and Equation 2.20.

$$\frac{v}{R_S} = \frac{v}{R_1 + R_2 \cdots + R_n} \tag{2.20}$$

R_S in Equation 2.20 is the resistance of n resistors in series, which is the sum of all the resistances, as shown in Equation 2.21.

$$R_S = R_1 + R_2 + \cdots + R_n = \sum_{m=1}^{n} R_m \tag{2.21}$$

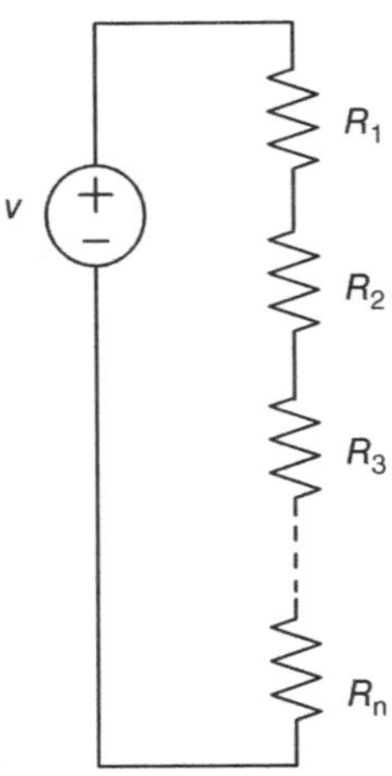

Figure 2.41 Circuit schematic with voltage source across n resistors in series.

2.5.3 Conductors in Series

Conductance G is the inverse of resistance R:

$$G = \frac{1}{R} \tag{2.22}$$

$$R = \frac{1}{G} \tag{2.23}$$

Equation 2.24 gives the formula for the equivalent series conductance G_S of two resistors (or you could say conductors) in series, as shown in Figure 2.42.

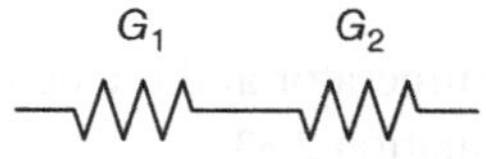

Figure 2.42 Two conductances in series.

Equation 2.25 states that the inverse of the series equivalent conductance, G_S, equals the sum of the inverses of the two conductances. Notice how these formulae in Equations 2.24 and 2.25, for conductances in series, look like the formulae for the equivalent resistance of two resistors in parallel.

$$G_S = \frac{G_1 G_2}{G_1 + G_2} \tag{2.24}$$

$$\frac{1}{G_S} = \frac{1}{G_1} + \frac{1}{G_2} \tag{2.25}$$

The formula for equivalent series conductance of n resistors (or conductors) in series, as shown in Figure 2.43, is that the inverse of the series equivalent conductance equals the sum of the inverses of the n resistors/conductors, as shown in Equation 2.26.

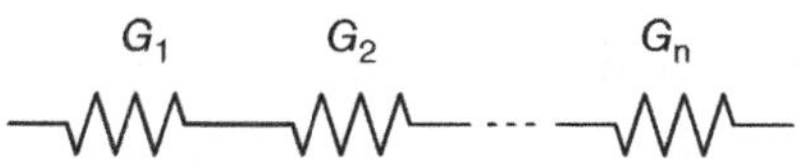

Figure 2.43 n conductances in series.

$$\frac{1}{G_S} = \frac{1}{G_1} + \frac{1}{G_2} + \cdots \frac{1}{G_n} = \sum_{m=1}^{n} \frac{1}{G_m} \tag{2.26}$$

2.5.3.1 Admittance is the Inverse of Impedance

$$Y = \frac{1}{Z} \tag{2.27}$$

These conductance formulae also apply to admittance Y, shown in Equation 2.27, which includes reactance in addition to conductance.

2.5.3.2 First Derivation of the Formula for the Conductance of Two Resistors in Series

This section derives the formula, in Equations 2.24 and 2.28, for two conductors in series, shown in Figure 2.44.

$$G_S = \frac{G_1 G_2}{G_1 + G_2} \tag{2.28}$$

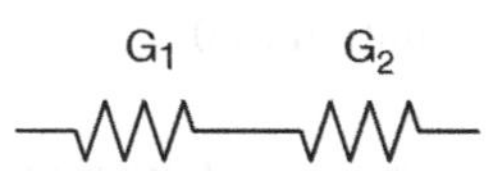

Figure 2.44 Two conductances in series.

The derivation starts from Equation 2.29, the formula for two resistors in series.

$$R_S = R_1 + R_2 \tag{2.29}$$

Conductance equals one over resistance, as stated in Equation 2.30.

$$G_S = \frac{1}{R_S} \tag{2.30}$$

From Equation 2.30, substitute in R_1 plus R_2 into Equation 2.30 for R_S to obtain Equation 2.31.

$$G_S = \frac{1}{R_S} = \frac{1}{R_1 + R_2} \tag{2.31}$$

From Equation 2.23, substitute $1/G_1$, for R_1 and $1/G_2$, for R_2, into Equation 2.31 to obtain Equation 2.32.

$$G_S = \frac{1}{R_S} = \frac{1}{\frac{1}{G_1} + \frac{1}{G_2}} \tag{2.32}$$

Multiply the numerator and the denominator of Equation 2.32 by G_1G_2 to get Equation 2.33.

$$G_S = \frac{1}{\frac{1}{G_1} + \frac{1}{G_2}} \frac{(G_1G_2)}{(G_1G_2)} \tag{2.33}$$

In Equation 2.34, multiply through some of the terms to distribute G_1G_2 in the denominator, obtaining Equation 2.34.

$$G_S = \frac{(G_1G_2)}{\frac{(G_1G_2)}{G_1} + \frac{(G_1G_2)}{G_2}} \tag{2.34}$$

In Equation 2.34, cancel some terms, as shown in Equation 2.35, to obtain the final result Equation 2.36.

$$G_S = \frac{(G_1G_2)}{\frac{(\cancel{G_1}G_2)}{\cancel{G_1}} + \frac{(G_1\cancel{G_2})}{\cancel{G_2}}} \tag{2.35}$$

Equation 2.36 is the equivalent series conductance G_S of two conductors (or resistors).

$$G_S = \frac{G_1G_2}{G_2 + G_1} \tag{2.36}$$

2.5.3.3 Second Derivation of the Formula for the Conductance of Two Resistors in Series

For two resistors in series, shown in Figure 2.45, this section calculates Equation 2.37, for the conductance of two resistors in series, from Equation 2.38, the formula for the inverse of two conductances in series.

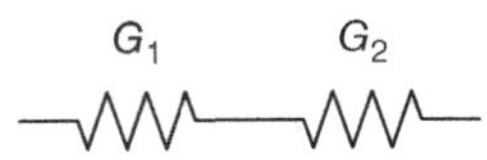

Figure 2.45 Two conductances in series.

$$G_S = \frac{G_1G_2}{G_1 + G_2} \tag{2.37}$$

$$\frac{1}{G_S} = \frac{1}{G_1} + \frac{1}{G_2} \tag{2.38}$$

Multiply both sides of Equation 2.38 by G_1G_2 to obtain Equation 2.39.

$$(G_1 \cdot G_2)\frac{1}{G_S} = (G_1 \cdot G_2)\frac{1}{G_1} + \frac{1}{G_2} \tag{2.39}$$

Multiply out the right hand side of Equation 2.39 to obtain Equation 2.40.

$$\frac{(G_1 \cdot G_2)}{G_S} = \frac{(G_1 \cdot G_2)}{G_1} + \frac{(G_1 \cdot G_2)}{G_2} \tag{2.40}$$

Cancel terms in Equation 2.40, as shown in Equation 2.41, to obtain Equation 2.42.

$$\frac{(G_1 \cdot G_2)}{G_S} = \frac{(\cancel{G_1} \cdot G_2)}{\cancel{G_1}} + \frac{(G_1 \cdot \cancel{G_2})}{\cancel{G_2}} \tag{2.41}$$

Multiply both sides of Equation 2.42 by G_S as shown in Equation 2.43.

$$\frac{(G_1 \cdot G_2)}{G_S} = G_2 + G_1 \tag{2.42}$$

$$\frac{(G_1 \cdot G_2) \cdot G_S}{G_S} = (G_2 + G_1) \cdot G_S \tag{2.43}$$

Cancel G_S terms on the left side of Equation 2.43, as shown in Equation 2.44, to obtain Equation 2.45.

$$\frac{(G_1 \cdot G_2) \cdot \cancel{G_S}}{\cancel{G_S}} = (G_2 + G_1) \cdot G_S \tag{2.44}$$

Divide both sides by of Equation 2.45 by G_1 plus G_2.

$$G_1 \cdot G_2 = (G_2 + G_1) \cdot G_S \tag{2.45}$$

$$\frac{1}{(G_2 + G_1)} \cdot G_1 \cdot G_2 = \frac{1}{(G_2 + G_1)} \cdot (G_2 + G_1) \cdot G_S \tag{2.46}$$

Cancel terms in Equation 2.46, as shown in Equation 2.27, to obtain the final result in Equation 2.48.

$$\frac{1}{(G_2 + G_1)} \cdot G_1 \cdot G_2 = \frac{1}{\cancel{(G_2 + G_1)}} \cdot \cancel{(G_2 + G_1)} \cdot G_S \tag{2.47}$$

Equation 2.48 matches Equation 2.38 for the conductance G_S of two conductors in series.

$$G_S = \frac{G_1 G_2}{G_1 + G_2} \tag{2.48}$$

2.5.4 Impedances in Series

Impedance Z is a generalization of resistance; impedance includes reactance of capacitors and inductors that can store energy. Resistance is a simple case of impedance where no energy is stored and the reactance X is zero.

$$Z = R + jX \tag{2.49}$$

Equation 2.49 states that impedance Z equals resistance R plus j multiplied by reactance X.

Note that in electronic and electrical engineering, j is the same complex operator as i, which is more commonly used in physics and mathematics to denote the complex operator (the square root of minus 1).

Impedances in series (Figures 2.46 and 2.47) follow the same fomulae as resistance is series; they add together (Equations 2.50 and 2.51).

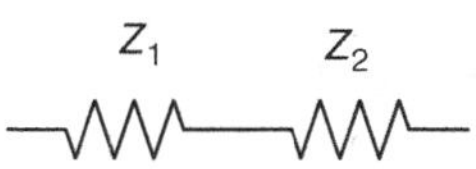

Figure 2.46 Two resistive impedances in series.

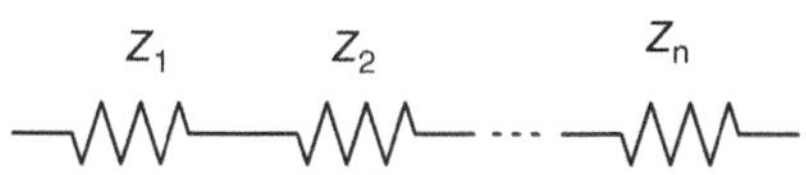

Figure 2.47 *n* resistive impedances in series.

Z_S in Equation 2.50 is the impedance of two circuit elements in series.

$$Z_S = Z_1 + Z_2 \tag{2.50}$$

Z_S in Equation 2.51 is the impedance of n circuit elements in series.

$$Z_S = Z_1 + Z_2 + \cdots + Z_n \tag{2.51}$$

2.5.4.1 Impedance *Z* in Either *s* or *jω* Frequency Domains

Both the Laplace s-domain and the steady-state omega frequency domain include the impedance concept and all of the formulae for resistance in this chapter also apply to impedances in both domains.

Z(s) $Z(s)$ indicates impedance in the Laplace s-domain.

Z(jω) $Z(j\omega)$ indicates impedance in the steady-state frequency domain.

2.5.4.2 Susceptance B

Susceptance B is the inverse of reactance X, as shown in Equation 2.52.

$$\mathrm{B} = \frac{1}{X} \tag{2.52}$$

B(s) $\mathrm{B}(s)$ indicates susceptance in the Laplace s-domain.

B(jω) $\mathrm{B}(j\omega)$ indicates susceptance in the steady-state frequency domain.

2.6 Voltage Dividers

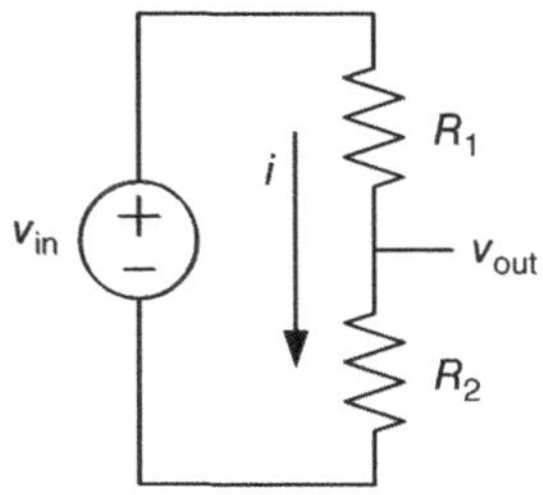

Figure 2.48 Resistive voltage divider circuit.

This circuit in Figure 2.48 is called a voltage divider because it looks at the voltage v_{out} divided between the two resistors. The voltage v_{in} is across the two resistors R_1 and R_2.

$$\frac{v_{out}}{v_{in}} = \frac{R_2}{R_1 + R_2} \quad (2.53)$$

Equation 2.53 is the voltage divider formula.

The derivation of the voltage divider formula, in Equation 2.53, makes use of the formulae for series resistances.

2.6.1 Calculation of the Voltage Divider Output Voltage v_{out}

$$v_{out} = iR_2 \quad (2.54)$$

By Ohm's law, the output voltage v_{out} in Equation 2.54 equals the current i times R_2.

$$i = \frac{v_{in}}{R_1 + R_2} \quad (2.55)$$

The current i is the input voltage v_{in} divided by the quantity R_1 plus R_2, the total combined series resistance.

Substitute Equation 2.55, the expression for i, into the Equation 2.54 for v_{out}, obtaining Equation 2.56.

$$v_{out} = iR_2 = \frac{v_{in}}{R_1 + R_2} R_2 = \frac{R_2 v_{in}}{R_1 + R_2} = \frac{R_2}{R_1 + R_2} v_{in} \quad (2.56)$$

The divided voltage v_{out} equals v_{in} times the ratio of R_2 divided by the quantity R_1 plus R_2.

Divide both sides of the Equation 2.56 by v_{in}, completing the derivation.

$$\frac{v_{out}}{v_{in}} = \frac{R_2}{R_1 + R_2} \quad (2.57)$$

The voltage divider formula in Equation 2.57 states that the ratio of v_{out} over v_{in} equals the ratio of R_2 divided by the quantity R_1 plus R_2.

2.6.2 Voltage Divider for Generalized Impedance

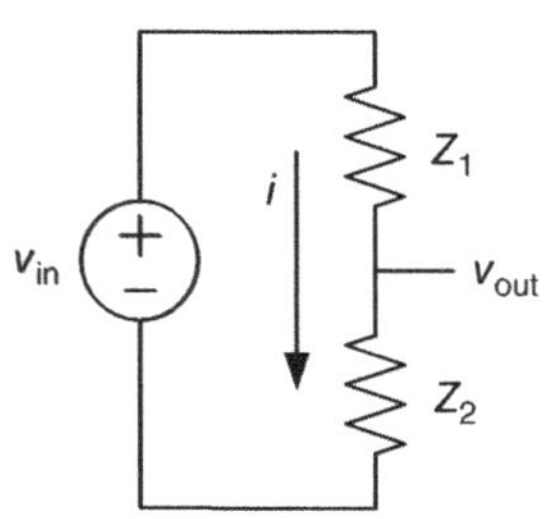

Figure 2.49 Resistive impedance voltage divider circuit.

Going beyond purely resistive circuits and considering circuits with reactive elements (capacitors and inductors), the voltage divider principle applies to generalized impedances.

$$\frac{\mathbf{v}_{out}}{\mathbf{v}_{in}} = \frac{Z_2}{Z_1 + Z_2} \quad (2.58)$$

$$\frac{v_{out}(s)}{v_{in}(s)} = \frac{Z(s)_2}{Z(s)_1 + Z(s)_2} \quad (2.59)$$

The voltage divider circuit in Figure 2.49 is rewritten from Figure 2.48, with z for impedance and boldface v for phasor notation, corresponding to Equation 2.58. Equation 2.59 corresponds to Laplace s-domain.

2.7 Parallel Circuit Element Formulae

Equation 2.60 shows the formula for the equivalent resistance R_P of two resistors in parallel, shown in Figure 2.50.

$$R_P = \frac{R_1 R_2}{R_1 + R_2} \tag{2.60}$$

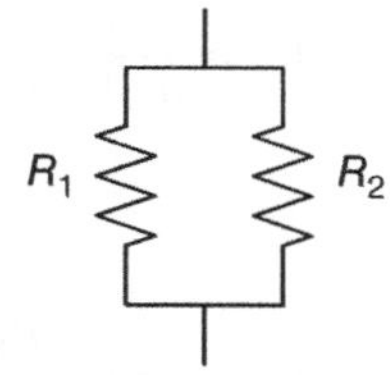

Figure 2.50 Two resistors in parallel.

Equation 2.61 shows the formula for the equivalent resistance R_P of n resistors in parallel, as shown in Figure 2.51.

$$\frac{1}{R_P} = \frac{1}{R_1} + \frac{1}{R_2} + \cdots + \frac{1}{R_n} = \sum_{m=1}^{n} \frac{1}{R_m} \tag{2.61}$$

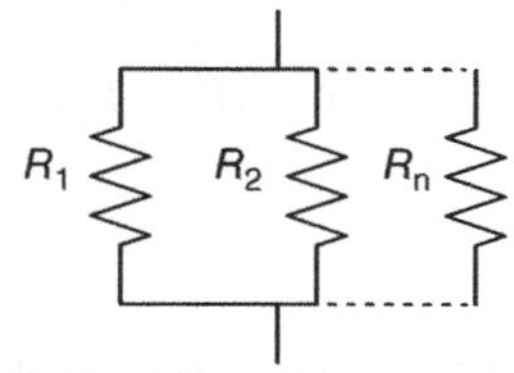

Figure 2.51 *n* resistors in parallel.

Invert Equation 2.61 to solve for R_P directly in Equation 2.62.

$$R_P = \frac{1}{\frac{1}{R_1} + \frac{1}{R_2} + \cdots + \frac{1}{R_n}} = \frac{1}{\sum_{m=1}^{n} \frac{1}{R_m}} \tag{2.62}$$

Equations 2.60, 2.61, and 2.62, for parallel resistance, apply also for parallel impedances Z_P.

The resistance of any parallel combination of resistors will be less than the resistance of any of the separate individual resistors, because the parallel combination gives current alternative paths to flow, reducing the overall parallel resistance relative to either resistance by itself.

2.7.1 Derivation of the Formulae for Two Resistors in Parallel

Use Kirchhoff's laws and Ohm's law to derive the formulae for two resistors in parallel.

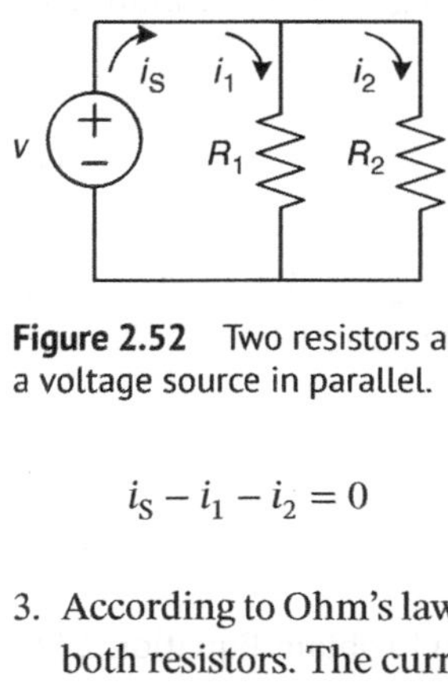

Figure 2.52 Two resistors and a voltage source in parallel.

1. Put a voltage source across two resistors, R_1 and R_2, connected in parallel, as shown in Figure 2.52. The resistors R_1 and R_2 might not have equal values of resistance.
2. According to KCL the currents entering and leaving a node add up to zero. Sum the currents entering the top node (noticing that i_1 and i_2 are defined as positive leaving the node) to obtain Equations 2.63a and 2.63b.

$$i_S - i_1 - i_2 = 0 \tag{2.63a}$$

$$i_S = i_1 + i_2 \tag{2.63b}$$

3. According to Ohm's law, the current in each resistor is v over R. (Remember that the voltage v is the same across both resistors. The currents i_1 and i_2 do not need to be the same.)

$$R_P = \frac{v}{i_T} = \frac{v}{(i_1 + i_2)} \tag{2.64}$$

Applying Ohm's law, the resistance of the parallel combination, R_P equals the voltage divided by the total current i_T, which is i_1 plus i_2. In Equation 2.64, substitute i_1 plus i_2 for i_T.

$$R_P = \frac{v}{(i_1 + i_2)} = \frac{v}{\left(\frac{v}{R_1} + \frac{v}{R_2}\right)} \tag{2.65}$$

In Equation 2.64, apply Ohm's law to the i_1 and i_2 terms on the right-hand side to obtain Equation 2.65 with currents expressed in terms of voltage over resistance.

$$R_P = \frac{vR_1R_2}{\left(\frac{v}{R_1} + \frac{v}{R_2}\right)R_1R_2} = \frac{vR_1R_2}{(vR_2 + vR_1)} \tag{2.66}$$

In Equation 2.65, multiply the numerator and denominator both by R_1 times R_2 to obtain Equation 2.66.

$$R_P = \frac{R_1R_2}{(R_2 + R_1)} \tag{2.67}$$

Cancel the voltage terms in Equation 2.66 to obtain Equation 2.64 for R_P in terms of resistances R_1 and R_2.

R_P in Equation 2.64 is the equivalent resistance for two resistors in parallel.

2.7.2 Derivation of the Formula for *n* Resistors in Parallel

Use Kirchhoff's laws and Ohm's law to derive the formulae for n resistors in parallel.

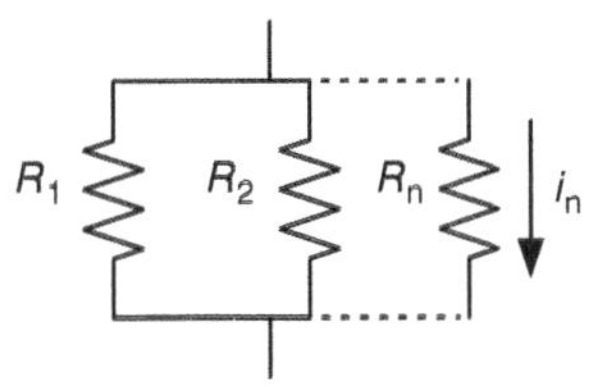

Figure 2.53 *n* resistors in parallel.

The same approach, for deriving the equivalent parallel resistance of two resistors, could be applied to any number of resistors n in parallel, since the voltage is the same across all n resistors.

In this derivation R_P is the resistance of n resistors in parallel, shown in Figure 2.53.

1. The voltage v across all n resistors is the same. The resistor values may be different from one another and the current through each resistor may be different.
2. The current through each individual resistor is given by Ohm's law, using the voltage v and the individual resistance R_1, R_2, etc., up to R_n, as shown in Equation 2.68.

$$i_1 = \frac{v}{R_1} \quad i_2 = \frac{v}{R_2} \quad \cdots\cdots\cdots \quad i_n = \frac{v}{R_n} \tag{2.68}$$

3. The total current i_T is the sum of each individual current in each resistor, as shown in Equation 2.69.

$$i_T = i_1 + i_2 + \cdots i_n = \frac{v}{R_1} + \frac{v}{R_2} + \cdots \frac{v}{R_n} \tag{2.69}$$

4. Factor out the common voltage v and introduce the equivalent parallel resistance R_P, as shown in Equation 2.70.

$$i_T = v\left(\frac{1}{R_1} + \frac{1}{R_2} + \cdots + \frac{1}{R_n}\right) = \frac{v}{R_P} \tag{2.70}$$

5. Cancel the voltage terms from Equation 2.70 and add the sigma summation notation to obtain Equation 2.71.

$$\frac{1}{R_P} = \frac{1}{R_1} + \frac{1}{R_2} + \cdots + \frac{1}{R_n} = \sum_{m=1}^{n} \frac{1}{R_m} \tag{2.71}$$

Equation 2.71 states that one over the equivalent parallel resistance equals the sum of one over each individual resistance for n resistors connected in parallel.

2.7.3 Conductance in Parallel

Conductance G is the inverse of resistance R, as shown in Equations 2.72 and 2.73.

$$G = \frac{1}{R} \quad (2.72)$$

$$R = \frac{1}{G} \quad (2.73)$$

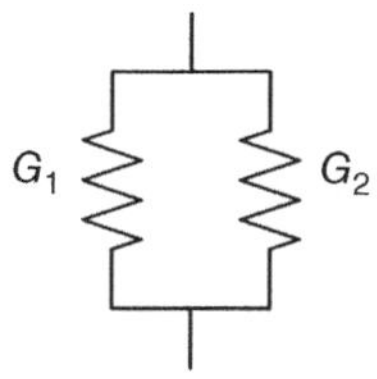

Figure 2.54 Two conductances in parallel.

The formula for equivalent parallel conductance G_P of two resistors (or conductors) in parallel, as shown in Figure 2.54, is the sum of the two conductances G_1 and G_2, shown in Equation 2.74.

$$G_P = G_1 + G_2 \quad (2.74)$$

The formula for equivalent parallel conductance G_P of n resistors (or conductors) in parallel is the sum of the n conductances, as shown in Equation 2.75.

$$G_P = G_1 + G_2 + \cdots G_n = \sum_{m=1}^{n} G_m \quad (2.75)$$

2.7.3.1 Admittance in Parallel

Admittance Y is the inverse of impedance Z. Like conductances, the inverse of resistances, admittances in parallel add, as shown in Equations 2.76 and 2.77.

$$Y_P = \frac{1}{Z_P} \quad (2.76)$$

$$Y_P = Y_1 + Y_2 + L + Y_n = \sum_{m=1}^{n} Y_m \quad (2.77)$$

2.7.4 Impedance in Parallel

Impedance includes both resistance and reactance (inductance or capacitance). Impedances in parallel behave like resistances in parallel.

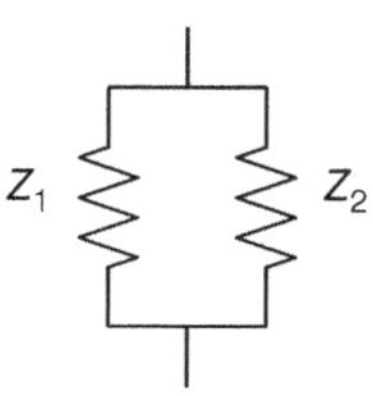

Figure 2.55 Two resistive impedances in parallel.

$$Z_P = \frac{Z_1 Z_2}{Z_1 + Z_2} \quad (2.78)$$

In Equation 2.78, Z_P is the impedance of two circuit elements in parallel, as shown in Figure 2.55.

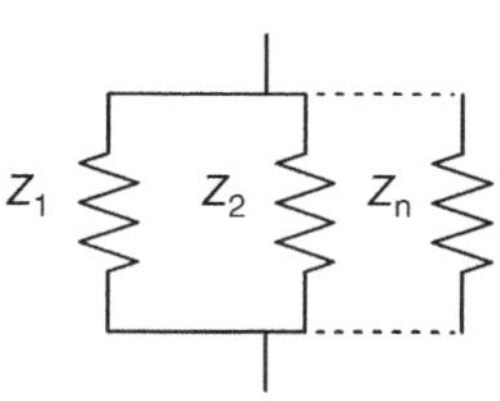

Figure 2.56 n resistive impedances in parallel.

$$\frac{1}{Z_P} = \frac{1}{Z_1 + Z_2 + \cdots Z_n} \quad (2.79)$$

In Equation 2.79, Z_P is the impedance of n circuit elements in parallel, as shown in Figure 2.56.

2.8 Current Dividers

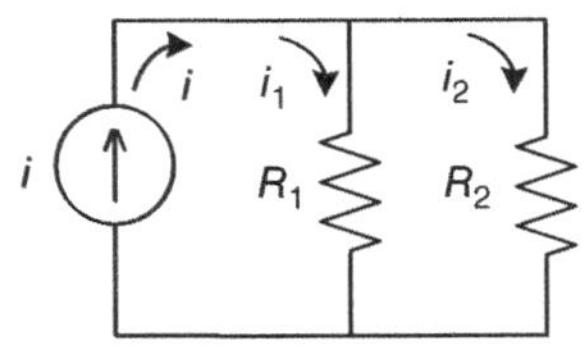

Figure 2.57 Current divider circuit.

The circuit in Figure 2.57 is called a current divider because the parallel resistors, R_1 and R_2, split the current i from the current source between into i_1 and i_2.

$$i_1 = i\frac{R_2}{(R_1 + R_2)} \quad (2.80) \qquad i_2 = i\frac{R_1}{(R_1 + R_2)} \quad (2.81)$$

Equations 2.80 and 2.81 are current divider formulae for the two-resistor circuit in Figure 2.57. If R_1 and R_2 are not the same value of resistance, then i_1 and i_2 won't be equal.

2.8.1 Derivation of Current Divider Formula for Two Resistors

Derive the current divider formula for two resistors, utilizing the fact that the voltage is the same across both resistors.

$$i_1 = v/R_1 \quad (2.82)$$

Apply Ohm's law to resistor R_1 in Figure 2.57. Equation 2.82 shows that the current i_1 equals the voltage divided by R_1.

$$v = iR_p = \frac{iR_1R_2}{(R_1 + R_2)} \quad (2.83)$$

Apply the formula for resistors in parallel, from Equation 2.67, to Equation 2.82, as shown in Equation 2.83.

$$i_1 = v/R_1 = \frac{v}{R_1} = \frac{iR_p}{R_1} = \frac{1}{R_1}\frac{iR_1R_2}{(R_1 + R_2)} \quad (2.84)$$

Apply Ohm's law to the voltage in Equation 2.83, by dividing by R_1, to obtain an expression for i_1, the current in R_1, in terms of resistors R_1 and R_2 in Equation 2.84.

$$i_1 = \frac{iR_p}{(R_1 + R_2)} = i\frac{R_2}{(R_1 + R_2)} \quad (2.85)$$

Cancel the R_1 terms in the numerator and denominator of Equation 2.84, and rearrange terms, to obtain Equation 2.85 for current i_1.

2.8.1.1 Current Divider Formulae

$$i_1 = i\frac{R_2}{(R_1 + R_2)} \quad (2.86)$$

$$i_2 = i\frac{R_1}{(R_1 + R_2)} \quad (2.87)$$

Equations 2.86 and 2.87 are current divider formulae for two resistors in parallel.

2.8.2 Alternate Derivation of Current Divider Formula

In the schematic in Figure 2.58, the voltage is the same across both resistors R_1 and R_2. The currents through each resistor are not required to be the same; use this fact to derive a current divider formula with a different approach than the first derivation.

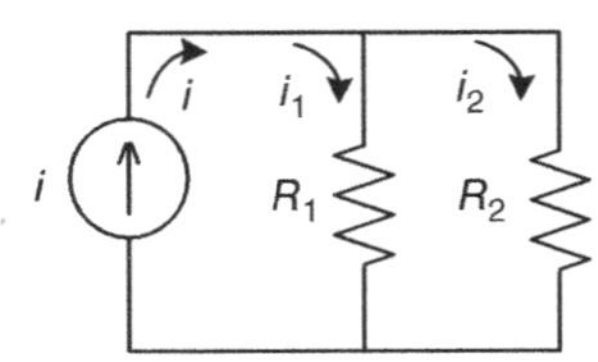

Figure 2.58 Current divider circuit.

$$v = i_1 R_1 = i_2 R_2 \quad (2.88)$$

Apply Ohm's law to either resistor or both resistors R_1 and R_2. The voltage v equals either branch current i_1 or i_2, times the resistance R_1 or R_2 in that same branch, as shown in Equation 2.88.

$$i_2 = i_1 \frac{R_1}{R_2} \quad (2.89)$$

In Equation 2.88, rearrange terms to express i_2 as a function of i_1 and the resistances as shown in Equation 2.88.

$$i_S = i_1 + i_2 \quad (2.90)$$

The total current i_S, through the source, equals the sum of the currents i_1 and i_2 in each branch, as indicated in Equation 2.90.
From Equation 2.89, substitute i_2 into Equation 2.90 for i_S, to express i_S in terms of i_1 and the resistances, as shown in Equation 2.91.

$$i_S = i_1 + i_1 \frac{R_1}{R_2} \quad (2.91)$$

In Equation 2.91, multiply the middle i_1 term by R_2 divided by R_2, in order to get the same R_2 denominator as the other term, as shown in Equation 2.92.

$$i_S = i_1 \frac{R_2}{R_2} + i_1 \frac{R_1}{R_2} \quad (2.92)$$

In Equation 2.92, factor out the common multiplier term, i_1, to obtain Equation 2.93.

$$i_S = i_1 \left(\frac{R_2}{R_2} + \frac{R_1}{R_2} \right) \quad (2.93)$$

In Equation 2.93, group the R_1 and R_2 over the same denominator R_2, as shown in Equation 2.94.

$$i_S = i_1 \left(\frac{R_2 + R_1}{R_2} \right) \quad (2.94)$$

Divide both sides of Equation 2.94 by the resistance term in the parentheses to solve for i_1 and obtain Equation 2.95.

$$\frac{i_S}{\left(\frac{R_2 + R_1}{R_2} \right)} = i_1 \quad (2.95)$$

Rearrange terms in Equation 2.92, inverting the denominator, to obtain the final result in Equation 2.95.

2.8.2.1 Current Divider Formula

$$i_1 = i \frac{R_2}{(R_1 + R_2)} \quad (2.96)$$

Equation 2.96, like Equation 2.87, is the current divider formula for two resistors in parallel.

2.8.3 The Current Divider Formula Generalizes to Impedances

2.8.3.1 Current Divider Formulae

$$i_1 = i \frac{Z_2}{(Z_1 + Z_2)} \quad (2.97)$$

$$i_2 = i \frac{Z_1}{(Z_1 + Z_2)} \quad (2.98)$$

Equations 2.97 and 2.98 are current divider formulae for two impedances in parallel, similar to the resistive counterparts.

2.9 Current and Voltage Intuitions

When first approaching and solving simple circuits, it is valuable to revisit and reinforce these two basic properties until they become intuitive.

2.9.1 Only One Current in a Series Circuit

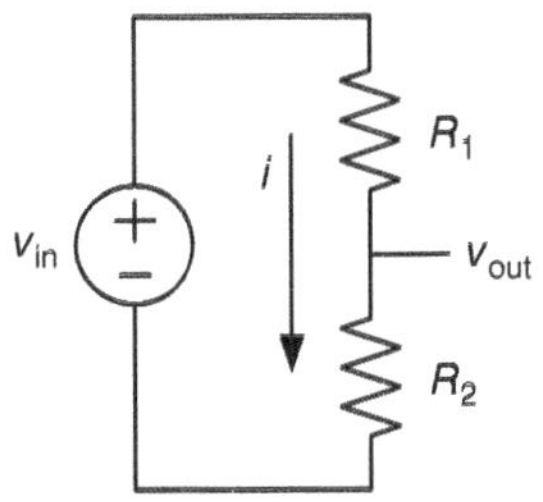

Figure 2.59 Voltage divider circuit.

The current is the same through two resistors in series (Figure 2.59).

The current is the same through all the circuit elements of a series loop, including the voltage source. This circuit has only one current.

This intuition comes from Kirchhoff's current law, which requires currents in and out of any node to sum to zero.

2.9.2 Only One Voltage in a Parallel Circuit

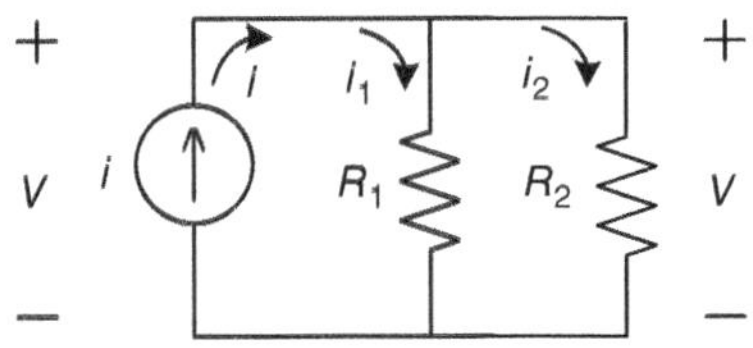

Figure 2.60 Current divider circuit.

The voltage is the same across the two resistors in parallel in Figure 2.60.

The voltage is the same across all the circuit elements of a parallel circuit, including the current source. The circuit in Figure 2.60 has only one voltage.

This intuition comes from Kirchhoff's voltage law, which requires voltages around loops to sum to zero.

3

Opamp Models and Resistive Circuits

3.1 Introduction and Ideal Opamp Model Results Overview

Operational amplifiers (opamps) are common circuit building blocks in electronics. Opamps are generally integrated circuits, like the well-known and widely used 741 opamp. This chapter focuses on some of the most simple and common opamp models.

The ideal opamp model says that the two opamp input pins, plus (+) and minus (−), are at the same voltage. At the same time, no current flows in or out of either pin.

The ideal opamp modeling notion of zero voltage between two pins, with no current flowing, defies our normal electronics notion of a short circuit that allows any amount of current to flow. In an ideal opamp model, this zero voltage is in fact not a short circuit. This potential confusion makes the careful study of basic opamp model circuits valuable.

3.1.1 Ideal Opamp Inverting Amplifier Circuit Results Summary

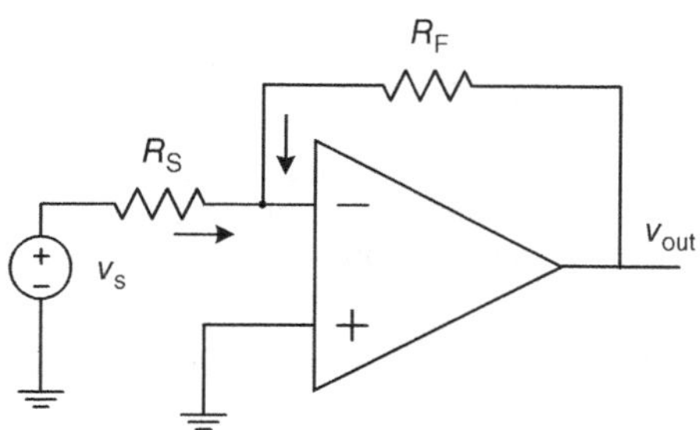

Figure 3.1 Inverting amplifier schematic.

$$\frac{v_{out}}{v_S} = \frac{-R_F}{R_S} \tag{3.1}$$

The gain of the inverting amplifier, shown in Figure 3.1, is minus R_F over R_S, as shown in Equation 3.1.

In the ideal opamp model, no current flows into the opamp through the plus and minus pins.

The minus pin on the opamp, connected to both R_S and R_F, is at zero voltage because the plus pin of the opamp is grounded.

First and Second Order Circuits and Equations: Technical Background and Insights, First Edition. Robert O'Rourke.

3.1.1.1 Ideal Opamp Non-Inverting Amplifier Results Summary

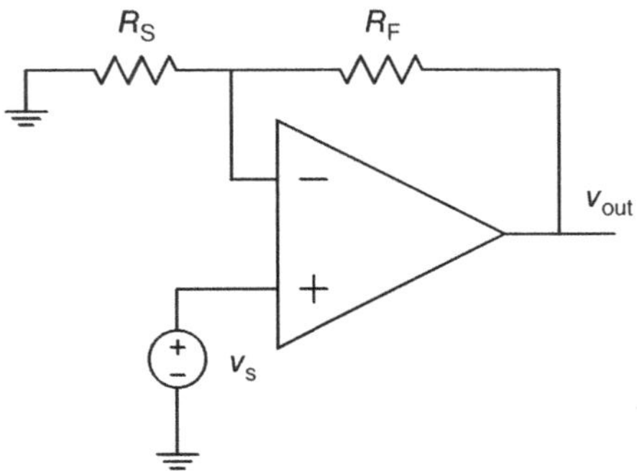

Figure 3.2 Non-inverting amplifier schematic.

$$\frac{v_{out}}{v_S} = \frac{R_F + R_S}{R_S} = \frac{R_F}{R_S} + 1 \tag{3.2}$$

The gain of the inverting amplifier is R_F over R_S plus 1.

No current flows into the opamp pins in the ideal model.

The minus pin on the opamp, connected to both R_S and R_F, is at voltage V_S because the plus pin of the opamp is at the same voltage as the minus pin.

There are various models for op-amps, each modeling containing different amounts or levels of details about the inner workings of the amplifier, such as input or output resistance. These basic circuits, shown in Figures 3.1 and 3.2, use an ideal opamp model which includes no details of the inner workings.

3.1.2 Opamp Models

3.1.2.1 Standard Opamp Model

The controlled voltage source, in the standard opamp model in Figure 3.3, multiplies the voltage, $v+$ minus $v-$, across R_I, by the gain A. R_I will be a large value and R_O will be a small value. There are finite input and output currents.

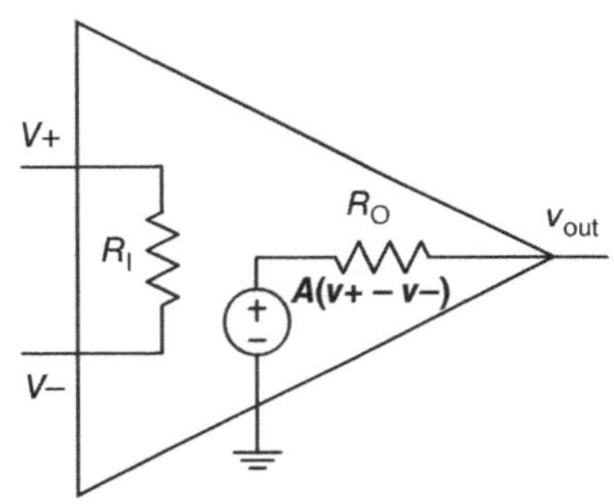

Figure 3.3 Standard opamp model with finite input resistance and finite output resistance.

3.1.2.2 Voltage-Controlled Voltage Source Opamp Model

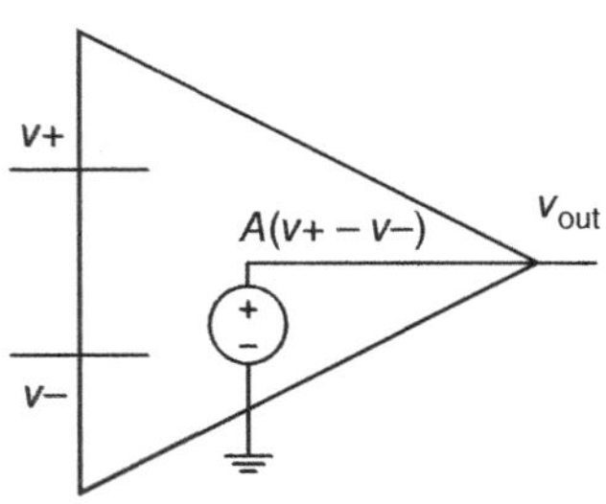

Figure 3.4 Opamp model showing controlled source modeling approach.

An operational amplifier, shown in Figure 3.4, can be modeled as a controlled voltage source. The output voltage is the input voltage multiplied by a gain A.

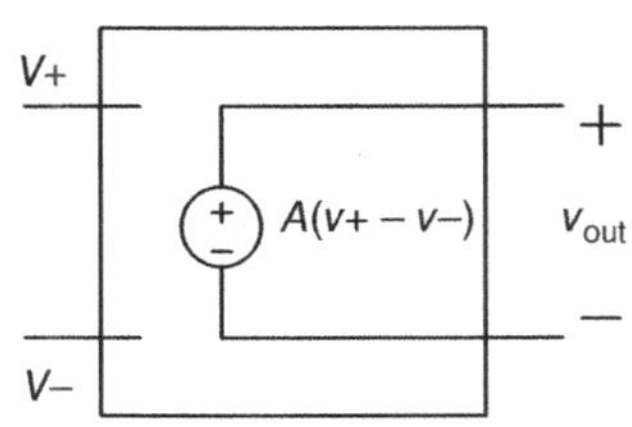

Figure 3.5 Voltage-controlled voltage source.

This controlled-source opamp model can be implemented in SPICE, or other lumped element circuit simulation using a voltage-controlled voltage source circuit element, as shown in Figure 3.5.

3.1.2.3 Ideal Opamp Model

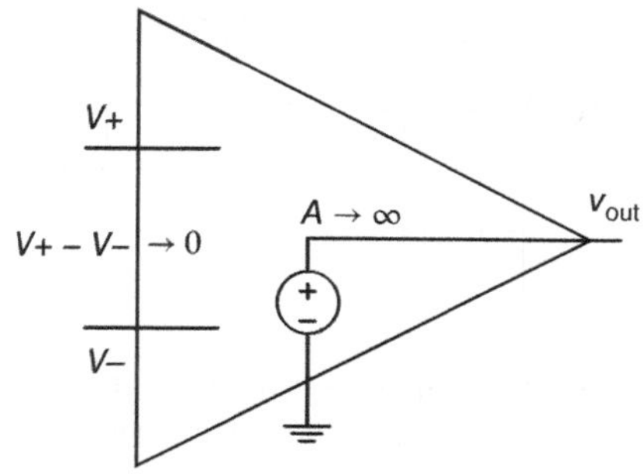

Figure 3.6 Ideal opamp model with vanishing input voltage and gain approaching infinity.

In the ideal opamp model, the input voltage ($v+$ minus $v-$) goes to zero. This implies that the gain goes to infinity in order to produce a finite voltage at the output. There is no input current into either input terminal $v+$ or $v-$ (in spite of the zero voltage across those terminals) (Figure 3.6).

A different way to describe the ideal opamp model, and its infinite gain, is that the output will be whatever it needs to be in order to keep the input voltage zero.

3.2 Ideal Opamp Resistive Amplifier Circuits

3.2.1 Inverting Amplifier Circuit Ideal Opamp Analysis

Since the input voltage difference is zero in the ideal op-amp model, the connection to the minus (−) terminal of the opamp is at the same voltage as the plus (+) terminal, which is grounded to zero, as shown in Figure 3.7.

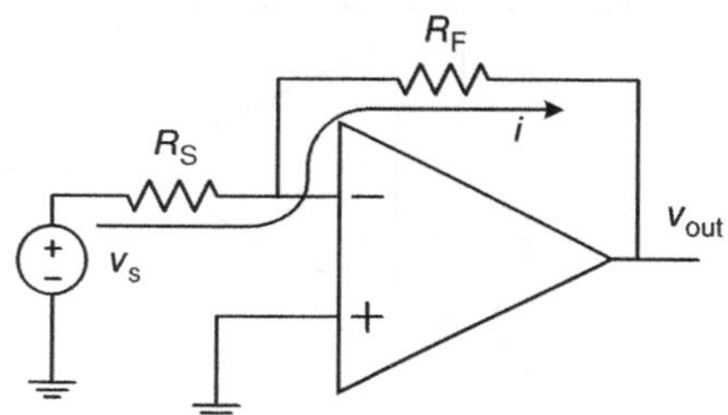

Figure 3.7 Inverting amplifier schematic.

Since no current flows into the $v-$ input terminal to the op-amp, in Figure 3.7 we can picture a current running from the voltage source v_S+ terminal, through R_S, past the $v-$ op-amp terminal, and through R_F to the output v_{out}.

If v_S is a positive voltage (with respect to ground), then current flowing past ground potential and to the output means that the output is at a negative voltage.

You can picture this like a voltage divider in Figure 3.8 where the middle is always at ground, indicated by the dashed ground symbol. If v_S goes negative, then v_{out} would be positive because the current would flow from ground potential to the output (Figure 3.7). Since the output v_{out} must always swing to the opposite polarity than v_S, Figure 3.7 is an inverting amplifier.

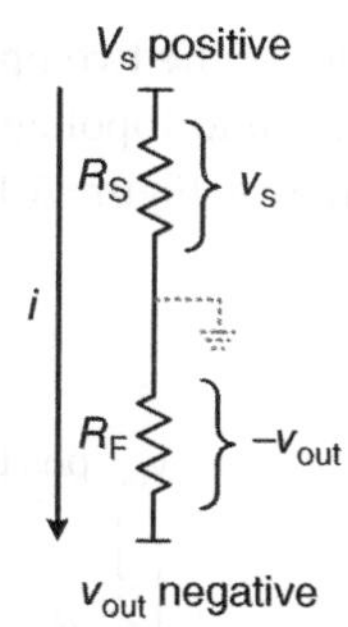

Figure 3.8 Voltage divider around opamp negative input terminal.

No current flows into or out of the $v-$ input terminal of the op-amp (Figure 3.9). The KCL (Kirchhoff's current law) Equation 3.3 sums the currents flowing into the $v-$ node (connecting the two resistors). This KCL equation only has two terms, one for each resistor current.

$$\frac{v_S - 0}{R_S} + \frac{v_{out} - 0}{R_F} = 0 \tag{3.3}$$

$$\frac{v_S}{R_S} = \frac{-v_{out}}{R_F} \tag{3.4}$$

Solve Equation 3.4 for v_{out} over v_S, as shown in Equation 3.5.

$$\frac{-R_F}{R_S} = \frac{v_{out}}{v_S} \tag{3.5}$$

Equation 3.5 characterizes the inverting amplifier. The ratio of output voltage to input voltage is the negative of the ratio of the output resistor to the input resistor.

Using the ideal opamp model, the inverting amplifier would work regardless of which opamp terminal was connected to the resistors. But this is not the case using an opamp model with finite input voltage.

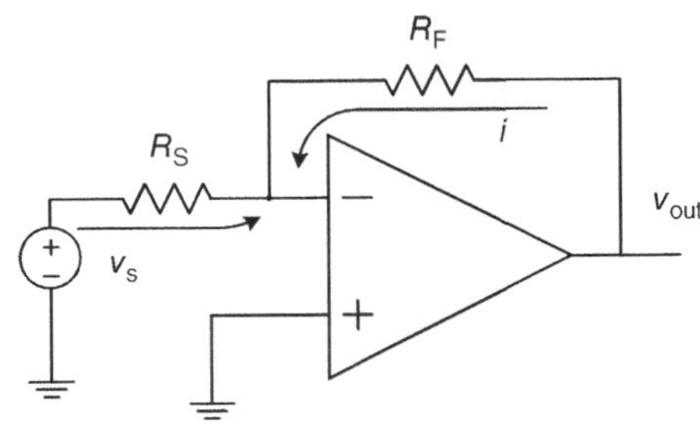

Figure 3.9 Inverting amplifier schematic showing no input current into opamp negative terminal.

3.2.2 Non-Inverting Amplifier Circuit Ideal Opamp Analysis

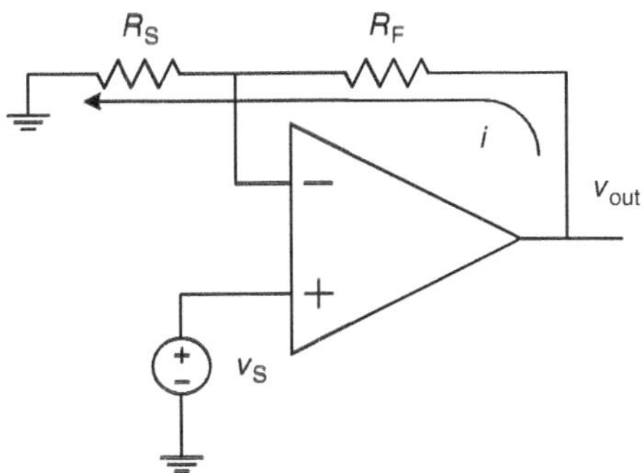

Figure 3.10 Non-inverting amplifier schematic drawn with negative terminal up.

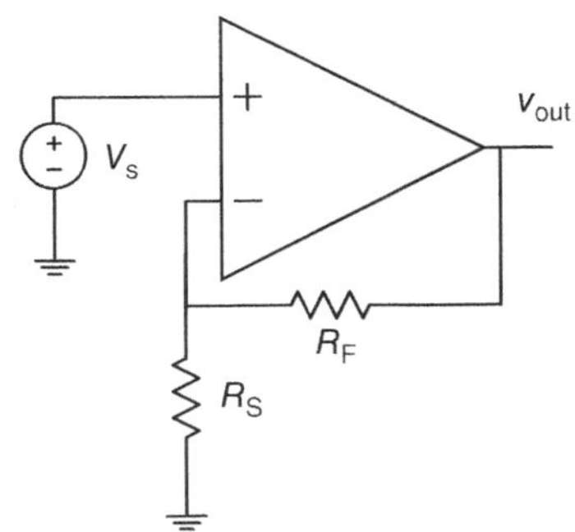

Figure 3.11 Non-inverting amplifier drawn with negative terminal down.

Both of the two opamp circuits, in Figures 3.10 and 3.11, have the same topology. Figure 3.10 has the $v-$ terminal on top, whereas Figure 3.11 has the $v+$ terminal on top.

The Ideal Op-amp Model

Input voltage ($v+ - v-$) is zero.
Input current is zero.
The output voltage is whatever it needs to be to keep the input voltage equal to zero.

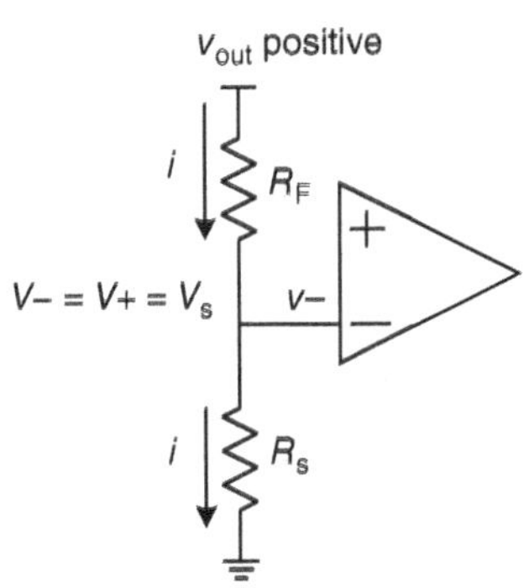

Figure 3.12 Non-inverting amplifier schematic with resistors drawn to indicate voltage divider.

Since the input voltage difference is zero using the ideal op-amp model, the connection to the minus (−) terminal of the opamp is at the same voltage as the plus (+) terminal, which is driven by the source v_S.

Since no current flows into or out of the opamp through the $v-$ input terminal, we can picture a current running from the output v_{out} through R_F, past the $v-$ op-amp terminal, and through R_F to ground. If you picture this like a voltage divider, as shown in Figure 3.12, then v_{out} will have the same polarity as v_S, but v_{out} will be a different size than v_S, scaled by R_F and R_S.

$$\frac{v_S - 0}{R_S} + \frac{v_S - v_{out}}{R_F} = 0 \tag{3.6}$$

In Equation 3.6 partition the v_S minus v_{out} term into two fractions, obtaining Equation 3.7.

$$\frac{v_S}{R_S} + \frac{v_S}{R_F} = \frac{v_{out}}{R_F} \tag{3.7}$$

Divide both sides of Equation 3.7 by v_S.

$$\frac{1}{R_S} + \frac{1}{R_F} = \frac{v_{out}}{v_S R_F} \tag{3.8}$$

Multiply both sides of Equation 3.8 by R_F.

$$\frac{R_F}{R_S} + \frac{R_F}{R_F} = \frac{v_{out} R_F}{v_S R_F} \tag{3.9}$$

Cancel R_F terms in Equation 3.9 to obtain Equation 3.10.

$$\frac{R_F}{R_S} + 1 = \frac{v_{out}}{v_S} \tag{3.10}$$

Equations 3.10 and 3.11 describe the gain of the ideal opamp non-inverting amplifier.

$$\frac{v_{out}}{v_S} = \frac{R_F}{R_S} + 1 \tag{3.11}$$

4

Reactive Circuit Elements

4.1 Capacitor and Inductor Comparison Summary

Capacitor

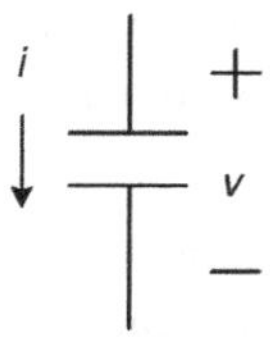

Figure 4.1a Capacitor schematic symbol with passive sign convention polarity marked.

A capacitor stores electric charge.
Capacitance is the ratio of stored charge q to the voltage v across the capacitor. A capacitor schematic symbol with sign convention is shown in Figure 4.1a

$$C = \frac{q}{v} \tag{4.1a}$$

$$q = CV \tag{4.2a}$$

q is the charge on the capacitor. C is the capacitance.

$$i = C\frac{dV}{dt} \tag{4.3a}$$

i is current through the capacitor.

The unit of capacitance is the farad, which equals one coulomb per volt.

$$1\ \mathrm{F} = 1\ \frac{\mathrm{C}}{\mathrm{V}} \tag{4.4a}$$

Inductor

An inductor stores magnetic flux.
Inductance is the ratio of stored flux ϕ (phi) to the current i through the inductor. An inductor schematic symbol with sign convention is shown in Figure 4.1b.

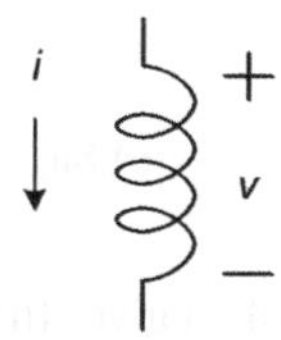

Figure 4.1b Inductor schematic symbol with passive sign convention polarity marked.

ϕ is the flux in the inductor. L is the inductance.

$$L = \frac{\phi}{i} \tag{4.1b}$$

$$\phi = Li \tag{4.2b}$$

v is the voltage across the inductor.

$$v = L\frac{di}{dt} \tag{4.3b}$$

The unit of inductance is the henry, which equals a volt-second per amp.

$$1\ \mathrm{H} = 1\ \frac{\mathrm{V\,s}}{\mathrm{A}} \tag{4.4b}$$

First and Second Order Circuits and Equations: Technical Background and Insights, First Edition. Robert O'Rourke.

Capacitor

If the capacitance is constant with respect to time, Equation 4.5a applies.

$$v_C = \frac{1}{C}\int_0^t i_C(\tau)\,d\tau + v_0 \tag{4.5a}$$

The voltage across a capacitor is the integral of the current through the capacitor with respect to time plus the initial voltage across the capacitor.

$$\Delta i \neq C\frac{\Delta v}{dt} \tag{4.6a}$$

Equation 4.6a indicates that the voltage across a capacitor cannot change instantaneously. An ideal capacitor can't have a finite change in voltage in an infinitesimally small time *dt*.

$$Z_C = \frac{1}{sC} \tag{4.7a}$$

Equation 4.7a gives capacitor impedance in the Laplace *s*-domain.

$$Z_C = \frac{1}{j\omega C} \tag{4.8a}$$

Equation 4.8a gives the impedance of a capacitor, one over *j* omega ($j\omega$) times capacitance in sinusoidal steady state.

The ω term in the denominator of Equation 4.8a indicates that a capacitor looks like a short circuit (low impedance) to high frequencies. A capacitor looks like an open circuit (high impedance) to low frequencies.

Voltage phase lags current phase for a capacitor when plotted in the impedance plane, shown in Figure 4.2a. This matches Equation 4.6a intuitively. Current can be applied to a capacitor instantaneously, but voltage across a capacitor must build up over time.

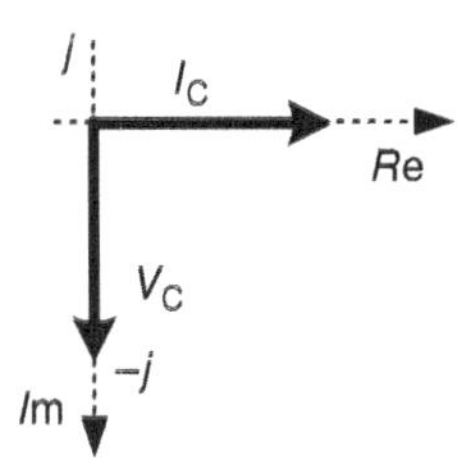

Figure 4.2a
Capacitor voltage and current phasors.

Inductor

If the inductance is constant with respect to time, Equation 4.5b applies.

$$i_L = \frac{1}{L}\int_0^t v(\tau)\,d\tau + I_0 \tag{4.5b}$$

The current through an inductor is the integral of the voltage across the inductor with respect to time plus the initial current in the inductor.

$$\Delta v \neq L\frac{\Delta i}{dt} \tag{4.6b}$$

Equation 4.6b indicates that the current through an inductor cannot change instantaneously. An ideal inductor can't have a finite change in current in an infinitesimally small time *dt*.

$$Z_L = sL \tag{4.7b}$$

Equation 4.7b gives inductor impedance in the Laplace *s*-domain.

$$Z_L = j\omega L \tag{4.8b}$$

Equation 4.8b gives the impedance of an inductor, *j* omega ($j\omega$) times inductance in sinusoidal steady state.

The ω term in the numerator of Equation 4.8b indicates that an inductor looks like a short circuit (low impedance) to low frequencies. An inductor looks like an open circuit (high impedance) to high frequencies.

Voltage phase leads current phase for an inductor when plotted in the impedance plane, shown in Figure 4.2b. This matches Equation 4.6b intuitively. Voltage can be applied to an inductor instantaneously, but current through an inductor must build up over time.

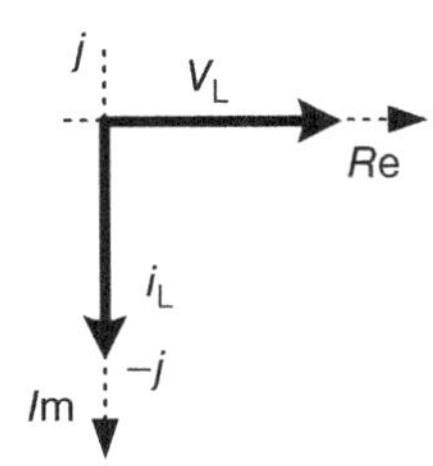

Figure 4.2b
Inductor voltage and current phasors.

Capacitor

Capacitance adds directly for capacitors connected in parallel, as shown in Figure 4.3a and Equation 4.9a.

$$C_{\text{Parallel}} = C_1 + C_1 + \cdots C_N \qquad (4.9a)$$

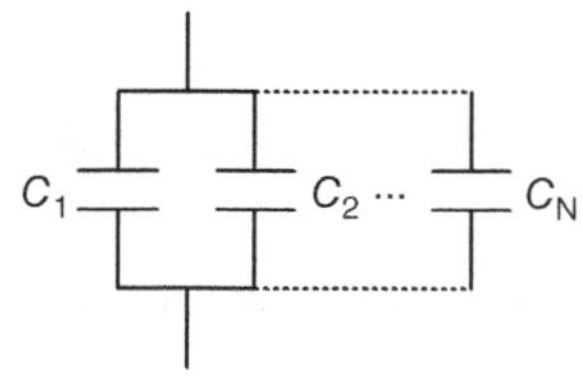

Figure 4.3a *N* capacitors in parallel.

Inductor

Inductance adds directly for inductors connected in series, as shown in Figure 4.3b and Equation 4.9b.

$$L_{\text{Series}} = L_1 + L_1 + \cdots L_N \qquad (4.9b)$$

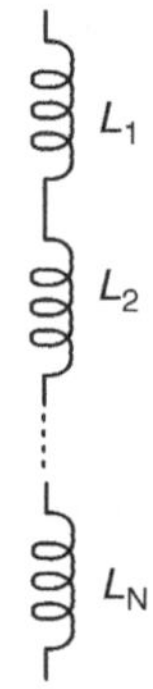

Figure 4.3b *N* inductors in series.

4.2 Capacitors

4.2.1 The Capacitor Circuit Element

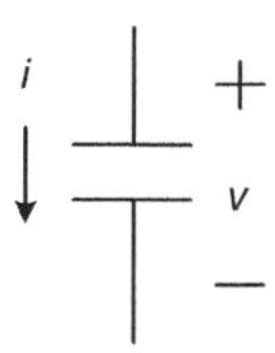

Figure 4.4 Capacitor schematic symbol with passive sign convention polarity marked.

A capacitor is a circuit element, shown in Figure 4.4, that stores energy in an electric field, based on stored electric charge, shown in Figure 4.5

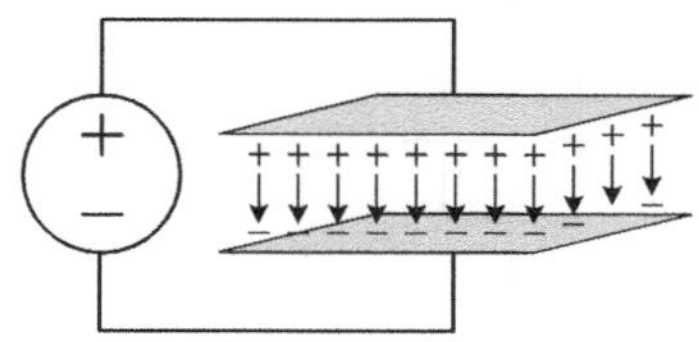

Figure 4.5 Parallel-plate capacitor drawing driven by lumped element voltage.

$$C = \frac{q}{v} \qquad (4.10)$$

From a circuit element point of view, we can define capacitance as the ratio of stored charge q to the voltage v across the capacitor, as shown in Equation 4.10.

A capacitor consists of two conductors and an insulating dielectric between them. The dielectric can be free space, air, or other dielectric materials with a wide range of permittivities.

From Equation 4.10, capacitance equals q divided by v. You can think of capacitance as a capacity, how much charge can you store in relation to the voltage.

4.2.1.1 Capacitor Branch Relationships

$$q = Cv \qquad (4.11)$$

The charge stored on one plate of an ideal capacitor equals the capacitance times the voltage across the capacitor as shown in Equation 4.11.

$$i = \frac{dq}{dt} = \frac{d(Cv)}{dt} = C\frac{dv}{dt} \quad (4.12)$$

Current is the derivative of charge with respect to time, as shown in Equation 4.12. If the capacitance is constant, the current through a capacitor equals the capacitance times the derivative of the voltage across the capacitor with respect to time, as shown in Equation 4.13.

$$i_C = C\frac{dv_C}{dt} \quad (4.13)$$

As a branch relationship for an ideal capacitor, Equation 4.13 relates the current through a capacitor to the voltage across the capacitor.

$$v = \frac{1}{C}\int_0^T i dt + v_0 \quad (4.14)$$

The voltage across a capacitor is the integral of the current through the capacitor, as shown in Equation 4.14, which is also a branch relationship.

Note how the integral starts at time zero, but the initial condition voltage v-nought (v_0) might represent what happened to the capacitor before time zero.

4.2.2 Capacitors in Series and Parallel

4.2.2.1 Parallel Capacitances

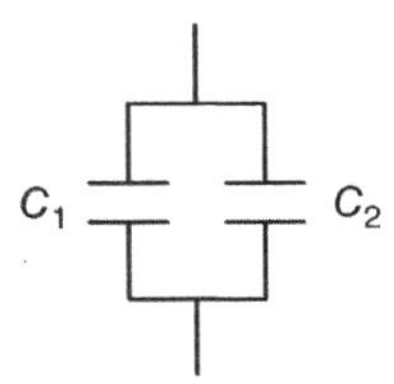

Figure 4.6 Two capacitors in parallel.

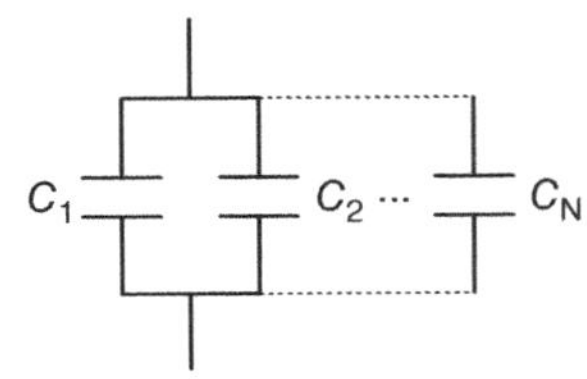

Figure 4.7 *N* capacitors in parallel.

$$C_{\text{Parallel}} = C_1 + C_2 \quad (4.15)$$

In Equation 4.15, C_{Parallel} is the equivalent parallel capacitance for two capacitors C_1 and C_2, as shown in Figure 4.6.

$$C_{\text{Parallel}} = C_1 + C_1 + \cdots C_N \quad (4.16)$$

In Equation 4.16, C_{Parallel} is the equivalent parallel capacitance for N capacitors shown in Figure 4.7.

4.2.2.2 Intuitive Memory Aid for Parallel Capacitance

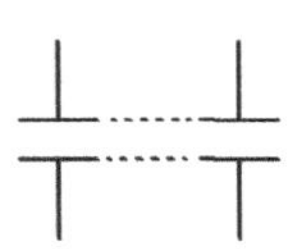

Figure 4.8 Capacitor schematic symbol dotted line visualization of plate size.

Picture parallel-plate capacitors, where a parallel connection of two capacitors can seem like simply joining the plates into one larger capacitor, as shown in Figure 4.8. This visual cue reminds us that the capacitances of parallel-plate capacitors simply add. The voltage is the same across both capacitors, but the charge in the joined capacitors is the combined amount of charge from each capacitor.

4.2.2.3 Capacitor Series Connections

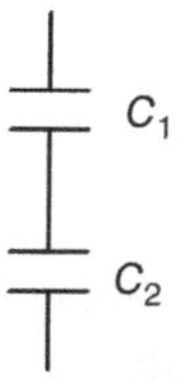

Figure 4.9 Two capacitors in series.

Figure 4.9 shows two capacitors in series. Their equivalent capacitance is given in Equation 4.17.

$$C_{\text{Series}} = \frac{C_1 C_2}{C_1 + C_2} \tag{4.17}$$

$$\frac{1}{C_{\text{Series}}} = \frac{1}{C_1} + \frac{1}{C_2} + \cdots + \frac{1}{C_N} \tag{4.18}$$

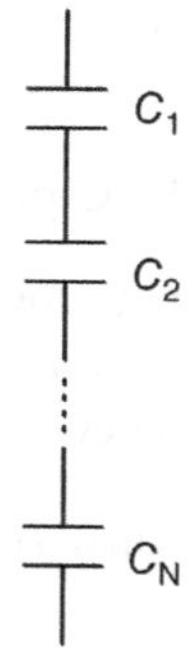

Figure 4.10 *N* capacitors in series.

Figure 4.10 shows an arbitrary number *N* of capacitors in parallel. Their equivalent capacitance is given in Equation 4.18.

4.2.2.4 Intuitive Memory Aid for Series Capacitance

The plus and minus charges on the plates between the two capacitors in Figures 4.9 and 4.10 would cancel one another somewhat, reducing the overall capacitance.

4.2.3 Capacitor Time Domain Behavior

$$i = C\frac{dv}{dt} \tag{4.19}$$

$$i \neq C\frac{\Delta v}{dt} \tag{4.20}$$

You can't change the voltage across a capacitor instantaneously. (That would require an infinite current.)

The voltage across a capacitor cannot change a finite amount in an infinitesimally small amount of time (instantaneously). Rather, the voltage across the capacitor changes gradually enough over time for charge to accumulate on (or dissipate from) the capacitor and change the voltage. The inequality in Equation 4.20 reminds us of this limitation.

Since the amount of current that flows into (or out of) a capacitor equals the capacitance times the derivative of the voltage across the capacitor with respect to time, shown in Equation 4.19, the larger the current, the larger the rate of change of voltage across the capacitor. In order to effect an instantaneous change in the voltage, an infinite amount of current would be required.

The mathematical expressions "*dv*" and "*dt*" represent infinitesimals. They are infinitely small, related to the limit in the definition of the derivative. By comparison, Δv and Δt are finite differences.

Equation 4.21 shows the definition of the derivative of voltage $v = f(t)$ with respect to time t.

$$\frac{dv}{dt} = \lim_{\Delta t \to 0} \frac{\Delta v}{\Delta t} = \frac{f(t + \Delta t) - f(t)}{\Delta t} \tag{4.21}$$

The derivative takes the limit as the finite Δt goes to zero; Δt becomes the infinitesimal dt, and finite Δv becomes infinitesimal dv in the derivative. An instantaneous change in voltage would imply a finite Δv in an infinitesimally small amount of time dt, or $i = C\,(\Delta v/dt)$. But, in a quotient $\Delta v/dt$, if the denominator (dt) goes to zero, while the numerator (Δv) remains finite sized, the whole expression (the current) increases to infinity. Hence the not equals sign in Equation 4.20.

4.2.3.1 Hydraulic Analogy for Capacitance

Filling a capacitor with charge is like filling up a container with water. The faster you want to fill the container, the more water you need in a short amount of time. But if the container has a finite size, like 10 l, then to fill it instantaneously would require a rate of water flow into the container increasing to infinity. Capacitance can represent elasticity in the sides of the container that exert a force on the water corresponding to the voltage on the capacitor.

4.2.4 Capacitor Voltage Phase Lags Current Phase

Periodic function signals, like sine and cosine, are common in electronics. Look at the phase relationship of voltage and current in circuit elements to develop concepts like phase lead and lag and impedance for use in steady-state frequency domain analysis.

$$v_C = \sin(\omega t) \tag{4.22}$$

Consider a voltage across a capacitor, where the voltage varies in time with the function sin of omega times t, as shown in Equation 4.22.

Calculate the current through the capacitor to understand the phase relationship with the voltage across the capacitor.

From Equation 4.22, substitute $\sin(\omega t)$ into Equation 4.19, the equation for the current in a capacitor, to obtain Equation 4.23.

$$i_C = C\frac{dv_C}{dt} = C\frac{d}{dt}[\sin(\omega t)] \tag{4.23}$$

The derivative of sine is cosine. Since ωt is also a function, the derivative in Equation 4.23 will use the chain rule and have an ω multiplying cosine ωt, as shown in Equation 4.24.

$$C\frac{d}{dt}[\sin(\omega t)] = C\omega\cos(\omega t) \tag{4.24}$$

In Equation 4.24, replace $\cos(\omega t)$ with $\sin(\omega t)$ plus π over 2 to obtain Equation 4.25.

$$i_C = C\frac{dv_C}{dt} = \omega C \sin\left(\omega t + \frac{\pi}{2}\right) \tag{4.25}$$

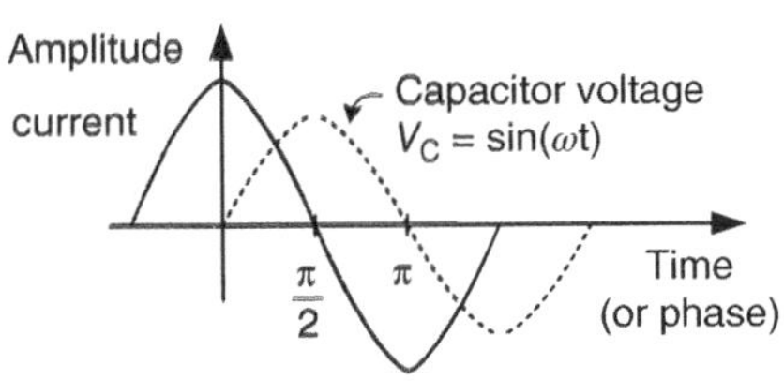

Figure 4.11 Capacitor voltage sinusoid lagging current sinusoid in phase.

Equation 4.25 expresses current through the capacitor, i_C, in terms of sine in order to compare the phase of the current through the capacitor to the phase of the voltage across the capacitor, as shown in Figure 4.11.

The voltage across the capacitor lags the current through the capacitor by 90° or pi over 2 rad, in Equation 4.25 and Figure 4.11. The vertical axis amplitudes in Figure 4.11 are not quantified.

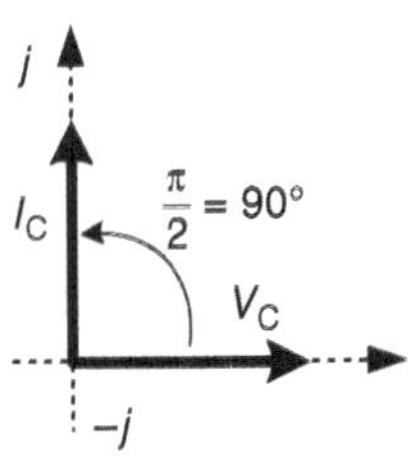

Figure 4.12 Capacitor phasor diagram with capacitor voltage as reference.

In the same way that sine and cosine function phasors can be mapped in a complex plane, we can plot the capacitor voltage and current phase relationship in the complex plane. The curved arrow in Figure 4.12 shows positive angle progression. The phase of the voltage across the capacitor, on the horizontal axis in Figure 4.12, lags the phase of the current through the capacitor, on the vertical axis, by 90°.

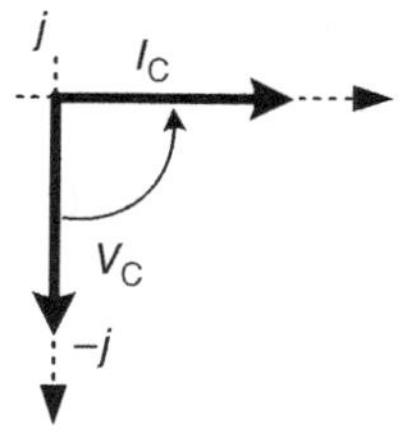

Figure 4.13 Capacitor phasor diagram with current as reference.

The plot in Figure 4.13, of the phase of the voltage across the capacitor lagging the phase of the current through the capacitor, shows the same 90° or pi over two-phase difference as Figures 4.11 and 4.12 show. Sometimes current is taken as a reference and plotted on the horizontal axis, which makes the voltage vector point down.

4.3 Inductors

4.3.1 The Inductor Circuit Element

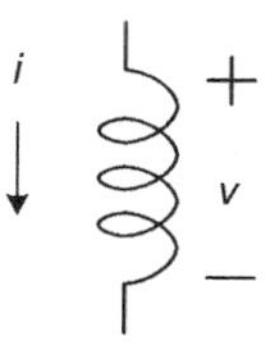

Figure 4.14 Inductor schematic symbol with passive sign convention polarity marked.

An inductor, shown in Figures 4.14 and 4.15, is a circuit element that stores energy in a magnetic field. Inductance produces induced voltage when the magnetic flux varies.

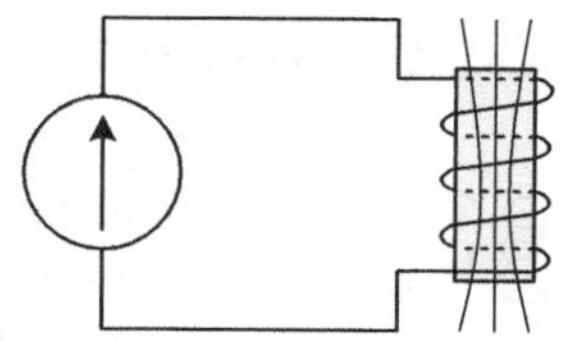

Figure 4.15 Inductor diagram with coil and fields and driven by a lumped element current source.

$$L = \frac{\phi}{i} \tag{4.26}$$

From a lumped circuit element point of view, inductance is the ratio of the total magnetic flux ϕ to the current i through the inductor, as shown in Equation 4.26. Inductance of a single conductor is also referred to as self-inductance to distinguish it from mutual inductance involving adjacent conductors.

An inductor consists of a metal conductor that typically has windings or some type of coil around a core. The core might be ferrite, metal, air, or substrate material. An individual straight wire has inductance, which becomes more important as signal frequency goes up, for example where the wire is a metal signal trace on a printed circuit board or integrated circuit.

4.3.1.1 Inductor Branch Relationships

The magnetic flux in an ideal inductor equals the inductance times the current through the inductor, as shown in Equation 4.27.

$$\phi = Li \tag{4.27}$$

Voltage across an inductor is the derivative of flux in the inductor with respect to time, as shown in Equation 4.28.

$$v = \frac{d\phi}{dt} \tag{4.28}$$

$$v = \frac{d(Li)}{dt} \quad (4.29)$$

$$V = L\frac{di}{dt} \quad (4.30)$$

If the inductance is constant with respect to time, then voltage in Equation 4.29 equals Equation 4.30.

Equation 4.30 states that the voltage across an inductor equals the inductance times the derivative of the current through the inductor with respect to time. This is an inductor branch relationship that relates the current through the inductor to the voltage across the inductor.

$$i = \frac{1}{L}\int_0^T v\,dt + i_0 \quad (4.31)$$

Equation 4.31 states that the current through an inductor is the integral of the voltage across the inductor. Equation 4.31 is also a statement of an inductor branch relationship.

Note in Equation 4.31 how the integral of the voltage starts at time zero, but the initial condition current *i*-nought (i_0) might represent what happened before time zero.

4.3.2 Inductors in Series and Parallel

4.3.2.1 Inductor Series Connections

Figure 4.16 Two inductors in series.

Figure 4.16 shows two inductances in series. Their equivalent inductance is given in Equation 4.32.

$$L_{\text{Series}} = L_1 + L_1 \quad (4.32)$$

Figure 4.17 and Equation 4.33 apply to an arbitrary number *N* of inductors in series.

$$L_{\text{Series}} = L_1 + L_1 + \cdots L_N \quad (4.33)$$

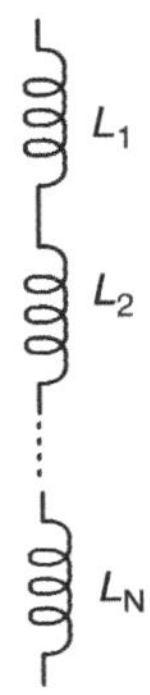

Figure 4.17 *N* inductors in series.

4.3.2.2 Intuitive Memory Aid

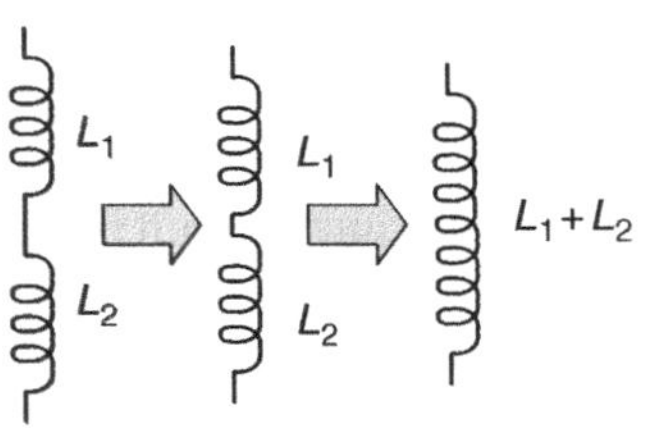

Figure 4.18 Inductor symbol joining to visualize inductance adding in series.

Think of an inductor as coils in a wire, where more coils on one wire makes more inductance. As Figure 4.18 shows, two inductors in series are like twice the number of coils on one inductor. This suggests that the inductances in series simply add. The current is the same through both inductors in series, but the overall flux is increased for two inductors in series.

4.3.2.3 Inductor Parallel Connections

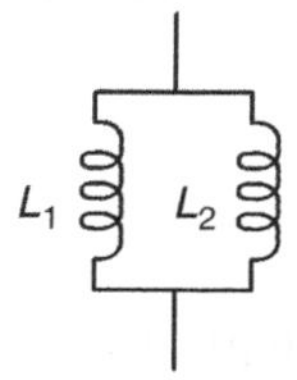

Figure 4.19 Two inductors in parallel.

Figure 4.19 shows two inductances in parallel. Their equivalent inductance is given in Equation 4.34.

$$L_{\text{Parallel}} = \frac{L_1 L_2}{L_1 + L_2} \tag{4.34}$$

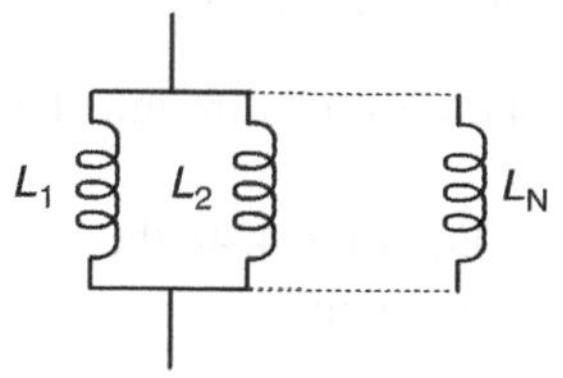

Figure 4.20 *N* inductors in parallel.

Figure 4.20 shows an arbitrary number N of inductors in parallel. Their equivalent inductance is given in Equation 4.35.

$$\frac{1}{L_{\text{Parallel}}} = \frac{1}{L_1} + \frac{1}{L_2} + \cdots \frac{1}{L_N} \tag{4.35}$$

4.3.3 Inductor Time Domain Behavior

$$v = L\frac{di}{dt} \tag{4.36}$$

$$v \neq L\frac{\Delta i}{dt} \tag{4.37}$$

You can't change the current flow into an inductor instantaneously. (That would require an infinite voltage.)

The current flow through an inductor cannot change a finite amount in an infinitesimally small amount of time (instantaneously). Rather, the current through the inductor changes gradually enough over time for magnetic flux to accumulate in (or dissipate from) the inductor and change the current flow. The inequality in Equation 4.37 reminds us of this limitation.

Since the voltage across an inductor equals the inductance times the derivative of the current through the inductor with respect to time, as shown in Equation 4.36, the larger the voltage, the larger the rate of change of current through the inductor. In order to effect an instantaneous change in the current, an infinite amount of voltage across the inductor would be required.

The expressions "*di*" and "*dt*" represent infinitesimals. They are infinitely small, related to the limit in the definition of the derivative. By comparison, Δi and Δt are finite differences.

Equation 4.38 is the definition of the derivative of current i equals $f(t)$ with respect to time t.

$$\frac{di}{dt} = \lim_{\Delta t \to 0} \frac{\Delta i}{\Delta t} = \frac{f(t + \Delta t) - f(t)}{\Delta t} \tag{4.38}$$

The derivative takes the limit as the finite Δt goes to zero; Δt becomes the infinitesimal dt, and finite Δi becomes di in the derivative. An instantaneous change would imply a finite Δi in an infinitesimally small amount of time dt. But, in a quotient $\Delta i/dt$, if the denominator (dt) goes to zero, while the numerator (Δi) remains finite sized, the whole expression (voltage) increases to infinity.

4.3.3.1 Hydraulic Analogy for Inductance

Current flowing in an ideal inductor is like water flowing in a frictionless pipe. The inductance corresponds to the inertia of the flowing water. Voltage corresponds to water pressure. Acceleration of water flow corresponds to the rate of change of the current. Changing the voltage on the inductor can change the current flow, and it is like changing the water pressure to change fluid flow. But the rate of current flow in an inductor changes gradually, even with a sudden change in in voltage. To instantaneously change the current flow would require an infinite amount of voltage.

4.3.4 Time and Frequency – Inductor Voltage Phase Leads Current Phase

Periodic function signals, like sine and cosine, are common in electronics. Look at the phase relationship of voltage and current in circuit elements to develop concepts like phase lead and lag and impedance for use in steady-state frequency domain analysis.

$$i_L = \sin(\omega t) \tag{4.39}$$

Consider a current through an inductor, where the current varies in time with the function sin of omega times t, as shown in Equation 4.39.

Calculate the voltage across the inductor to understand the phase relationship with the current through the inductor.

$$v_L = L\frac{di_L}{dt} = L\frac{d}{dt}[\sin(\omega t)] \tag{4.40}$$

From Equation 4.39, substitute sin(ωt) into Equation 4.40, the equation for the voltage across an inductor, to obtain Equation 4.41.

The derivative of sine is cosine. Since ωt is also a function, the derivative in Equation 4.40 will use the chain rule and have an ω multiplying cosine ωt, as shown in Equation 4.41.

$$L\frac{d}{dt}[\sin(\omega t)] = L\omega\cos(\omega t) \tag{4.41}$$

In Equation 4.41, replace cos(ωt) with sin(ωt) plus π over 2 to obtain Equation 4.42.

$$v_L = L\frac{di_L}{dt} = \omega L \sin\left(\omega t + \frac{\pi}{2}\right) \tag{4.42}$$

Equation 4.42 now expresses voltage across the inductor, v_L, in terms of sine in order to compare the phase of the voltage the inductor to the phase of the current through the inductor.

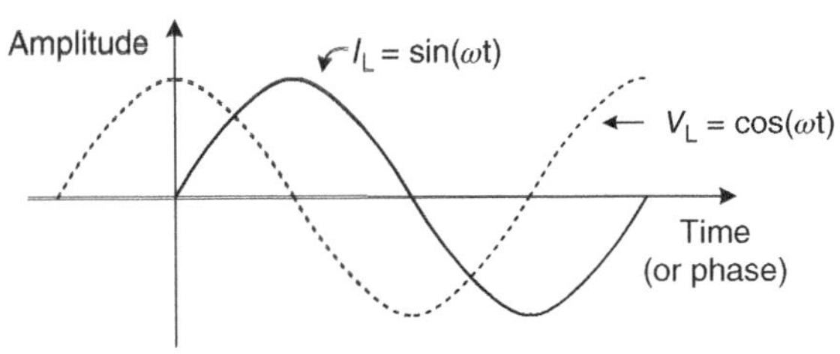

Figure 4.21 Inductor voltage sinusoid leading current sinusoid in phase.

The current through the inductor lags the voltage across the inductor by 90° or pi over 2 rad, as shown in Equation 4.42 and Figure 4.21. The vertical axis amplitudes in Figure 4.21 are not quantified.

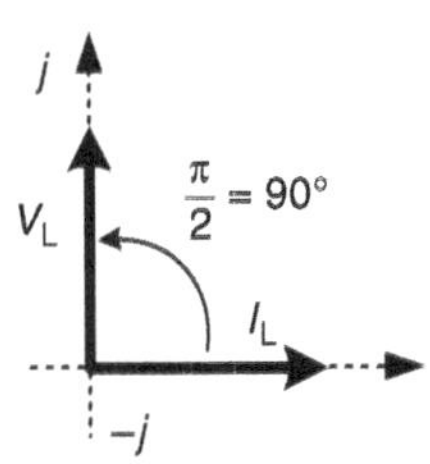

Figure 4.22 Inductor phasor diagram with inductor current as reference.

In the same way that sine and cosine function phasors can be mapped in a complex plane, we can plot the inductor voltage and current phase relationship in the complex plane. The curved arrow in Figure 4.22 shows positive angle progression going counterclockwise. The phase of the current through the inductor, on the horizonal axis in Figure 4.22, lags the phase of the voltage across the inductor, on the vertical axis, by 90°.

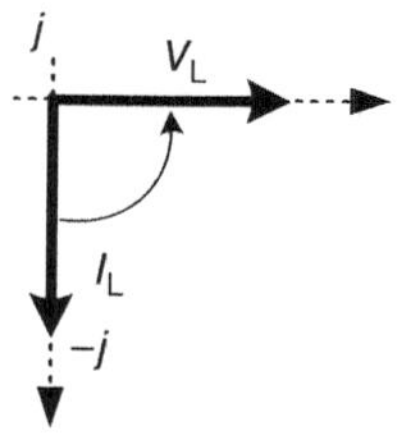

Figure 4.23 Inductor phasor diagram with inductor voltage as reference.

The plot in Figure 4.23, of the phase of the current through the inductor lagging the phase of the voltage across the inductor, shows the same 90° or pi over two phase difference as Figure 4.22 shows. Sometimes voltage is taken as a reference and plotted on the horizontal axis, which makes the current vector point down.

Part 2

First-Order Circuits

5

First-Order RC and RL Circuits Introduction

This chapter introduces first-order circuits in both time and frequency domain. The time domain response from initial conditions is called natural or zero-input response (ZIR). The time domain response from a step function input is a form of zero-state response (ZSR). The frequency domain response analyzed here is the sinusoidal steady-state response.

5.1 What are First-Order Circuits?

5.1.1 A Capacitor or an Inductor Makes a First-Order Circuit

First-order circuits have one reactive circuit element, such as a capacitor or inductor. Capacitors and inductors can store energy.

First-order circuits can make frequency selective filters in the frequency domain.

First-order circuits are described by first-order differential equations.

In the time domain, first-order circuit responses are often rising or decaying exponential functions. (Other responses, like integrators or delta functions, are possible for some topologies and inputs.)

5.1.2 First-Order Circuits Can Filter in the Frequency Domain

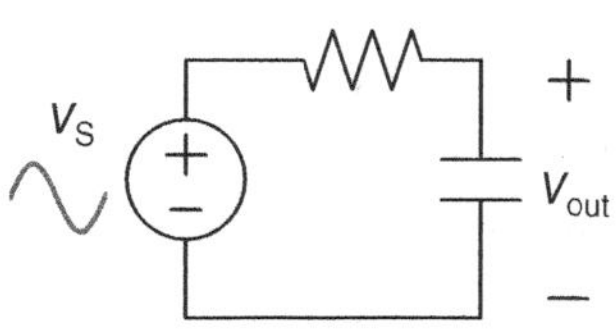

Figure 5.1a Series RC circuit driven by AC voltage.

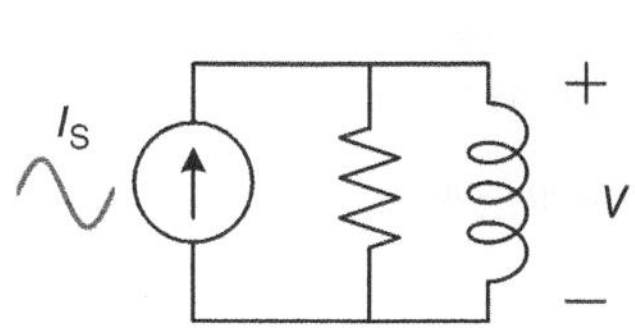

Figure 5.1b Parallel RL circuit driven by AC current.

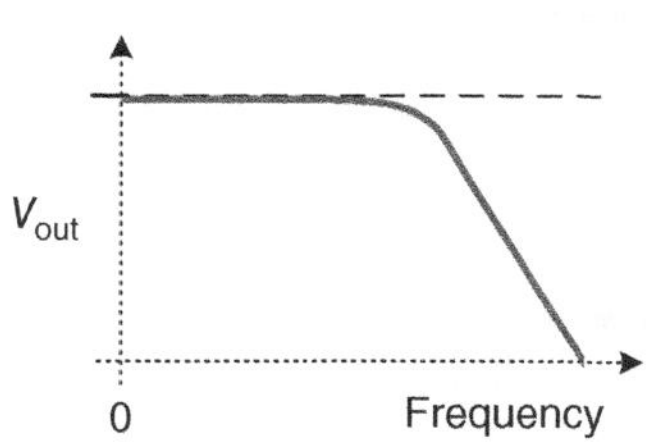

Figure 5.2a RC low-pass filter response.

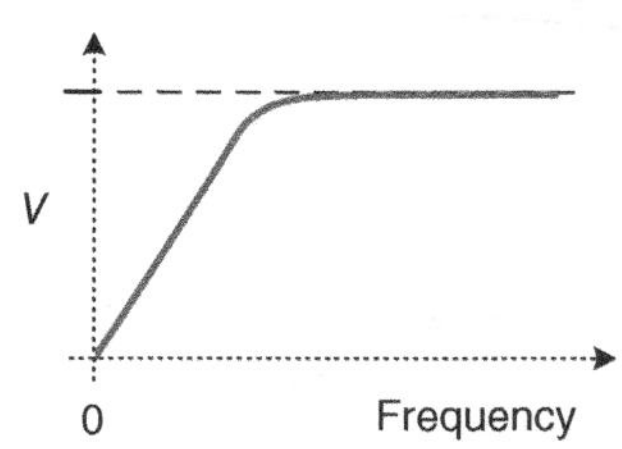

Figure 5.2b RL high-pass filter response.

Figure 5.1a shows a shunt capacitor low-pass filter schematic. The one-pole low-pass frequency response is shown in Figure 5.2a.

Figure 5.1b shows a shunt inductor high-pass filter schematic. The one-pole high-pass frequency response is shown in Figure 5.2b.

First and Second Order Circuits and Equations: Technical Background and Insights, First Edition. Robert O'Rourke.

5.1.3 First-Order Circuits Are Described by First-Order Differential Equations

Ordinary differential equations, with respect to time, describe lumped element circuits, and first-order differential equations describe first-order RC and RL circuits in the time domain. These equations come from applications of Kirchhoff's current and voltage laws.

$$\frac{V_{out} - V_S}{R} + C\frac{dV_{out}}{dt} = 0 \quad (5.1a)$$

Equation 5.1a comes from Kirchhoff's current law (KCL) applied to the resistor–capacitor (RC) circuit in Figures 5.1a and 5.3a.

$$\frac{L}{R}\frac{di_L(t)}{dt} + i_L(t) - I_S = 0 \quad (5.1b)$$

Equation 5.1b comes from Kirchhoff's current law (KCL) applied to the resistor–inductor (RL) circuit in Figures 5.1b and 5.3b.

5.1.4 First-Order Circuit Exponential Step Responses in Time Domain

First-order RC and RL circuits are shown in Figures 5.3a,b. Figures 5.4a and 5.4b show the voltage and current input step functions. The time domain step responses, exponentially decaying signals, are shown in Figures 5.5a,b.

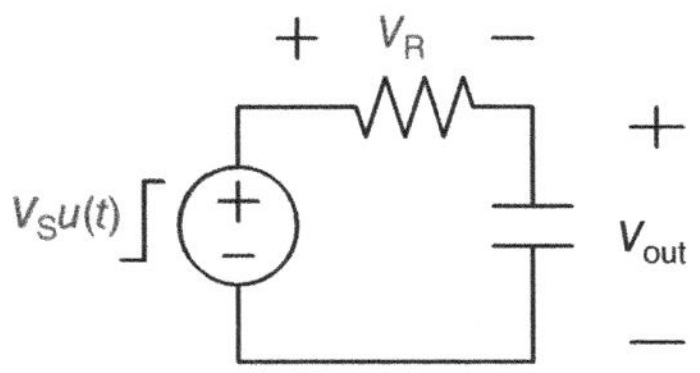

Figure 5.3a Series RC circuit driven by voltage step.

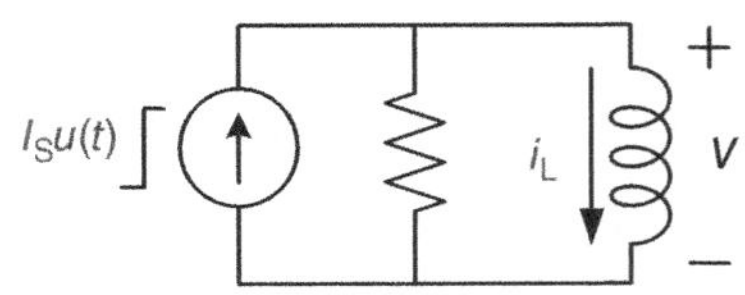

Figure 5.3b Parallel RL circuit driven by current step.

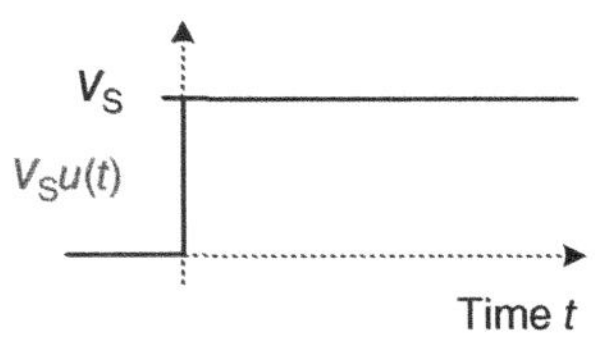

Figure 5.4a Voltage step input function.

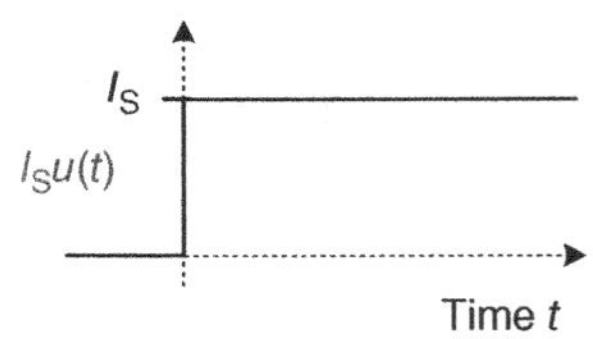

Figure 5.4b Current step input function.

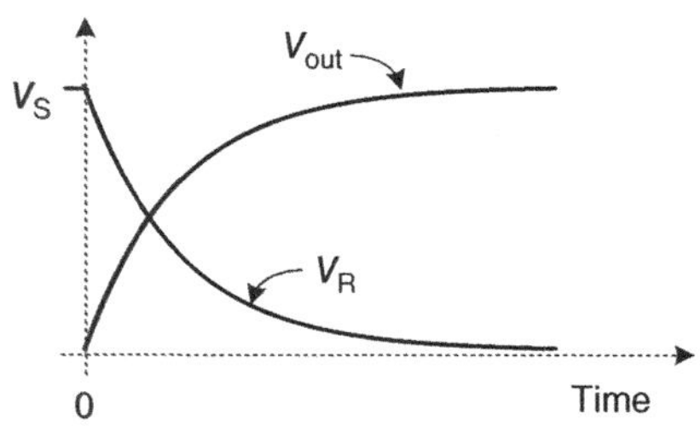

Figure 5.5a RC circuit step voltage responses.

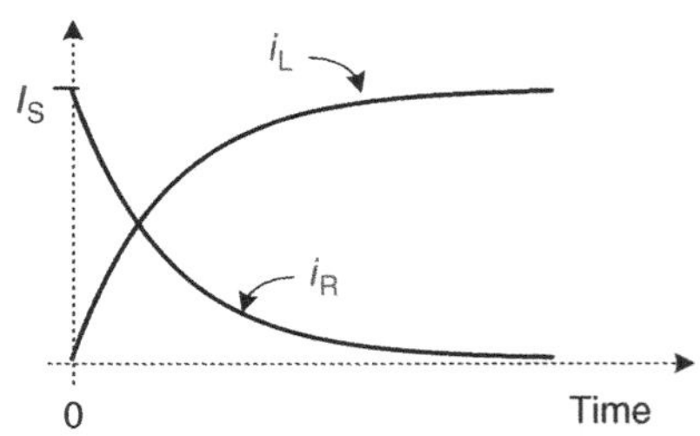

Figure 5.5b RL circuit step current responses.

5.2 Intuitive First-Order Circuit Frequency Domain Examples

Considering the impedance of inductors and capacitors at high and low frequencies, one can determine by inspection whether these first order topologies are low pass or high pass filters in the sinusoidal steady state frequency domain.

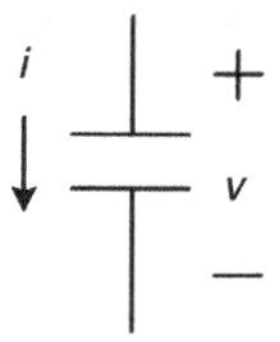

Figure 5.6 Capacitor schematic symbol marked with reference polarity.

$$Z_C = \frac{1}{j\omega C} = \frac{j}{j \cdot j\omega C} = \frac{j}{-\omega C} = -j\frac{1}{\omega C} \tag{5.2}$$

For the capacitor shown in Figure 5.6, the impedance equals one over j omega C $(1/j\omega C)$ and the reactance of the capacitor is minus one over ωC, as shown in Equation 5.2. At high frequencies, capacitor impedance goes down.

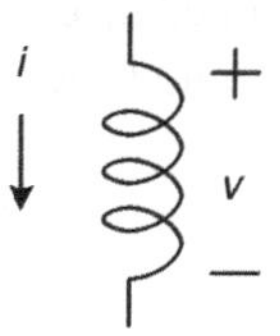

Figure 5.7 Inductor schematic symbol marked with reference polarity.

$$Z_L = j\omega L \tag{5.3}$$

For the inductor shown in Figure 5.7, the impedance is j omega times inductance $(j\omega L)$, as shown in Equation 5.3. At low frequencies, the impedance goes down.

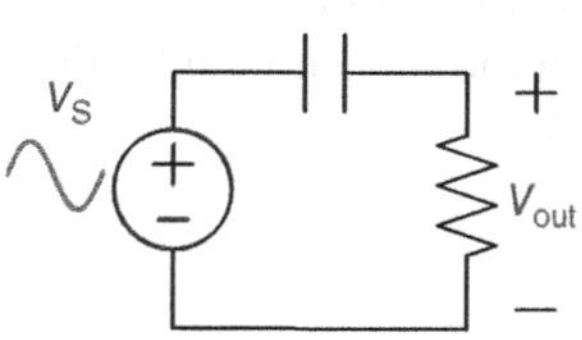

Figure 5.8a RC high-pass filter schematic.

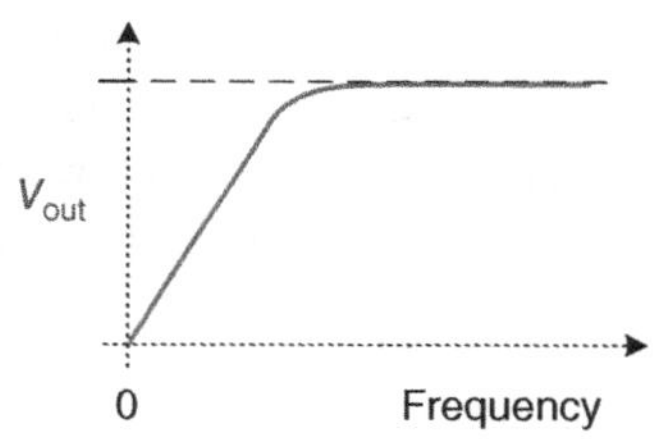

Figure 5.8b RC high-pass filter frequency response.

In Figure 5.8a, the series capacitor is low impedance at high frequency, allowing high frequencies to easily pass from the voltage source to the output voltage V_{out} across the resistor in Figure 5.8b.

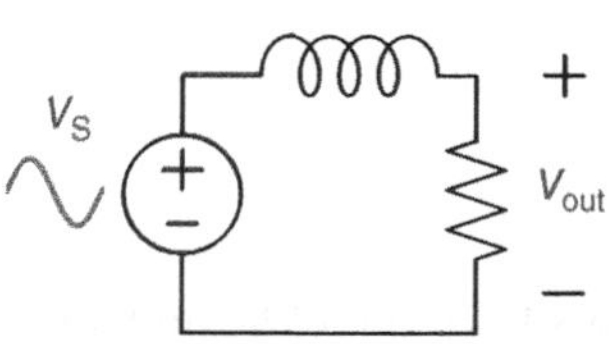

Figure 5.9a RL low-pass filter schematic.

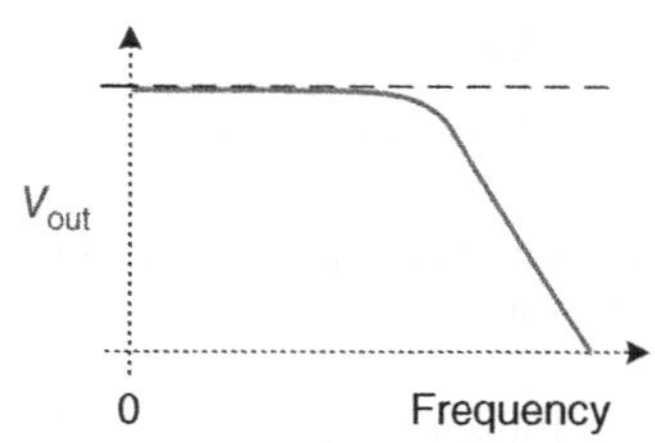

Figure 5.9b RL low-pass filter frequency response.

In Figure 5.9a, the series inductor is low impedance at low frequency, allowing low frequencies to easily pass from the voltage source to the output voltage V_{out} across the resistor in Figure 5.9b.

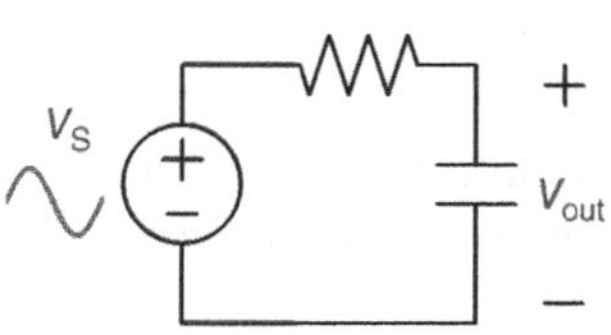

Figure 5.10a RC low-pass filter schematic.

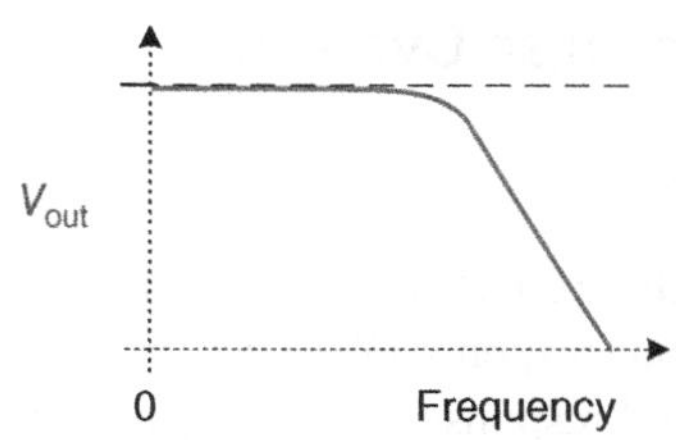

Figure 5.10b RC low-pass filter frequency response.

In Figure 5.10a the shunt capacitor has high impedance at low frequencies and voltage V_{out} can drop across the capacitor as shown in Figure 5.9b. At high frequencies, where the capacitor is low impedance, current flows and little voltage drops across V_{out}.

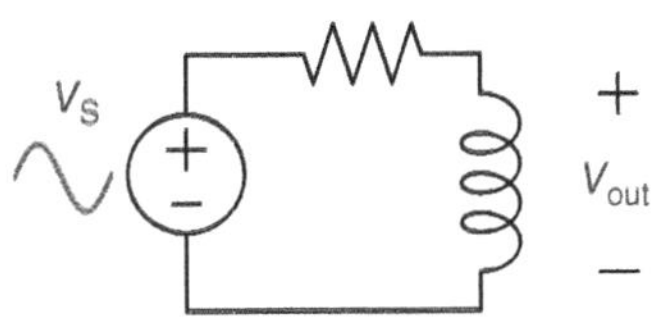

Figure 5.11a RL high-pass filter schematic.

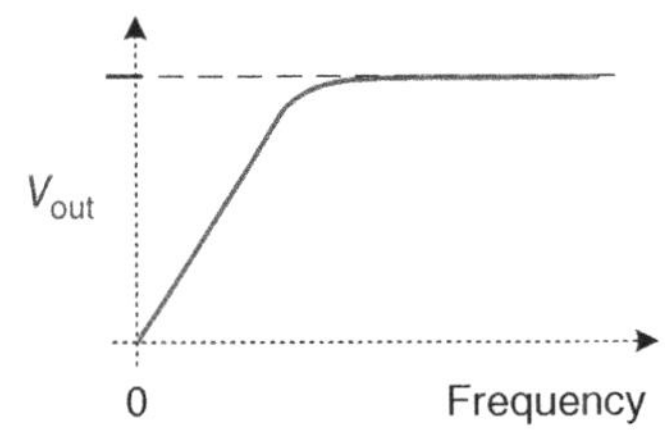

Figure 5.11b RL low-pass filter frequency response.

In Figure 5.11a the shunt inductor has high impedance at high frequencies and voltage V_{out} can drop across the inductor as shown in Figure 5.11b. At low frequencies, where the inductor is low impedance, current flows and little voltage drops across V_{out}.

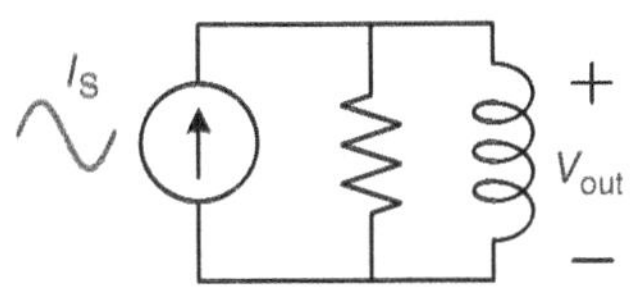

Figure 5.12a RL high-pass filter schematic.

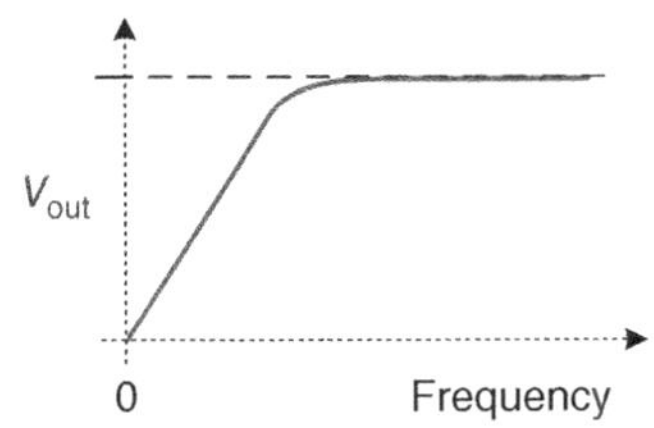

Figure 5.12b RL high-pass filter frequency response.

In Figure 5.12a the shunt inductor has high impedance at high frequencies and voltage V_{out} can drop across the inductor and resistor as shown in Figure 5.12b. At low frequencies, where the inductor is low impedance, current flows through the inductor and little voltage drops across V_{out}.

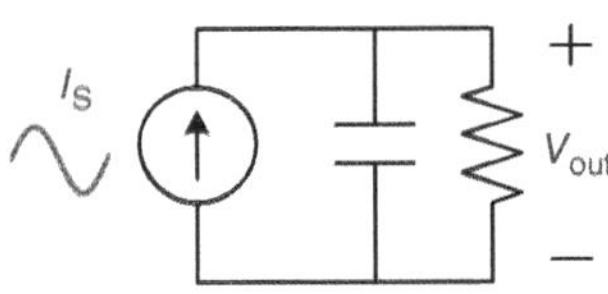

Figure 5.13a RC low-pass filter schematic.

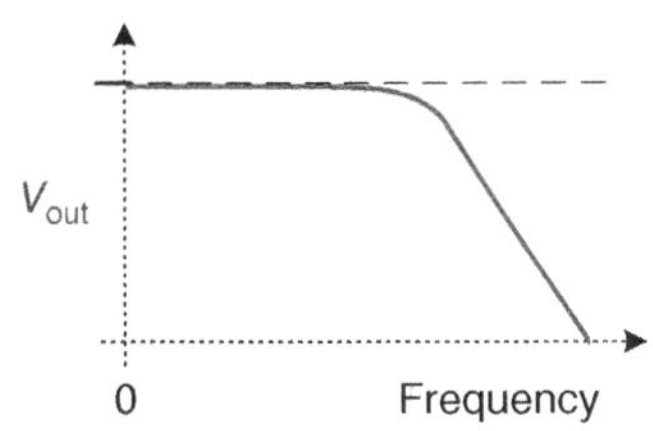

Figure 5.13b RC low-pass filter frequency response.

In Figure 5.13a the shunt capacitor has high impedance at low frequencies and voltage V_{out} can drop across the capacitor and resistor as shown in Figure 5.13b. At high frequencies, where the capacitor is low impedance, current flows through the capacitor and little voltage drops across V_{out}.

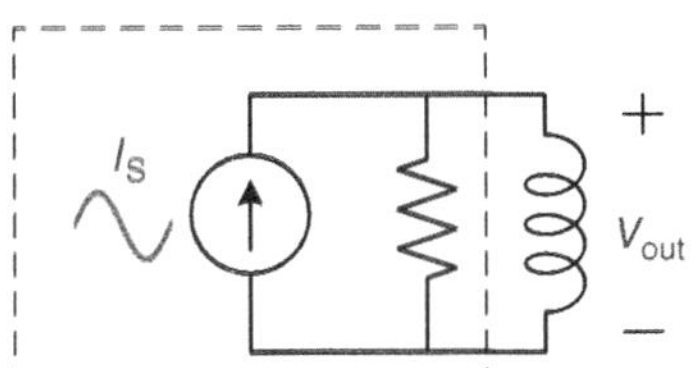

Figure 5.14a Parallel RL schematic driven by AC current source.

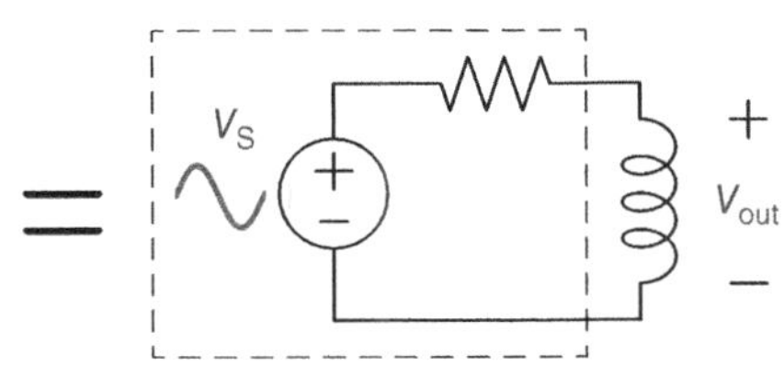

Figure 5.14b Series RL schematic driven by AC voltage source.

From Thévenin's and Norton's theorems we know that these two topologies in Figure 5.14a (and 5.12a) and 5.14b (5.11a) are equivalent. Both are high-pass filters.

5.3 First-Order Natural and Step Response Overview

5.3.1 First-Order RC and RL Natural Response

Natural response is the circuit response behavior that reflects only the nature of the circuit, and not the nature of any input. Natural response is also called zero-input response (ZIR) because there is no input signal driving the circuit. The differential equations describing natural response are homogeneous.

RC Natural Response

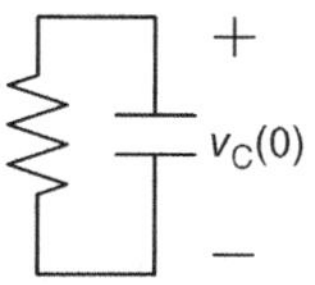

$$C\frac{dv(t)}{dt} + \frac{v(t)}{R} = 0 \quad (5.4a)$$

Figure 5.15a Parallel RC circuit schematic.

$$v(t) = v_C(0)e^{\frac{-t}{RC}} \quad (5.5a)$$

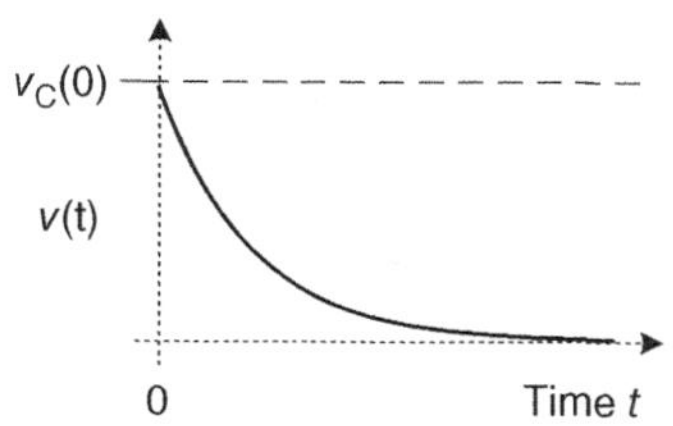

Figure 5.16a RC natural response voltage.

RL Natural Response

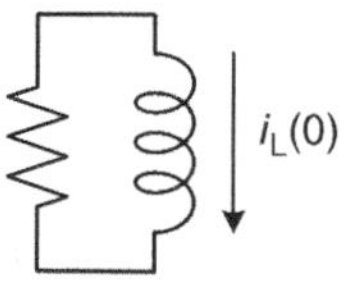

$$L\frac{di(t)}{dt} + i(t)R = 0 \quad (5.4b)$$

Figure 5.15b Parallel RL circuit schematic.

$$i(t) = i_L(0)e^{\frac{-tR}{L}} \quad (5.5b)$$

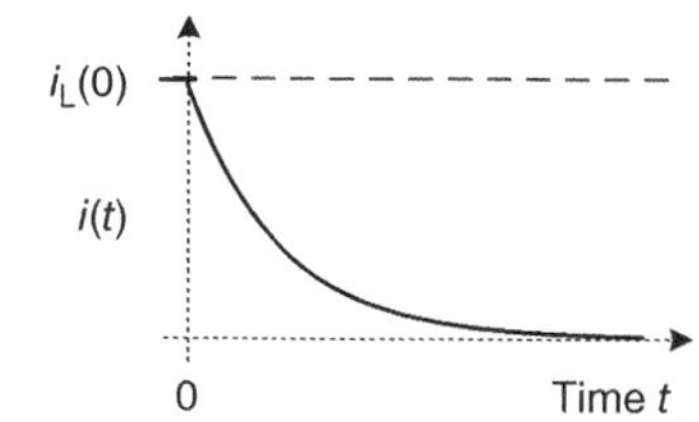

Figure 5.16b RL natural response current.

First-order natural response, RC or RL, comes from an initial condition, an initial capacitor voltage $v_C(0)$ or an initial inductor current $i_L(0)$, decaying to zero over time. Figures 5.15a and 5.15b show RC and RL circuits. The RC and RL natural responses are shown in Figures 5.16a and 5.16b.

Step response is the response to a step function voltage or current input to the circuit, as shown in Section 5.1.4. In contrast to natural response, step response is a zero-state response (ZSR) where there are no initial conditions in the circuit. The differential equations describing step response are non-homogeneous. The differential equations have a driving function corresponding to the voltage or current source in the circuit schematic.

Complete response is both ZIR and ZSR, the response of a circuit to an input signal and any initial conditions which may be present. Complete response equations can be arranged in terms of ZIR plus ZSR or alternatively in terms of transient and steady-state components.

5.3.2 Intuitive Analysis of First-Order RC and RL Step Response

From Chapter 4, we know that for the capacitor branch relationship in Equation 5.6 (and Equation 4.18), an instantaneous voltage cannot be dropped across a capacitor, as indicated by the inequality in Equation 5.7 (and 4.19).

$$i = C\frac{dv}{dt} \quad (5.6)$$

$$i \neq C\frac{\Delta v}{dt} \quad (5.7)$$

When a step voltage is applied instantaneously to the RC circuit in Figure 5.17, all the voltage must initially drop across the resistor as V_{out}, as shown in Figure 5.18. The voltage across the capacitor builds up from zero. As the capacitor voltage builds up, the resistor voltage decreases, and the current decreases in proportion, as shown in Figure 5.19.

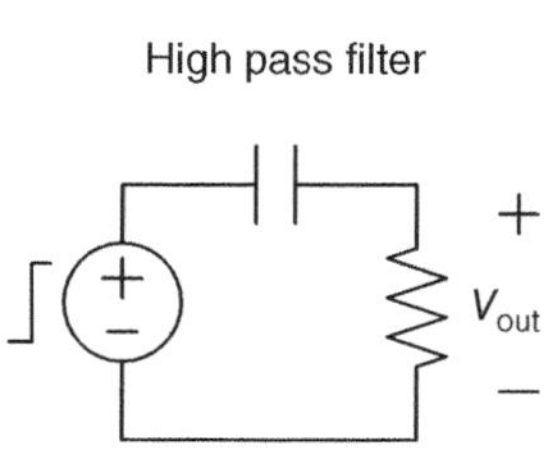

Figure 5.17 Series RC circuit with voltage step input.

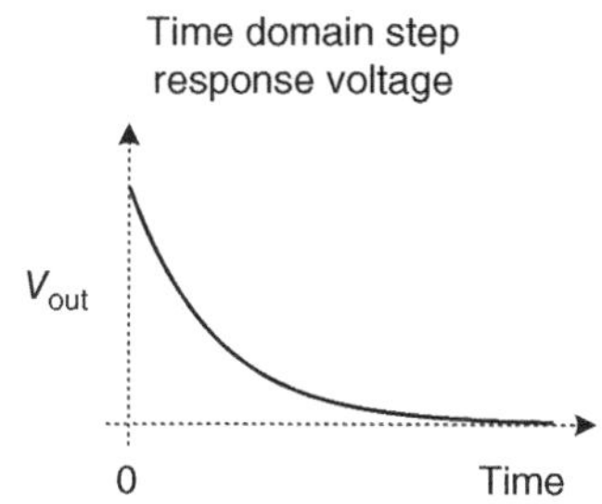

Figure 5.18 Series RC circuit step response resistor voltage.

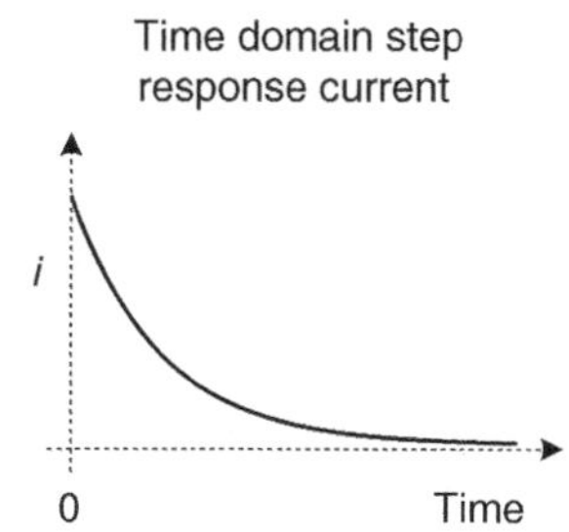

Figure 5.19 Series RC circuit step response loop current.

From Chapter 4, we know that for the inductor branch relationship in Equation 5.8 (and 4.35), an instantaneous current cannot be put through an inductor, indicated by Equation 5.9 (and 4.36).

$$v = L\frac{di}{dt} \tag{5.8}$$

$$v \neq L\frac{\Delta i}{dt} \tag{5.9}$$

When a step voltage is applied instantaneously to the RL circuit in Figure 5.20, all the voltage must initially drop across the inductor, because there can be no instantaneous current, which means there can be no current through the resistor and therefore no instantaneous V_{out}, as shown in Figure 5.21. The loop current builds up current from zero, as shown in Figure 5.22. As the current builds up, the resistor voltage V_{out} increases, as shown in Figure 5.21.

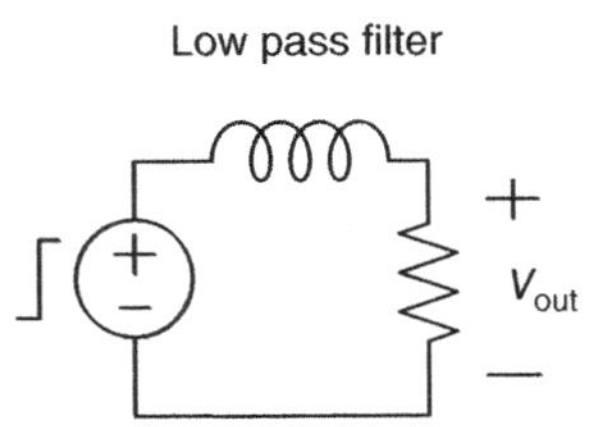

Figure 5.20 Series RL circuit with voltage step input.

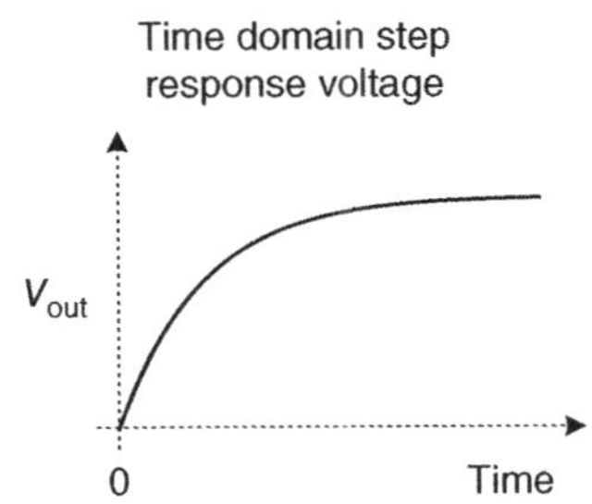

Figure 5.21 Series RL circuit step response resistor voltage.

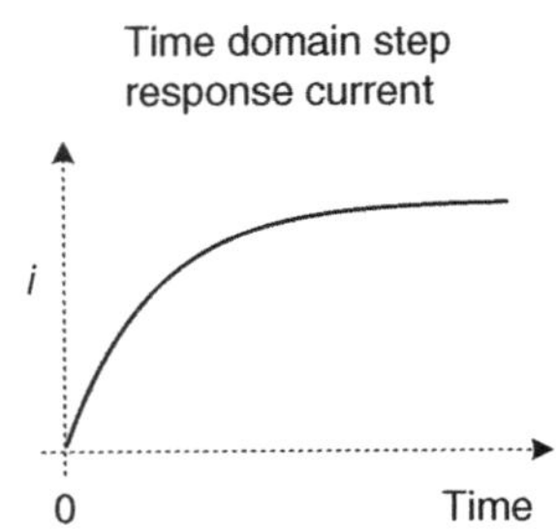

Figure 5.22 Series RL circuit step response loop current.

6

First-Order Frequency Domain Response

6.1 First-Order Frequency Response Overview

This chapter explores first-order RC and RL circuits in the steady-state frequency domain, where phasors and transfer functions are useful to understand circuit frequency response to steady-state sinusoidal input signals like sine or cosine of ωt.

6.1.1 Frequency Response Notations and Complex Magnitude

$$H(j\omega) = \frac{\mathbf{V}_{\text{out}}}{\mathbf{V}_{\text{S}}} \tag{6.1}$$

$H(j\omega)$ is a standard notation for steady-state frequency domain transfer functions in terms of angular frequency ω, as shown in Equation 6.1. Bold typeface is used to express phasors.

$$|H(j\omega)| = \left|\frac{\mathbf{V}_{\text{out}}}{\mathbf{V}_{\text{S}}}\right| \tag{6.2}$$

Equation 6.2 shows notations for the magnitude of the voltage transfer function $H(j\omega)$.

$$H(j\omega) = \frac{\overline{V}_{\text{out}}}{\overline{V}_{\text{S}}} \tag{6.3}$$

Equation 6.3 shows a common notation for a voltage transfer function $H(j\omega)$, using bars over the variables to indicate phasor quantities.

6.1.1.1 Complex Magnitude

The magnitude of a complex number C, in Equation 6.4, is the square root of the sum of the squares of the real and imaginary parts shown in Equation 6.5.

$$C = a + jb \tag{6.4}$$

$$|C| = \sqrt{a^2 + b^2} \tag{6.5}$$

The magnitude of a complex function $G(\omega)$, in Equation 6.6, is the square root of the sum of the squares of the real and imaginary parts, shown in Equation 6.7.

$$G(\omega) = R(\omega) + jX(\omega) \tag{6.6}$$

$$|G(\omega)| = \sqrt{[R(\omega)]^2 + [X(\omega)]^2} \tag{6.7}$$

Equations 6.6 and 6.7 are general to all complex numbers and functions. The variable ω has been chosen here because the steady-state frequency domain context is radian frequency ω equals $2\pi f$, where f in frequency in Hertz.

First and Second Order Circuits and Equations: Technical Background and Insights, First Edition. Robert O'Rourke.

Many first-order filter transfer functions are the quotient of two complex functions, where the numerator may have only a real component, only an imaginary $j\omega$ term, or both real and imaginary terms.

$$\left|\frac{A}{B}\right| = \frac{|A|}{|B|} \quad (6.8)$$

The magnitude of the ratio of two complex numbers, A and B in Equation 6.8, or two complex functions $A(\omega)$ and $B(\omega)$ in Equation 6.9, equals the ratio of the magnitudes of those same two complex numbers or functions.

$$\left|\frac{A(\omega)}{B(\omega)}\right| = \frac{|A(\omega)|}{|B(\omega)|} \quad (6.9)$$

6.1.2 Frequency Domain Transfer Function Magnitude Bode Plots

The magnitudes of steady-state frequency domain transfer functions are often plotted versus frequency in Bode plots, as shown in Figure 6.1.

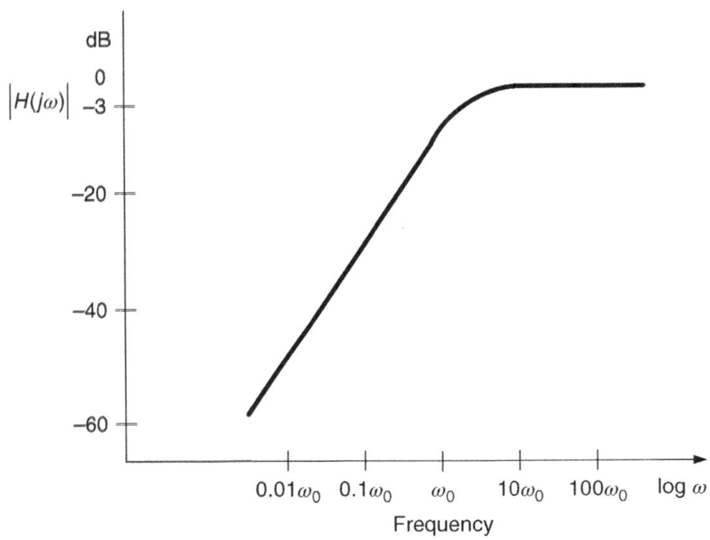

Figure 6.1 Example Bode magnitude plot of first-order high-pass filter.

Figure 6.1 shows angular frequency $\omega = 2\pi f$ on the horizontal scale. Horizontal frequency scales often use f. The graph in Figure 6.1 is a first-order high-pass filter.

The vertical scale of Bode diagrams is logarithmic in decibels (dB), as shown in Figure 6.1 and Equations 6.10 and 6.11.

$$H \text{ in dB} = 20 \log_{10} H \quad (6.10)$$

The number H expressed in decibels is 20 times the logarithm to the base 10 of H, as shown in Equation 6.10.

$$|H(j\omega)|_{\text{dB}} = 20 \log |H(j\omega)| \quad (6.11)$$

The magnitude of the transfer function $H(\omega)$ expressed decibels, in Equation 6.11, is 20 times the logarithm to the base 10 of the magnitude of $H(\omega)$.

The horizontal scale of the Bode diagram is log frequency. Notice in Figure 6.1 how the equally spaced tick marks on the horizontal axis represent successive powers of ten in frequency.

6.1.3 Frequency Domain Transfer Function Phase Calculation

$$H(j\omega) = R(\omega) + jX(\omega) \quad (6.12)$$

$$\theta(\omega) = \tan^{-1}\frac{X(\omega)}{R(\omega)} \quad (6.13)$$

For a complex transfer function like Equation 6.12, the phase $\theta(\omega)$ is the arctangent of the quotient of the reactive term $X(\omega)$ over the real term $R(\omega)$, as shown in Equation 6.13.

$$H(j\omega) = \frac{F(j\omega)}{G(j\omega)} \quad (6.14)$$

Transfer functions can have a complex function $F(\omega)$ in the numerator and a complex function $G(\omega)$ in the denominator, as shown in Equation 6.14.

For calculating the phase response of transfer functions like Equation 6.14, apply the arctangent formula for the quotient of two complex numbers or functions in Equation 6.14.

$$H(j\omega) = \frac{F(j\omega)}{G(j\omega)} = \frac{|F|\angle\theta_F}{|G|\angle\theta_G} = \frac{|F|}{|G|}\angle(\theta_F - \theta_G) \quad (6.15)$$

The quotient of two complex functions, like Equations 6.14 and 6.15, equals the quotient of the magnitudes and the difference in phase as shown in Equation 6.15.

$$\angle H(j\omega) = \angle F(j\omega) - \angle G(j\omega) \quad (6.16)$$

$$\angle H(j\omega) = \tan^{-1}\frac{\mathrm{Im}(F)}{Re(F)} - \tan^{-1}\frac{\mathrm{Im}(G)}{Re(G)} \quad (6.17)$$

The phase of the transfer function in Equation 6.14 is the phase of the numerator minus the phase of the denominator, as shown in Equation 6.16.

Combining Equations 6.13 and 6.16, the phase of the transfer functions, in Equations 6.14 and 6.15, is the arctangent of the imaginary part of the numerator over the real part of the numerator minus the arctangent of the imaginary part of the denominator over the real part of the denominator, as expressed in Equation 6.17.

The phase of steady-state frequency domain transfer functions is often plotted versus frequency in Bode plots, as shown in Figure 6.2. The vertical scale is in degrees.

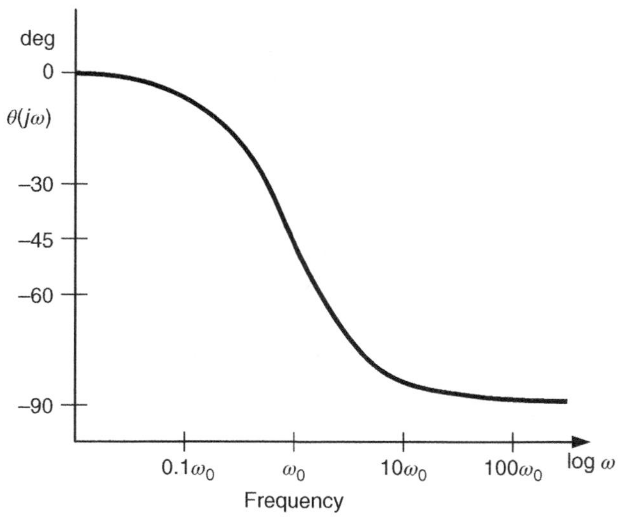

Figure 6.2 Example Bode phase plot.

6.1.4 Frequency Domain Response Topology Examples

The topology matrix in Table 6.1 shows the dB magnitude response and the phase response of some first-order RC and RL circuit topologies.

Table 6.1 Common first-order circuit topologies and the shapes of their frequency domain transfer function magnitudes and phases.

Circuit topology	Frequency domain magnitude response	Frequency domain phase response
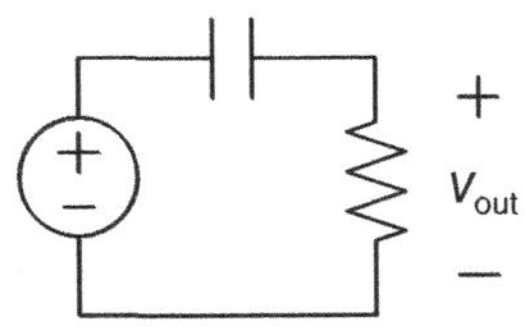 High-pass filter	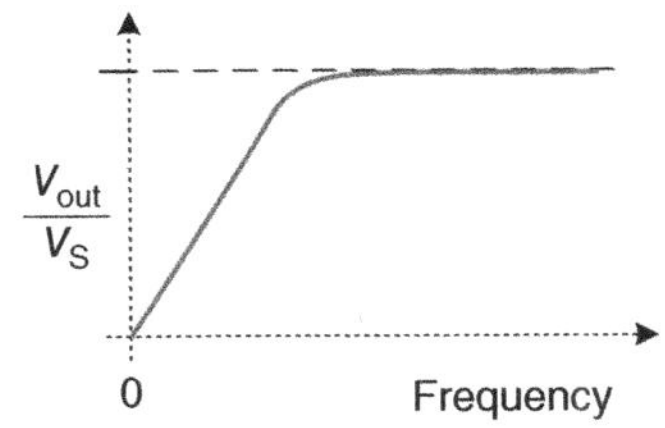	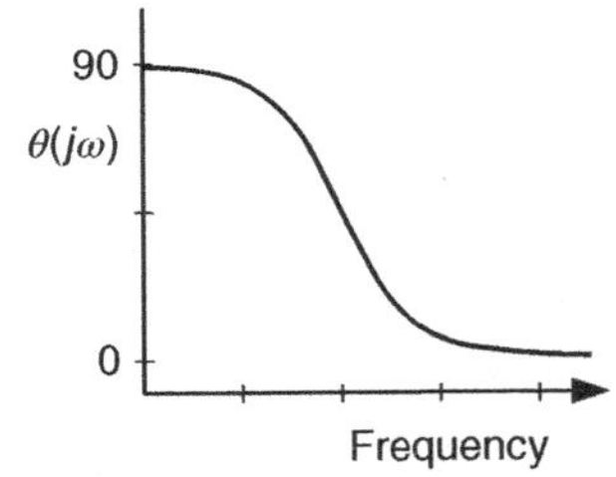
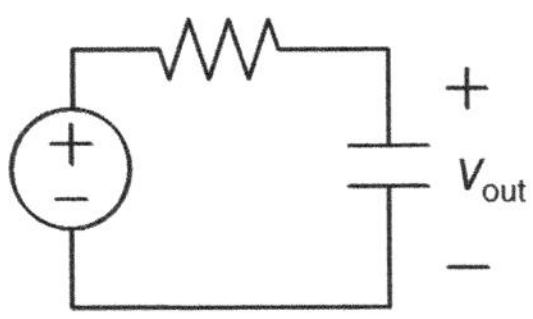 Low-pass filter	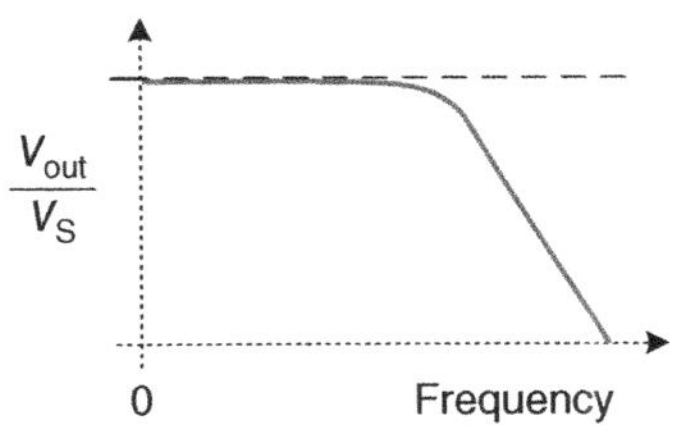	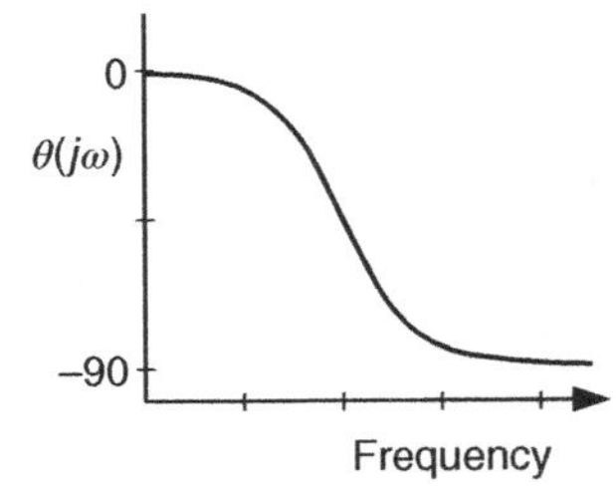
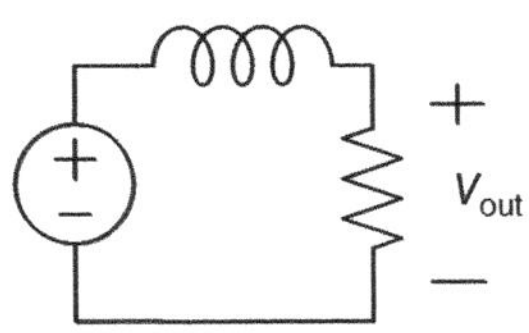 Low-pass filter	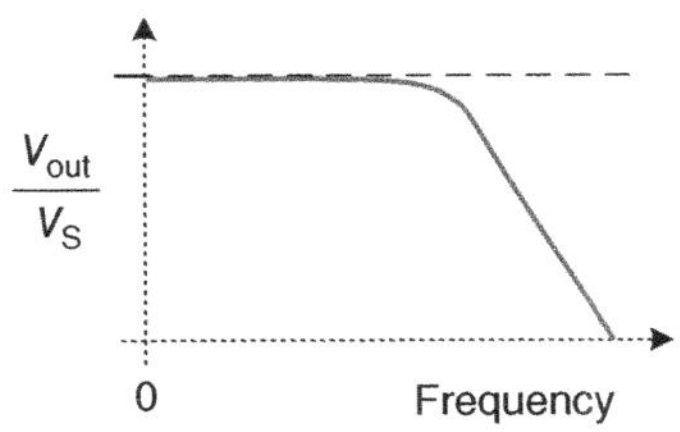	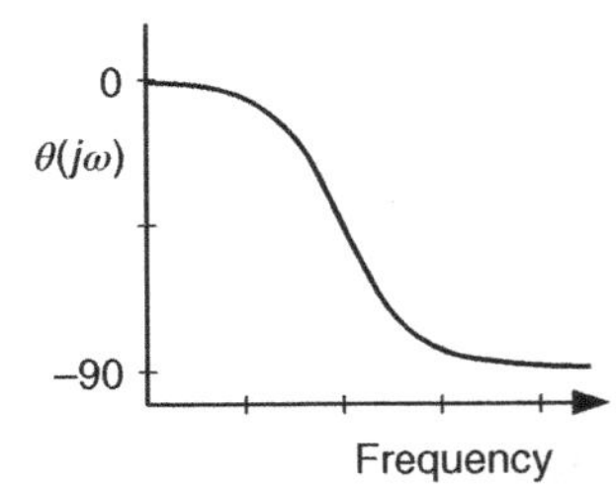
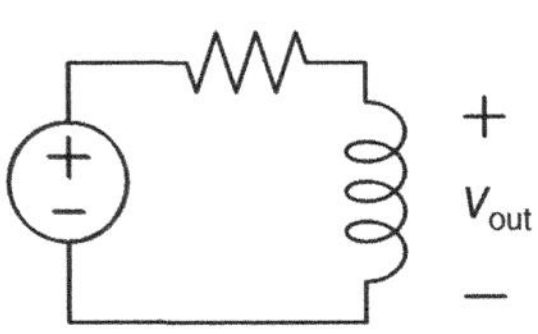 High-pass filter	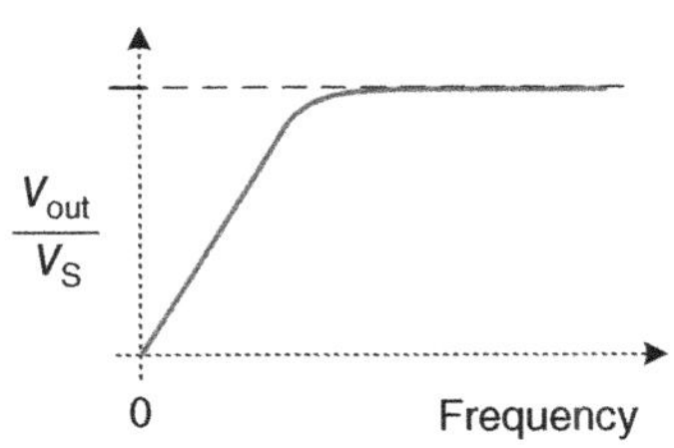	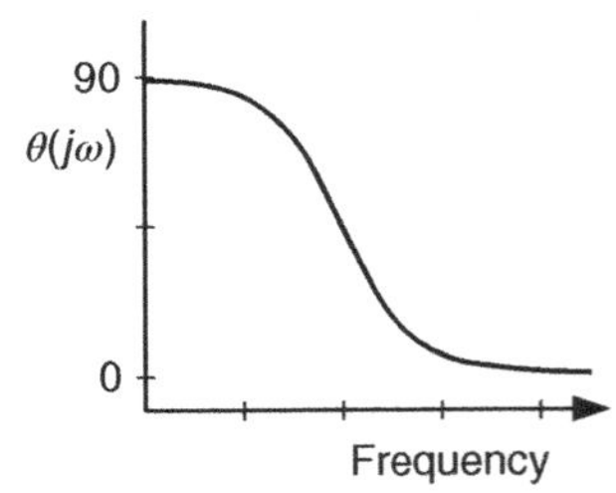
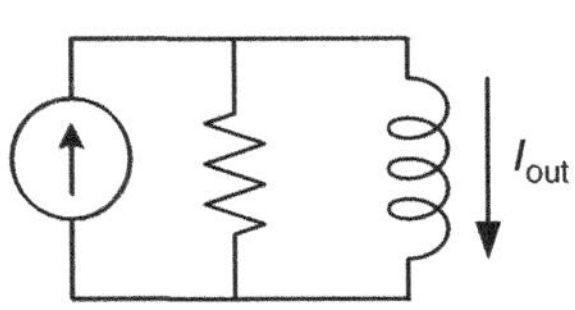 Low-pass filter	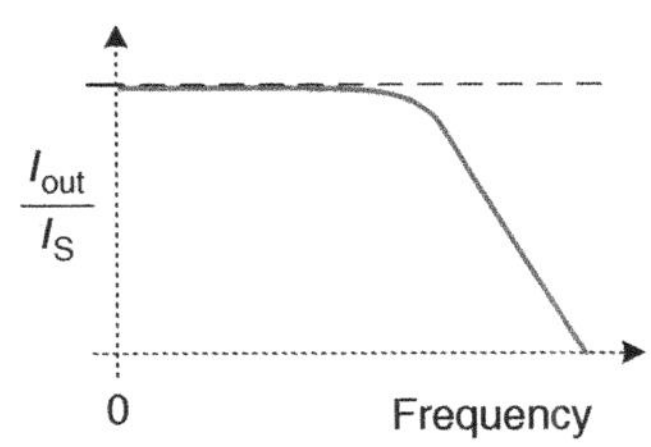	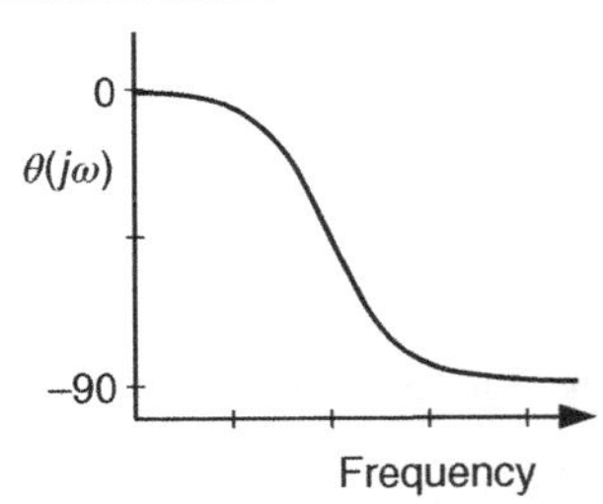

Table 6.1 (Continued)

Circuit topology	Frequency domain magnitude response	Frequency domain phase response
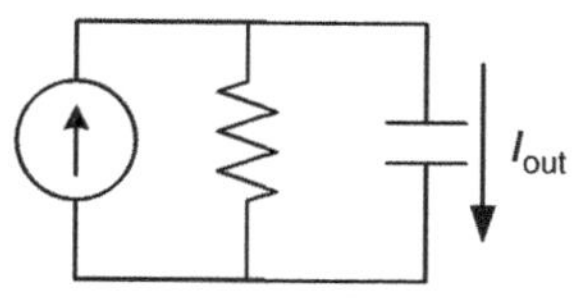 High-pass filter	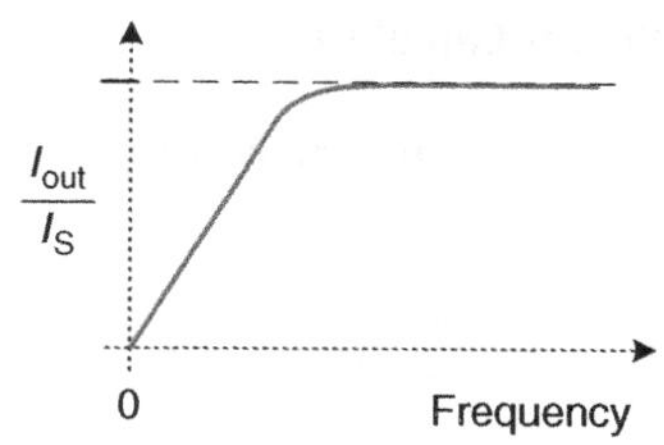	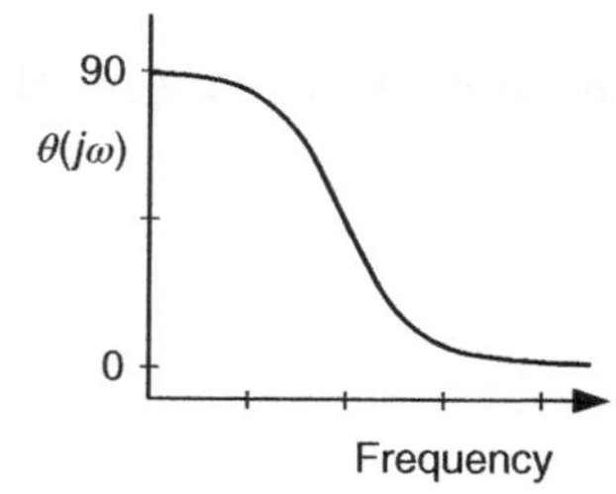

6.1.5 First-Order Frequency Response Formulae with ω_0

$$\omega_0 = \frac{1}{\tau} \qquad (6.18)$$

In Equation 6.18, the corner frequency ω_0, in the frequency domain, equals one over τ (tau), the time constant in time domain.

Low-pass filters

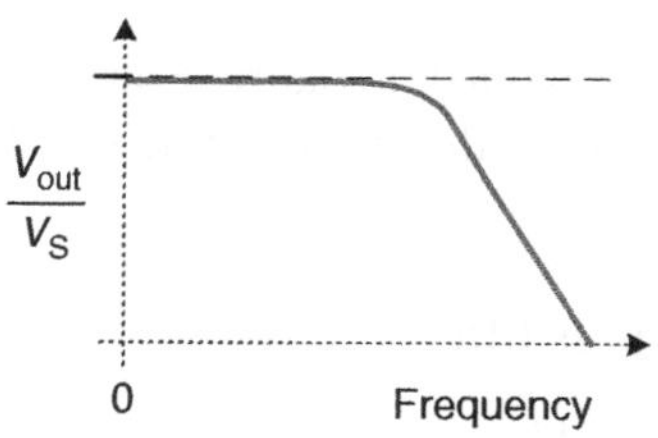

Figure 6.3a Example Bode plot of first-order low-pass filter.

$$\frac{V_{out}}{V_S} = \frac{1}{1 + j\frac{\omega}{\omega_0}} \qquad (6.19a)$$

Equation 6.19a is in the general formula for a first-order low-pass filter transfer function in terms of angular frequency ω and corner frequency ω_0, plotted in Figure 6.3a.

$$|H(j\omega)| = \left|\frac{V_{out}}{V_S}\right| = \frac{1}{\sqrt{1 + (\omega/\omega_0)^2}} \qquad (6.20a)$$

Equation 6.20a is in the general format for low-pass filter transfer function magnitude in terms of angular frequency ω and break frequency ω_0.

$$\theta = \angle H(j\omega) = \angle\frac{V_{out}}{V_S} = -\tan^{-1}\left(\frac{\omega}{\omega_0}\right) \qquad (6.21a)$$

High-pass filters

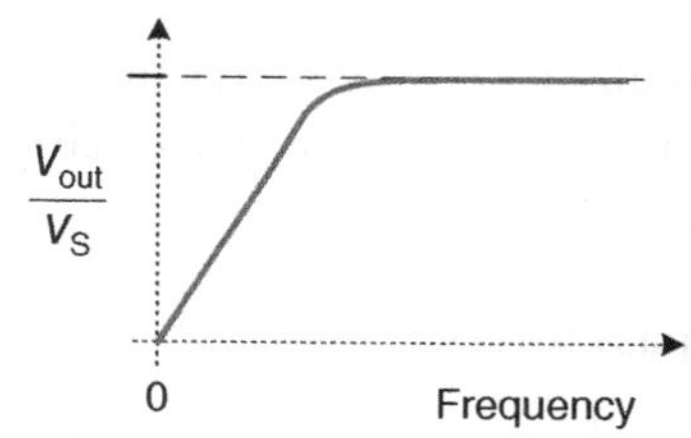

Figure 6.3b Example Bode plot of first-order high-pass filter.

$$\frac{V_{out}}{V_S} = \frac{j\frac{\omega}{\omega_0}}{1 + j\frac{\omega}{\omega_0}} \qquad (6.19b)$$

Equation 6.19b is in the general formula for a first-order high-pass filter transfer function in terms of angular frequency ω and corner frequency ω_0, plotted in Figure 6.3b.

$$|H(j\omega)| = \left|\frac{V_{out}}{V_S}\right| = \frac{\omega/\omega_0}{\sqrt{1 + (\omega/\omega_0)^2}} \qquad (6.20b)$$

Equation 6.20b is in the general format for a first-order high-pass filter transfer function magnitude in terms of angular frequency ω and break frequency ω_0.

$$\angle H(j\omega) = \frac{\pi}{2} - \tan^{-1}(\omega/\omega_0) \qquad (6.21b)$$

Equation 6.21a is in the general format for low-pass filter phase response in terms of angular frequency ω and break frequency ω_0.

Equation 6.21b is in the general format for a first-order high-pass filter phase response in terms of angular frequency ω and break frequency ω_0.

6.1.6 First-Order Corner Frequency and Half-Power Calculations

Low-pass filters

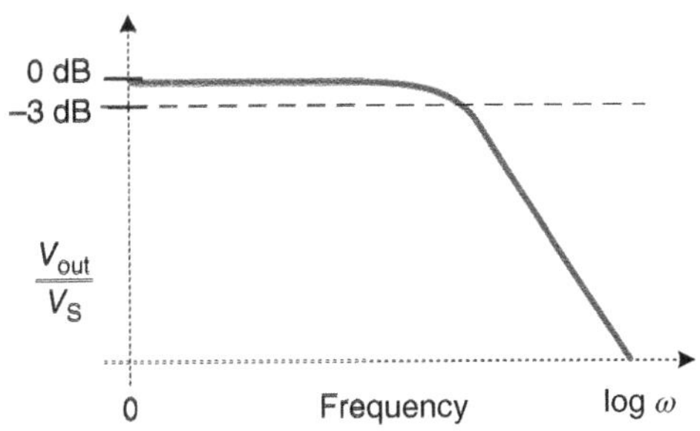

Figure 6.4a Bode plot of first-order low-pass filter with –3 dB point marked.

High-pass filters

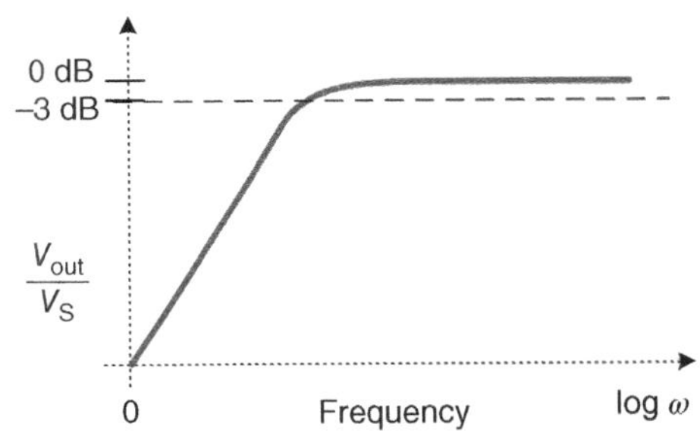

Figure 6.4b Bode plot of first-order high-pass filter with –3 dB point marked.

Use the standard formulae for low-pass filter and high-pass filter transfer functions, plotted in Figures 6.4a,b, in terms of ω_0, in Equations 6.22a (and 6.20a) and 6.22b (and 6.20b), to calculate the value of the transfer function at the corner frequency ω_0 as follows.

Low-pass filters

$$|H(j\omega)| = \left|\frac{\boldsymbol{V}_{\text{out}}}{\boldsymbol{V}_{\text{S}}}\right| = \frac{1}{\sqrt{1+(\omega/\omega_0)^2}} \tag{6.22a}$$

High-pass filter

$$|H(j\omega)| = \left|\frac{\boldsymbol{V}_{\text{out}}}{\boldsymbol{V}_{\text{S}}}\right| = \frac{\omega/\omega_0}{\sqrt{1+(\omega/\omega_0)^2}} \tag{6.22b}$$

In Equations 6.22a and 6.22b, substitute in ω_0 for the angular frequency ω, as shown in Equations 6.23a and 6.23b.

Low-pass filters

$$|H(j\omega)| = \left|\frac{\boldsymbol{V}_{\text{out}}}{\boldsymbol{V}_{\text{S}}}\right| = \frac{1}{\sqrt{1+(\omega_0/\omega_0)^2}} \tag{6.23a}$$

High-pass filter

$$|H(j\omega)| = \left|\frac{\boldsymbol{V}_{\text{out}}}{\boldsymbol{V}_{\text{S}}}\right| = \frac{\omega_0/\omega_0}{\sqrt{1+(\omega_0/\omega_0)^2}} \tag{6.23b}$$

Simplify Equations 6.23a and 6.23b to get Equation 6.24, which applies to both low-pass and high-pass filters.

$$|H(j\omega)| = \left|\frac{\boldsymbol{V}_{\text{out}}}{\boldsymbol{V}_{\text{S}}}\right| = \frac{1}{\sqrt{1+(1)^2}} = \frac{1}{\sqrt{2}} = 0.707 \tag{6.24}$$

Taking the logarithm of the transfer function magnitudes, at the corner frequency ω_0, and multiplying by 20 to convert to dB, results in the magnitude of about minus 3 dB, as shown in Equation 6.25.

$$|H(j\omega)|_{\text{dB}} = 20\log|H(j\omega)| = 20\log(0.707) = -3.01\ \text{dB} \tag{6.25}$$

The magnitude of the transfer function $H(j\omega)$ corresponds to either a voltage or a current. Power corresponds to voltage or current squared. Squaring the magnitude of the transfer function at the corner frequency ω_0 shows the factor of one-half related to power in Equation 6.26.

$$|H(j\omega)|^2 = \left|\frac{\mathbf{V}_{\text{out}}}{\mathbf{V}_{\text{S}}}\right|^2 = \left(\frac{1}{\sqrt{2}}\right)^2 = \frac{1}{2} \tag{6.26}$$

6.2 Series RC High-pass Filter Frequency Response

6.2.1 Series RC Frequency Domain Results Overview

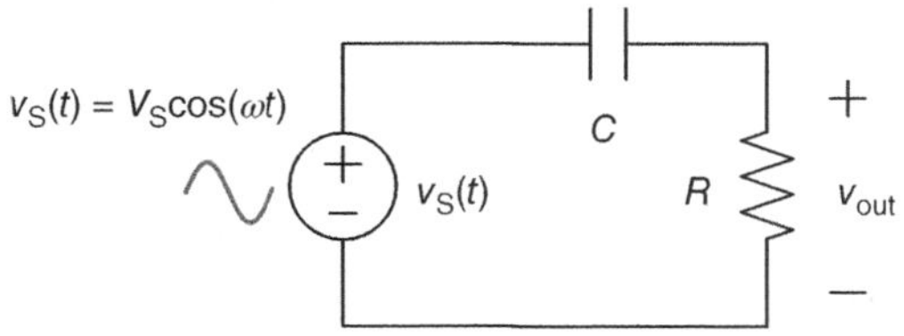

Figure 6.5 Series RC high-pass circuit driven by sinusoidal voltage source.

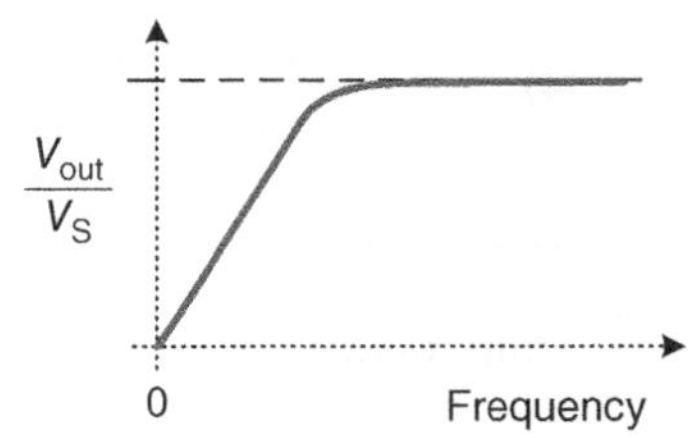

Figure 6.6 High-pass filter Bode magnitude plot.

Figure 6.5 is a high-pass filter because the series capacitor passes high frequencies; the high-frequency current appears across the resistor as v_{out}. At low frequencies, less current flows around the loop and more voltage drops across the capacitor instead of v_{out}.

$$\frac{\mathbf{V}_{\text{out}}}{\mathbf{V}_{\text{S}}} = \frac{j\omega/\omega_0}{1 + j\omega/\omega_0} \tag{6.27}$$

Equation 6.27 is in the general format for high-pass filter transfer functions. This form is useful for easily calculating specific transfer functions.

$$\omega_0 = \frac{1}{RC} \tag{6.28}$$

$$\frac{\mathbf{V}_{\text{out}}}{\mathbf{V}_{\text{S}}} = \frac{j\dfrac{\omega}{1/RC}}{1 + j\dfrac{\omega}{1/RC}} \tag{6.29}$$

From Equation 6.28, substitute ω_0 equals $1/RC$, the corner frequency, into the transfer function Equation 6.27 to get Equation 6.29.

$$\frac{\mathbf{V}_{\text{out}}}{\mathbf{V}_{\text{S}}} = \left(\frac{j\omega RC}{1 + j\omega RC}\right) = H(j\omega) \tag{6.30}$$

Equation 6.30 shows the steady-state frequency domain voltage transfer function, in phasor notation, for the series RC high-pass filter circuit in Figure 6.5.

Because the output voltage v_{out} over v_{S}, varies with frequency, $\mathbf{V}_{\text{out}}$ over $\mathbf{V}_{\text{S}}$ makes a convenient transfer function. The transfer function has the same frequency-dependent $j\omega RC$ term in the numerator and the denominator. As the frequency goes up, the 1 in the denominator becomes small, in comparison to the two $j\omega RC$ terms, and the magnitude of the transfer function approaches 1 (0 dB). At DC (zero frequency) the $j\omega RC$ term becomes zero and the magnitude of the transfer function goes to zero.

$$|H(j\omega)| = \left|\frac{\mathbf{V}_{\text{out}}}{\mathbf{V}_{\text{S}}}\right| = \frac{\omega/\omega_0}{\sqrt{1 + (\omega/\omega_0)^2}} \tag{6.31}$$

Equation 6.31 is the general formula for a first-order high-pass filter transfer function magnitude in terms of angular frequency ω and break frequency ω_0.

From Equation 6.28, substitute ω_0 equals $1/RC$, the corner frequency, into the transfer function magnitude Equation 6.31, to obtain Equation 6.32.

$$|H(j\omega)| = \left|\frac{V_{out}}{V_S}\right| = \frac{\omega RC}{\sqrt{1+(\omega RC)^2}} \quad (6.32)$$

Equation 6.32 shows the voltage transfer function magnitude, shown in Figure 6.6, for the series RC high-pass filter circuit in Figure 6.5.

$$\angle H(j\omega) = \frac{\pi}{2} - \tan^{-1}(\omega/\omega_0) \quad (6.33)$$

Equation 6.33 is in the general format for a first-order high-pass filter transfer function phase in terms of angular frequency ω and break frequency ω_0.

$$\frac{\pi}{2} = 90° \quad (6.34)$$

From Equation 6.28, substitute ω_0 equals 1 over RC, the corner frequency, into the transfer function phase Equation 6.33.

$$\angle H(j\omega) = \frac{\pi}{2} - \tan^{-1}(\omega RC) \quad (6.35)$$

Equation 6.35 shows the voltage transfer function phase, plotted in Figure 6.7, for the series RC high-pass filter circuit in Figure 6.5.

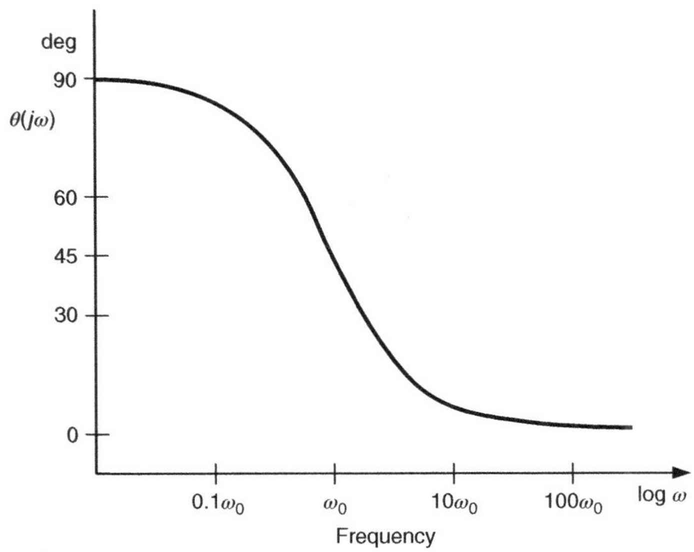

Figure 6.7 Series RC high-pass filter Bode phase plot.

6.2.2 Series RC Frequency Domain Example Intuitive Analysis

The capacitor, in the simulation schematic in Figure 6.8, looks like an open circuit (high impedance) to DC, and the capacitor looks like a short circuit (low impedance) to high frequency.

The circuit in Figure 6.8 is a high-pass filter because the series capacitor passes the high frequencies most easily from the input voltage source, labeled V_S and V_{in} in the simulation schematic, to the output, labeled V_{out} in the schematic. The high-pass filter response is shown in Figure 6.9.

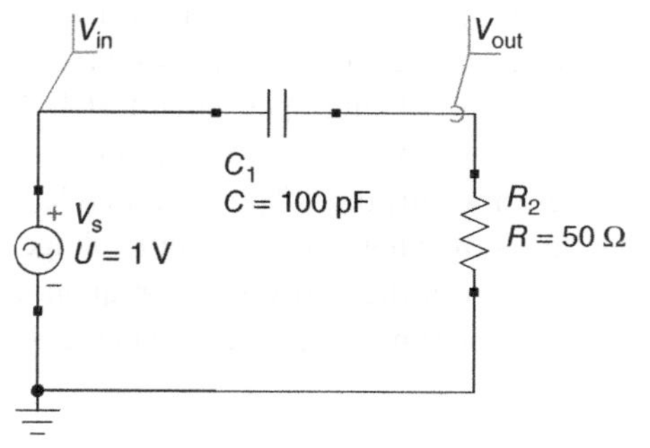

Figure 6.8 Series RC circuit simulation schematic driven by sinusoidal voltage source.

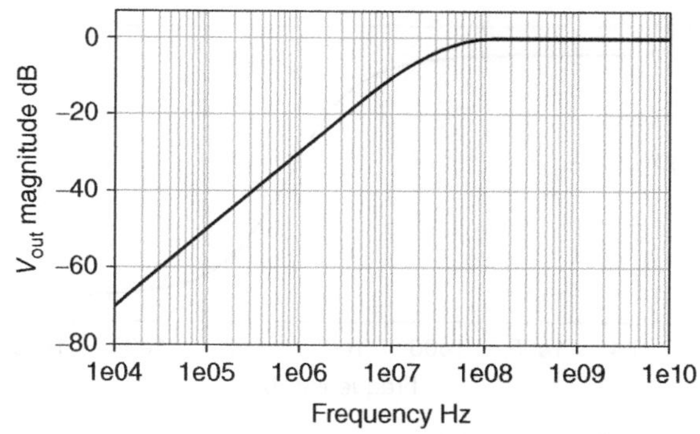

Figure 6.9 Series RC frequency domain simulation of output voltage magnitude.

As the signal (V_S) frequency goes up, the impedance of the capacitor, 1 over $j\omega C$, goes down and more of the source voltage V_S gets dropped across the output resistor (V_{out}), leaving less voltage across the capacitor ($V_{out} - V_{in}$), as shown in the plot in Figure 6.10. In the high-frequency limit, all of V_S can drop across the resistor and current becomes V_S over R (also V_{in} over R), as shown in the current plot Figure 6.11. $1\,\text{V}/50\,\Omega$ equals 20 mA.

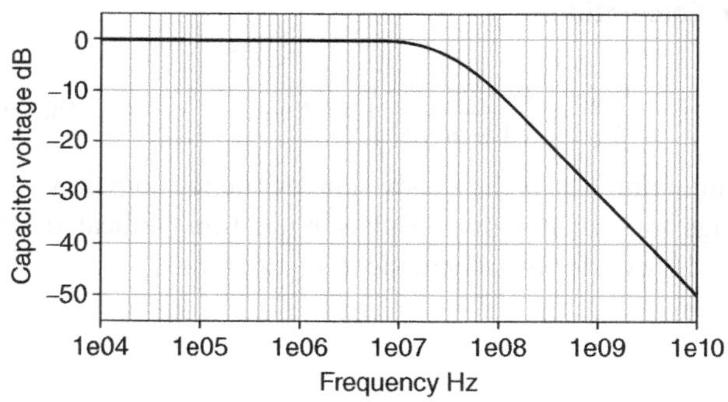

Figure 6.10 Series RC simulation of capacitor voltage frequency response.

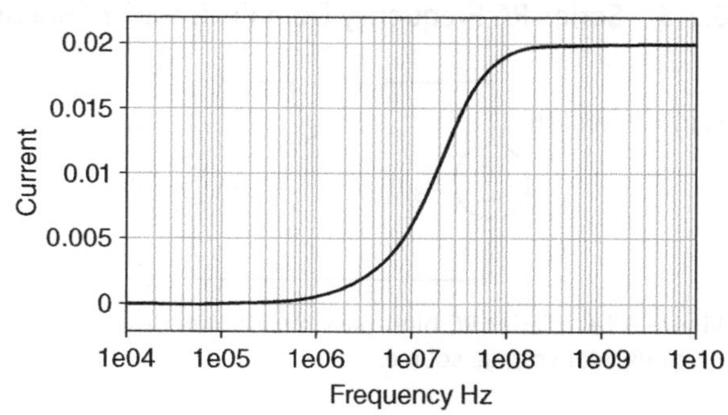

Figure 6.11 Series RC simulation of loop current over frequency.

As the signal (V_S) frequency goes down, the impedance of the capacitor ($1/j\omega C$) goes up and more of the source voltage V_S gets dropped across the capacitor, leaving less voltage across the resistor (V_{out}), as shown in Figure 6.10. In the low-frequency limit, all of V_S can drop across the capacitor and no current flows at all, as shown in Figure 6.11.

The impedance of the capacitor, which is a pure reactance $-1/\omega C$, changes with frequency while the impedance of the resistor (a pure resistance) does not. The total series impedance of the resistor and capacitor together varies, and this variation in impedance changes the loop current. The higher the loop current, the higher the output voltage V_{out} across the output resistor.

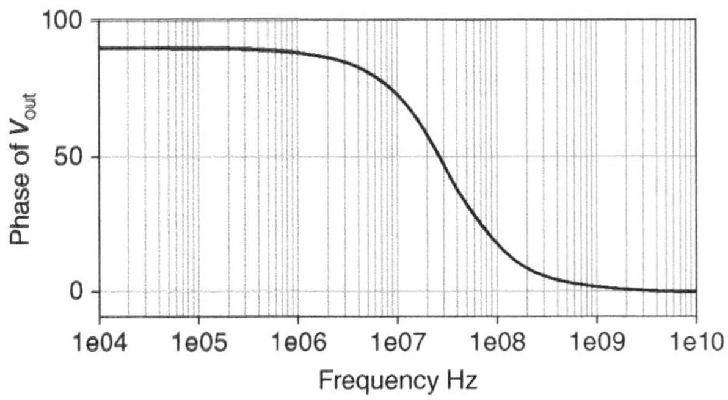

Figure 6.12 Series RC frequency domain simulation of output voltage phase.

At DC, the reactance $-1/\omega C$, blows up to minus infinity, making the impedance infinite, which corresponds to capacitors being open circuit at DC. Equation 6.35 shows that at DC, the arctangent term goes to zero and only the 90° phase term is left. Figure 6.12 shows the phase of the output voltage, which is taken across the resistor. As frequency goes up, the reactance and the phase shift decrease toward zero.

6.2.2.1 Lead Network

In a capacitor, the voltage lags the current in phase. The output of this circuit is the voltage across the resistor, which is always in phase with the current through the resistor. The current is also the reference for the phase. Since the output voltage has the phase of the current, which leads the voltage across the capacitor, this is a lead circuit. Figure 6.12 shows the phase starting at 90° at DC, consistent with a lead network.

6.2.3 Series RC Frequency Domain Transfer Function Calculation

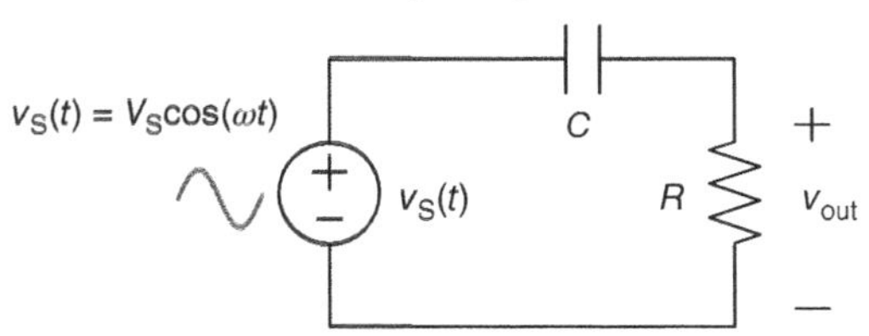

Figure 6.13 Series RC high-pass filter circuit driven by sinusoidal voltage source.

$$\frac{\mathbf{V}_{out}}{\mathbf{V}_S} = \left(\frac{j\omega RC}{1 + j\omega RC}\right) = H(j\omega) \tag{6.36}$$

Equation 6.36 is the steady-state frequency domain voltage transfer function in phasor notation (indicated by bold text), for the schematic in Figure 6.13.

6.2.3.1 Calculating the Transfer Function

$$\frac{\mathbf{V}_{out}}{\mathbf{V}_S} \tag{6.37}$$

We can use a voltage divider to calculate the steady-state frequency domain transfer function of the output voltage V_{out} over the input voltage V_S. Notice the use of j as the imaginary operator rather than "i" used in other mathematics and physics literature.

$$Z_C = \frac{1}{j\omega C} \tag{6.38}$$

Equation 6.38 is the impedance of an ideal capacitor.

$$Z_R = R \tag{6.39}$$

Equation 6.39 is the impedance of an ideal resistor.

$$Z_{RC} = \frac{1}{j\omega C} + R \tag{6.40}$$

The impedance of the series combination of the resistor and capacitor, in Equation 6.40, is R plus one over $j\omega C$.

$$I = \frac{V_S}{\frac{1}{j\omega C} + R} \tag{6.41}$$

Applying Ohm's law to the voltage source across both the resistor and capacitor in series, in Equation 6.41, the current in the whole loop equals the input voltage V_S divided by the impedance of the capacitor and the resistor in series, one over the quantity $j\omega C$ plus R.

$$V_{out} = \underbrace{\left(\frac{V_S}{\frac{1}{j\omega C} + R}\right)}_{\text{Current}} R \tag{6.42}$$

Applying Ohm's law to the output resistor, the output voltage V_{out} equals the current in the loop times the impedance (i.e. the resistance) of the resistor, shown in Equation 6.42.

$$\frac{V_{out}}{V_S} = \left(\frac{R}{\frac{1}{j\omega C} + R}\right) \tag{6.43}$$

In Equation 6.42, divide both sides of the equation by V_S, and move the R inside the brackets, to obtain Equation 6.43.

Multiply both top and bottom of the right-hand side of Equation 6.43 by $j\omega C$ to get Equation 6.44.

$$\frac{V_{out}}{V_S} = \left(\frac{j\omega RC}{1 + j\omega RC}\right) \tag{6.44}$$

Equation 6.44 is the steady-state frequency domain voltage transfer function of the series RC circuit in Figure 6.13.

The RC notation can be converted into the ω_0 notation.

$$\omega_0 = \frac{1}{RC} \tag{6.45}$$

Equation 6.45 is the corner frequency or break frequency of the RC circuit.

$$\frac{V_{out}}{V_S} = \left(\frac{j\omega/\omega_0}{1 + j\omega/\omega_0}\right) \tag{6.46}$$

Substitute Equation 6.45 into Equation 6.44 to get the transfer function in terms of ω and ω_0 in Equation 6.46.

$$\frac{V_{out}}{V_S} = \left(\frac{j\omega RC}{1 + j\omega RC}\right) \tag{6.47}$$

This notation in Equation 6.46 allows us to track the ratio of corner frequency (omega ω_0 equals $1/RC$) relative to the frequency ω.

$$\frac{V_{out}}{V_S} = \left(\frac{j\omega}{1/RC + j\omega}\right) \tag{6.48}$$

For an alternative notation, divide numerator and denominator of Equation 6.47 (same as 6.36) by RC to isolate the ω terms in Equation 6.48.

$$\frac{V_{out}}{V_S} = \left(\frac{j\omega}{\omega_0 + j\omega}\right) \tag{6.49}$$

Substitute Equation 6.45 into Equation 6.45 to get the transfer function in terms of ω and ω_0 in Equation 6.49.

6.2.4 Series RC Transfer Function Magnitude Calculation

$$\frac{V_{out}}{V_S} = \left(\frac{j\omega}{1/RC + j\omega}\right) = H(j\omega) \tag{6.50}$$

To plot the frequency response, we need to calculate the magnitude and phase of the transfer function of the first-order RC high-pass filter circuit in Figure 6.14.

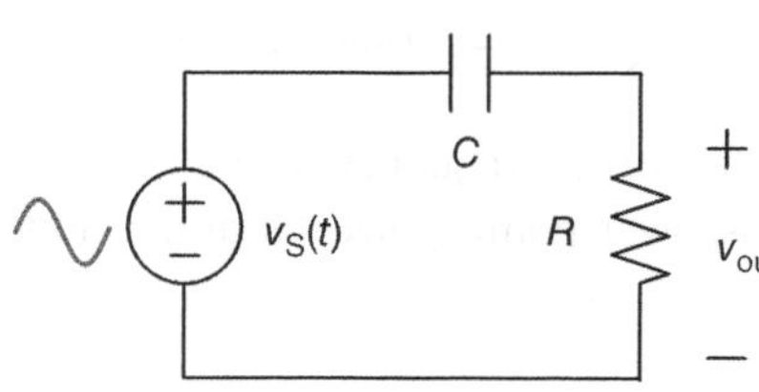

Figure 6.14 Series RC high-pass filter circuit driven by sinusoidal voltage source.

The magnitude of a complex function, $G(\omega)$ in Equation 6.51, is the square root of the sum of the squares of the real part $R(\omega)$ and the imaginary part $X(\omega)$, as shown in Equation 6.52.

$$G(\omega) = R(\omega) + jX(\omega) \tag{6.51}$$

$$|G(\omega)| = \sqrt{[R(\omega)]^2 + [X(\omega)]^2} \tag{6.52}$$

The series RC high-pass filter transfer function, in Equation 6.50, can be thought of as the quotient of two complex functions, where the numerator has no real component, only the imaginary $j\omega$ term.

The magnitude of the ratio of two complex functions $A(\omega)$ and $B(\omega)$, equals the ratio of the magnitudes of those same two complex functions, as shown in Equation 6.53.

$$\left|\frac{A(\omega)}{B(\omega)}\right| = \frac{|A(\omega)|}{|B(\omega)|} \tag{6.53}$$

Starting with Equation 6.53, calculate the magnitude of the RC high-pass filter transfer function, in Equation 6.50, using the square root of the sum of the squares, shown in Equation 6.52. Notice that the imaginary operator j does not get squared.

$$\frac{\boldsymbol{V}_{\text{out}}}{\boldsymbol{V}_{\text{S}}} = \frac{\boldsymbol{V}_{\text{out}}}{\boldsymbol{V}_{\text{S}}} = \frac{|j\omega|}{|(1/RC) + j\omega|} = \frac{\sqrt{\omega^2}}{\sqrt{(1/RC)^2 + \omega^2}} = \frac{\omega}{\sqrt{(1/RC)^2 + \omega^2}} \tag{6.54}$$

$$\left|\frac{\boldsymbol{V}_{\text{out}}}{\boldsymbol{V}_{\text{S}}}\right| = \frac{\omega}{\sqrt{(1/RC)^2 + \omega^2}} = |H(j\omega)| \tag{6.55}$$

Equation 6.55 is the magnitude of the RC high-pass filter transfer function $H(j\omega)$ of the circuit in Figure 6.14.

6.2.4.1 Transfer Function Magnitude at the Corner Frequency

$$\left|\frac{\boldsymbol{V}_{\text{out}}}{\boldsymbol{V}_{\text{S}}}\right| = \frac{\omega}{\sqrt{{\omega_0}^2 + \omega^2}} = |H(j\omega)| \tag{6.56}$$

Equation 6.56 is the magnitude of $H(j\omega)$ in an alternate notation.

Calculate the magnitude of the series RC high-pass filter at the corner frequency $\omega_0 = 1/RC$, shown in Equation 6.58.

$$\omega_0 = \frac{1}{RC} \tag{6.57}$$

$$\left|\frac{\boldsymbol{V}_{\text{out}}}{\boldsymbol{V}_{\text{S}}}\right| = \frac{\omega_0}{\sqrt{{\omega_0}^2 + {\omega_0}^2}} = \frac{\omega_0}{\sqrt{2{\omega_0}^2}} = \frac{1}{\sqrt{2}}\frac{\omega_0}{\sqrt{{\omega_0}^2}} = \frac{1}{\sqrt{2}} = 0.707 \tag{6.58}$$

Equation 6.58 shows that when $\omega = \omega_0$, the denominator of the transfer function magnitude becomes the square root of 2.

$$\left|\frac{\boldsymbol{V}_{\text{out}}}{\boldsymbol{V}_{\text{S}}}\right|_{\text{dB}} = 20\log(\text{mag}) = 20\log(0.707) = 20(-0.15058) \cong\sim 3\text{ dB} \tag{6.59}$$

The corner frequency is often referred to as the -3 dB point or 3 dB down. This corner frequency is also the half-power point; power is proportional to voltage squared. The square of one-over-the-square-root-of-two is one half.

6.2.4.2 Alternate Form of Magnitude of the Transfer Function

$$\frac{V_{out}}{V_S} = \left(\frac{j\omega RC}{1 + j\omega RC}\right) \quad (6.60)$$

Starting from Equation 6.60 for the transfer function, calculate the magnitude keeping an ωRC term in the numerator.

Apply Equation 6.52, the square root of the sum of the squares, to Equation 6.60 to get Equation 6.61.

$$\frac{V_{out}}{V_S} = \frac{|j\omega RC|}{|1 + j\omega RC|} = \frac{\sqrt{(\omega RC)^2}}{\sqrt{1^2 + (\omega RC)^2}} = \frac{\omega RC}{\sqrt{1 + (\omega RC)^2}} \quad (6.61)$$

$$\left|\frac{V_{out}}{V_S}\right| = |H|(j\omega) = \frac{\omega RC}{\sqrt{1 + (\omega RC)^2}} \quad (6.62)$$

Equation 6.62 is the magnitude of the transfer function $H(j\omega)$, of the circuit in Figure 6.14, in an alternate notation from Equation 6.55.

6.2.5 Series RC Magnitude Bode Diagram Calculation

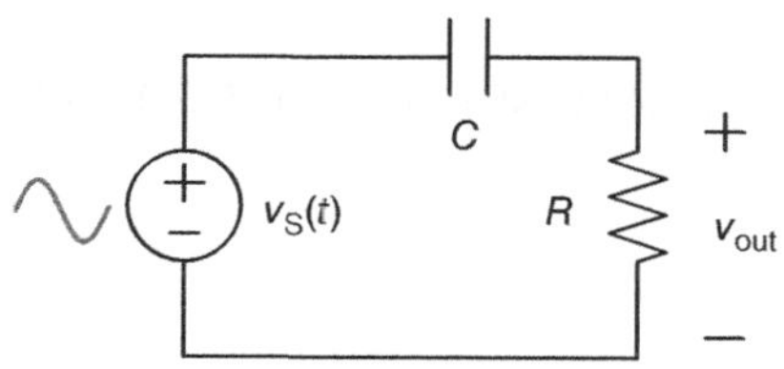

Figure 6.15 Series RC high-pass filter circuit driven by sinusoidal voltage source.

$$\frac{V_{out}}{V_S} = \frac{j\omega}{1/RC + j\omega} = H(j\omega) \quad (6.63)$$

Equation 6.63 is the transfer function of the series RC high-pass filter.

From the magnitude of the transfer function in Equation 6.63, we'll generate the exact equations used to create the Bode plots for the circuit in Figure 6.15. This plot will have dB magnitude on the vertical scale and radian frequency ω on the horizontal axis. $\omega = 2\pi f$.

$$|H(j\omega)| = \left|\frac{V_{out}}{V_S}\right| = \frac{\omega}{\sqrt{(1/RC)^2 + \omega^2}} \quad (6.64)$$

Equation 6.64 shows the magnitude of the transfer function $H(\omega)$ of the RC circuit as derived previously in Equation 6.55.

$$|H(j\omega)| = \omega[(1/RC)^2 + \omega^2]^{-1/2} \quad (6.65)$$

Restate the square root in the denominator of Equation 6.64 as an exponent of minus one half, as shown in Equation 6.65.

The vertical scale of the Bode diagram is logarithmic in decibels (dB).

$$H \text{ in dB} = 20 \log_{10} H \quad (6.66)$$

The number H expressed in decibels is 20 times the logarithm to the base 10 of H.

$$|H(j\omega)|_{dB} = 20 \log |H(j\omega)| \quad (6.67)$$

The magnitude of the transfer function $H(\omega)$ expressed decibels, in Equation 6.67, is 20 times the logarithm to the base 10 of the magnitude of $H(\omega)$.

$$|H(j\omega)|_{dB} = 20 \log \omega[(1/RC)^2 + \omega^2]^{-1/2} \quad (6.68)$$

From Equation 6.65, substitute the transfer function magnitude $|H(j\omega)|$ into Equation 6.67 to obtain Equation 6.68.

Equation 6.68 is the logarithm of a product. Use a logarithmic identity, in Equation 6.69, to separate the Equation 6.68 ω product term from the "square root" (bracketed) expression, resulting in Equation 6.71. The A and B terms for Equation 6.69 are given in Equation 6.70.

$$\log(A \cdot B) = \log(A) + \log(B) \qquad (6.69)$$

$$A = \omega \qquad B = [\omega^2 + (1/RC)^2]^{-1/2} \qquad (6.70)$$

$$|H(j\omega)|_{\mathrm{dB}} = 20\log\omega + 20\log[\omega^2 + (1/RC)^2]^{-1/2} \qquad (6.71)$$

The application of Equation 6.69 to Equation 6.68 yields the sum of two logarithm terms in Equation 6.71.

The right-hand log term in Equation 6.71 is the logarithm of an exponential term, the bracketed term taken to the minus one half-power.

$$\log_a G^k = k\log_a G \qquad (6.72)$$

Simplify the second term of Equation 6.71, using the log Equation 6.72 with $k = -1/2$, to bring down the minus one half, to get Equation 6.73. Twenty times minus one-half equals minus 10.

$$|H(j\omega)|_{\mathrm{dB}} = 20\log\omega + -10\log[\omega^2 + (1/RC)^2] \qquad (6.73)$$

For the RC filter schematic in Figure 6.15, Equation 6.73 is the magnitude of the transfer function $H(j\omega)$ in dB, expressed in the Bode plot in Figure 6.16.

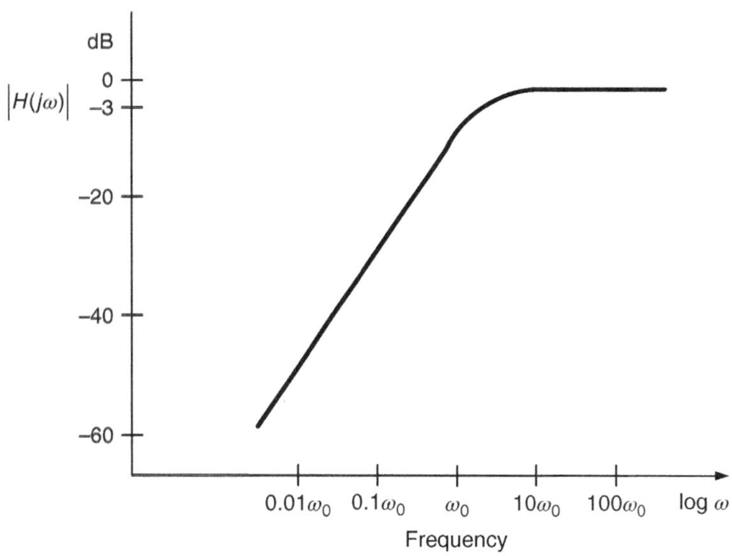

Figure 6.16 Series RC high-pass filter Bode magnitude plot.

In the Bode plot in Figure 6.16, the vertical axis is dB scale, as expressed in Equation 6.67.

In order to manually construct the Bode plot graph in Figure 6.16, and to gain insight, plot the asymptotes of the curve in high- and low-frequency ranges.

Additionally, plot the corner frequency ω_0 where the slope of the curve changes. The Bode plot curve can then be drawn manually using these three pieces of information.

6.2.6 Series RC Magnitude Bode Diagram Plot

6.2.6.1 Calculate the High-Frequency 0 dB Asymptote

Starting with Equations 6.73 and 6.74, for the magnitude of the transfer function, calculate the high-frequency asymptote line. As frequency goes up, far beyond the corner frequency, the magnitude frequency response of the *RC* transfer function will asymptotically approach this line.

$$|H(j\omega)|_{\text{dB}} = 20\log\omega + -10\log[\omega^2 + (1/RC)^2] \quad (6.74)$$

The magnitude of the RC high-pass filter transfer function H in dB.

At large enough values of ω, the $1/RC$ squared term (corner frequency ω_0 squared) becomes small in comparison to ω. As shown in Equation 6.75, remove the $1/RC$ term.

$$|H(j\omega)|_{\text{dB}} = 20\log\omega + -10\log[\omega^2 + \cancel{(1/RC)^2}] \quad (6.75)$$

$$|H(j\omega)|_{\text{dB}} \approx 20\log\omega - 10\log\omega^2 \quad (6.76)$$

In Equation 6.76, simplify the exponent in the minus 10 log term using the log Equation 6.72. Bring down the exponent 2 to get Equation 6.77.

$$|H(j\omega)|_{\text{dB}} \approx 20\log\omega - 20\log\omega \quad (6.77)$$

$$|H(j\omega)|_{\text{dB}} \approx 0\,\text{dB} \quad (6.78)$$

As the frequency ω goes up, the magnitude of the transfer function at high frequencies approaches 0 dB or unity gain, as shown in Equations 6.77 and 6.78.

6.2.6.2 Plot the High-Frequency 0 dB Asymptote

The high-frequency asymptote, of the RC high-pass filter magnitude Bode diagram, is plotted in Figure 6.17 for frequencies above $10\omega_0$.

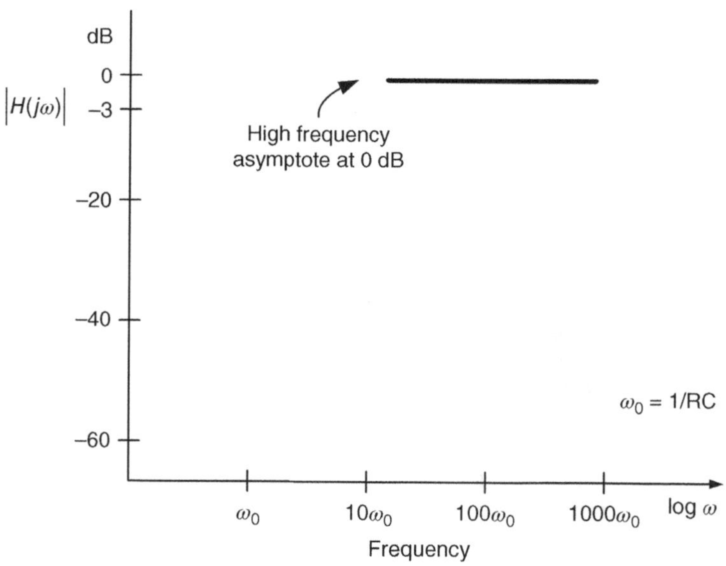

Figure 6.17 Series RC Bode magnitude plot high-frequency asymptote.

6.2.6.3 Calculate the Low-Frequency Asymptote

At low frequencies, the Bode plot of the magnitude of the series RC high-pass filter transfer function will asymptotically approach the line that we calculate in Equation 6.84.

$$|H(j\omega)|_{dB} = 20\log\omega + -10\log[\omega^2 + (1/RC)^2] \quad (6.79)$$

Start with the magnitude of the RC circuit transfer function $H(j\omega)$ in dB in Equation 6.79.

$$|H(j\omega)|_{dB} = 20\log\omega + -10\log[\cancel{\omega^2} + (1/RC)^2] \quad (6.80)$$

As the frequency ω goes below ω_0, the omega squared term becomes small in comparison to the $1/RC$ corner frequency squared term. Cross out the ω^2 term in Equation 6.80 simplifying it to obtain Equation 6.81.

$$|H(j\omega)|_{dB} \approx 20\log\omega - 10\log(1/RC)^2 = 20\log\omega - 10\log\omega_0^2 \quad (6.81)$$

6.2.6.4 Plot the Low-Frequency Asymptote

Plot the low-frequency asymptote of the magnitude of the transfer function from Equation 6.81, using Equation 6.82 as the 0 dB intercept. The asymptote increases with frequency, or rolls up, at a rate 20 dB per decade ($\log\omega$). The high-frequency asymptote and the corner frequency behavior are still missing from the plot in Figure 6.18.

$$|H(j\omega)|_{dB} \approx 20\log\omega - 20\log\omega_0 \quad (6.82) \qquad \omega_0 = 1/RC \quad (6.83)$$

Equation 6.82 is a line with a slope of 20 dB per decade crossing through the point 0 dB at $\omega = \omega_0$. In terms of the line plotting formula $y = mX + B$, $m = 20$ and X is $\log\omega$, as shown in Equation 6.84.

$$20\log\omega - 20\log\omega_0 = 20\log\frac{\omega}{\omega_0} \quad (6.84) \qquad \omega = \omega_0 \qquad 20\log(1) = 0\text{ dB} \quad (6.85)$$

The coordinates of the top of the low-frequency asymptote line in Figure 6.18 are 0 dB, on the vertical scale, and ω_0 on the horizontal scale, as illustrated in Equations 6.84 and 6.85.

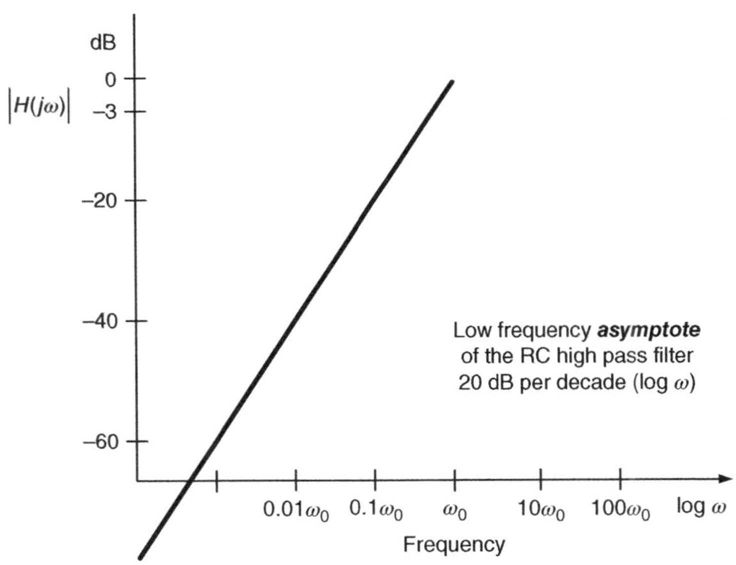

Figure 6.18 Series RC Bode magnitude plot low-frequency asymptote.

The low-frequency asymptote line in Figure 6.19 is intentionally drawn going below the horizontal X axis to illustrate that the X axis does not represent a linear scale 0 for the y axis. The vertical axis can go down further. On the dB scale in Figure 6.18, 0 dB is near the top of the graph.

Similarly, the X axis in Figure 6.18 is also a log scale and the vertical Y axis may not represent zero frequency for the horizontal frequency axis.

6.2.6.5 20 dB per Decade Roll Up Illustration

$$|H(j\omega)|_{\text{dB}} \approx 20\log\omega \tag{6.86}$$

Explore intuition around the roll up at a rate of 20 dB per decade, shown in Equation 6.86, by calculating the magnitude of the transfer function at 10 times ω, as shown in Equation 6.87.

$$|H(j\omega)|_{\text{dB}\ 10\omega} \approx 20\log 10\omega \tag{6.87}$$

10ω is a product and we can again invoke a logarithmic identity, shown in Equation 6.86. Apply Equation 6.88 to Equation 6.87, using 10 for A and ω for B, to obtain Equation 6.89.

$$\log(A \cdot B) = \log(A) + \log(B) \tag{6.88}$$

$$A = 10 \qquad B = \omega$$

$$|H(j\omega)|_{\text{dB}\ 10\omega} \approx 20(\log 10 + \log\omega) \tag{6.89}$$

Multiply out Equation 6.89 and remove the brackets to obtain Equation 6.90.

$$|H(j\omega)|_{\text{dB}\ 10\omega} \approx 20\log 10 + 20\log\omega \tag{6.90}$$

The logarithm to the base 10 of a number is the exponent to which you raise 10 in order to get that number. $\log_{10} = 1$. Simplify Equation 6.90 to obtain 6.91.

$$|H(j\omega)|_{\text{dB}\ 10\omega} \approx 20 + 20\log\omega \tag{6.91}$$

The magnitude of the transfer function at 10ω, one decade higher in frequency than ω, is 20 dB higher than it is at ω. This demonstrates how our transfer function rolls up, increases in value, at the rate of 20 dB per decade of increasing frequency.

6.2.6.6 Corner Frequency

The corner frequency ω_0 is where the magnitude of the transfer function is −3 dB down (from its maximum 0 dB). This half-power point occurs at the frequency $\omega_0 = 1/RC$. ω_0 is sometimes called a break frequency. Along with the asymptotes, the corner frequency ω_0 is useful in manually plotting the magnitude Bode plot of the transfer function.

6.2.6.7 Transfer Function dB Magnitude at the Corner Frequency

$$|H(j\omega)|_{\text{dB}} = 20\log - 10\log[\omega^2 + (1/RC)^2] \tag{6.92}$$

To calculate and verify the −3 dB value of the transfer function magnitude at the corner frequency ω_0, substitute the corner frequency 1 over RC into Equation 6.92 for ω, to obtain Equation 6.93.

$$|H(j\omega)|_{\text{dB}} = 20\log(1/RC) - 10\log[(1/RC)^2 + (1/RC)^2] \tag{6.93}$$

In Equation 6.93, combine terms inside the square bracket, yielding Equation 6.94.

$$|H(j\omega)|_{dB} = 20\log(1/RC) - 10\log[2(1/RC)^2] \tag{6.94}$$

Two times one over RC squared is a product and we can again invoke the logarithmic identity in Equation 6.95 to split the product term into a sum, as shown in Equation 6.96.

$$\log(A \cdot B) = \log(A) + \log(B) \qquad (6.95) \qquad\qquad A = 2 \qquad B = 1/RC$$

$$|H(j\omega)|_{dB} = 20\log(1/RC) - 10\log(2) - 10\log[(1/RC)^2] \tag{6.96}$$

In Equation 6.96, simplify the exponent using the log Equation 6.72, $\log_a G^k$ equals $k\log_a G$. Applying Equation 6.72 to the right side, with k equals 2, "brings down" the exponent 2 to obtain Equation 6.97.

$$|H(j\omega)|_{dB} = 20\log(1/RC) - 10\log(2) - 20\log(1/RC) \tag{6.97}$$

The two 20 log terms in Equation 6.97 cancel one another, resulting in Equation 6.98.

$$|H(j\omega)|_{dB} = -10\log(2) =\sim 3 \text{ dB} \tag{6.98}$$

At the corner frequency 1 over RC the magnitude of the transfer function is said to be minus three dB down.

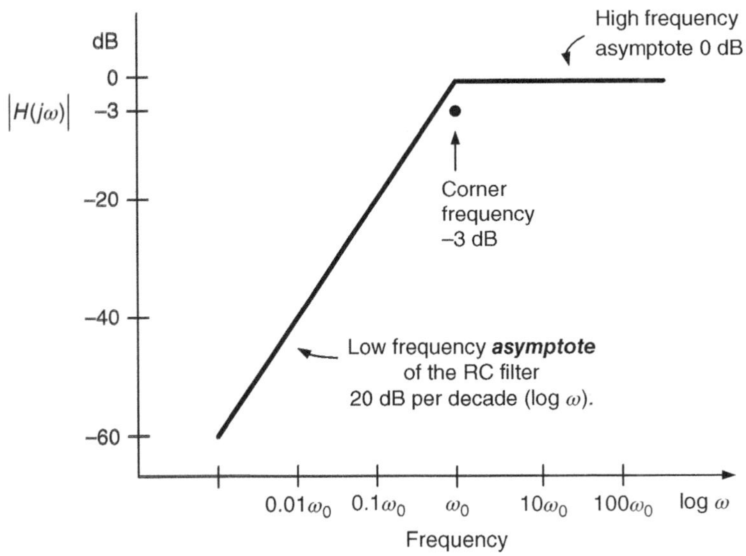

Figure 6.19 Series RC Bode magnitude plot corner frequency.

With all three pieces needed for the Bode plot, as shown in Figure 6.19, the high-frequency asymptote, the low-frequency asymptote, and the corner frequency, one can manually draw the RC high-pass filter transfer function magnitude Bode plot, using the corner frequency for the transition between the two asymptotes, as shown in Figure 6.20.

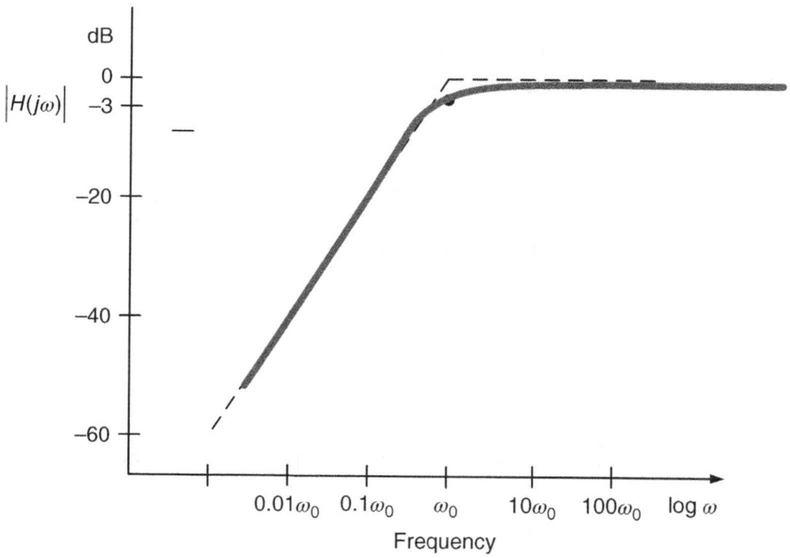

Figure 6.20 Series RC high-pass filter finished Bode magnitude plot.

6.2.7 Series RC Transfer Function Phase Calculation

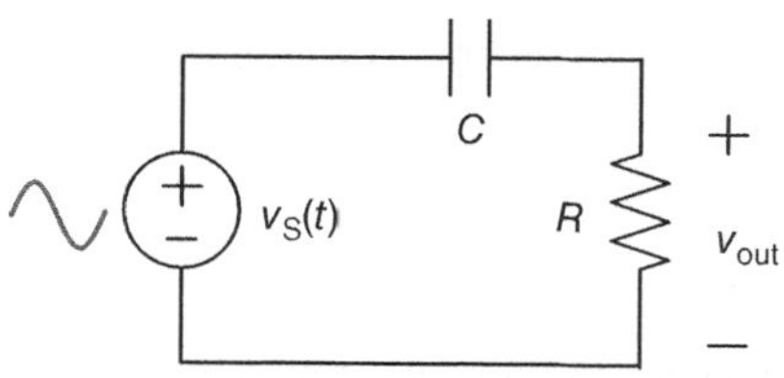

Figure 6.21 Series RC high-pass filter circuit driven by sinusoidal voltage source.

$$\frac{\boldsymbol{V}_{\text{out}}}{\boldsymbol{V}_{\text{S}}} = \left(\frac{j\omega RC}{1 + j\omega RC}\right) = H(j\omega) \tag{6.99}$$

Equation 6.99 is the transfer function of the RC high-pass filter in Figure 6.21.

$$\angle H(j\omega) = \frac{\pi}{2} - \tan^{-1}(\omega RC) \tag{6.100}$$

$$\angle H(j\omega) = \frac{\pi}{2} - \tan^{-1}(\omega/\omega_0) \tag{6.101}$$

Equations 6.100 and 6.101 show the phase response of the RC high-pass filter in Figure 6.21.

$\omega_0 = 1/RC$ $\quad \frac{\pi}{2} = 90°$

Equations 6.102 and 6.103 summarize the calculation of the phase response of the transfer function $H(j\omega)$.

$$\angle H(j\omega) = \angle z_1(j\omega) - \angle z_2(j\omega) = \tan^{-1}\left[\frac{X}{R}\right]\text{numerator} - \tan^{-1}\left[\frac{X}{R}\right]\text{denominator} \tag{6.102}$$

$$\angle H(j\omega) = \angle(j\omega RC) - \angle(j\omega + 1/RC) = 90° - \tan^{-1}(\omega RC) \tag{6.103}$$

6.2.7.1 Calculating the Phase Response of a Complex Transfer Function

The angle theta of a complex function, like the transfer function $H(j\omega)$ in Equation 6.99, is the arctangent of the imaginary part $X(\omega)$ over the real part $R(\omega)$, as shown in Equation 6.104.

$$\theta(\omega) = \angle H(\omega) = \tan^{-1}\frac{X(\omega)}{R(\omega)} \tag{6.104}$$

The first-order RC high-pass filter transfer function in Equation 6.99 is a quotient of two complex expressions $F(\omega)$ and $G(\omega)$ as shown in Equation 6.105.

$$H(\omega) = \frac{\boldsymbol{V}_{\text{out}}}{\boldsymbol{V}_{\text{S}}} = \frac{j\omega RC}{1 + j\omega RC} = \frac{F(\omega)}{G(\omega)} \tag{6.105}$$

$$F = 0 + j\omega RC \tag{6.106}$$

$$G = 1 + j\omega RC \tag{6.107}$$

Equation 6.108 shows how the phase angle of a quotient of complex functions is the phase angle of the numerator minus the phase angle of the denominator.

$$\angle H(j\omega) = \angle\frac{F(j\omega)}{G(j\omega)} = \frac{\angle\theta_F}{\angle\theta_G} = \angle(\theta_F) - \angle(\theta_G) \tag{6.108}$$

Into Equation 6.108, substitute the numerator, F from Equation 6.106, and the denominator G, from Equation 6.107, of the RC high-pass transfer function $H(j\omega)$ to obtain Equation 6.109.

$$\angle H(j\omega) = \angle(j\omega RC) - \angle(j\omega + 1/RC) \tag{6.109}$$

Apply Equation 6.104, the arctangent of the imaginary part X over the real part R, separately to the numerator and denominator terms in Equation 6.109.

Numerator F

$$\angle(j\omega RC) = \tan^{-1}[X/R] = \tan^{-1}[\omega RC/0] = \tan^{-1}(\infty) = 90° = \frac{\pi}{2} \tag{6.110}$$

The phase angle of the numerator is 90° or pi over two because tangent approaches 90° as omega goes to infinity.

Denominator G

$$\angle(1 + j\omega RC) = \tan^{-1}[\omega RC/1] = \tan^{-1}(\omega RC) \tag{6.111}$$

The phase angle of the denominator is arctangent of omega times R times C.

Substitute Equation 6.110 and 6.111 both into Equation 6.109 for the phase angle of $H(j\omega)$ to obtain Equation 6.112.

$$\angle H(j\omega) = \tan^{-1}(\infty) - \tan^{-1}[\omega RC/1] = 90° - \tan^{-1}(\omega RC) \tag{6.112}$$

$$\angle H(j\omega) = \frac{\pi}{2} - \tan^{-1}(\omega RC) \tag{6.113}$$

Equation 6.113 is the phase response of the RC high pass filter transfer function in Equation 6.105.

This phase shift, as a function of angular frequency ω, of the transfer function $H(j\omega)$ for the series RC high-pass filter circuit can be expressed in terms of the break frequency $\omega_0 = 1/RC$.

Insert Equation 6.114a, the corner frequency ω_0, into Equation 6.113 for the phase response of the series RC high-pass filter to obtain Equation 6.114b.

$$\omega_0 = \frac{1}{RC} \quad (6.114a) \qquad \theta = \angle H(j\omega) = \angle \frac{\boldsymbol{V}_{out}}{\boldsymbol{V}_S} = 90° - \tan^{-1}\left(\frac{\omega}{\omega_0}\right) \quad (6.114b)$$

6.2.8 Series RC Phase Bode Diagram Calculation

$$\theta = \angle H(j\omega) = 90° - \tan^{-1}(\omega RC) = \frac{\pi}{2} - \tan^{-1}\left(\frac{\omega}{\omega_0}\right) \quad (6.115)$$

For the RC high-pass filter in Figure 6.21, Equation 6.115 shows the phase response from Equation 6.113.

As a guide to drawing the phase Bode diagram curve by hand, plot several straight lines that will approximate the curve. To do this, calculate a low-frequency asymptote line, a high-frequency asymptote line, and plot the phase of the corner frequency ω_0.

6.2.8.1 Phase as ω Approaches 0

As the frequency approaches zero, the phase from Equation 6.115 approaches 90°, as shown in Equation 6.116b.

$$\theta = \angle H(j\omega) \Rightarrow \frac{\pi}{2} - \tan^{-1}(0) \quad (6.116a)$$

$$\theta = \angle H(j\omega) = \frac{\pi}{2} - 0 = \frac{\pi}{2} = 90° \quad (6.116b)$$

The arctangent of zero degrees is zero.

Draw a flat straight line from 0 to $0.1\omega_0$ frequency for this low-frequency asymptote.

6.2.8.2 Phase as ω Approaches Infinity

As the frequency increases, the argument of arctangent, in the phase Equation 6.115, blows up to infinity, and the arctangent approaches −90°, as shown in Equation 6.117a. The phase at high frequencies approaches 0°, as shown in Equation 6.117a.

$$\theta = \angle H(j\omega) \Rightarrow \frac{\pi}{2} - \tan^{-1}(\infty) \quad (6.117a)$$

$$\theta = \angle H(j\omega) = \frac{\pi}{2} - 90° = 90° - 90° = 0 \quad (6.117b)$$

Draw a flat straight line at 0° starting at $10\omega_0$.

6.2.8.3 Phase at Corner (Break) Frequency $\omega = \omega_0 = 1/RC$

From the phase Equation 6.115, calculate the phase at the corner frequency ω_0, as shown in Equations 6.118 and 6.120.

$$\theta = \angle H(j\omega) \Rightarrow 90° - \tan^{-1}\left(\frac{\omega}{\omega_0}\right) = 90° - \tan^{-1}(1) \quad (6.118) \qquad \omega_0 = \frac{1}{RC} \quad (6.119)$$

$$\theta = \angle H(j\omega) = 90° - \tan^{-1}(1) = 90° - 45° = 45° = 45° = \frac{\pi}{4} \quad (6.120)$$

Mark this 45° value on the graph at $1\omega_0$ on the horizontal axis. The graph goes through this point.

Figure 6.22 shows the two asymptote lines and a slanted line drawn through the corner frequency at 45°.

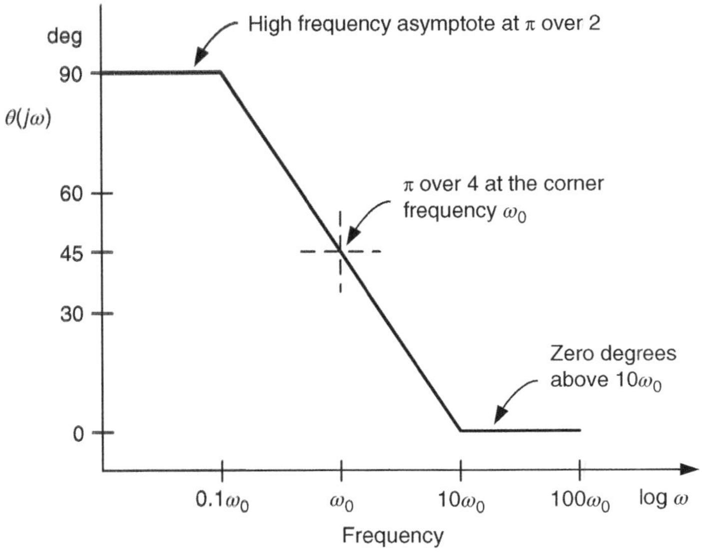

Figure 6.22 Series RC Bode phase plot construction.

6.2.9 Series RC Phase Bode Diagram Plot

Sketch in the phase response plot of the RC filter transfer function, as shown in Figure 6.23.

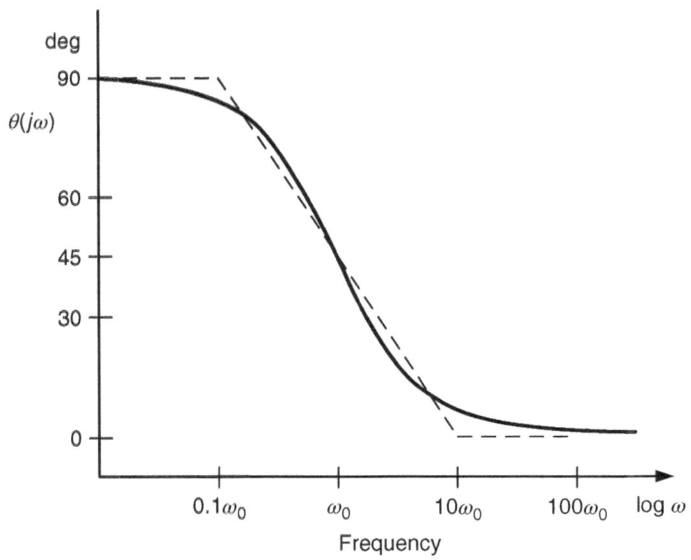

Figure 6.23 Series RC Bode phase plot finished.

$$\theta = \angle H(j\omega) = \frac{\pi}{2} - \tan^{-1}\left(\frac{\omega}{\omega_0}\right) \quad (6.121)$$

$$\angle H(j\omega) = \frac{\pi}{2} - \tan^{-1}(\omega RC) \quad (6.122)$$

Equations 6.121 and 6.122 show the phase response of the RC high-pass filter circuit, in terms of ω and ω_0 in Equation 6.121, and in terms of RC in Equation 6.122.

6.2.10 Series RC High-pass Filter Phase Relationships

Phasor diagrams are useful tools for performing circuit calculations and understanding relationships between voltages and phase angles in circuits. Phasor diagrams use head-to-tail graphical addition to calculate values. Although phasors are complex numbers and not vector quantities, phasor diagrams add voltages and currents like vectors. For additional information on phasor diagrams, please see Chapter 16.

6.2.10.1 Phasor Diagram at Corner Frequency

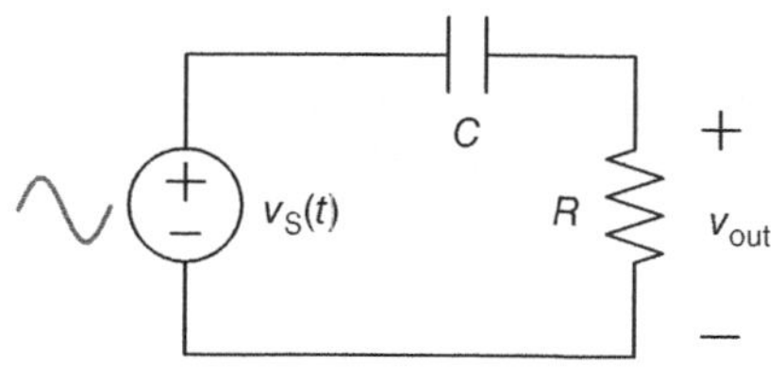

Figure 6.24 Series RC high-pass filter circuit driven by sinusoidal voltage source.

$$\angle H(j\omega) = \frac{\pi}{2} - \tan^{-1}(\omega RC) = \angle \frac{\mathbf{V}_{out}}{\mathbf{V}_S} \tag{6.123}$$

Equation 6.123 shows the phase of the sinusoidal steady-state transfer function, for the series RC high-pass filter in Figure 6.24.

For the series RC high-pass filter in Figure 6.24, calculate the phase angle of V_{out} at the corner frequency ω_0, in Equation 6.124, using phasor diagrams.

$$\omega_0 = \frac{1}{RC} \tag{6.124}$$

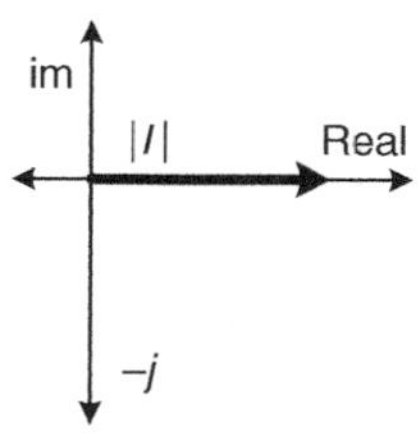

Figure 6.25 Series RC high-pass filter current phasor diagram.

Use current as the reference for zero phase, because in a series circuit, the current is the same through all the circuit elements. We know how the phase of the voltage across the resistor and the capacitor relates to the phase of the current through the resistor and capacitor.

The magnitude of the current phasor |**I**| gets plotted along the positive real axis in Figure 6.25 because the phase of the reference is zero. The unit of the current magnitude is amps.

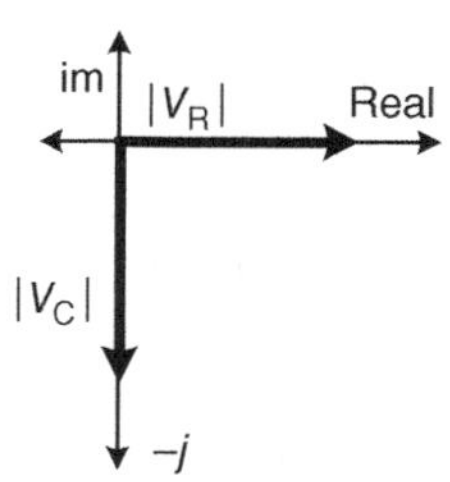

Figure 6.26 Series RC resistor and capacitor voltages phasor diagram.

Plot the magnitude of the resistor voltage $\mathbf{V}_{out}$ along the positive real axis in Figure 6.26 as $|\mathbf{V}_R|$ because the phase of the voltage across a resistor, relative to the current through the resistor, is zero.

Plot the magnitude of the capacitor voltage phasor $|\mathbf{V}_C|$ along the negative imaginary axis in Figure 6.26 because the current through a capacitor leads the voltage by 90° and the direction of positive rotation is counterclockwise.

The length of the two phasors in Figure 6.26 is the same because the magnitude of the resistance and the capacitive reactance are the same at the corner frequency ω_0, as shown in Equation 6.126.

$$Z_C = \frac{1}{j\omega C} \tag{6.125}$$

Equation 6.125 shows the impedance of the capacitor.

$$Z_{\text{C-corner}} = \frac{1}{j\omega_0 C} = \frac{RC}{jC} = jR \tag{6.126}$$

Equation 6.126 shows the impedance of the capacitor at the *t* the corner frequency $\omega_0 = 1/RC$.

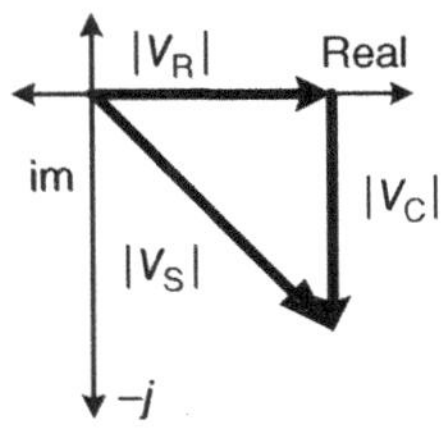

Figure 6.27 Series RC phasor diagram calculating source phasor.

Set the capacitance phasor head-to-tail to the resistor voltage phasor for vector addition, as shown in Figure 6.27. Connect the two phasors to get the magnitude and phase of the source voltage.

The direction of the $\boldsymbol{V}_S$ phasor in Figure 6.27 reflects a Kirchhoff's voltage law expression in the form of Equation 6.127.

$$|\boldsymbol{V}_S| = |\boldsymbol{V}_R| + |\boldsymbol{V}_C| \tag{6.127}$$

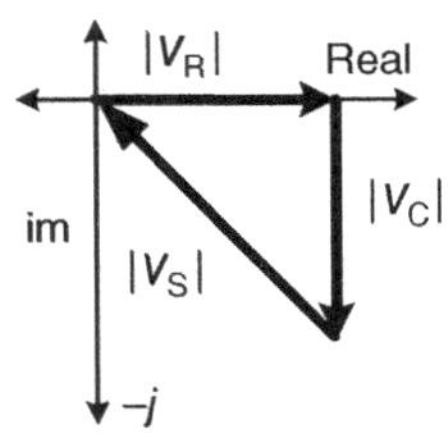

Figure 6.28 Series RC phasor diagram with opposite orientation of source phasor.

The direction of the V_S phasor in Figure 6.28, the opposite of Figure 6.27, reflects a Kirchhoff's voltage law in the form of Equation 6.128, but the result is the same.

$$|\boldsymbol{V}_R| + |\boldsymbol{V}_C| - |\boldsymbol{V}_S| = 0 \tag{6.128}$$

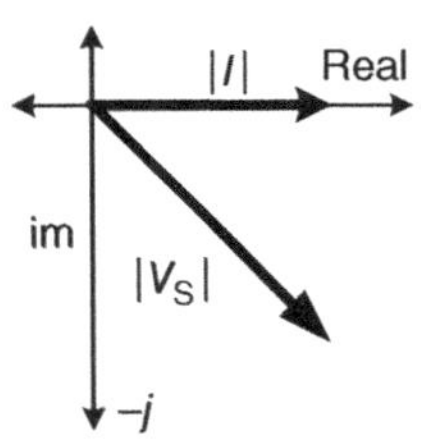

Figure 6.29 Series RC phasor diagram with source phasor at −45°.

Figure 6.29 shows that the V_S phasor lags the current phasor by 45° in phase.

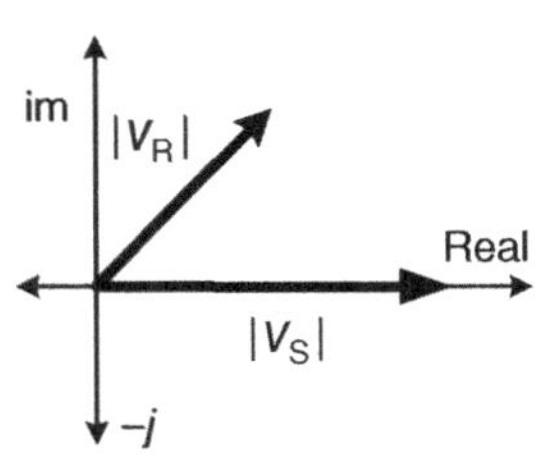

Figure 6.30 Series RC phasor diagram with source phasor on real axis.

The voltage across the resistor is V_{out}, which is in phase with the current. To complete the calculation of the phase of the output voltage at the corner frequency, rotate the phasors counterclockwise (in the positive phase direction) by 45°, as shown in Figure 6.30. This rotation puts the $\boldsymbol{V}_S$ phasor on the real axis, with zero phase, making $\boldsymbol{V}_S$ the reference.

Figure 6.30 shows that relative to V_S, the phase of the output voltage V_{out} is 45°.

The output voltage leads the input voltage in phase, making this a lead network.

6.3 Series RL Low-pass Filter Frequency Response

6.3.1 Series RL Frequency Domain Results Overview

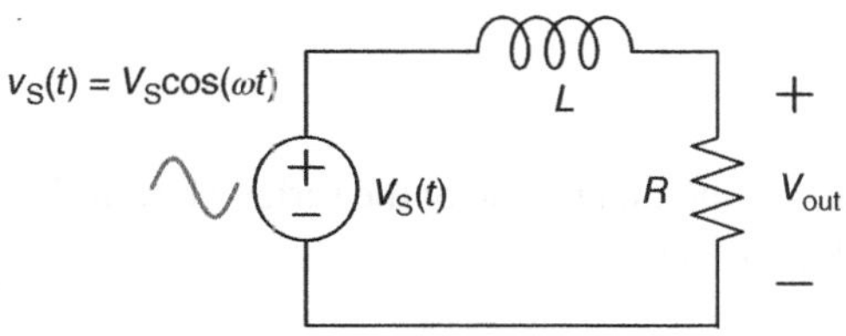

Figure 6.31 Series RL low-pass filter circuit driven by sinusoidal voltage source.

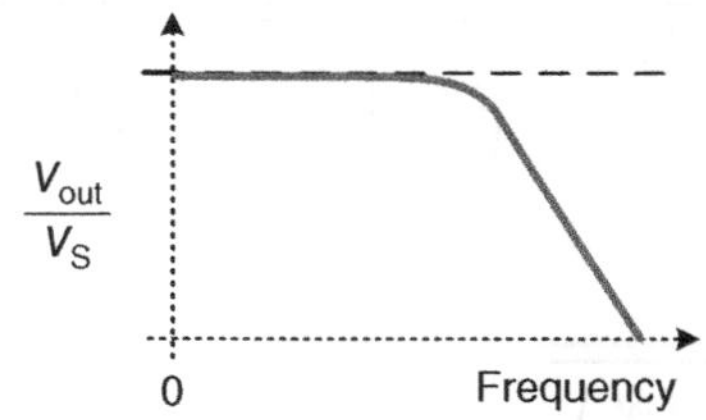

Figure 6.32 example Bode magnitude plot of first-order low-pass filter.

At low frequency, the inductor impedance is low and current can flow, causing voltage to drop across the resistor as V_{out}. At high frequencies, little current flows and V_{out} goes down. From this we know that the RL circuit in Figure 6.31 is a low-pass filter, as plotted in Figure 6.32.

$$\frac{\mathbf{V}_{out}}{\mathbf{V}_S} = \frac{1}{1+j\frac{\omega}{\omega_0}} \quad (6.129)$$

Equation 6.129 is in the general formula for low-pass filter transfer functions. This form is useful for easily calculating specific transfer functions.

$$\omega_0 = R/L \quad (6.130)$$

For this series RL circuit in Figure 6.31, substitute ω_0 equals R/L, the corner frequency from Equation 6.130, into the general transfer function formula Equation 6.129, to obtain Equation 6.131.

$$\frac{\mathbf{V}_{out}}{\mathbf{V}_S} = \frac{1}{j\omega\frac{L}{R}+1} = H(j\omega) \quad (6.131)$$

Multiply the top and bottom of the transfer function Equation 6.131 by R to get Equation 6.132.

$$\frac{\mathbf{V}_{out}}{\mathbf{V}_S} = \frac{R}{j\omega L+R} = H(j\omega) \quad (6.132)$$

Equation 6.132 shows the steady-state frequency domain voltage **transfer function**, in phasor notation, of the RL low-pass circuit in Figure 6.31.

The output voltage varies with frequency, and the source is a voltage, so $\mathbf{V}_{out}/\mathbf{V}_S$ makes a convenient transfer function. The denominator of the transfer function has a frequency-dependent $j\omega L$ term, which means that as frequency goes up, the magnitude of the transfer function decreases. At DC (zero frequency) the $j\omega L$ term becomes zero and the magnitude of the transfer function is 1 (R/R) in magnitude and 0 dB.

$$|H(j\omega)| = \left|\frac{\mathbf{V}_{out}}{\mathbf{V}_S}\right| = \frac{1}{\sqrt{1+(\omega/\omega_0)^2}} \quad (6.133)$$

Equation 6.133 is in the general format for low-pass filter transfer function magnitude in terms of angular frequency ω and break frequency ω_0.

For this series RL circuit, substitute $\omega_0 = R/L$ from Equation 6.130, into the general transfer function magnitude Equation 6.133, to obtain 6.135.

$$|H(j\omega)| = \left|\frac{\boldsymbol{V}_{\text{out}}}{\boldsymbol{V}_{\text{S}}}\right| = \frac{1}{\sqrt{1+\left(\frac{\omega}{\omega_0}\right)^2}} = \frac{1}{\sqrt{1+\left(\frac{\omega L}{R}\right)^2}} \quad (6.134)$$

$$\lvert H(j\omega)\rvert = \frac{1}{\sqrt{1+\left(\frac{\omega L}{R}\right)^2}}$$	(6.135)	Equation 6.135 is the magnitude of the series RL transfer function in Equation 6.132.

$$\theta = \angle H(j\omega) = \angle\frac{\boldsymbol{V}_{\text{out}}}{\boldsymbol{V}_{\text{S}}} = -\tan^{-1}\left(\frac{\omega}{\omega_0}\right) \quad (6.136)$$

Equation 6.136 is in the general format for low-pass filter phase response in terms of angular frequency ω and break frequency ω_0.

For this series RL circuit, substitute $\omega_0 = R/L$ from Equation 6.130, into the general phase response Equation 6.136, to obtain 6.138.

$$\theta = \angle H(j\omega) = \angle\frac{\boldsymbol{V}_{\text{out}}}{\boldsymbol{V}_{\text{S}}} - \tan^{-1}\left(\frac{\omega}{R/L}\right) = -\tan^{-1}(\omega L/R) \quad (6.137)$$

$\angle H(j\omega) = -\tan^{-1}(\omega L/R)$	(6.138)	Equations 6.138 and Figure 6.33 are the phase response of the series RL transfer function in Equation 6.132.

As indicated by the negative sign on the phase in Equation 6.138, and the corresponding phase plot in Figure 6.33, this series RL circuit in Figure 6.31 is a **lag network**. The phase of the output voltage lags the phase of the input voltage.

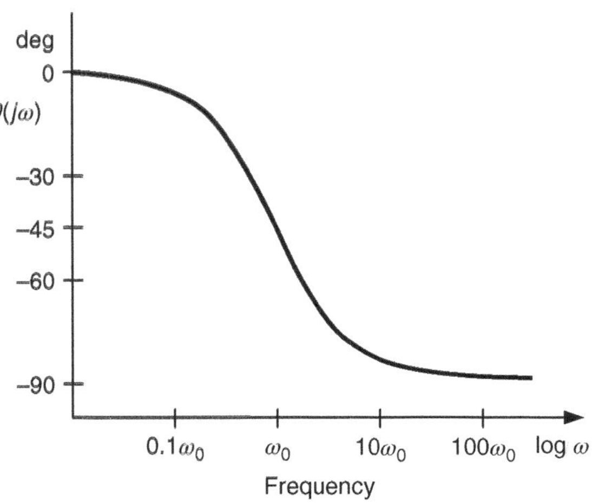

Figure 6.33 Series RL Bode phase plot.

6.3.2 Series RL Frequency Domain Example Intuitive Analysis

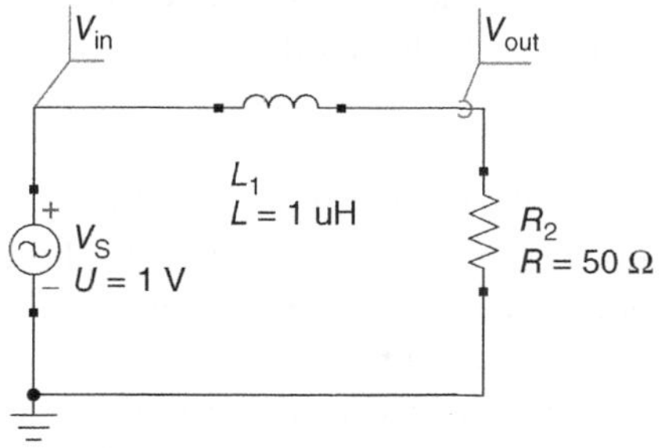

Figure 6.34 Series RL circuit simulation schematic driven by sinusoidal voltage source.

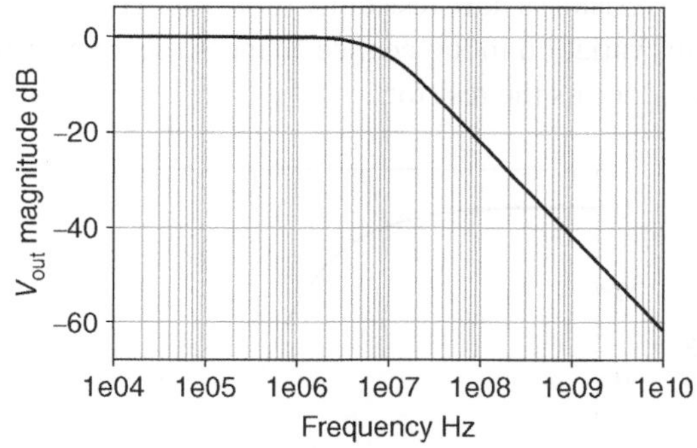

Figure 6.35 Series RL frequency domain simulation of output voltage magnitude.

In the schematic in Figure 6.34, the inductor looks like a short circuit (low impedance) to DC, and the inductor looks like an open circuit (high impedance) to high frequency.

This circuit in Figure 6.34 is a low-pass filter because the series inductor passes the low frequencies most easily from the input voltage source, labeled V_S and V_{in} to the output, labeled V_{out} on the simulation schematic. The frequency response is shown in Figure 6.35.

As the frequency of the input voltage V_S goes down, the inductor impedance $Z = j\omega L$ decreases toward zero. The output voltage V_{out} approaches V_S because the voltage across the inductor, V_S minus V_{out}, decreases toward zero. Figure 6.35, the V_{out} magnitude plot, shows 0 dB at lower frequencies, indicating a ratio of V_{out} over V_{in} near 1. The decreasing size of the inductor voltage also appears in Figure 6.36, which goes down to about −60 dB. At zero frequency (DC), the current in the loop, shown in Figure 6.37 is V_S over R (also labeled V_{in} over R). One volt over 50 Ω equals the 20 mA.

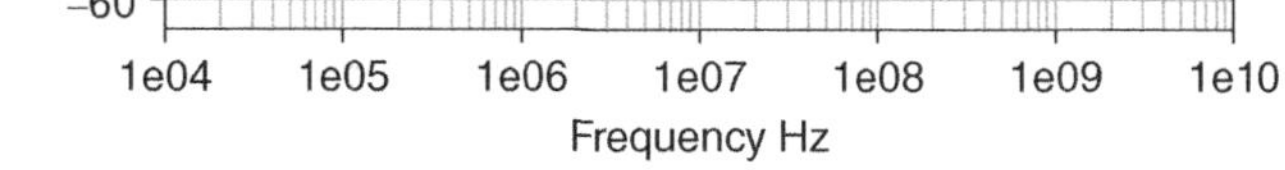

Figure 6.36 Series RL simulation of inductor voltage frequency response .

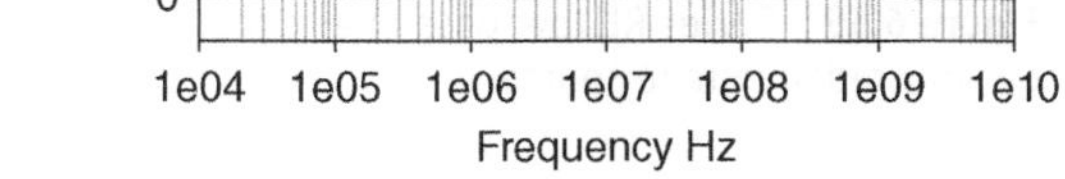

Figure 6.37 Series RL low-pass filter simulation of current.

As the signal (V_S) frequency goes up, the impedance of the inductor ($j\omega L$) goes up and more of the source voltage V_S gets dropped across the inductor, as shown in Figure 6.36, leaving less voltage across the resistor (V_{out}), as shown in Figure 6.35. In the high-frequency limit, all of V_S can drop across the inductor and no current flows at all, as shown in Figure 6.37.

The impedance of the inductor, which is a pure reactance ωL, changes with frequency while the impedance of the resistor (a pure resistance) does not. The total series impedance of the resistor and inductor together varies, and this variation in impedance changes the loop current. The higher the loop current, the higher the output voltage V_{out} across the resistor.

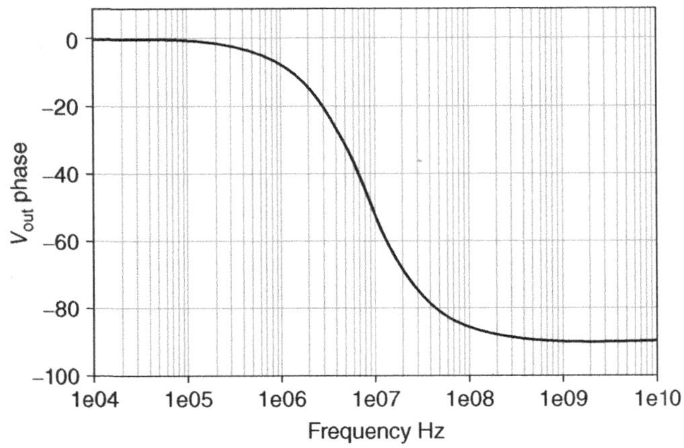

Figure 6.38 Series RL simulation of output voltage phase response.

At DC the reactance $j\omega L$ goes to zero, making the inductor effectively a short circuit. With the reactance zero, there is no phase shift at DC. As frequency goes up, the reactance increases and at high frequency makes the inductor effectively open circuit. Figure 6.38 shows the phase of the output voltage, which is taken across the resistor. From the phase Equation 6.138, we see that the phase shift is negative.

6.3.2.1 Lag Network

In an inductor, the voltage leads the current in phase. The output of this circuit is the voltage across the resistor, which is always in phase with the current through the resistor. The current is also the reference for the phase. Since the output voltage has the phase of the current, which lags the voltage across the inductor, this is a lag circuit. Figure 6.38 shows the phase starting at zero decreasing in the negative direction as frequency goes up, consistent with a lag network. At higher frequencies, the output voltage lags the input voltage in phase.

6.3.3 Series RL Frequency Domain Transfer Function Calculation

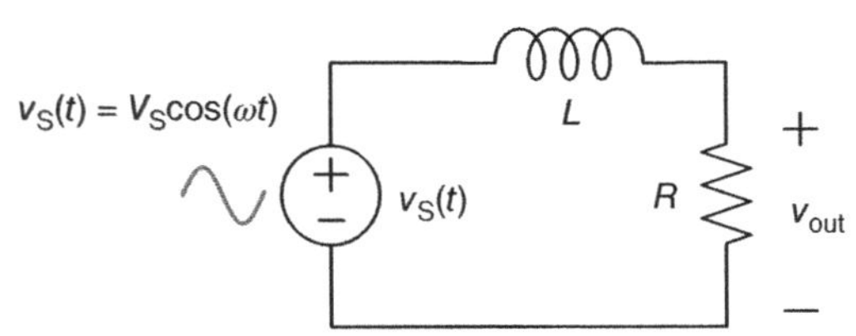

Figure 6.39 Series RL low-pass filter circuit driven by sinusoidal voltage source.

$$\frac{\mathbf{V}_{out}}{\mathbf{V}_S} = \frac{R}{j\omega L + R} = H(j\omega) \tag{6.139}$$

Equation 6.139 is the steady-state frequency domain voltage transfer function, in phasor notation, of the RL low-pass filter in Figure 6.39.

6.3.3.1 Calculating the Transfer Function $\frac{V_{out}}{V_S}$

We can use a voltage divider to calculate the steady-state frequency domain transfer function of the output voltage V_{out} over the input voltage V_S. Notice the use of j as the imaginary operator rather than i used in other mathematics and physics writing.

$$Z_L = j\omega L \tag{6.140}$$

Equation 6.140 is the impedance of an ideal inductor.

$$Z_R = R \tag{6.141}$$

Equation 6.141 is the impedance of an ideal resistor.

$$\boldsymbol{V}_{\text{out}} = \left(\underbrace{\frac{V_S}{j\omega L + R}}_{\text{current}} \right) R \quad (6.142)$$

The output voltage equals the current in the loop multiplied by the impedance (resistance) of the resistor. The current in the loop equals the input voltage V_S divided by the impedance of the inductor and resistor in series ($j\omega L$ plus R). This is shown in Equation 6.142.

$$\frac{\boldsymbol{V}_{\text{out}}}{\boldsymbol{V}_S} = \frac{R}{j\omega L + R} \quad (6.143)$$

Divide both sides of Equation 6.142 by V_S and move the R term inside the brackets to obtain Equation 6.143.

Divide top and bottom of the right-hand side of Equation 6.143 by L to obtain Equation 6.144.

$$\frac{\boldsymbol{V}_{\text{out}}}{\boldsymbol{V}_S} = \frac{R/L}{j\omega + R/L} \quad (6.144)$$

Equation 6.144 is the steady-state frequency domain voltage transfer function, in phasor notation, of the RL circuit in Figure 6.39.

Another commonly seen form for the transfer function is obtained by multiplying Equation 6.144 by L over R yielding Equation 6.145.

$$\frac{\boldsymbol{V}_{\text{out}}}{\boldsymbol{V}_S} = \frac{1}{j\omega L/R + 1} = H(j\omega) \quad (6.145)$$

Equation 6.145 is the steady-state frequency domain voltage transfer function, in phasor notation, of the circuit in Figure 6.39.

$$\frac{\boldsymbol{V}_{\text{out}}}{\boldsymbol{V}_S} = \frac{1}{j\omega \frac{L}{R} + 1} = H(j\omega) \quad (6.146)$$

$$\omega_0 = R/L \quad (6.147)$$

From Equation 6.147, substitute $\omega_0 = R/L$, the corner frequency that makes the L/R term 1, into the transfer function Equation 6.146, to obtain Equation 6.148.

$$\frac{\boldsymbol{V}_{\text{out}}}{\boldsymbol{V}_S} = \frac{1}{1 + j\frac{\omega}{\omega_0}} \quad (6.148)$$

Equation 6.148 is in the general format for low-pass filters. This form is useful for easily calculating the magnitude and phase of the transfer function.

6.3.3.2 Calculating the Magnitude of the Transfer Function – Summarized

$$\frac{\boldsymbol{V}_{\text{out}}}{\boldsymbol{V}_S} = \frac{1}{j\omega \frac{L}{R} + 1} = H(j\omega) \quad (6.149)$$

Equation 6.149 is the transfer function for the series RL low-pass filter circuit in Figure 6.39.

$$|H(j\omega)| = \left| \frac{1}{1 + j\omega \frac{L}{R}} \right| = \frac{|1|}{\left|1 + j\omega \frac{L}{R}\right|} \quad (6.150)$$

The magnitude of a quotient of two complex functions is the quotient of the magnitudes of the complex functions, as shown in Equation 6.150. In this case the numerator is just the number 1.

The magnitude of a complex function $H(\omega)$, in Equation 6.150, is the square root of the sum of the squares of the real and imaginary parts, as shown in Equation 6.151.

$$|H(j\omega)| = \frac{1}{\sqrt{(1)^2 + \left(\omega \frac{L}{R}\right)^2}} = \frac{1}{\sqrt{1 + \left(\omega \frac{L}{R}\right)^2}} = \frac{1}{\sqrt{1 + \left(\frac{\omega}{\omega_0}\right)^2}} \quad (6.151)$$

$$|H(j\omega)| = \frac{1}{\sqrt{1+\left(\omega\frac{L}{R}\right)^2}} \quad (6.152)$$

Equation 6.152 is the **magnitude** of the transfer function of the series RL low-pass filter circuit in Figure 6.39.

6.3.3.3 Phase of the RL Low-pass Filter Transfer Function

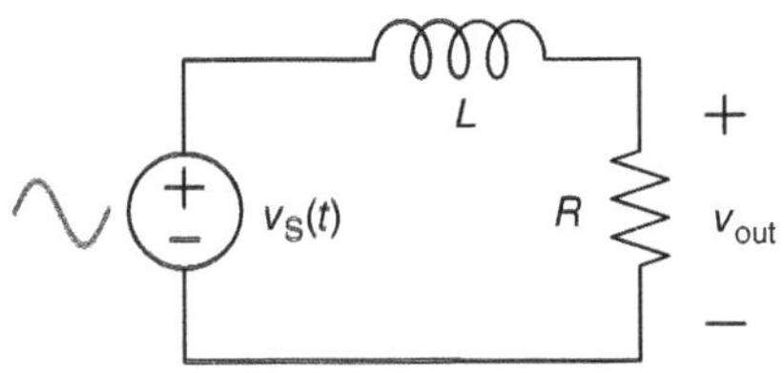

Figure 6.40 Series RL low-pass filter circuit driven by sinusoidal voltage source.

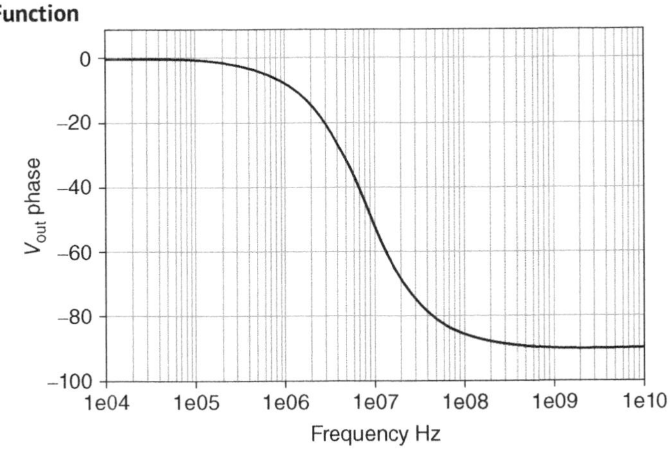

Figure 6.41 Series RL simulation of output voltage phase response.

6.3.3.4 Calculating the Phase of the Transfer Function – Summarized

$$\frac{\mathbf{V}_{out}}{\mathbf{V}_S} = \frac{1}{j\omega\frac{L}{R}+1} = H(j\omega) \quad (6.153)$$

Equation 6.153 is the transfer function for the series RL low-pass filter circuit in Figure 6.40. A simulation of the phase is shown in Figure 6.41.

$$\phi(\omega) = \tan^{-1}\frac{X(\omega)}{R(\omega)} \quad (6.154)$$

Equation 6.154 shows that the phase of the voltage transfer function is the arctangent of the imaginary component $X(\omega)$ divided by the real component $R(\omega)$.

Equation 6.155 shows the phase angle of a quotient of complex functions is the phase angle of the numerator $F(j\omega)$ minus the phase angle of the denominator $G(j\omega)$.

$$\angle H(j\omega) = \angle\frac{F(j\omega)}{G(j\omega)} = \frac{\angle\theta_F}{\angle\theta_G} = \angle(\theta_F) - \angle(\theta_G) \quad (6.155)$$

$$\theta(\omega) = \angle H(j\omega) = \angle\left(\frac{1}{j\omega\frac{L}{R}+1}\right) = \tan^{-1}(1) - \tan^{-1}(\omega L/R) = 0 - \tan^{-1}(\omega L/R) \quad (6.156)$$

$$\theta(\omega) = \angle H(j\omega) = -\tan^{-1}(\omega L/R) \quad (6.157)$$

Equation 6.157 is the **phase** response, as shown in Figure 6.41, of the series RL low-pass filter circuit in Figure 6.40.

$$\theta(\omega) = \angle H(j\omega) = -\tan^{-1}(\omega/\omega_0) \quad (6.158)$$

Equation 6.158 is the general form of the phase response in terms of ω_0.

$$\omega_0 = R/L \tag{6.159}$$

In Equation 6.159, ω_0 is the corner or break frequency of the series RL low-pass filter.

6.3.4 Series RL Low-pass Filter Phase Relationships

Phasor diagrams are useful tools for performing circuit calculations and understanding relationships between voltages and phase angles in circuits. Phasor diagrams use head-to-tail graphical addition to calculate values. Although phasors are complex numbers and not vector quantities, phasor diagrams add voltages and currents like vectors. For additional information on phasor diagrams, please see Chapter 16.

6.3.4.1 Phasor Diagram at Corner Frequency

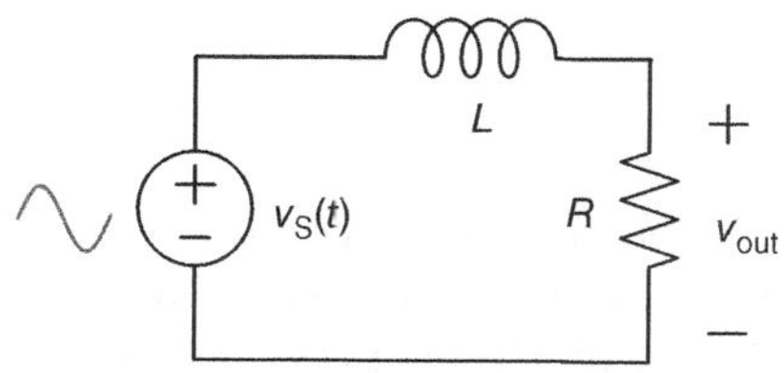

Figure 6.42 Series RL low-pass filter circuit driven by sinusoidal voltage source.

$$\theta = \angle H(j\omega) = -\tan^{-1}\left(\frac{\omega L}{R}\right) = \angle \frac{\boldsymbol{V}_{\text{out}}}{\boldsymbol{V}_{\text{S}}} \tag{6.160}$$

Equation 6.160 shows the phase of the sinusoidal steady-state transfer function for the series RL low-pass filter in Figure 6.42.

For the series RL low-pass filter circuit in Figure 6.42, use phasor diagrams to calculate the phase angle of V_{out} at the corner frequency ω_0.

$$\omega_0 = \frac{R}{L} \tag{6.161}$$

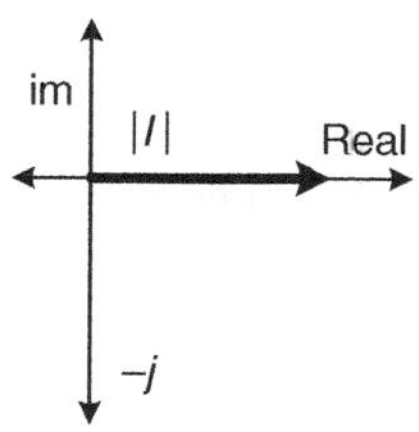

Figure 6.43 Series RL low-pass filter current phasor diagram .

Use the current phasor **I** as the reference for zero phase, because in a series circuit, the current is the same through all the circuit elements.

The magnitude of the current phasor |**I**| gets plotted along the positive real axis in Figure 6.43 because the phase of the reference is zero. The unit of the current magnitude is amps.

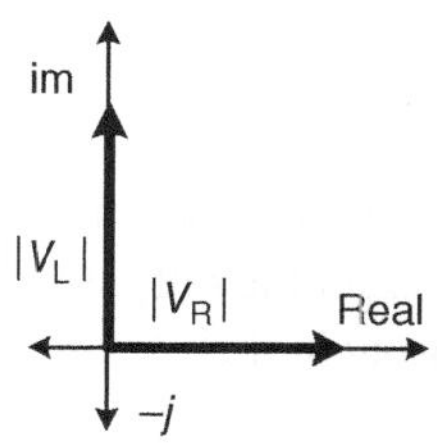

Figure 6.44 Series RL resistor and inductor voltages phasor diagram.

Plot the magnitude of the resistor voltage Vout as $|\mathbf{V}_{\text{R}}|$ along the positive real axis in Figure 6.44 because the phase of the voltage across a resistor, relative to the current through the resistor, is zero.

Plot the magnitude of the inductor voltage phasor $|\mathbf{V}_{\text{L}}|$ along the positive imaginary axis in Figure 6.44 because the current through an inductor lags the voltage by 90°.

The length, of the two phasors in Figure 6.44, is the same, as shown in Equation 6.164. This is because the magnitude of the resistance and inductive reactance are the same at the corner frequency ω_0, which is shown in Equation 6.163. Equation 6.162 is the impedance of an ideal inductor, which is purely reactive. A reactive component of the same value as the resistive component in the circuit means that the two magnitudes are equal.

$$Z_L = j\omega L \qquad (6.162)$$

$$\omega_0 = \frac{R}{L} \qquad (6.163)$$

$$Z_{\text{L-corner}} = j\omega_0 L = j\frac{R}{L}L = jR \qquad (6.164)$$

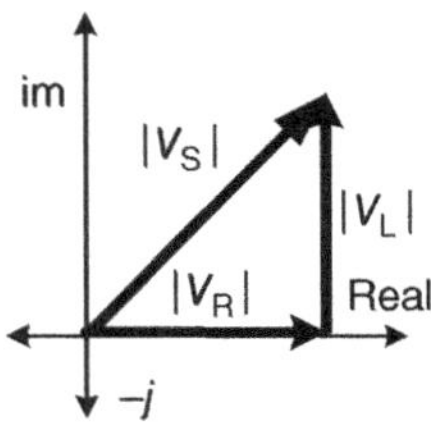

Figure 6.45 Series RL phasor diagram calculating source phasor.

Set the inductor voltage phasor head-to-tail with the resistor voltage phasor for vector addition. Connect the two phasors to get the magnitude and phase of the source voltage V_S, as shown in Figure 6.45.

The direction of the $\boldsymbol{V}_S$ phasor in Figure 6.45 reflects a Kirchhoff's voltage law in the form of Equation 6.165.

$$|\boldsymbol{V}_S| = |\boldsymbol{V}_R| + |\boldsymbol{V}_L| \qquad (6.165)$$

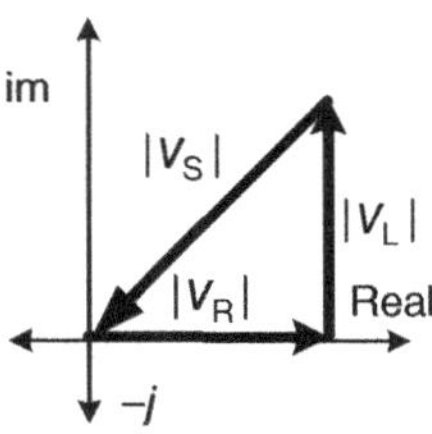

Figure 6.46 Series RL phasor diagram with opposite orientation of source phasor.

The opposite direction of the $\boldsymbol{V}_S$ phasor in Figure 6.46, in comparison to Figure 6.45, reflects a Kirchhoff's voltage law in the form of equation 6.166, but the magnitude and phase angle result is the same as in Figure 6.45.

$$|\boldsymbol{V}_R| + |\boldsymbol{V}_L| - |\boldsymbol{V}_S| = 0 \qquad (6.166)$$

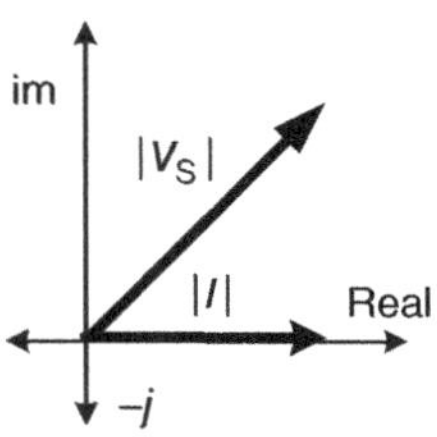

Figure 6.47 Series RL phasor diagram with source phasor at plus 45°.

Figure 6.47 shows that the V_S phasor leads the current phasor by 45°.

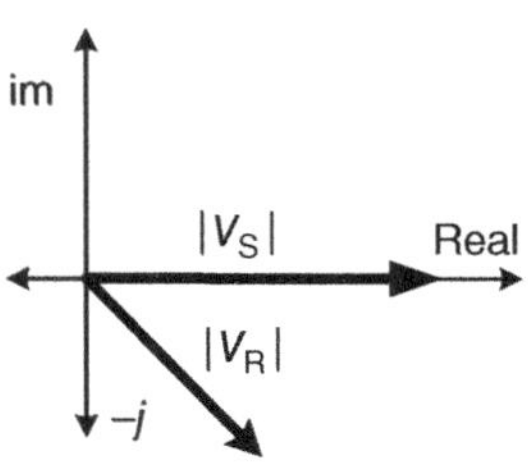

Figure 6.48 Series RL phasor diagram with source phasor on real axis.

The voltage across the resistor is V_{out}, which is in phase with the current. To complete the calculation of the phase of the output voltage at the corner frequency, rotate the phasors in the clockwise (negative phase) direction by 45°, as shown in Figure 6.48. This rotation puts the V_S phasor on the real axis, with zero phase, making V_S the reference. Relative to V_S, the phase of the output voltage $V_{out} = V_R$ is −45°.

6.3.5 Series RL Transfer Function Magnitude – Detailed Calculation

$$\frac{\boldsymbol{V}_{\text{out}}}{\boldsymbol{V}_{\text{S}}} = \frac{R/L}{j\omega + R/L} = H(j\omega) \qquad (6.167)$$

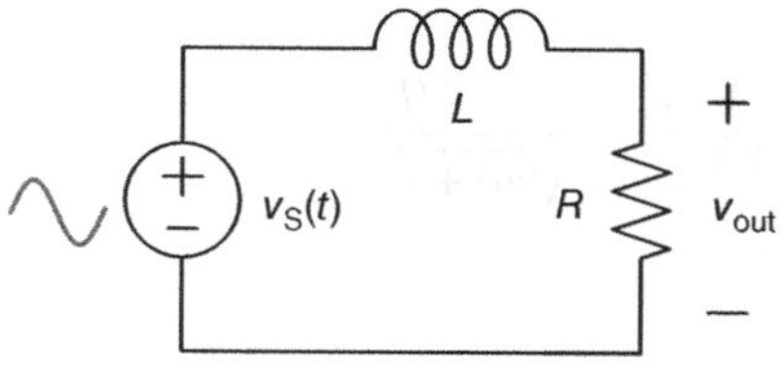

Figure 6.49 Series RL low-pass filter circuit driven by sinusoidal voltage source.

To plot the frequency response of the first-order RL low-pass filter circuit in Figure 6.49, we need to calculate the magnitude and phase of the transfer function in Equation 6.167.

The magnitude of a complex function, $G(\omega)$ in Equation 6.168, is the square root of the sum of the squares of the real part $R(\omega)$ and the imaginary part $X(\omega)$, as shown in Equation 6.169.

$$G(\omega) = R(\omega) + jX(\omega) \qquad (6.168)$$

$$|G(\omega)| = \sqrt{[R(\omega)]^2 + [X(\omega)]^2} \qquad (6.169)$$

The series RL low-pass filter transfer function, in Equation 6.167, can be thought of as the quotient of two complex functions, where the numerator is a real-valued constant, R. The numerator has no imaginary component.

The magnitude of the ratio of two complex functions $A(\omega)$ and $B(\omega)$ equals the ratio of the magnitudes of those same two complex functions, as shown in Equation 6.170.

$$\left|\frac{A(\omega)}{B(\omega)}\right| = \frac{|A(\omega)|}{|B(\omega)|} \qquad (6.170)$$

Starting with Equation 6.170, calculate the magnitude of the RL low-pass filter transfer function, in Equation 6.167, using the square root of the sum of the squares, as shown in Equation 6.171. Notice that the j does not get squared.

$$\left|\frac{\boldsymbol{V}_{\text{out}}}{\boldsymbol{V}_{\text{S}}}\right| = \frac{V_{\text{out}}}{V_{\text{S}}} = \frac{|R/L|}{|j\omega + R/L|} = \frac{R/L}{\sqrt{(\omega L)^2 + (R/L)^2}} \qquad (6.171)$$

$$|H(j\omega)| = \left|\frac{\boldsymbol{V}_{\text{out}}}{\boldsymbol{V}_{\text{S}}}\right| = \frac{R/L}{\sqrt{\omega^2 + (R/L)^2}} \qquad (6.172)$$

Equation 6.172 is the **magnitude** of the RL lowpass filter transfer function, $H(j\omega)$, of the circuit in Figure 6.49.

6.3.6 Series RL Magnitude Bode Diagram Calculation

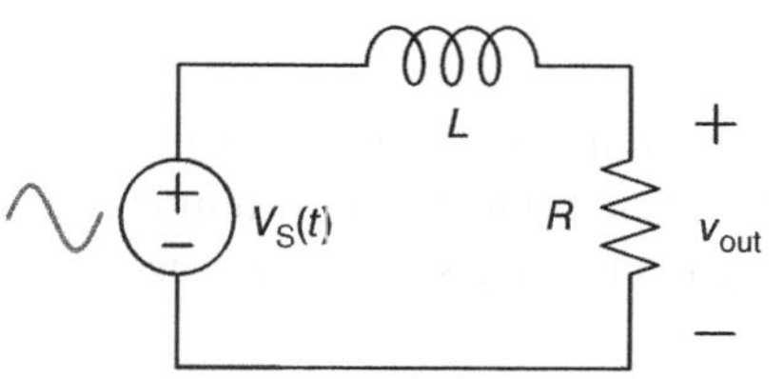

Figure 6.50 Series RL low-pass filter circuit driven by sinusoidal voltage source.

$$\frac{\boldsymbol{V}_{\text{out}}}{\boldsymbol{V}_{\text{S}}} = \frac{R/L}{j\omega + R/L} = H(j\omega) \qquad (6.173)$$

Equation 6.173 states the steady-state frequency domain voltage transfer function, in phasor notation, of the RL circuit in 6.50.

From the magnitude of the transfer function, we'll generate the exact equations used to create the Bode plots. This plot will have dB magnitude on the vertical scale and radian frequency ω on the horizontal axis. $\omega = 2\pi f$.

$$|H(j\omega)| = \left|\frac{\boldsymbol{V}_{\text{out}}}{\boldsymbol{V}_{\text{S}}}\right| = \frac{R/L}{\sqrt{\omega^2 + (R/L)^2}} \tag{6.174}$$

Equation 6.174 is the magnitude of the transfer function $H(j\omega)$ of the RL circuit in Figure 6.50, from Equation 6.172.

From equation 6.174, take the R/L term from the numerator to the front as a multiplicative constant, as shown in Equation 6.175.

$$|H(j\omega)| = \left|\frac{\boldsymbol{V}_{\text{out}}}{\boldsymbol{V}_{\text{S}}}\right| = \frac{R}{L}\frac{1}{\sqrt{\omega^2 + (R/L)^2}} \tag{6.175}$$

Restate the square root in the denominator of Equation 6.175 as an exponent of minus one half.

$$|H(j\omega)| = \frac{R}{L}[\omega^2 + (R/L)^2]^{-1/2} \tag{6.176}$$

The vertical scale of the Bode diagram is logarithmic in decibels (dB).

$$H \text{ in dB} = 20\log_{10}H \tag{6.177}$$

The number H, expressed in decibels in Equation 6.177, is 20 times the logarithm to the base 10 of H.

$$|H(j\omega)|_{\text{dB}} = 20\log|H(j\omega)| \tag{6.178}$$

The magnitude of the transfer function $H(\omega)$ expressed in decibels, in Equation 6.178, is 20 times the logarithm to the base 10 of the magnitude of $H(\omega)$.

Substitute the transfer function magnitude $|H(j\omega)|$, from Equation 6.176, into Equation 6.178, to obtain Equation 6.179.

$$|H(j\omega)|_{\text{dB}} = 20\log\frac{R}{L}[\omega^2 + (R/L)^2]^{-1/2} \tag{6.179}$$

Equation 6.179 is the logarithm of a product. Use a logarithmic identity, in Equation 6.180, to split this product term of R over L times the "square root" (bracketed) expression in Equation 6.179, to obtain Equation 6.183.

$$\log(A\cdot B) = \log(A) + \log(B) \tag{6.180}$$

$$A = R/L \tag{6.181}$$

$$B = [\omega^2 + (R/L)^2]^{-1/2} \tag{6.182}$$

When we apply the logarithmic rule for the log of a product, Equation 6.180, we get this sum of two logarithm terms in Equation 6.183.

$$|H(j\omega)|_{\text{dB}} = 20\log(R/L) + 20\log[\omega^2 + (R/L)^2]^{-1/2} \tag{6.183}$$

The right-hand log term in Equation 6.183 is the logarithm of an exponential term, the bracketed term taken to the minus one half-power.

$$\log_a G^k = k\log_a G \tag{6.184}$$

Simplify the second term of Equation 6.183, using the log Equation 6.185 with k equals $-1/2$, to bring down the minus one half, to obtain Equation 6.185. Twenty times minus one-half equals minus 10.

$$|H(j\omega)|_{\text{dB}} = 20\log(R/L) - 10\log[\omega^2 + (R/L)^2] \tag{6.185}$$

Equation 6.185 is the magnitude of the transfer function $H(j\omega)$ in dB, plotted in the Figure 6.51 Bode plot.

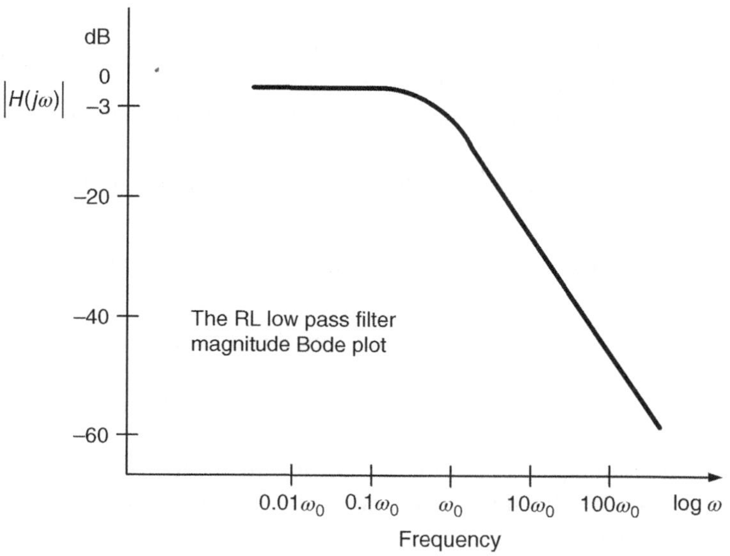

Figure 6.51 Series RL low-pass filter Bode magnitude plot.

In the Bode plot in Figure 6.51, the vertical axis is dB scale, as defined in Equation 6.178.

6.3.7 Series RL Magnitude Bode Diagram Plot

In order to manually construct the Bode plot graph in Figure 6.51, and to gain technical familiarity with the circuit behavior, plot the asymptotes of the curve in high- and low-frequency ranges. Additionally, plot the corner frequency where the slope of the curve changes. The Bode plot curve is then plotted using these pieces of information.

6.3.7.1 Calculate the High-Frequency Asymptote

At high frequencies, the Bode plot of the magnitude of the RL transfer function will approach the asymptotic line in Equation 6.188, calculated as follows. Start with the magnitude of the transfer function from Equation 6.185 and here in Equation 6.186.

$$|H(j\omega)|_{\text{dB}} = 20\log(R/L) - 10\log[\omega^2 + (R/L)^2] \quad (6.186)$$

At large enough values of ω, the R/L terms from Equation 6.186 become small in comparison to ω, as shown in Equation 6.187.

$$|H(j\omega)|_{\text{dB}} = \cancel{20\log(R/L)} - 10\log[\omega^2 + \cancel{(R/L)^2}] \quad (6.187)$$

Removing terms from Equation 6.187 yields Equation 6.188.

$$|H(j\omega)|_{\text{dB}} \approx -10\log\omega^2 \quad (6.188)$$

Simplify Equation 6.188, using the log Equation 6.184 with $k = 2$. Bring down the 2, to get Equation 6.189.

$$|H(j\omega)|_{\text{dB}} \approx -20\log\omega \quad (6.189)$$

Equation 6.189 represents the part of the magnitude of the transfer function, at high frequencies, that declines, or "rolls off" at a rate of −20 dB per decade. A decade is a change by a factor of 10 up or down.

The next step is to plot this transfer function magnitude asymptote in Equation 6.189 on a graph with a vertical axis (y) in dB and a horizontal axis (x) in log ω.

6.3.7.2 Plot the High-Frequency Asymptote

The asymptotes of the Bode plot diagram are approximations used to construct or sketch a Bode plot by hand. But the asymptotes do also represent the actual transfer function pretty well in their frequency ranges.

The high-frequency asymptote in Equation 6.189 is a line which the transfer function approaches asymptotically as the frequency goes up.

To plot the high-frequency asymptote in Equation 6.189, we also need to know an intercept in order to locate the asymptote on a logarithmic graph.

$$\log(1) = 0 \tag{6.190}$$

Where the transfer function magnitude equals 1, the quantity in dB is 0 dB.

Going back to Equation 6.186, with $\omega = 0$, R/L terms on the right-hand side cancel and the magnitude of the transfer function equals 0 dB. The R/L term can be used as the DC intercept frequency and labeled ω_0. At low frequency, the high-frequency asymptote hits 0 dB on the vertical scale when it is at the corner frequency ω_0 on the horizontal frequency scale.

$$\omega_0 = \frac{R}{L} \tag{6.191}$$

$\omega_0 =$ in Equation 6.191 is R/L, the low-frequency 0 dB intercept. The top of the high-frequency asymptote starts at 0 dB at frequency ω_0, as shown in Figure 6.52.

6.3.7.3 Graph of the High-Frequency Asymptote

Plot the high-frequency asymptote of the magnitude of the transfer function from Equation 6.189, using Equation 6.191 as the 0 dB intercept. The asymptote declines, or rolls off, at a rate −20 dB per decade ($\log \omega$). The low-frequency asymptote and the corner frequency are still missing from the plot in Figure 6.52.

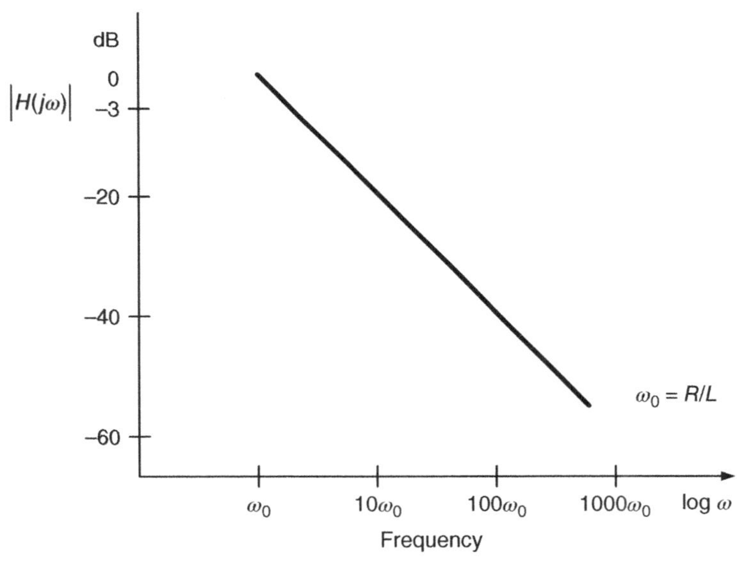

Figure 6.52 Series RL Bode magnitude plot high-frequency asymptote.

Figure 6.52 shows the high-frequency asymptote of the RL filter with $\log \omega$ on the horizontal axis.

6.3.7.4 Illustrating the −20 dB per Decade Roll Off

$$|H(j\omega)|_{\text{dB}} \approx -20 \log \omega \tag{6.192}$$

Let's develop some intuition around the roll off at a rate of −20 dB per decade on the logarithmic graph in Figure 6.52.

$$Y = -20x \quad (6.193)$$

Substituting y and x into our high-frequency Equation 6.192 for the transfer function, we can see a slope in the familiar terms of x and y graphs. This transfer function magnitude has a slope of minus 20.

$$|H(j\omega)|_{\text{dB}\ 10\omega} \approx -20\log 10\omega \quad (6.194)$$

Let's illustrate the "per decade" roll off by calculating the magnitude of the transfer function at 10 times ω.

$$\log(A \cdot B) = \log(A) + \log(B)$$
$$A = 10 \qquad B = \omega \quad (6.195)$$

In Equation 6.194, 10ω is a product. Invoke the logarithmic identity in Equation 6.195 to separate the product term, to obtain Equation 6.196.

$$|H(j\omega)|_{\text{dB}\ 10\omega} \approx -20(\log 10 + \log\omega) \quad (6.196)$$

In Equation 6.196, multiply out the 20 and remove the brackets to obtain Equation 6.197.

$$|H(j\omega)|_{\text{dB}\ 10\omega} \approx -20\log 10 - 20\log\omega \quad (6.197)$$

The logarithm to the base 10 of a number is the exponent to which you raise 10 in order to get that number. $\log 10$ equals 1 for example. In Equation 6.197 the $-20\log 10$ term becomes -20 in Equation 6.198.

$$|H(j\omega)|_{\text{dB}\ 10\omega} \approx -20 - 20\log\omega \quad (6.198)$$

The magnitude of the transfer function at 10ω, one decade higher in frequency than ω, is -20 dB lower than the magnitude is at ω. This demonstrates how the RL low-pass filter transfer function magnitude rolls off at the rate of -20 dB per decade of increasing frequency.

6.3.7.5 Low-Frequency Asymptote for the Magnitude Bode Diagram

Calculate the Low-Frequency 0 dB Asymptote Another step in the manual construction of the Bode plot is to calculate a low-frequency asymptotic line.

$$|H(j\omega)|_{\text{dB}} = 20\log(R/L) - 10\log[\omega^2 + (R/L)^2] \quad (6.199)$$

Start with Equation 6.199, the magnitude of the RL circuit transfer function $H(j\omega)$ in dB, from Equation 6.186.

At low enough values of ω, the L/R terms in Equation 6.199 become large in comparison to ω. Set $\omega = 0$ to obtain Equation 6.200.

$$|H(j\omega)|_{\text{dB}} = 20\log(R/L) - 10\log(R/L)^2 \quad (6.200)$$

$$\log_a G^k = k\log_a G \quad (6.201)$$

In Equation 6.200, to simplify the exponent 2, on the R over L quantity in the bracket, apply the log Equation 6.201 to the right side of Equation 6.200 to bring down the exponent 2. k equals 2. Two times minus 10 equals minus 20, as shown in Equation 6.202.

$$|H(j\omega)|_{\text{dB}} = 20\log(R/L) - 20\log(R/L) \quad (6.202)$$

$$|H(j\omega)|_{\text{dB}} = 0 \quad (6.203)$$

At DC (zero frequency) the magnitude of the transfer function is 0 dB or magnitude 1, as shown in Equation 6.203. This is the low-frequency asymptote for the magnitude Bode diagram of the RL low-pass filter circuit in Figure 6.50.

The dB magnitude, of $H(j\omega)$ equal to 0 in Equation 6.203, isn't the plot of the magnitude of the transfer function. Rather, Equation 6.203 is the value that the magnitude of the transfer function asymptotically approaches. At even small frequencies above DC (zero frequency), the magnitude of the transfer function starts to decline below the 0 dB unity gain asymptote in Equation 6.203.

At DC the magnitude of the transfer function is 0 dB or magnitude of 1, this means that all the voltage gets through from the input to the output. This makes sense, because at DC there is no impedance in the inductor opposing the current going through it.

Plot the Low-Frequency 0 dB Asymptote To plot the low-frequency asymptote of the RL circuit transfer function magnitude, draw a line at 0 dB on the vertical axes, extending from the far left to the angular frequency value of ω_0, as shown in Figure 6.53.

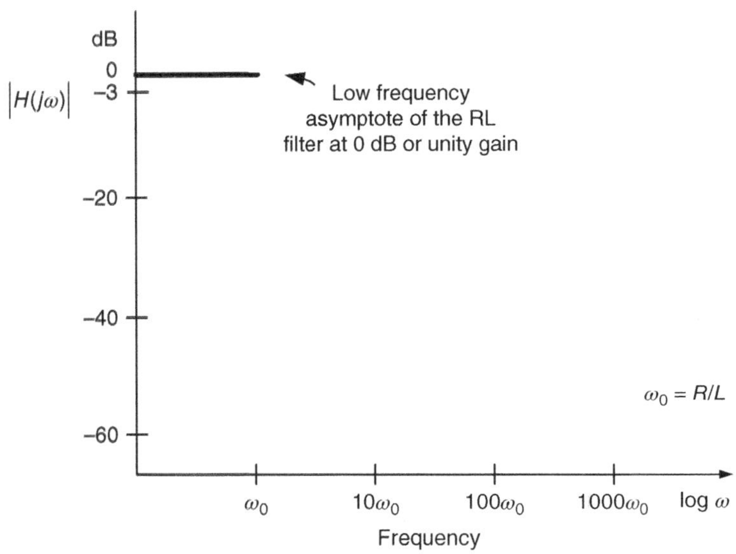

Figure 6.53 Series RL Bode magnitude plot low-frequency asymptote.

6.3.7.6 Corner Frequency

$$\omega_0 = \frac{R}{L} \tag{6.204}$$

The corner frequency is where the magnitude of the transfer function is −3 dB. In the RL low-pass filter circuit, this occurs at the frequency ω equals R over L in Equation 6.204.

$$|H(j\omega)|_{\text{dB}} = 20\log(R/L) - 10\log[\omega^2 + (R/L)^2] \tag{6.205}$$

To calculate or derive the value of the magnitude of the transfer function at the corner frequency ω_0, substitute the frequency ω equals L over R, from Equation 6.204, into the dB transfer function Equation in 6.205, to obtain Equation 6.206.

$$|H(j\omega)|_{\text{dB}} = 20\log(R/L) - 10\log[(R/L)^2 + (R/L)^2] \tag{6.206}$$

In Equation 6.206, gather terms in the square bracket yielding Equation 6.207.

$$|H(j\omega)|_{dB} = 20\log(R/L) - 10\log[2(R/L)^2] \tag{6.207}$$

In Equation 6.207, two times R over L is a product. Invoke the logarithmic identity in Equation 6.208 to separate the product term, to obtain Equation 6.209.

$$\log(A \cdot B) = \log(A) + \log(B) \qquad (6.208) \qquad A = 2 \quad B = R/L$$

$$|H(j\omega)|_{\text{dB}} = 20\log(R/L) - 10\log(2) - 10\log[(R/L)^2] \tag{6.209}$$

$$\log_a G^k = k\log_a G \tag{6.210}$$

In Equation 6.209, simplify the exponent 2, in the right-hand bracketed term, using the log formula in Equation 6.210 with k equal to 2, which brings down the exponent 2, resulting in Equation 6.211. Two times minus 10 equals minus 20.

$$|H(j\omega)|_{\text{dB}} = 20\log(R/L) - 10\log(2) - 20\log(R/L) \tag{6.211}$$

In Equation 6.211, the two 20log terms cancel one another, yielding Equation 6.212.

$$|H(j\omega)|_{\text{dB}} = -10\log(2) = \sim 3\text{ dB} \tag{6.212}$$

$$\omega_0 = \frac{R}{L} \tag{6.213}$$

At the corner frequency ω_0 equals R over L, in Equation 6.213, the magnitude of the transfer function in Equation 6.211, is said to be minus three dB down.

Figure 6.54 plots the corner frequency ω_0, along with the two calculated asymptotic lines that approximate the magnitude of the RL low pass filter circuit transfer function in Equation 6.205.

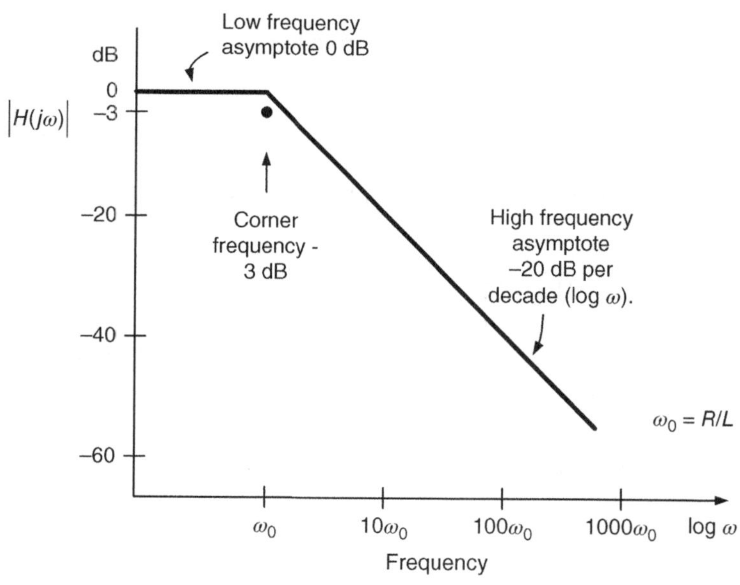

Figure 6.54 Series RL Bode magnitude plot corner frequency.

In Figure 6.55, the corner frequency ω_0 point is used to draw the transfer function magnitude Bode plot by hand, along with the two calculated asymptotic lines.

The previous calculation, of the corner frequency at ω equals L over R, started with the dB equation for the transfer function from Equation 6.205. Alternatively, we can start with the (non-dB) equation for the transfer function magnitude in Equation 6.214, taken from Equation 6.174.

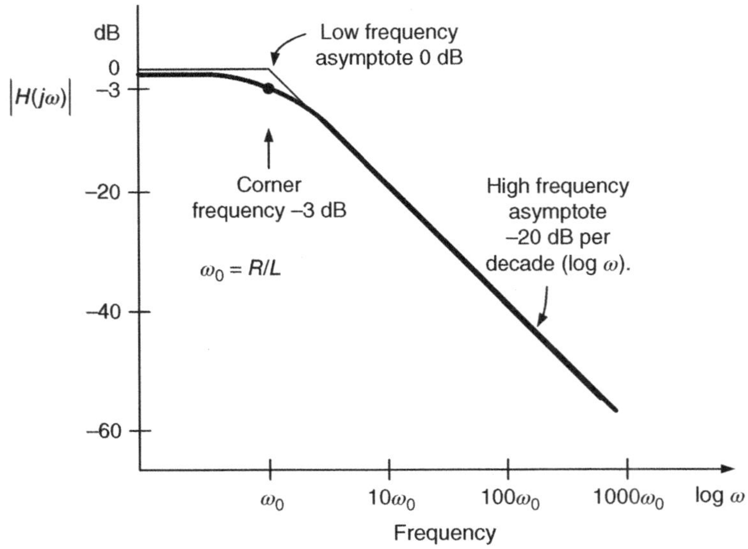

Figure 6.55 Series RL low-pass filter finished Bode magnitude plot.

$$|H(j\omega)| = \left|\frac{\boldsymbol{V}_{\text{out}}}{\boldsymbol{V}_{\text{S}}}\right| = \frac{R/L}{\sqrt{\omega^2 + (R/L)^2}} \tag{6.214}$$

Equation 6.214 is the magnitude of the transfer function of the RL series low-pass filter circuit in Figures 6.50 and 6.31.

Calculate the corner frequency ω equals R over L from Equation 6.213, as shown in Equation 6.215.

$$\left|H\left(j\frac{R}{L}\right)\right| = \frac{R/L}{\sqrt{(R/L)^2 + (R/L)^2}} = \frac{R/L}{\sqrt{2(R/L)^2}} = \frac{R/L}{(R/L)}\frac{1}{\sqrt{2}} = \frac{1}{\sqrt{2}} \simeq 0.707 \tag{6.215}$$

Equation 6.215 says that the magnitude of the transfer function at the corner frequency is 0.707 times the maximum value of the transfer function.

$$\left|H\left(j\frac{L}{R}\right)\right|_{\text{dB}} = 20\log\left|H\left(j\frac{L}{R}\right)\right| = 20\log\left(\frac{1}{\sqrt{2}}\right) \simeq 20\log(0.707) \simeq -3\ \text{dB} \tag{6.216}$$

The magnitude, of the RL low-pass circuit transfer function at the corner frequency L over R, is 3 dB down from the maximum value of the transfer function, as shown in Equation 6.216.

6.3.8 Series RL Transfer Function Phase – Detailed Calculation

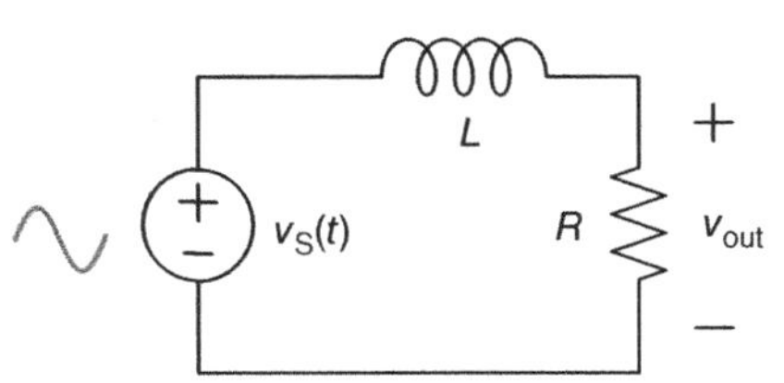

Figure 6.56 Series RL low-pass filter circuit driven by sinusoidal voltage source.

$$\frac{\boldsymbol{V}_{\text{out}}}{\boldsymbol{V}_{\text{S}}} = \frac{R/L}{j\omega + R/L} = H(j\omega) \tag{6.217}$$

Equation 6.217 is the steady-state frequency domain voltage transfer function, in phasor notation, for the series RL low-pass filter circuit in Figure 6.56.

The angle theta of a complex function, like the transfer function $H(j\omega)$ in Equation 6.217, is the arctangent of the imaginary part $X(\omega)$ over the real part $R(\omega)$, as shown in Equation 6.218.

$$\theta(\omega) = \angle H(\omega) = \tan^{-1}\frac{X(\omega)}{R(\omega)} \tag{6.218}$$

A first-order filter transfer function, shown in Equation 6.219, can be a quotient of two complex expressions $F(\omega)$ and $G(\omega)$.

$$H(\omega) = \frac{\boldsymbol{V}_{\text{out}}}{\boldsymbol{V}_{\text{S}}} = \frac{R/L}{j\omega + R/L} = \frac{F(\omega)}{G(\omega)} \tag{6.219}$$

$$F = R/L \tag{6.220}$$

$$G = j\omega + R/L \tag{6.221}$$

Equation 6.222 shows how the phase angle of a quotient of complex functions is the phase angle of the numerator minus the phase angle of the denominator.

$$\angle H(j\omega) = \angle\frac{F(j\omega)}{G(j\omega)} = \frac{\angle\theta_F}{\angle\theta_G} = \angle(\theta_F) - \angle(\theta_G) \tag{6.222}$$

$$\angle H(j\omega) = \angle F(j\omega) - \angle G(j\omega) \tag{6.223}$$

Equation 6.223 expresses Equation 6.222 in a slightly different notation.

Into Equation 6.222, substitute the numerator, $\boldsymbol{F}$ from Equation 6.220, and the denominator $\boldsymbol{G}$, from Equation 6.221, of the RL high-pass transfer function $H(j\omega)$ to get Equation 6.224.

The phase of the transfer function will be the phase of the numerator, R over L minus the phase of the denominator, the quantity j-omega plus R over L.

$$\angle H(j\omega) = \angle(R/L) - \angle(j\omega + R/L) \tag{6.224}$$

Apply Equation 6.218, the arctangent of the imaginary part X over the real part R, separately to the numerator and denominator terms in Equation 6.224, as shown in Equations 6.225 and 6.226.

Numerator
F

$$\angle(R/L) = \tan^{-1}[X/R] = \tan^{-1}[0/R/L] = \tan^{-1}(0) = 0 \tag{6.225}$$

Equation 6.225 says that the phase angle of the numerator is zero because the numerator is a purely real number.

Denominator
G

$$\angle(j\omega + R/L) = \tan^{-1}[\omega/(R/L)] = \tan^{-1}(\omega L/R) \tag{6.226}$$

Equation 6.226 says that the phase angle of the denominator is the arctangent of omega times L over R.

Substitute Equations 6.225 for the numerator $\boldsymbol{F}$ and 6.226 for the denominator $\boldsymbol{G}$ into Equation 6.224 for the phase angle of H, to obtain Equation 6.227.

$$\angle H(j\omega) = \angle(R/L) - \angle(j\omega + R/L) = 0 - \tan^{-1}(\omega L/R) \tag{6.227}$$

$$\angle H(j\omega) = -\tan^{-1}(\omega L/R) \tag{6.228}$$

Equation 6.228 and Figure 6.57 are the phase response of the series RL low-pass filter circuit in Figure 6.56.

Substitute the corner or break frequency ω_0, in Equation 6.229, into Equation 6.228 for the series RL lowpass filter circuit phase angle as a function of ω, to get a general form for the phase angle of low-pass filters, shown in Equation 6.230.

$$\omega_0 = R/L \quad (6.229)$$

$$\theta = \angle H(j\omega) = -\tan^{-1}\left(\frac{\omega}{\omega_0}\right) = \angle\frac{\boldsymbol{V}_{\text{out}}}{\boldsymbol{V}_{\text{S}}} \quad (6.230)$$

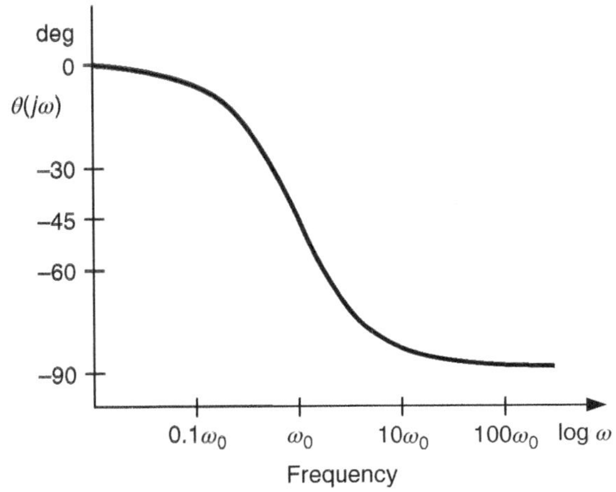

Figure 6.57 Series RL circuit output voltage phase response.

6.3.8.1 Constructing the Transfer Function Phase Bode Diagram

$$\angle H(j\omega) = -\tan^{-1}(\omega L/R) \quad (6.231)$$

Equation 6.231 is the phase response of the RL low-pass filter.

As a guide to manually drawing the phase response Bode diagram curve, plot several straight lines that will approximate the curve.

Calculate asymptotes for low and high frequencies.

6.3.8.2 Phase as ω Approaches 0

$$\text{As } \omega \Rightarrow 0: \quad \theta = \angle H(j\omega) \Rightarrow -\tan^{-1}(0) = 0 \quad (6.232)$$

Draw a flat straight line from 0 to $0.1\omega_0$ frequency for this low-frequency asymptote, as shown in the graph in Figure 6.58.

6.3.8.3 Phase as ω Approaches Infinity

As $\omega \Rightarrow$ infinity, the argument of arctan blows up to infinity, and the phase approaches −90°.

In the graph in Figure 6.58, draw a flat straight line at −90° starting at $10\omega_0$.

6.3.8.4 Phase at Corner (Break) Frequency $\omega = \omega_0 = R/L$

Corner/break frequency $\omega = \omega_0 = R/L$:

$$\theta = \angle H(j\omega) \Rightarrow -\tan^{-1}\left(\frac{\omega}{\omega_0}\right) = -\tan^{-1}(\omega L/R) \quad (6.233)$$

$$\theta = \angle H(j\omega) = -\tan^{-1}\left(\frac{R}{L}\frac{L}{R}\right) = -\tan^{-1}(1) = -45 \text{ deg} = -\frac{\pi}{4} \quad (6.234)$$

Mark this −45° value on the graph at ω_0. The graph goes through this point.

Figure 6.58 shows the straight lines used to approximate the phase response of the RL filter transfer function from Equation 6.217.

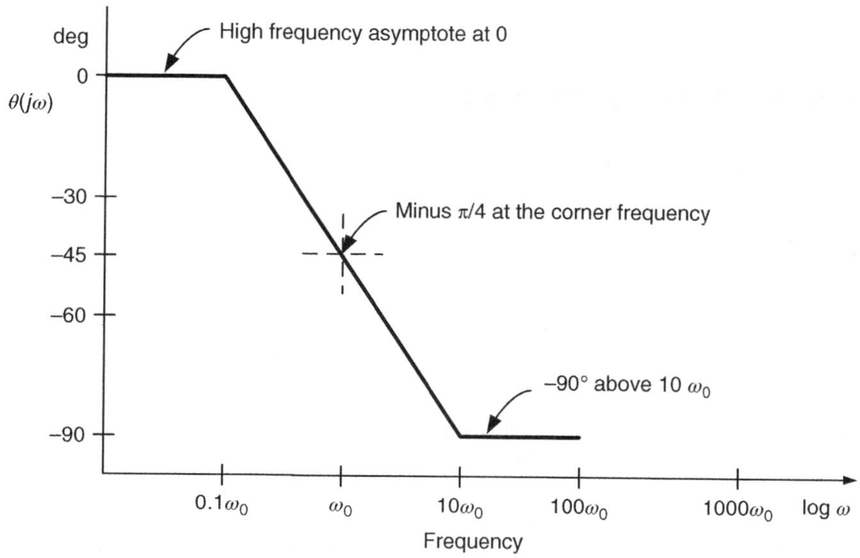

Figure 6.58 Series RL circuit output voltage phase response construction.

Sketch in the phase response plot of the RL filter transfer function, using the asymptotes and the corner frequency at 45°, as shown in Figure 6.59.

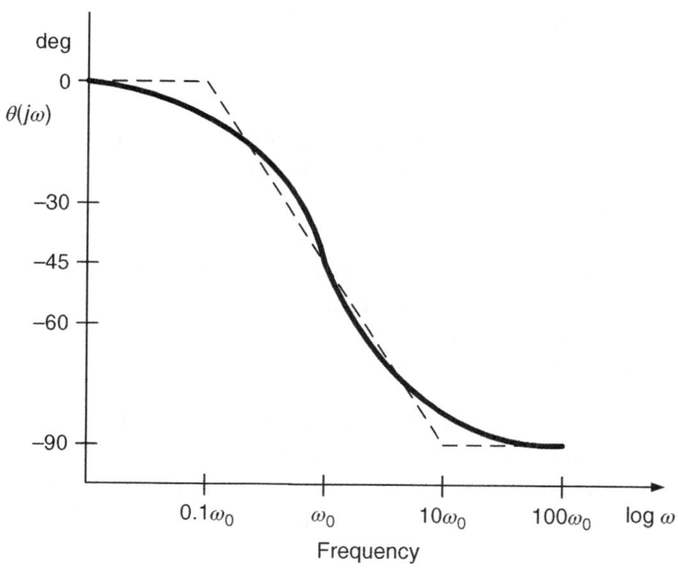

Figure 6.59 Series RL circuit output voltage phase response construction finished.

$$\theta = \angle H(j\omega) = -\tan^{-1}\left(\frac{\omega}{\omega_0}\right) = -\tan^{-1}\left(\frac{\omega L}{R}\right) \quad (6.235)$$

Because the phase of the output over the input is negative, the output voltage phase **lags** the input voltage phase.

6.4 Series RC Low-pass Filter Frequency Response

6.4.1 Series RC Frequency Domain Results Overview

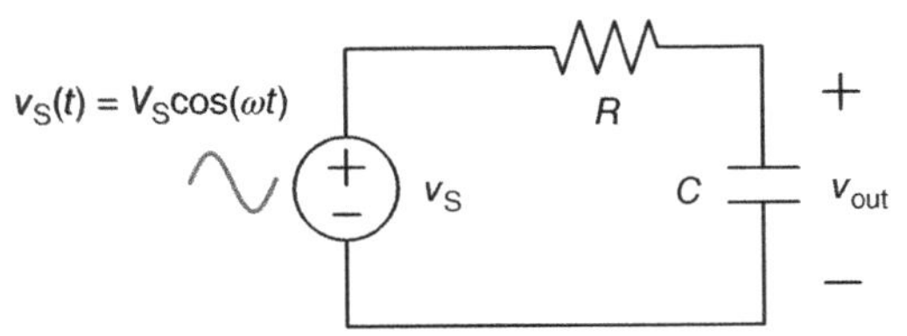

Figure 6.60 Series RC low-pass circuit driven by sinusoidal voltage source.

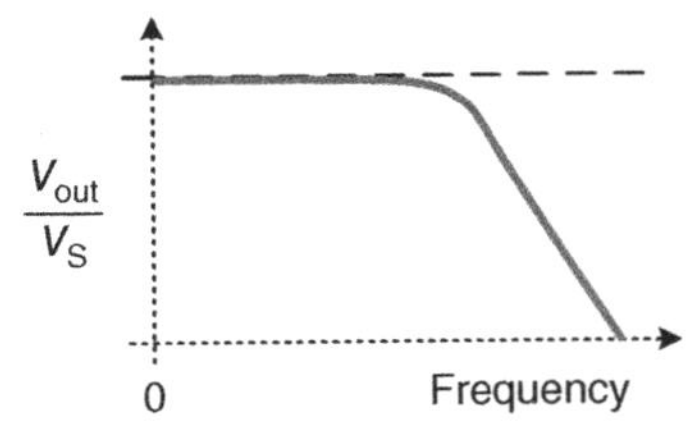

Figure 6.61 Low-pass filter Bode magnitude plot.

At low frequencies, where the capacitive impedance 1 over $j\omega C$ is high, little current flows and little voltage drops across the resistor. Instead, the output voltage V_{out} drops across the output capacitor. Low current means little difference between source voltage v_S and capacitor voltage V_{out}. This makes the RC circuit in Figure 6.60 a low-pass filter, as shown in the example Bode diagram in Figure 6.61. As frequency goes up, more current flows and more voltage drops across the resistor rather than the capacitor.

$$\frac{\boldsymbol{V}_{out}}{\boldsymbol{V}_S} = \frac{1}{1 + j\frac{\omega}{\omega_0}} \quad (6.236)$$

Equation 6.236 is the general formula for a first-order low-pass filter transfer function. This formula is useful for easily calculating specific transfer functions.

$$\omega_0 = \frac{1}{RC} \quad (6.237)$$

$$\frac{\boldsymbol{V}_{out}}{\boldsymbol{V}_S} = \frac{1}{1 + j\omega RC} = H(j\omega) \quad (6.238)$$

From Equation 6.237, substitute ω_0 equals 1 over RC, the corner frequency, into the transfer function Equation 6.236 to obtain Equations 6.238 and 6.239.

$$\frac{\boldsymbol{V}_{out}}{\boldsymbol{V}_S} = \frac{1}{j\omega RC + 1} = H(j\omega) \quad (6.239)$$

Equation 6.239 shows the steady-state frequency domain voltage **transfer function,** in phasor notation, for the RC circuit in Figure 6.60.

The output voltage varies with frequency s and the source is a voltage, so $\boldsymbol{V}_{out}/\boldsymbol{V}_S$ makes a convenient transfer function for the RC circuit in Figure 6.60. The denominator of the transfer function has a frequency-dependent $j\omega RC$ term, which means that as frequency goes up, the magnitude of the transfer function decreases. At DC (zero frequency) the $j\omega RC$ term becomes zero and the magnitude of the transfer function in Equation 6.239 is 1 or 0 dB.

$$|H(j\omega)| = \left|\frac{V_{out}}{V_S}\right| = \frac{1}{\sqrt{1+\left(\frac{\omega}{\omega_0}\right)^2}} \quad (6.240)$$

Equation 6.240 is the general formula for a first-order low-pass filter transfer function magnitude, in terms of angular frequency ω and break frequency ω_0.

For this series RC circuit in Figure 6.61, substitute the corner frequency ω_0 equals 1 over RC, from Equation 6.237, into the general transfer function magnitude Equation 6.240, to obtain Equation 6.241, the magnitude of the transfer function in terms of RC.

$$|H(j\omega)| = \left|\frac{V_{out}}{V_S}\right| = \frac{1}{\sqrt{1+\left(\frac{\omega}{1/RC}\right)^2}} = \frac{1}{\sqrt{1+(\omega RC)^2}} \quad (6.241)$$

$$|H(\omega)| = \left|\frac{V_{out}}{V_S}\right| = \frac{1}{\sqrt{1+(\omega RC)^2}} \quad (6.242)$$

Equation 6.242 is the **magnitude** of the series RC low-pass filter transfer function in Equation 6.239.

$$\theta = \angle H(j\omega) = \angle\frac{V_{out}}{V_S} = -\tan^{-1}\left(\frac{\omega}{\omega_0}\right) \quad (6.243)$$

Equation 6.243 is in the general formula for first-order low-pass filter phase response in terms of angular frequency ω and break frequency ω_0.

From Equation 6.38, substitute in ω_0 equals 1 over RC, the corner frequency, into the general phase response Equation 6.243 to obtain Equation 6.244.

$$\theta(\omega) = \angle H(j\omega) = -\tan^{-1}\left(\frac{\omega}{1/RC}\right) \quad (6.244)$$

$$\theta(\omega) = \angle H(j\omega) = -\tan^{-1}(\omega RC) \quad (6.245)$$

Equation 6.245 and Figure 6.62 are the **phase** of the series RC low-pass filter transfer function in Equation 6.239.

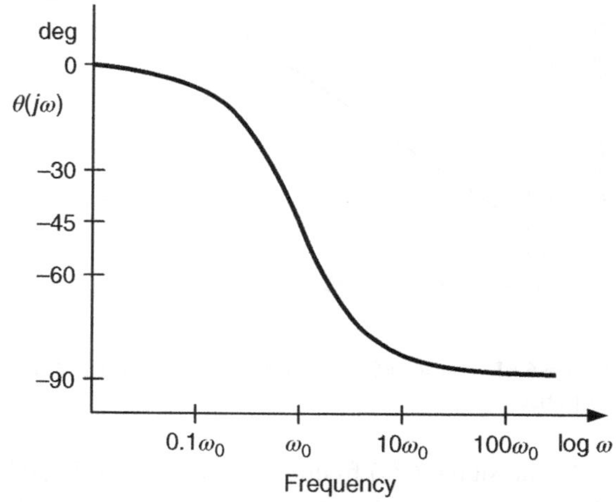

Figure 6.62 Series RC low-pass filter Bode phase plot.

Because of the minus sign in Equation 6.245, and the phase in Figure 6.62, descending from zero to negative values as frequency increases, the series RC circuit in Figure 6.60 is a **lag network**.

6.4.2 Series RC Frequency Domain Example Intuitive Analysis

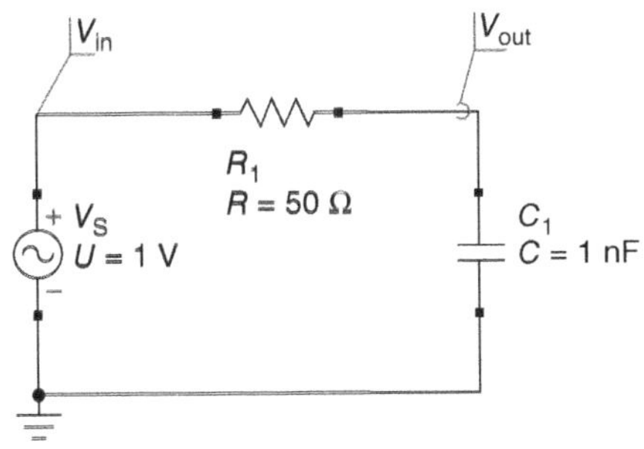

Figure 6.63 Series RC circuit simulation schematic driven by sinusoidal voltage source.

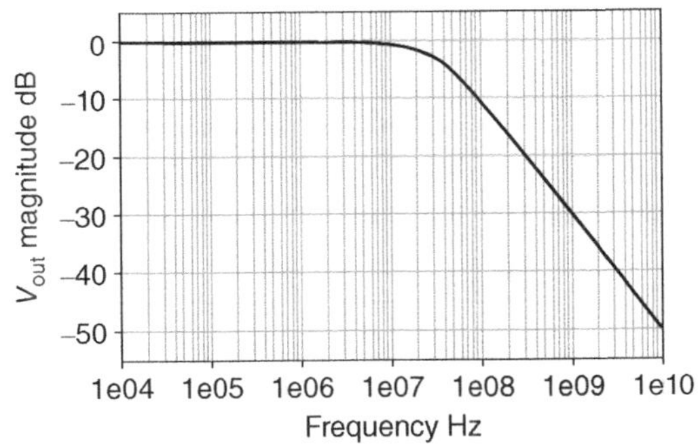

Figure 6.64 Series RC frequency domain simulation of output voltage magnitude.

In the schematic in Figure 6.63, the capacitor looks like a short circuit (low impedance) to high frequency, and the capacitor looks like an open circuit (high impedance) to low frequency and DC.

This RC circuit in Figure 6.63 is a low-pass filter, as shown in Figure 6.64, because the shunt capacitor blocks the low frequencies from flowing through the capacitor, leaving more voltage across the capacitor as output voltage V_{out}.

As the frequency of the input voltage, labeled V_S and V_{in} in the simulation schematic, goes down, the capacitor impedance $Z = 1/j\omega C$ increases toward infinity (open circuit). The output voltage, labeled V_{out} in the schematic and in the frequency response plot, approaches the input voltage level V_S because the voltage across the resistor, V_{in}-V_{out} in the simulation schematic, as shown in Figure 6.65, decreases toward zero. At zero frequency (DC), the current in the loop, shown in Figure 6.66, goes to zero.

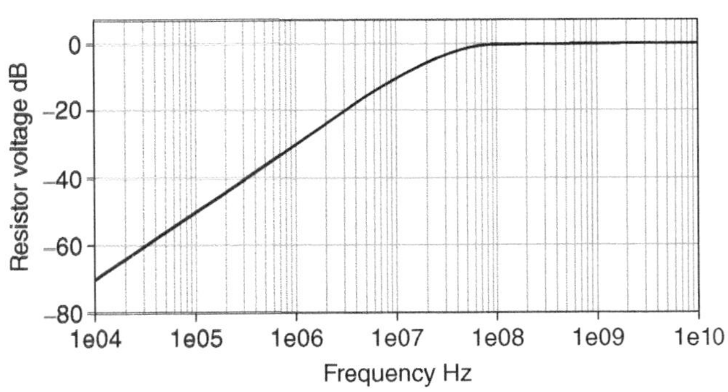

Figure 6.65 Series RC simulation of resistor voltage frequency response.

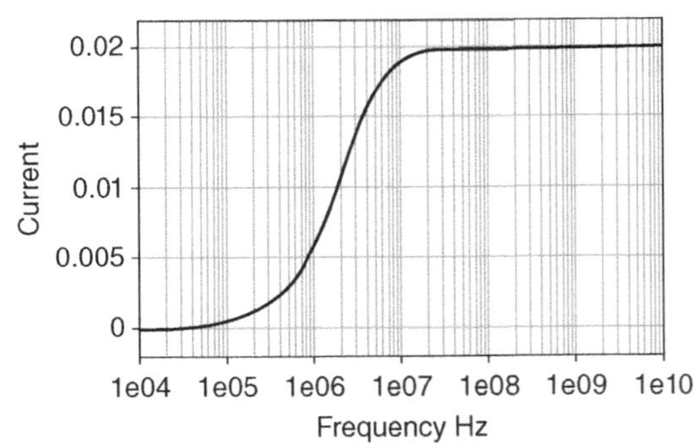

Figure 6.66 Series RC simulation of loop current over frequency.

As the signal (V_S) frequency goes up, the impedance of the capacitor (1 over $j\omega C$) goes down and more of the source voltage V_S gets dropped across the resistor (V_{out} minus V_S), shown in Figure 6.65, leaving less output voltage V_{out} across the capacitor. In the high-frequency limit, all of V_S can drop across the resistor and the loop current goes to V_S/R, shown in Figure 6.66.

The impedance of the capacitor, 1 over $j\omega C$, a pure reactance -1 over ωC, changes with frequency while the impedance of the resistor (pure resistance) does not. As the impedance of the capacitor changes with frequency, the total series impedance of the resistor and capacitor together varies, and this change in impedance varies the loop current. The higher the loop current, the lower the output voltage V_{out}, which Figures 6.65 and 6.66 together illustrate.

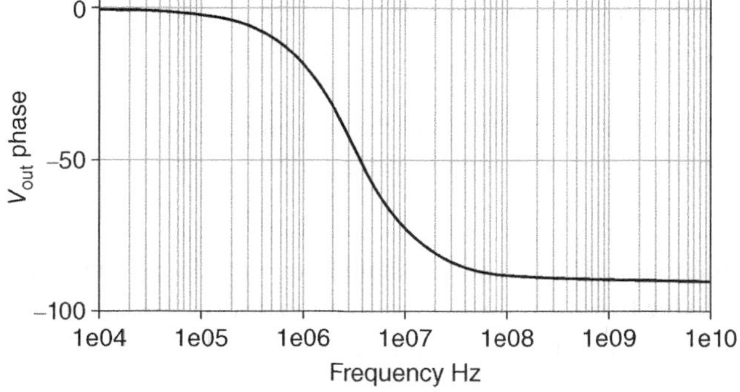

Figure 6.67 Series RC frequency domain simulation of output voltage phase.

$$\theta(\omega) = \angle H(j\omega) = -\tan^{-1}(\omega RC) \tag{6.246}$$

Equation 6.246 shows the phase shift in this RC series low-pass filter circuit. As frequency goes to zero, the argument of arctangent goes to zero and the phase goes to zero. As frequency increase to infinity, the arctangent goes to 90°, resulting in a phase of −90°, as shown in Figure 6.67.

6.4.2.1 Lag Network

In a capacitor, the current leads the voltage in phase. The output of this circuit is the voltage across the capacitor, so the phase of the output voltage will lag the loop current. The loop current is the phase reference for the voltage source, so the output voltage lags the input voltage, making the RC circuit in Figure 6.60 a lag network.

Figure 6.67 shows the phase, starting at zero, decreasing in the negative direction as frequency goes up, consistent with a lag network. At frequencies above DC, the output voltage lags the input voltage in phase.

6.4.3 Series RC Frequency Domain Transfer Function Calculation

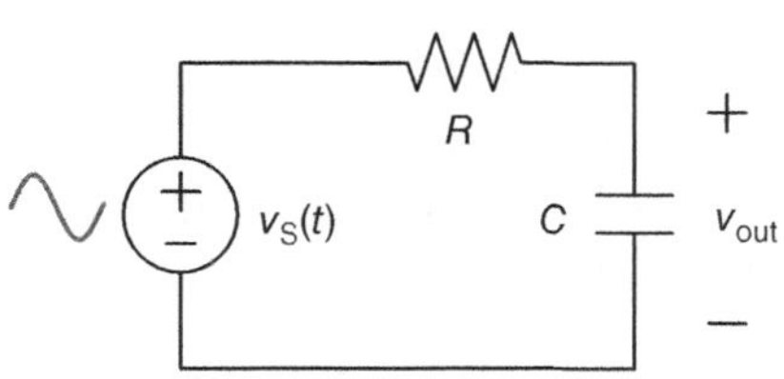

Figure 6.68 Series RC low-pass filter circuit driven by sinusoidal voltage source.

$$H(\omega) = \frac{\boldsymbol{V}_{out}}{\boldsymbol{V}_S} = \frac{1}{j\omega RC + 1} \tag{6.247}$$

Equation 6.247 is the steady-state frequency domain voltage transfer function, in phasor notation, for the RC low-pass filter in Figure 6.68.

6.4.3.1 Calculating the Transfer Function

$$H(\omega) = \frac{\boldsymbol{V}_{out}}{\boldsymbol{V}_S} \tag{6.248}$$

We can use a voltage divider to calculate the Equation 6.247 steady-state frequency domain transfer function of the output voltage V_{out} over the input voltage V_S. Notice the use of j as the imaginary operator rather than i used in other mathematics and physics literature.

$$Z_C = \frac{1}{j\omega C} \tag{6.249}$$

Equation 6.249 is the impedance of an ideal capacitor.

$$Z_R = R \tag{6.250}$$

Equation 6.250 is the impedance of an ideal resistor. The impedance is a pure resistance.

$$\boldsymbol{V}_{\text{out}} = \underbrace{\left(\frac{\boldsymbol{V}_{\text{S}}}{R + \frac{1}{j\omega C}}\right)}_{\text{current}} \frac{1}{j\omega C} \tag{6.251}$$

The output voltage $\boldsymbol{V}_{\text{out}}$ equals the current in the loop multiplied by the impedance of the capacitor. The current in the loop equals the input voltage v_s divided by the impedance of the capacitor and resistor in series, R plus 1-over-$j\omega C$, as shown in Equation 6.251.

$$\frac{\boldsymbol{V}_{\text{out}}}{\boldsymbol{V}_{\text{S}}} = \left(\frac{1}{R + \frac{1}{j\omega C}}\right) \frac{1}{j\omega C} \tag{6.252}$$

Divide both sides of Equation 6.251 by $\boldsymbol{V}_{\text{S}}$, to obtain Equation 6.252.

Multiply both the numerator and the denominator of Equation 6.252 by $j\omega C$ to obtain Equation 6.253.

$$\frac{\boldsymbol{V}_{\text{out}}}{\boldsymbol{V}_{\text{S}}} = \left(\frac{1}{j\omega RC + 1}\right) \tag{6.253}$$

Equation 6.253 is the steady-state frequency domain **voltage transfer function**, in phasor notation, for the series RC low-pass filter circuit in Figure 6.68.

Substitute ω_0, from Equation 6.237 and 6.254, into the transfer function in Equation 6.253, to get Equation 6.255, a general form for low-pass filter transfer functions.

$$\omega_0 = \frac{1}{RC} \tag{6.254}$$

$$H(\omega) = \frac{\boldsymbol{V}_{\text{out}}}{\boldsymbol{V}_{\text{S}}} = \frac{1}{j\frac{\omega}{\omega_0} + 1} \tag{6.255}$$

Obtain another form for the transfer function in Equation 6.253 by dividing the numerator and the denominator of Equation 6.253 by RC, resulting in Equation 6.256.

$$\frac{\boldsymbol{V}_{\text{out}}}{\boldsymbol{V}_{\text{S}}} = \frac{1/RC}{j\omega + 1/RC} = H(j\omega) \tag{6.256}$$

Equation 6.256 is the steady-state frequency domain **voltage transfer function**, in phasor notation, for the series RC low-pass filter circuit in Figure 6.68.

$$\frac{\boldsymbol{V}_{\text{out}}}{\boldsymbol{V}_{\text{S}}} = \frac{\omega_0}{j\omega + \omega_0} = H(j\omega) \tag{6.257}$$

From Equation 6.254, substitute in ω_0 equals 1 over RC into Equation 6.256 to obtain Equation 6.257, another common form that is especially useful when considering the corner frequency, the minus three dB half-power frequency.

6.4.4 Series RC Transfer Function Magnitude Calculation

$$H(\omega) = \frac{\boldsymbol{V}_{\text{out}}}{\boldsymbol{V}_{\text{S}}} = \frac{1}{j\omega RC + 1} \tag{6.258}$$

Equation 6.258 is the steady-state frequency domain voltage transfer function of the circuit in Figure 6.60.

$$\left|\frac{A(\omega)}{B(\omega)}\right| = \frac{|A(\omega)|}{|B(\omega)|} \tag{6.259}$$

The magnitude of a quotient of complex functions, $A(\omega)$ and $B(\omega)$, is the quotient of magnitudes of those complex functions, as shown in Equation 6.259.

$$\left|\frac{\boldsymbol{V}_{\text{out}}}{\boldsymbol{V}_{\text{S}}}\right| = \frac{\boldsymbol{V}_{\text{out}}}{\boldsymbol{V}_{\text{S}}} = \frac{|1|}{|j\omega CR + 1|} \tag{6.260}$$

In Equation 6.260, the denominator is a complex function, while the numerator is a purely real number 1.

$$|G(\omega)| = \sqrt{[R(\omega)]^2 + [X(\omega)]^2} \tag{6.261}$$

The magnitude of a complex function is the square root of the sum of the squares of the real and imaginary components, $R(\omega)$ and $X(\omega)$, as shown in Equation 6.261.

Apply Equation 6.261 to Equation 6.260 to calculate the magnitude of the denominator of Equation 6.260, resulting in Equation 6.262.

$$\left|\frac{V_{out}}{V_S}\right| = \frac{1}{\sqrt{1 + (\omega RC)^2}} \quad (6.262)$$

Equation 6.262 is the **magnitude** of the transfer function of the series RC low-pass filter circuit in Figure 6.60.

6.4.4.1 Calculating the Transfer Function Magnitude in dB

In many plots of transfer functions, such as in circuit simulation or test and measurement, the vertical axis will be in dB. The vertical scale of the Bode diagrams is logarithmic in decibels (dB). The subsection calculates the transfer function in dB.

$$|H(\omega)| = \frac{1}{\sqrt{1 + (\omega RC)^2}} = [1 + (\omega RC)^2]^{-1/2} \quad (6.263)$$

Express Equation 6.262, for the transfer function magnitude, in terms of an exponent of minus one half, as shown in Equation 6.263.

$$H \text{ in dB} = 20 \log_{10} H \quad (6.264)$$

The number H expressed in decibels is 20 times the logarithm to the base 10 of H, as shown in Equation 6.264.

$$|H(\omega)|_{dB} = 20 \log |H(\omega)| \quad (6.265)$$

The magnitude of the transfer function $H(\omega)$, expressed decibels, is 20 times the logarithm to the base 10 of the magnitude of $H(\omega)$ as shown in Equation 6.265

Substitute the transfer function $H(j\omega)$ magnitude, from Equation 6.263, into Equation 6.265, to obtain Equation 6.266.

$$|H(j\omega)|_{dB} = 20 \log[1 + (\omega RC)^2]^{-1/2} \quad (6.266)$$

The right-hand log term in Equation 6.266 is the logarithm of an exponentiated term, the bracketed term taken to the minus one half-power.

$$\log_a G^k = k \log_a G \quad (6.267)$$

Using the log formula in Equation 6.267, with k equal to minus one half, simplify the second term of Equation 6.266. Bring down the minus one half, to obtain Equation 6.268. Twenty times minus one-half equals minus 10.

$$|H(j\omega)|_{dB} = -10 \log[1 + (\omega RC)^2] \quad (6.268)$$

Equation 6.268 is the dB magnitude of the transfer function of the series RC in Figure 6.68.

From Equation 6.268, calculate the value of the dB transfer function magnitude, in Equation 6.268, at the corner frequency ω_0. Start by substituting ω_0 into Equation 6.268 for ω, to obtain 6.269.

$$|H(j\omega)|_{dB} = -10 \log\left[1 + (\omega_0 RC)^2\right] \quad (6.269)$$

$$|H(j\omega)|_{dB} = -10 \log\left[1 + \left(\frac{RC}{RC}\right)^2\right] \quad (6.270)$$

$$|H(j\omega)|_{dB} = -10 \log[2] \cong -3 \text{ dB} \quad (6.271)$$

In Equation 6.269, substitute 1 over RC for ω_0, to obtain Equation 6.270.

In 6.270, cancel the RC terms, to obtain Equation 6.271 with the calculated value of about minus 3 dB.

The low-pass filter transfer function Bode plot, in terms of ω_0, is shown in Figure 6.69.

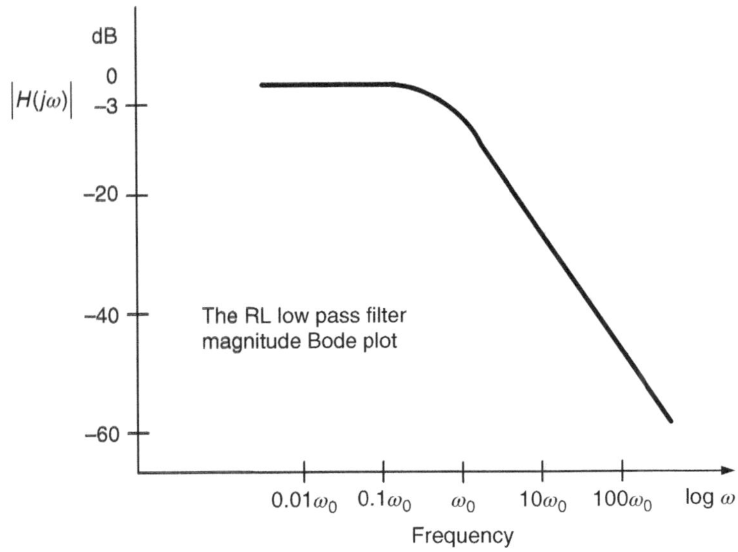

Figure 6.69 Series RC low-pass filter Bode magnitude plot.

At ω equals 0, in Figure 6.69, the logarithm of 1 is 0 dB.

At ω equals ω_0, in Figure 6.69, the magnitude equals −0.301 dB.

6.4.5 Series RC Transfer Function Phase Response Calculation

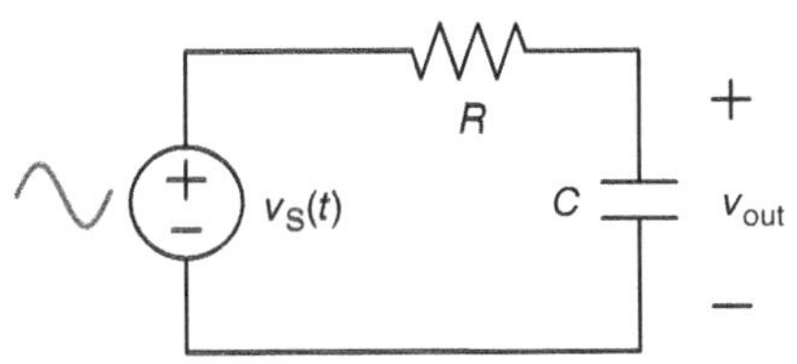

Figure 6.70 Series RC low-pass filter circuit driven by sinusoidal voltage source.

$$H(j\omega) = \frac{\mathbf{V}_{\text{out}}}{\mathbf{V}_{\text{S}}} = \frac{1}{j\omega RC + 1} \tag{6.272}$$

Equation 6.272 is the steady-state frequency domain voltage transfer function of the series RC low-pass filter in Figures 6.70 and 6.68.

$$\theta(\omega) = \angle H(\omega) = \tan^{-1}\frac{X(\omega)}{R(\omega)} \tag{6.273}$$

The angle theta of a complex function, like the transfer function $H(j\omega)$ in Equation 6.272, is the arctangent of the imaginary part $X(\omega)$ over the real part $R(\omega)$, as shown in Equation 6.273.

$$H(\omega) = \frac{\mathbf{V}_{\text{out}}}{\mathbf{V}_{\text{S}}} = \frac{1}{j\omega RC + 1} = \frac{F(\omega)}{G(\omega)} \tag{6.274}$$

A first-order filter transfer function, shown in Equation 6.274 can be a quotient of two complex expressions $F(\omega)$ and $G(\omega)$ in Equations 6.275a and 6.275b.

$F = 1$ (6.275a) $G = j\omega RC + 1$ (6.275b)

In the case of the RC high-pass filter transfer function $H(j\omega)$ in Equation 6.272, the numerator is the real number 1.

Equation 6.276 shows how the phase angle of a quotient of complex functions is the phase angle of the numerator minus the phase angle of the denominator.

$$\angle H(j\omega) = \angle\frac{F(j\omega)}{G(j\omega)} = \frac{\angle\theta_F}{\angle\theta_G} = \angle(\theta_F) - \angle(\theta_G) \tag{6.276}$$

Into Equation 6.276, substitute the numerator and denominator from Equation 6.274, of the RC high-pass transfer function $H(j\omega)$ to get Equation 6.277.

$$\angle H(j\omega) = \angle\frac{1}{j\omega RC+1} = \frac{\angle 1}{\angle(j\omega RC+1)} = \angle(1) - \angle(j\omega RC+1) \tag{6.277}$$

Apply the formula, in Equation 6.273, for the phase of a complex function, to both terms separately in Equation 6.277, the numerator and the denominator, as shown in Equations 6.278 and 6.279.

Numerator

F

$$\angle(1) = \tan^{-1}[X/R] = \tan^{-1}[0/1] = \tan^{-1}(0) = 0 \tag{6.278}$$

The phase angle the numerator is zero because the numerator is a purely real number.

Denominator

G

$$\angle(j\omega RC+1) = \tan^{-1}[\omega RC/1] = \tan^{-1}(\omega RC) \tag{6.279}$$

The phase angle of the denominator is arctangent of omega times RC.

Combine the results from Equations 6.278 and 6.279, back into Equation 6.277 to get Equation 6.280 for the phase angle response of the series RC filter circuit.

$$\theta(\omega) = \angle H(j\omega) = \angle\left(\frac{1}{j\omega RC+1}\right) = 0 - \tan^{-1}(\omega RC) \tag{6.280}$$

$$\angle H(\omega) = \angle\left(\frac{1}{j\omega RC+1}\right) = -\tan^{-1}(\omega RC) \tag{6.281}$$

Equation 6.281 and Figure 6.71 are the phase response of the series RC low-pass filter circuit in Figure 6.70.

To express the phase in terms of the corner frequency, in Equation 6.282, substitute ω_0 equals 1 over RC into Equation 6.281 to obtain Equation 6.283.

$$\omega_0 = \frac{1}{RC} \tag{6.282}$$

$$\angle H(\omega) = \angle\left(\frac{1}{j\omega RC+1}\right) = -\tan^{-1}\left(\frac{\omega}{\omega_0}\right) \tag{6.283}$$

Equation 6.283 is the phase response of the series RC low-pass filter.

Figure 6.71 plots the series RC low-pass filter phase from Equation 6.281.

At ω equals 0, the arctangent of 0 is 0°, shown in Equation 6.284.

$$\angle H(\omega) = -\tan^{-1}(0) = 0^{\circ} \tag{6.284}$$

At the break frequency ω_0 equals 1 over RC, the phase of the transfer function in Equation 6.272 is −45°, as shown in Equations 6.285 and 6.286.

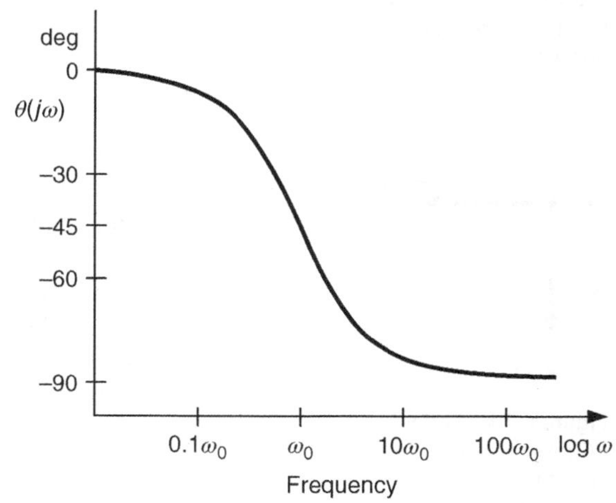

Figure 6.71 Series RC low-pass filter Bode magnitude plot.

$$\angle H(\omega) = -\tan^{-1}(\omega_0 RC) = -\tan^{-1}\left(\frac{RC}{RC}\right) = -\tan^{-1}(1) = -45^{\circ} \quad (6.285)$$

$$\angle H(\omega) = -\tan^{-1}\left(\frac{\omega}{\omega_0}\right) = -\tan^{-1}(1) = -45^{\circ} \quad (6.286)$$

6.4.6 Series RC Low-pass Filter Phase Relationships

Phasor diagrams are useful tools for performing circuit calculations and understanding relationships between voltages and phase angles in circuits. Phasor diagrams use head-to-tail graphical addition to calculate values. Although phasors are complex numbers and not vector quantities, phasor diagrams add voltages and currents like vectors. For additional information on phasor diagrams, please see Chapter 16.

6.4.6.1 Phasor Diagram at Corner Frequency

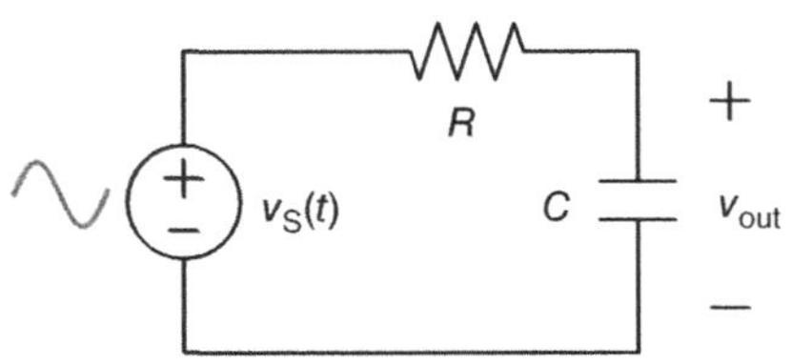

Figure 6.72 Series RC low-pass filter circuit driven by sinusoidal voltage source.

$$\angle H(\omega) = \angle\left(\frac{1}{j\omega RC + 1}\right) = -\tan^{-1}(\omega RC) \quad (6.287)$$

Equation 6.287 shows the phase of the sinusoidal steady-state transfer function for the series RC low-pass filter in Figure 6.72.

For the series RC low-pass filter, calculate the phase angle of $\boldsymbol{V}_{\text{out}}$ at the corner frequency ω_0, shown in Equation 6.288.

$$\omega_0 = \frac{1}{RC} \quad (6.288)$$

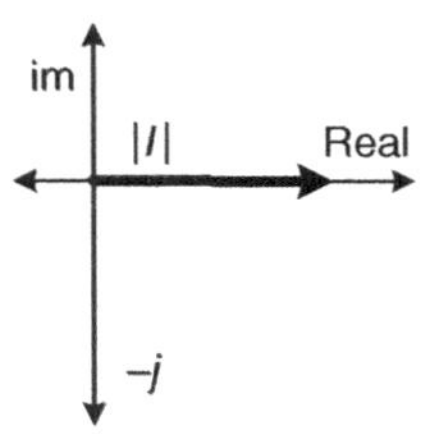

Figure 6.73 Series RC low-pass filter current phasor diagram.

Use current as the reference for zero phase, as shown in Figure 6.73, because in a series circuit, the current is the same through all the circuit elements.

The magnitude of the current phasor |I| gets plotted along the positive real axis in Figure 6.73 because the phase of the reference is zero. The unit of the current magnitude is amps.

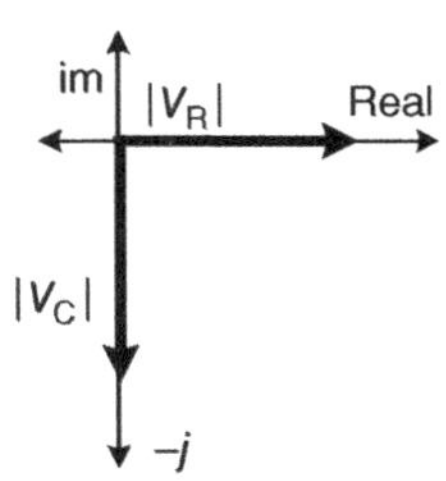

Figure 6.74 Series RC resistor and capacitor voltages phasor diagram.

Plot the magnitude of the resistor voltage, $\boldsymbol{V}_{\text{out}}$ minus $\boldsymbol{V}_{\text{S}}$, along the positive real axis in Figure 6.74 because the phase of the voltage across a resistor, relative to the current through the resistor, is zero.

Plot the magnitude of the capacitor voltage, $\boldsymbol{V}_{\text{out}}$, along the negative imaginary axis in Figure 6.74 because the voltage through a capacitor lags the current by 90°. Counterclockwise rotation is the positive phase direction.

The lengths of the two voltage phasors in Figure 6.74 are the same because the magnitude of the resistance and magnitude of the inductive reactance are the same at the corner frequency ω_0, as shown in Equation 6.291. Substituting in 1 over RC into the capacitive impedance Equation 6.289, in place of ω, shows that the impedance, a pure reactance, has the same magnitude as the resistance at ω equals ω_0.

$$Z_C = \frac{1}{j\omega C} \quad (6.289) \qquad \omega_0 = \frac{1}{RC} \quad (6.290) \qquad Z_{\text{C-corner}} = \frac{1}{j\omega_0 C} = \frac{RC}{jC} = -jR \quad (6.291)$$

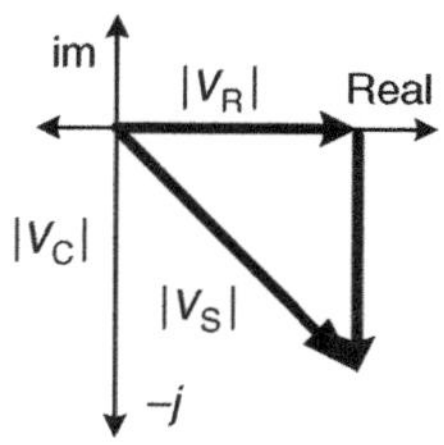

Figure 6.75 Series RC phasor diagram calculating source phasor.

Set the capacitor phasor head-to-tail with the resistor voltage phasor for vector addition in Figure 6.75. Connect the two phasors to get the magnitude and phase of the source voltage V_S.

The direction of the V_S phasors in Figure 6.75 reflects Kirchhoff's voltage law in the form of Equation 6.292.

$$|\boldsymbol{V}_S| = |\boldsymbol{V}_R| + |\boldsymbol{V}_C| \quad (6.292)$$

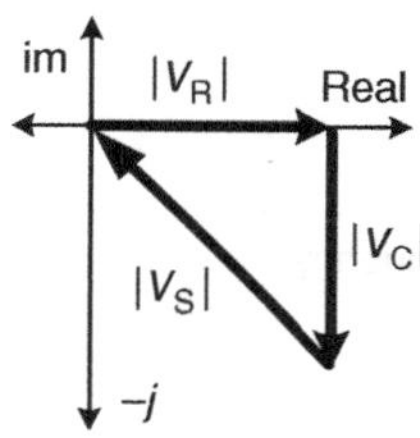

Figure 6.76 Series RC phasor diagram with opposite orientation of source phasor.

The opposite direction of the V_S phasor, in Figure 6.76, reflects a Kirchhoff's voltage law in the form of Equation 6.293, but the result is the same.

$$|\boldsymbol{V}_R| + |\boldsymbol{V}_C| - |\boldsymbol{V}_S| = 0 \quad (6.293)$$

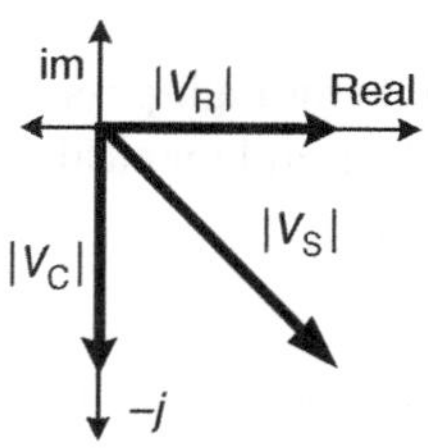

Figure 6.77 Series RC phasor diagram with source phasor at −45°.

Figure 6.77 shows that the V_S phasor lags the current phasor, in the resistor, by 45°.

The voltage across the capacitor is V_{out}, which lags the V_S by 45°. This series RC circuit is a lag network.

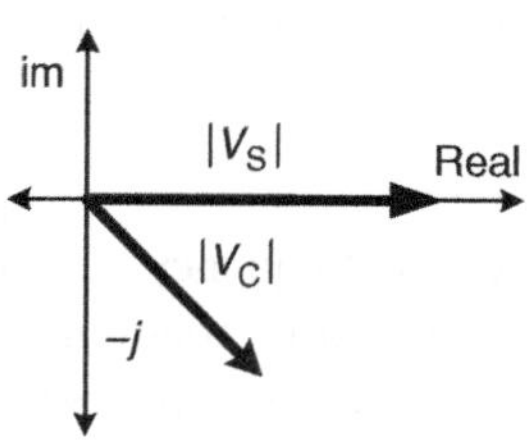

Figure 6.78 Series RC phasor diagram with source phasor on real axis.

To complete the calculation of the phase of the output voltage at the corner frequency, rotate the phasors in the counterclockwise (positive phase) direction by 45°, as shown in Figure 6.78. This rotation puts the V_S phasor on the real axis, with zero phase, making $\boldsymbol{V}_S$ the reference. Relative to $\boldsymbol{V}_S$, the phase of the output voltage $\boldsymbol{V}_{out} = \boldsymbol{V}_C$ is −45°.

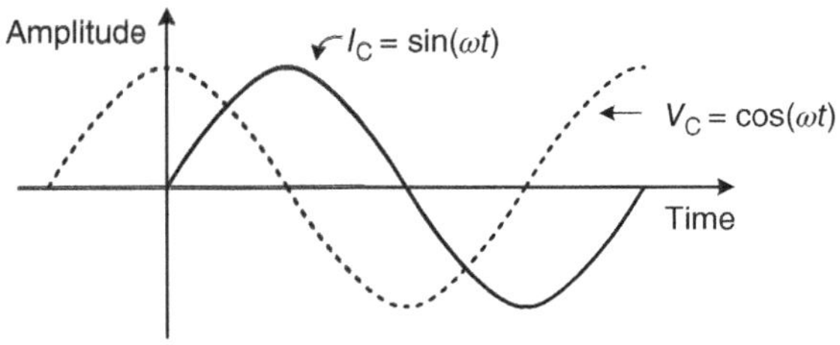

Figure 6.79 Sinusoids showing capacitor voltage lagging capacitor current.

Figure 6.79 shows how the phase of the voltage across an ideal capacitor lags the phase of the current across the capacitor.

6.5 Series RL High-pass Filter Frequency Response

6.5.1 Series RL Frequency Domain Results Overview

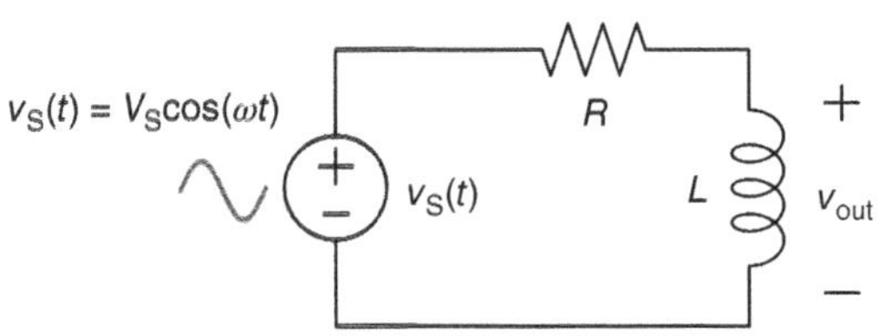

Figure 6.80 Series RL circuit schematic driven by sinusoidal voltage source.

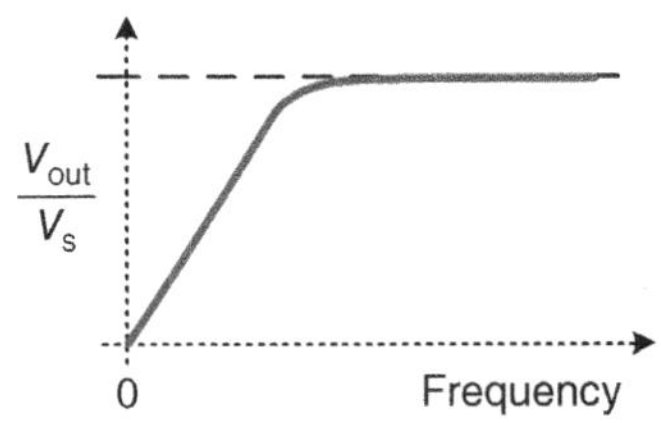

Figure 6.81 Series RL output voltage high-pass frequency response magnitude.

At high frequencies, the shunt inductor in Figure 6.80 won't pass significant current. Low loop current means little voltage across the resistor; all the voltage from the source gets dropped across the output inductor, making the filter high-pass, as shown in Figure 6.81. The schematic in Figure 6.80 is also a lead network.

$$\frac{\mathbf{V}_{\text{out}}}{\mathbf{V}_{\text{S}}} = \frac{j\omega/\omega_0}{1 + j\omega/\omega_0} = \frac{j\dfrac{\omega}{\omega_0}}{1 + j\dfrac{\omega}{\omega_0}} \tag{6.294}$$

Equation 6.294 is the general formula for high-pass filter transfer functions. This form is useful for easily calculating specific transfer functions.

$$\omega_0 = R/L \tag{6.295}$$

From Equation 6.295, substitute the corner frequency, ω_0, into the transfer function in Equation 6.294 to obtain Equation 6.296.

$$\frac{\mathbf{V}_{\text{out}}}{\mathbf{V}_{\text{S}}} = \frac{j\dfrac{\omega}{R/L}}{1 + j\dfrac{\omega}{R/L}} \tag{6.296}$$

Simplify the numerator and denominator of Equation 6.296 to generate Equation 6.297.

$$\frac{\mathbf{V}_{\text{out}}}{\mathbf{V}_{\text{S}}} = \left(\frac{j\omega L/R}{1 + j\omega L/R}\right) = H(j\omega) \tag{6.297}$$

Equation 6.297 shows the steady-state frequency domain voltage transfer function, in phasor notation, for the series RL high-pass filter circuit in Figure 6.80.

$$|H(j\omega)| = \left|\frac{V_{\text{out}}}{V_S}\right| = \frac{\omega/\omega_0}{\sqrt{1+(\omega/\omega_0)^2}} \quad (6.298)$$

Equation 6.298 is in the general format for a first-order high-pass filter transfer function magnitude in terms of angular frequency ω and break frequency ω_0.

From Equation 6.295, substitute R over L into the transfer function magnitude Equation 6.298 for ω_0, to obtain Equation 6.299.

$$|H(j\omega)| = \left|\frac{V_{\text{out}}}{V_S}\right| = \frac{\omega L/R}{\sqrt{1+(\omega L/R)^2}} \quad (6.299)$$

Equation 6.299 shows the voltage transfer function magnitude for the series RL high-pass filter circuit in Figure 6.80.

$$\angle H(j\omega) = \frac{\pi}{2} - \tan^{-1}(\omega/\omega_0) \quad (6.300)$$

Equation 6.300 is in the general format for a first-order high-pass filter transfer function phase in terms of angular frequency ω and break frequency ω_0.

$$\omega_0 = R/L \quad \frac{\pi}{2} = 90^\circ \quad (6.301)$$

From Equation 6.301, substitute R over L into the transfer function phase Equation 6.300 to obtain Equation 6.302.

$$\angle H(j\omega) = \frac{\pi}{2} - \tan^{-1}(\omega L/R) \quad (6.302)$$

Equation 6.302 shows the voltage transfer function phase for the series RL high-pass filter circuit in Figure 6.80.

The $\pi/2$ term in Equation 6.302 indicates that the phase is positive and that this series RL circuit, with the output taken across the inductor, is a **lead** network.

The $\pi/2$ term in Equation 6.302 accounts for the 90° phase shift at DC shown in Figure 6.82. The phase shift decreases to approach zero degrees at very high frequencies.

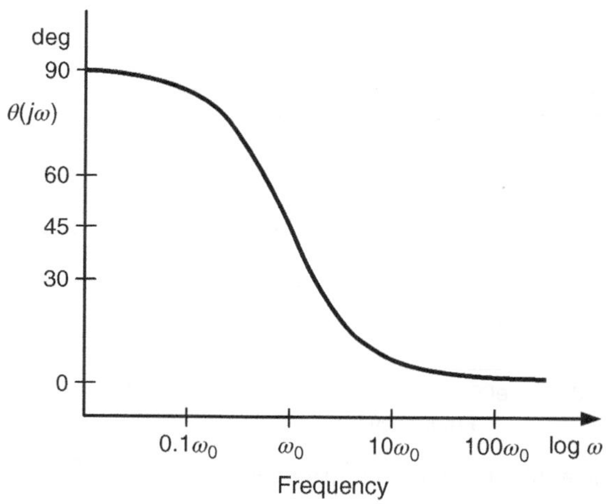

Figure 6.82 Series RL high-pass filter output voltage phase response.

6.5.2 Series RL Frequency Domain Example Intuitive Analysis

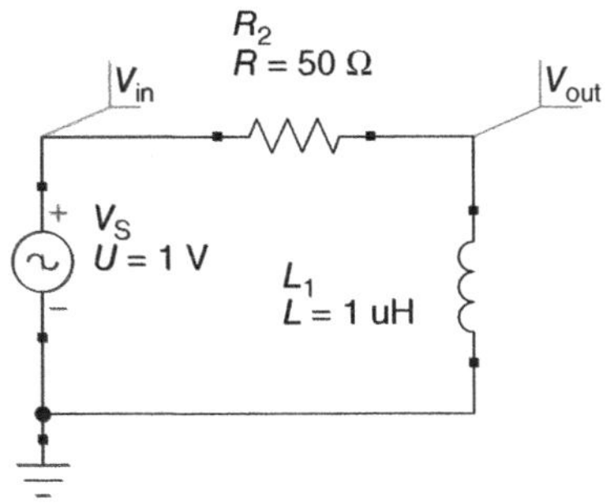

Figure 6.83 Series RL circuit simulation schematic driven by sinusoidal voltage source.

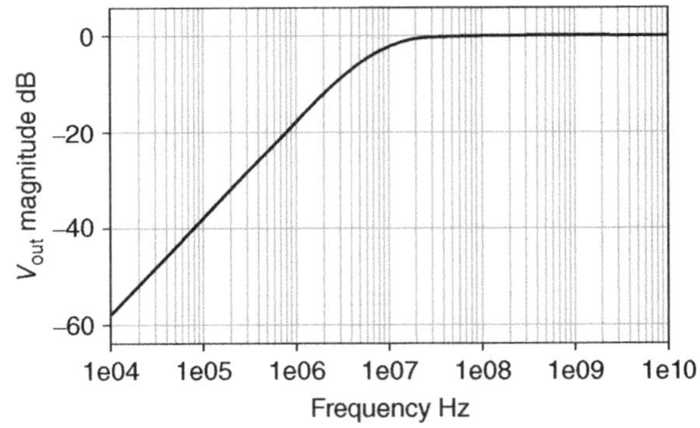

Figure 6.84 Series RL simulation of output voltage magnitude frequency response.

In the schematic in Figure 6.83, the inductor looks like an open circuit (high impedance) to high frequency, and the inductor looks like a short circuit (low impedance) to low frequency.

The circuit in Figure 6.83 is a high-pass filter, as shown in Figure 6.84, because the series inductor blocks high frequencies. Less current around the loop at high frequencies means less voltage across the resistor and more voltage, labeled V_{out} in the simulation schematic in Figure 6.83, across the output inductor.

As the frequency of the input signal, labeled V_S and V_{in} in the simulation schematic in Figure 6.83, goes down, the impedance of the inductor ($j\omega L$) goes down and more of the source voltage, V_S, gets dropped across the resistor, as shown in the simulation results plot in Figure 6.85. In the low-frequency limit, all the source voltage, V_S, drops across the resistor and the current becomes V_S over R equals 0.2 A (same thing as V_{in} over R), as shown in Figure 6.86.

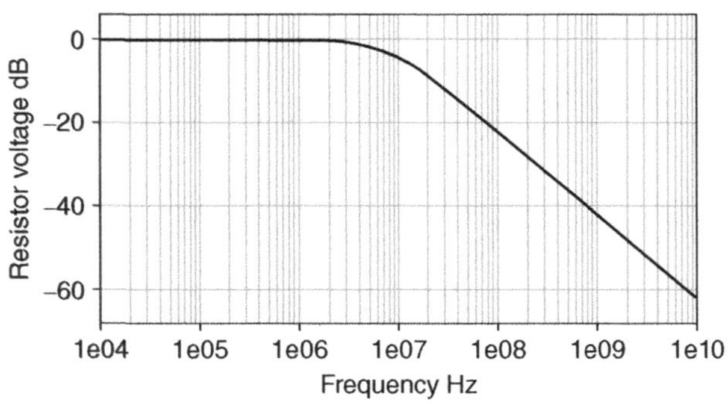

Figure 6.85 Series RL simulation of resistor output magnitude response.

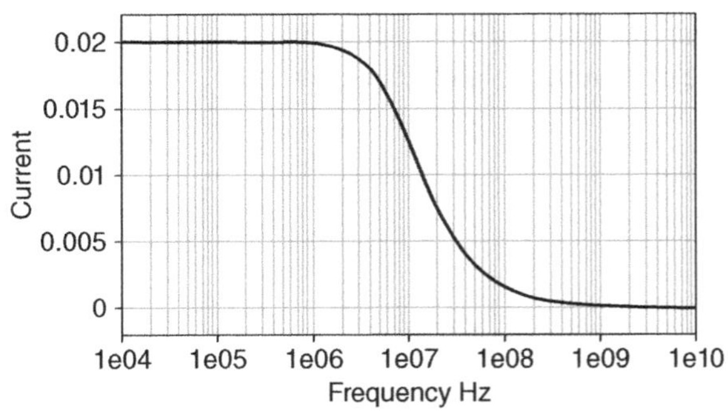

Figure 6.86 Series RL high-pass filter simulation of current.

The lower the impedance of the inductor, the lower the total impedance of the inductor and resistor in series, and the larger the current in the loop. Larger loop current means larger voltage across the resistor and smaller the output voltage V_{out} across the inductor.

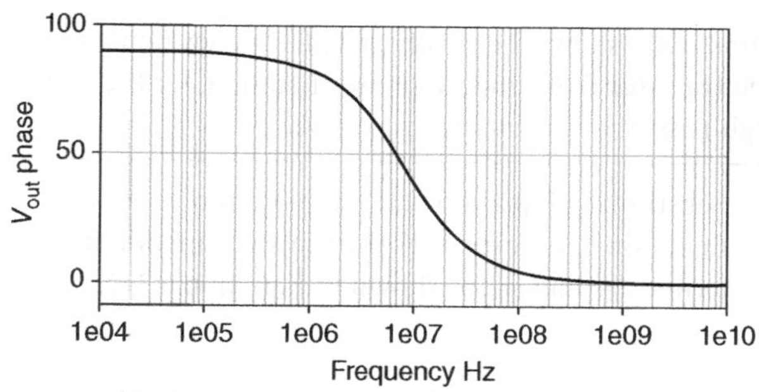

Figure 6.87 Series RL simulation of inductor output voltage phase response.

At DC, the reactance ωL is zero and the 90° phase value in Figure 6.87 represents the $\pi/2$ term in the phase Equations 6.302 and 6.303. As frequency increases, the second term, the arctangent of ωL over R term, grows and is subtracted from the first term. As ω grows to infinity, the arctangent term approaches 90°, making the phase of the transfer function approach zero.

$$\angle H(j\omega) = \frac{\pi}{2} - \tan^{-1}(\omega L/R) \qquad (6.303)$$

6.5.2.1 Lead Network

In the circuit simulation, the voltage source V_S is the reference for phase. The positive phase in Figure 6.87 indicates that this first-order high-pass filter is a lead network. The output voltage V_{out} leads the source voltage V_S in phase.

6.5.3 Series RL Frequency Domain Transfer Function Calculation

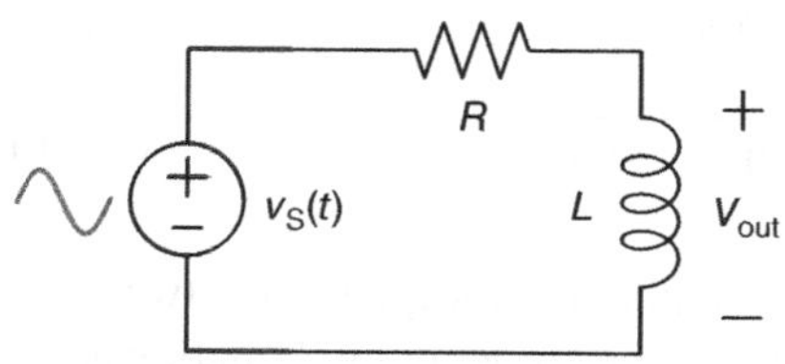

Figure 6.88 Series RL circuit schematic driven by sinusoidal voltage source.

$$\frac{\boldsymbol{V}_{out}}{\boldsymbol{V}_S} = \frac{j\omega L/R}{1 + j\omega L/R} = H(j\omega) \qquad (6.304)$$

Equation 6.304 is the steady-state frequency domain voltage transfer function for the circuit schematic in Figure 6.88.

6.5.3.1 Calculating the Transfer Function $\frac{\boldsymbol{V}_{out}}{\boldsymbol{V}_S}$

We can use a voltage divider approach to calculate the steady-state frequency domain transfer function of the output voltage V_{out} over the input voltage V_S.

$$Z_L = j\omega L \qquad (6.305)$$

Equation 6.305 is the impedance of an ideal inductor.

$$Z_R = R \qquad (6.306)$$

Equation 6.306 is the impedance of an ideal resistor.

$$\boldsymbol{I}_{out} = \left(\frac{\boldsymbol{V}_S}{R + j\omega L}\right) \qquad (6.307)$$

The loop current in Equation 6.307 equals the input voltage V_S divided by the series impedance, R plus $j\omega L$, that the voltage drops across.

$$\boldsymbol{V}_{out} = \left(\frac{\boldsymbol{V}_S}{R + j\omega L}\right) j\omega L \qquad (6.308)$$

Applying Ohm's law to Equation 6.307, the output voltage V_{out} equals the loop current I_{out} times the impedance of the output inductor $j\omega L$, as shown in Equation 6.308.

$$\frac{\boldsymbol{V}_{out}}{\boldsymbol{V}_S} = \frac{j\omega L}{R + j\omega L} \qquad (6.309)$$

Divide both sides of the Equation 6.308 by V_S and remove the brackets, to obtain Equation 6.309.

$$\frac{\boldsymbol{V}_{\text{out}}}{\boldsymbol{V}_{\text{S}}} = \frac{j\omega L/R}{1 + j\omega L/R} \quad (6.310)$$

Equation 6.310 shows the steady-state frequency domain voltage **transfer function**, in phasor notation, for the series RL high-pass filter circuit in Figure 6.88.

$$\omega_0 = R/L \quad (6.311)$$

$$\frac{\boldsymbol{V}_{\text{out}}}{\boldsymbol{V}_{\text{S}}} = \left(\frac{j\omega/\omega_0}{1 + j\omega/\omega_0}\right) \quad (6.312)$$

Substituting Equation 6.311, for ω_0, into Equation 6.310 expresses the transfer function in terms of ω and ω_0, as shown in Equation 6.312.

$$\frac{\boldsymbol{V}_{\text{out}}}{\boldsymbol{V}_{\text{S}}} = \frac{j\omega}{R/L + j\omega} = \frac{j\omega}{\omega_0 + j\omega} \quad (6.313)$$

For another convenient notation, multiply the numerator and denominator of Equation 6.310 by R over L to generate Equation 6.313.

6.5.4 Series RL Transfer Function Magnitude Calculation

$$H(j\omega) = \frac{j\omega}{R/L + j\omega} = \frac{\boldsymbol{V}_{\text{out}}}{\boldsymbol{V}_{\text{S}}} \quad (6.314)$$

To plot the frequency response, of the first-order RL high-pass filter, we need to calculate the magnitude and phase of the transfer function in Equation 6.314.

The magnitude of a complex function, $G(\omega)$ in Equation 6.315, is the square root of the sum of the squares of the real part $R(\omega)$ and the imaginary part $X(\omega)$, as shown in Equation 6.316.

$$G(\omega) = R(\omega) + jX(\omega) \quad (6.315)$$

$$|G(\omega)| = \sqrt{[R(\omega)]^2 + [X(\omega)]^2} \quad (6.316)$$

The series RL low-pass filter transfer function, in Equation 6.314, can be thought of as the quotient of two complex functions, where the numerator in this case is an imaginary valued term $j\omega$. The numerator in Equation 6.314 has no real component.

The magnitude of the ratio of two complex functions $A(\omega)$ and $B(\omega)$ equals the ratio of the magnitudes of those same two complex functions, as shown in Equation 6.317.

$$\left|\frac{A(\omega)}{B(\omega)}\right| = \frac{|A(\omega)|}{|B(\omega)|} \quad (6.317)$$

Starting with Equation 6.314, calculate the magnitude of the RL high-pass filter transfer function using the square root of the sum of the squares, shown in Equation 6.316. Notice that the j does not get squared. The result is Equations 6.318 and 6.319.

$$\left|\frac{\boldsymbol{V}_{\text{out}}}{\boldsymbol{V}_{\text{S}}}\right| = \frac{V_{\text{out}}}{V_{\text{S}}} = \frac{|j\omega|}{|R/L + j\omega|} = \frac{\sqrt{\omega^2}}{\sqrt{(R/L)^2 + \omega^2}} = \frac{\omega}{\sqrt{(R/L)^2 + \omega^2}} \quad (6.318)$$

$$\left|\frac{\boldsymbol{V}_{\text{out}}}{\boldsymbol{V}_{\text{S}}}\right| = \frac{\omega}{\sqrt{(R/L)^2 + \omega^2}} = |H(j\omega)| \quad (6.319)$$

Equation 6.319 expresses the **magnitude** of the RL high-pass filter transfer function $H(j\omega)$ for the circuit in Figures 6.88 and 6.89.

The corner frequency in Equation 6.320 can be used to transform the transfer function magnitude in Equation 6.319 into the ω and ω_0 form in Equation 6.321.

$$\omega_0 = \frac{R}{L} \quad (6.320)$$

$$\left|\frac{\boldsymbol{V}_{\text{out}}}{\boldsymbol{V}_{\text{S}}}\right| = \frac{\omega}{\sqrt{\omega_0^2 + \omega^2}} = |H(j\omega)| \quad (6.321)$$

6.5.4.1 Series RL Bode Plot Calculations

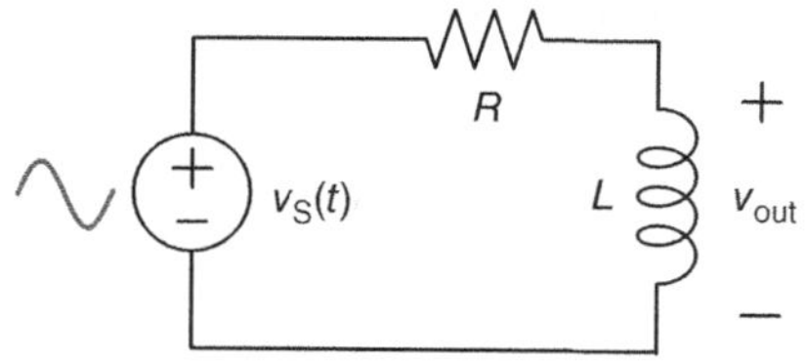

Figure 6.89 Series RL circuit schematic driven by sinusoidal voltage source.

$$\frac{V_{out}}{V_S} = \frac{j\omega}{R/L + j\omega} = H(j\omega) \qquad (6.322)$$

Equation 6.322 expresses the steady-state frequency domain voltage transfer function for the RL circuit schematic in Figure 6.89.

From the transfer function and magnitude of the transfer function, we'll generate the exact equations used to create the Bode plots. This plot will have dB magnitude on the vertical scale and radian frequency ω $(2\pi f)$ on the horizontal axis.

$$|H(j\omega)| = \left|\frac{V_{out}}{V_S}\right| = \frac{\omega}{\sqrt{(R/L)^2 + \omega^2}} \qquad (6.323)$$

Equation 6.323 is the magnitude of the transfer function $H(\omega)$ of the RL circuit previously derived in Equation 6.319.

$$|H(j\omega)| = \omega[\omega^2 + (R/L)^2]^{-1/2} \qquad (6.324)$$

Restate the square root in the denominator of Equation 6.323 as an exponent of minus one half, as shown in Equation 6.324.

The vertical scale of the Bode diagram is logarithmic in decibels (dB).

$$H \text{ in dB} = 20\log_{10} H \qquad (6.325)$$

Equation 6.325 states that the number H expressed in decibels is 20 times the logarithm to the base 10 of H.

$$|H(j\omega)|_{dB} = 20\log|H(j\omega)| \qquad (6.326)$$

The magnitude of the transfer function $H(j\omega)$ expressed in decibels is 20 times the logarithm, to the base 10, of the magnitude of $H(j\omega)$. From Equation 6.324 substitute the RL transfer function magnitude into Equation 6.326, to obtain 6.337.

$$|H(j\omega)|_{dB} = 20\log\omega[\omega^2 + (R/L)^2]^{-1/2} \qquad (6.327)$$

This expression in Equation 6.327 is the logarithm of a product. Use the logarithmic identity in Equation 6.328 to simplify the product term in the "square root" (bracketed) expression in Equation 6.327.

$$\log(A \cdot B) = \log(A) + \log(B) \qquad (6.328) \qquad A = \omega \qquad B = [\omega^2 + (R/L)^2]^{-1/2}$$

Apply the logarithmic rule, in Equation 6.328, for the log of a product, to Equation 6.327 to obtain the sum of two logarithm terms in Equation 6.329.

$$|H(j\omega)|_{dB} = 20\log(\omega) + 20\log[\omega^2 + (R/L)^2]^{-1/2} \qquad (6.329)$$

The right-hand log term in Equation 6.329 is the logarithm of an exponential term, where the exponent is the minus one-half on the brackets.

$$\log_a G^k = k\log_a G \qquad (6.330)$$

Apply Equation 6.330 to the right side of Equation 6.329, with k equals minus one half, to bring down the minus one half, to obtain Equation 6.331.

Twenty times minus one-half equals minus 10, as shown in Equation 6.331.

$$|H(j\omega)|_{dB} = 20\log(\omega) - 10\log[\omega^2 + (R/L)^2] \quad (6.331)$$

Equation 6.331 is the magnitude of the transfer function $H(j\omega)$ in dB for Bode plots.

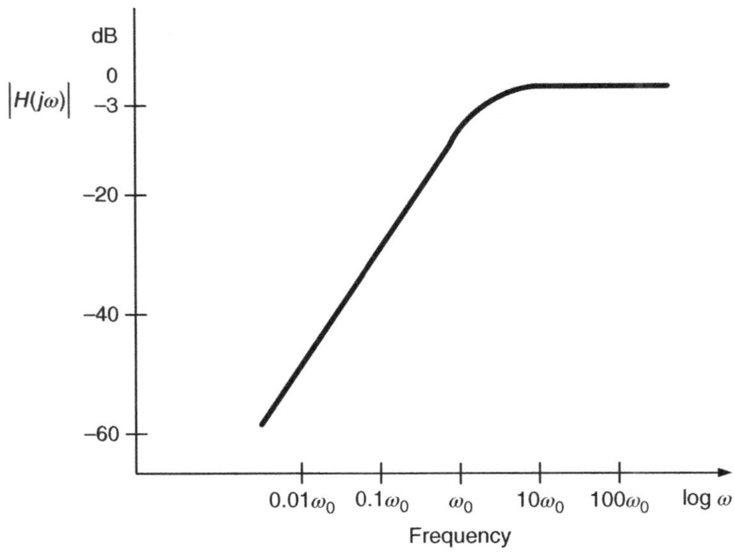

Figure 6.90 Series RL high-pass filter Bode magnitude plot.

Figure 6.90 is the RL high-pass filter magnitude Bode plot.

The vertical axis in Figure 6.90 is also log scale, but more specifically, it is a dB scale as defined in Equation 6.327.

6.5.5 Series RL Transfer Function Phase Response Calculation

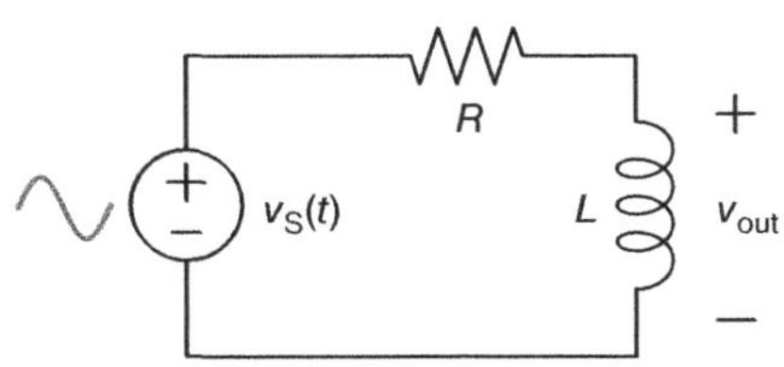

Figure 6.91 Series RL circuit schematic driven by sinusoidal voltage source.

$$\angle H(j\omega) = \frac{\pi}{2} - \tan^{-1}(\omega L/R) \quad (6.332)$$

Plotted in Figure 6.92, Equation 6.332 is the phase response of the series RL circuit in Figure 6.91.

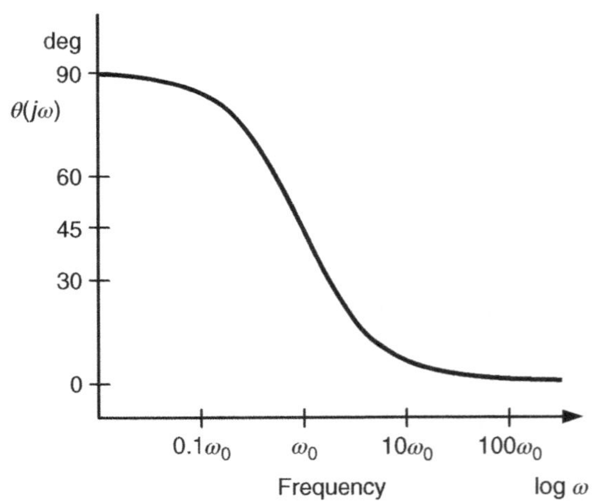

Figure 6.92 Series RL high-pass filter Bode phase plot.

For the transfer function in Equation 6.333, the phase in Equation 6.332 can be calculated from the general formula for the phase of a first-order high-pass filter in Equation 6.335.

$$\frac{\mathbf{V}_{out}}{\mathbf{V}_S} = \frac{j\omega}{R/L + j\omega} = H(j\omega) \quad (6.333)$$

$$\angle H(j\omega) = \frac{\pi}{2} - \tan^{-1}(\omega/\omega_0)$$
$$\omega_0 = R/L \quad \frac{\pi}{2} = 90^\circ \quad (6.334)$$

Equation 6.333 is the steady-state frequency domain voltage transfer function, in phasor notation, for the series RL high-pass filter in Figure 6.91.

6.5.5.1 Calculating the Phase Response of a Complex Transfer Function

The angle theta of a complex function, like the transfer function $H(j\omega)$ in Equation 6.333, is the arctangent of the imaginary part $X(\omega)$ over the real part $R(\omega)$, as shown in Equation 6.335.

$$H(j\omega) = R(\omega) + jX(\omega) = \frac{j\omega}{j\omega + R/L} \quad (6.335)$$

$$\theta(\omega) = \angle H(\omega) = \tan^{-1}\frac{X(\omega)}{R(\omega)} \quad (6.336)$$

The first-order RL high-pass filter transfer function in Equation 6.335 is a quotient of two complex expressions, as shown in Equations 6.337, 6.338, and 6.339.

$$H(\omega) = \frac{\boldsymbol{V}_{\text{out}}}{\boldsymbol{V}_{\text{S}}} = \frac{j\omega}{j\omega + R/L} = \frac{F(\omega)}{G(\omega)} \quad (6.337)$$

$$F = j\omega \quad (6.338) \qquad G = j\omega + R/L \quad (6.339)$$

Equation 6.337 shows the transfer function numerator expressed as the complex function $F(\omega)$ and the denominator expressed as $G(\omega)$.

$$\angle H(j\omega) = \angle\frac{F(j\omega)}{G(j\omega)} = \frac{\angle\theta_F}{\angle\theta_G} = \angle(\theta_F) - \angle(\theta_G) \quad (6.340)$$

Equation 6.340 shows how the phase angle of a quotient of complex functions is the phase angle of the numerator minus the phase angle of the denominator.

$$\angle H(j\omega) = \angle F(j\omega) - \angle G(j\omega) \quad (6.341)$$

Equation 6.341 expresses Equation 6.340 in a slightly different notation.

From Equations 6.338 and 6.339, insert the transfer function numerator and denominator into Equation 6.341 to get Equation 6.342.

The phase of the transfer function will be the phase of the numerator $j\omega$ term minus the phase of the denominator, the quantity $j\omega$ plus R over L, as shown in Equation 6.342.

$$\angle H(j\omega) = \angle(j\omega) - \angle(j\omega + R/L) \quad (6.342)$$

Apply Equation 6.336, the arctangent of the imaginary part $X(\omega)$ over the real part $R(\omega)$, separately to the numerator and denominator terms in Equation 6.342, as shown in Equations 6.343 and 6.344.

Numerator F

$$\angle(j\omega) = \tan^{-1}[X/R] = \tan^{-1}[\omega/0] = \tan^{-1}(\infty) = \frac{\pi}{2} = 90^\circ \quad (6.343)$$

Denominator G

$$\angle(j\omega + R/L) = \tan^{-1}[X/R] = \tan^{-1}[\omega/(R/L)] = \tan^{-1}(\omega L/R)$$

Substitute Equations 6.343 and 6.344 into Equation 6.342 for the phase angle of $H(j\omega)$ in Equation 6.345.

$$\angle H(j\omega) = \angle(j\omega) - \angle(j\omega + R/L) = 90^O - \tan^{-1}(\omega L/R) \quad (6.345)$$

$$\angle H(j\omega) = \frac{\pi}{2} - \tan^{-1}(\omega L/R) \quad (6.346)$$

Equation 6.346 is the **phase** response of the RL high-pass filter transfer function in Equation 6.333.

The phase response, from Equation 6.346, can be expressed in terms of the break frequency ω_0, as shown in Equation 6.347.

$$\theta = \angle H(j\omega) = -\tan^{-1}\left(\frac{\omega}{\omega_0}\right) = \angle\left|\frac{\boldsymbol{V}_{\text{out}}}{\boldsymbol{V}_{\text{S}}}\right| \quad (6.347) \qquad \omega_0 = R/L$$

6.5.6 Series RL Low-pass Filter Phase Relationships

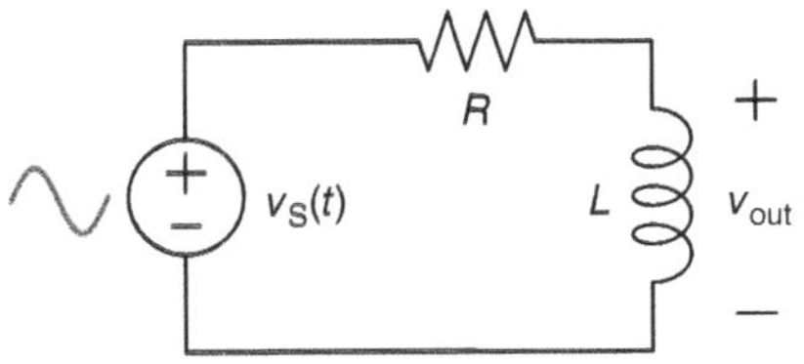

Figure 6.93 Series RL circuit schematic driven by sinusoidal voltage source.

Because the phase of the output over the input, shown in Equation 6.348, is positive, the output voltage phase leads the input voltage phase for the series RL circuit in Figure 6.93.

$$\angle H(j\omega) = \angle\left|\frac{\boldsymbol{V}_{\text{out}}}{\boldsymbol{V}_{\text{S}}}\right| = \theta(j\omega) = \frac{\pi}{2} - \tan^{-1}(\omega L/R) \tag{6.348}$$

The voltage across an ideal inductor is 90° out of phase with the current through the inductor at all frequencies, as shown in Figure 6.94.

The phase of the voltage across an inductor leads the phase of the current through the inductor. In Figure 6.94, the arrow indicates a counterclockwise rotation, putting the V_L phasor ahead of the I_L phasor in phase.

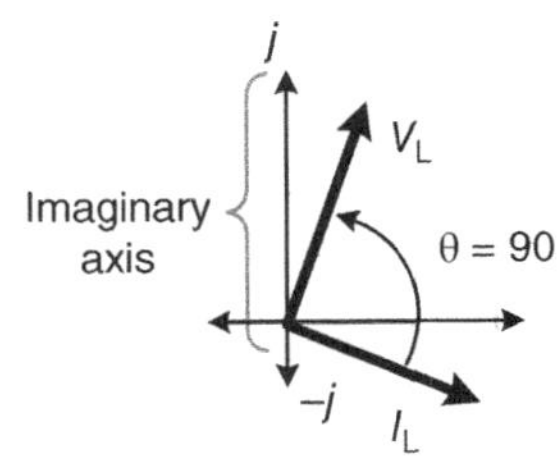

Figure 6.94 Phasor diagram showing voltage leading current for an inductor.

Shown in Figure 6.95, the impedance of the inductor, $j\omega L$, gets plotted at 90°, the positive imaginary direction, corresponding to the positive reactance ωL.

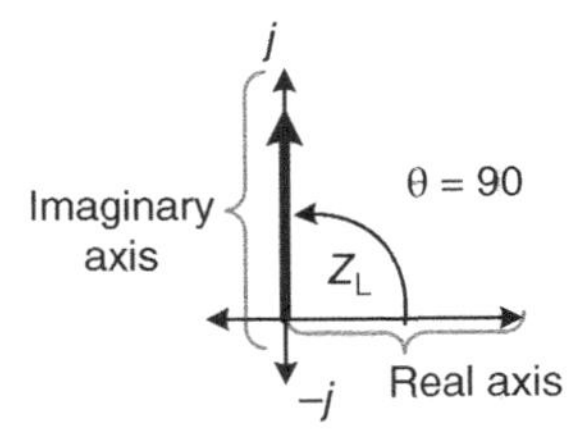

Figure 6.95 Phasor diagram of inductive impedance showing 90° phase angle.

The voltage across a resistor is always in phase with the current through the resistor. The impedance of an ideal resistor, a pure resistance, gets plotted on the positive real axis, as shown in Figure 6.96.

The voltage across a resistor is always in phase with the current through the resistor. The impedance of an ideal resistor, a pure resistance, gets plotted on the positive real axis, as shown in Figure 6.96.

The current in an independent voltage source can be whatever value it needs to be in order to supply the specified voltage of the source. This variability includes the phase of the current in the voltage source. The current can have whatever phase, relative to the voltage, that it needs in order to supply the specified voltage.

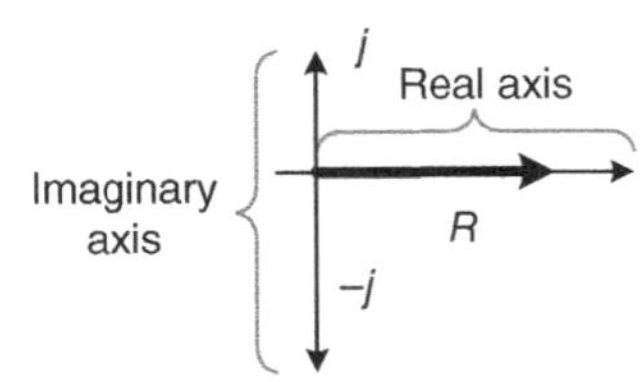

Figure 6.96 Phasor diagram of resistive impedance plotted on real axis.

KCL (Kirchhoff's current law) and KVL (Kirchhoff's voltage law) apply in the steady-state frequency domain. Currents and voltages add and subtract phasors like vector addition and subtraction. But currents and voltages are on different magnitude scales.

For analyzing the phasor relationships among the circuit elements in a first-order series circuit, use the current as a reference because it is the same through all three circuit elements. The current vector lines up on the positive real (horizontal) axis as shown in Figure 6.97. Note that current, being a different quantity than voltage, will have its own scale in a phasor diagram.

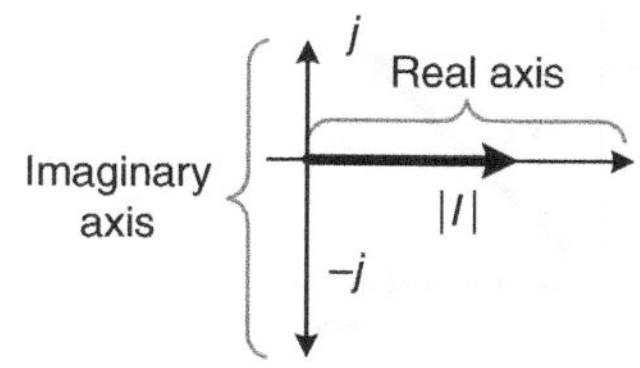

Figure 6.97 Phasor diagram of current on real axis as a reference.

The phase relationship of the voltage across the independent voltage source, relative to its current, changes with frequency.

The amount of current in the whole circuit also changes with frequency, changing the magnitudes of the voltage phasors.

Variations in the amount of current and the phase of the current, relative to the voltage of the independent voltage source, allow the phase of the output voltage over the input voltage to vary according to the arctangent function of omega times *L* over *R*.

6.5.6.1 Phase Relationships and Phasors at the Corner Frequency

Use phasors and vector addition to calculate the phase of the series RL network in Figure 6.93 at the corner frequency ω_0 equals *R* over *L*. The value of the inductor reactance and the value of the resistor's resistance are equal at the corner frequency.

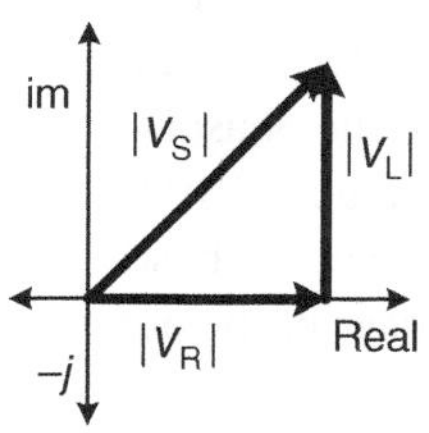

Figure 6.98 Series RL phasor diagram calculating source voltage phasor.

Set the inductance phasor $\boldsymbol{V}_L$ head-to-tail to the resistor voltage phasor $\boldsymbol{V}_R$ for vector addition, as shown in Figure 6.98. Connect the two phasors, $\boldsymbol{V}_R$ and $\boldsymbol{V}_L$, to get the magnitude and phase of the source voltage $\boldsymbol{V}_S$.

The direction of the $\boldsymbol{V}_S$ phasor in Figure 6.98 reflects Kirchhoff's voltage law in the form of Equation 6.349.

$$|\boldsymbol{V}_S| = |\boldsymbol{V}_R| + |\boldsymbol{V}_L| \tag{6.349}$$

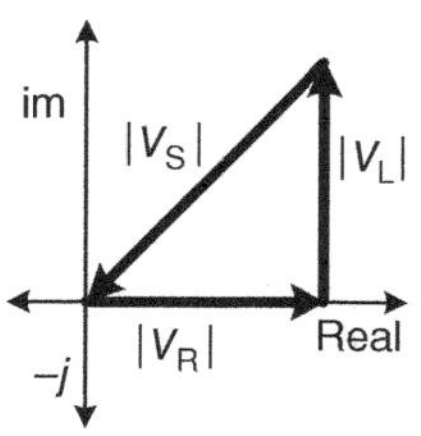

Figure 6.99 Series RL phasor diagram with opposite orientation of source phasor.

The direction of the $\boldsymbol{V}_S$ phasor in Figure 6.99 reflects Kirchhoff's voltage law in the form of Equation 6.350.

$$|\boldsymbol{V}_R| + |\boldsymbol{V}_L| - |\boldsymbol{V}_S| = 0 \tag{6.350}$$

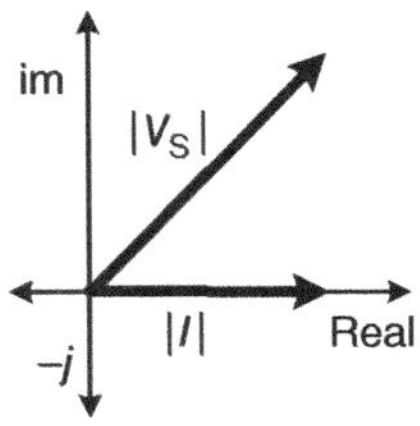

Figure 6.100 Series RL phasor diagram with source phasor at plus 45°.

Figure 6.100 shows that the voltage source $\boldsymbol{V}_S$ phasor leads the current phasor by 45°. Current has a different magnitude scale from voltage.

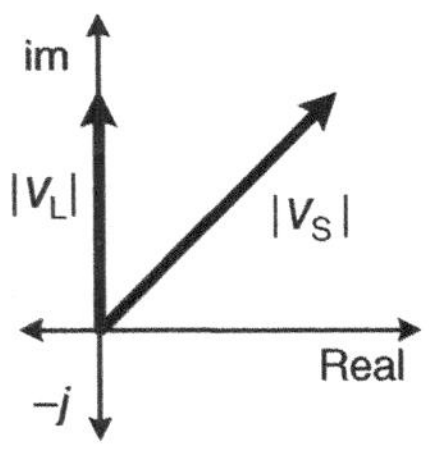

Figure 6.101 Series RL phasor diagram with inductor voltage phasor on imaginary axis.

In Figure 6.101, the $\boldsymbol{V}_R$ phasor is removed from the graph and the $\boldsymbol{V}_L$ phasor is moved to the vertical (imaginary) axis. The $\boldsymbol{V}_S$ phasor lags the $\boldsymbol{V}_L$ phasor by 45°.

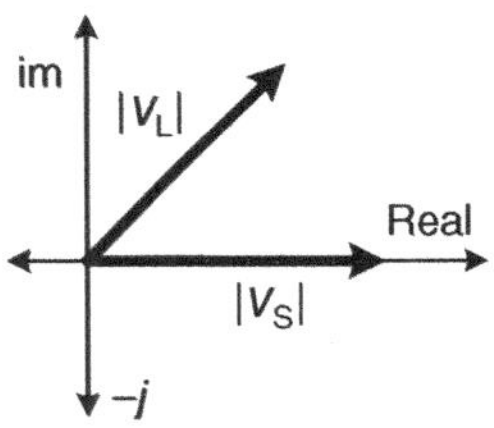

Figure 6.102 Series RL phasor diagram showing source voltage phasor on real axis.

The voltage across the inductor, $\boldsymbol{V}_L$, is also the output voltage $\boldsymbol{V}_{out}$. To complete the calculation of the phase of the voltage transfer function at the corner frequency, rotate the phasors clockwise (negative phase) by 45°, as shown in Figure 6.102. This rotation puts the $\boldsymbol{V}_S$ phasor on the real axis, with zero phase, making $\boldsymbol{V}_S$ the reference. Relative to V_S, the phase of the output voltage V_{out} is 45°.

6.6 Parallel RL Low-pass Filter Frequency Response

6.6.1 Parallel RL Frequency Domain Results Overview

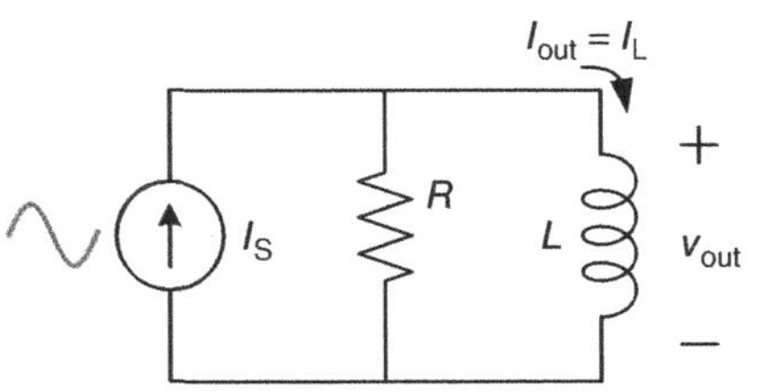

Figure 6.103 Parallel RL circuit schematic driven by sinusoidal current source.

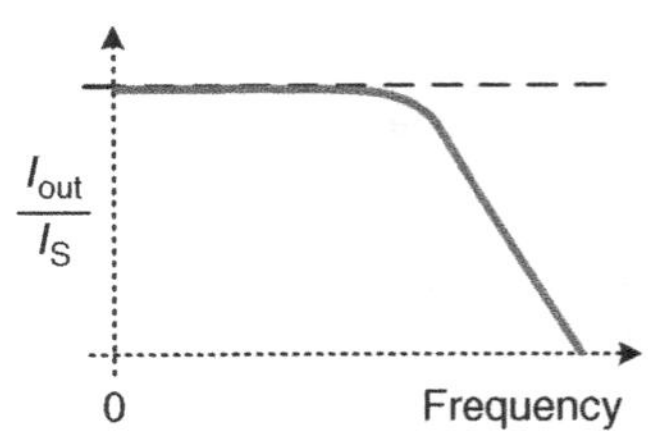

Figure 6.104 Parallel RL output current low-pass frequency response magnitude.

Equation 6.351 is the current transfer function for the parallel RL circuit in Figure 6.103. Figure 6.104 shows the plot of the magnitude of the **current** transfer function $\boldsymbol{I}_{out}/\boldsymbol{I}_S$, where $\boldsymbol{I}_S$ equals 1.

$$\frac{\boldsymbol{I}_{out}}{\boldsymbol{I}_S} = \frac{1}{(j\omega L/R + 1)} \tag{6.351}$$

At low frequency, where the inductor impedance $j\omega L$ is small, current flows easily through the output shunt inductor, making the **current** transfer function $\boldsymbol{I}_{out}/\boldsymbol{I}_S$ a low-pass filter. At high frequencies, the current goes through the resistor, making V_{out} large.

The input current is an independent variable and the output current varies with frequency, so $\boldsymbol{I}_{out}/\boldsymbol{I}_S$ makes a convenient transfer function.

$$H(j\omega) = \frac{1}{1 + j\omega/\omega_0} \quad (6.352)$$

Equation 6.352 is in the general format for first-order low-pass filter transfer functions.

$$\omega_0 = R/L \quad (6.353)$$

From Equation 6.353, substitute R over L into Equation 6.352, in place of for the corner frequency ω_0, to obtain Equation 6.354.

$$H(j\omega) = \frac{1}{1 + j\frac{\omega}{R/L}} \quad (6.354)$$

Rearrange the denominator j term in Equation 6.354 to obtain Equation 6.355.

$$H(j\omega) = \frac{\boldsymbol{I}_{out}}{\boldsymbol{I}_S} = \frac{1}{1 + j\omega L/R} \quad (6.355)$$

Equation 6.355 is the steady-state frequency domain **current transfer function** of Figure 6.104.

$$H(j\omega) = \frac{\boldsymbol{I}_{out}}{\boldsymbol{I}_S} = \frac{R}{R + j\omega L} \quad (6.356)$$

In Equation 6.355, multiply the numerator and the denominator by R to generate another common form of this low-pass filter transfer function, shown in 6.365.

$$|H(j\omega)| = \left|\frac{\boldsymbol{I}_{out}}{\boldsymbol{I}_S}\right| = \frac{1}{\sqrt{1 + (\omega/\omega_0)^2}} \quad (6.357)$$

Equation 6.357 is the general format for low-pass filter transfer function magnitude in terms of angular frequency ω and break frequency ω_0.

From Equation 6.353, substitute R over L into Equation 6.357, in place of the corner frequency ω_0, to obtain Equation 6.358.

$$|H(j\omega)| = \left|\frac{\boldsymbol{I}_{out}}{\boldsymbol{I}_S}\right| = \frac{1}{\sqrt{1 + \left(\frac{\omega}{R/L}\right)^2}} \quad (6.358)$$

Rearrange the denominator in Equation 6.358 to get Equation 6.359.

$$|H(j\omega)| = \left|\frac{\boldsymbol{I}_{out}}{\boldsymbol{I}_S}\right| = \frac{1}{\sqrt{1 + (\omega L/R)^2}} \quad (6.359)$$

Equation 6.359 is in the **magnitude** of the **current** transfer function for the low-pass filter in Figure 6.103.

$$\theta = \angle H(j\omega) = \angle\frac{\boldsymbol{I}_{out}}{\boldsymbol{I}_S} = -\tan^{-1}\left(\frac{\omega}{\omega_0}\right) \quad (6.360)$$

Equation 6.360 is in the general format for low-pass filter phase response in terms of angular frequency ω and break frequency ω_0.

From Equation 6.353, substitute R over L into the general phase response, Equation 6.360, in place of ω_0, to obtain Equation 6.361.

$$\theta = \angle H(j\omega) = \angle\frac{\boldsymbol{I}_{out}}{\boldsymbol{I}_S} = -\tan^{-1}\left(\frac{\omega}{R/L}\right) = -\tan^{-1}(\omega L/R) \quad (6.361)$$

$$\angle H(j\omega) = \angle\frac{\boldsymbol{I}_{out}}{\boldsymbol{I}_S} = -\tan^{-1}(\omega L/R) \quad (6.362)$$

Equations 6.361 and 6.362 are the **phase** of the parallel RL current transfer function in Equation 6.351.

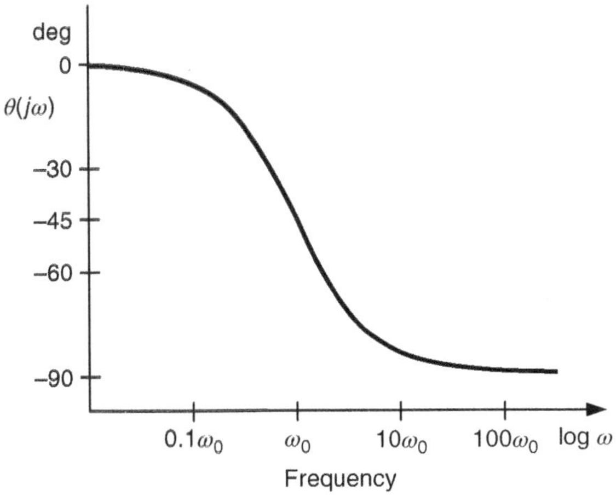

Figure 6.105 Parallel RL low-pass filter voltage phase response.

First-order phase response Equations 6.361 and 6.362 are plotted in Figure 6.105.

Because of the minus sign in Equation 6.362, and the phase in Figure 6.105 descending from zero to negative values, the parallel RL circuit in Figure 6.103 is a **lag network**.

6.6.2 Parallel RL Frequency Domain Example Intuitive Analysis

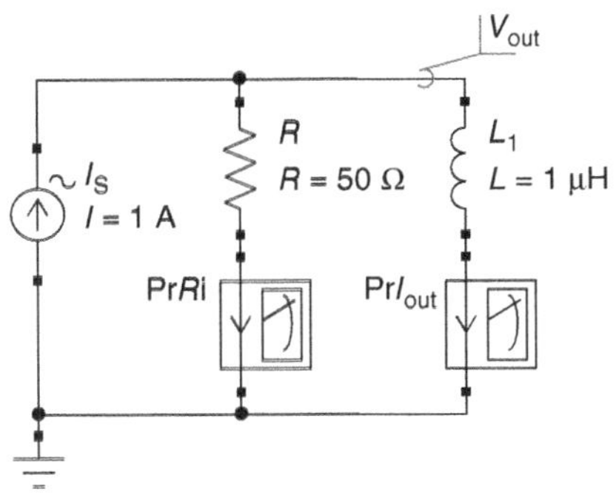

Figure 6.106 Parallel RL circuit simulation schematic driven by sinusoidal current source.

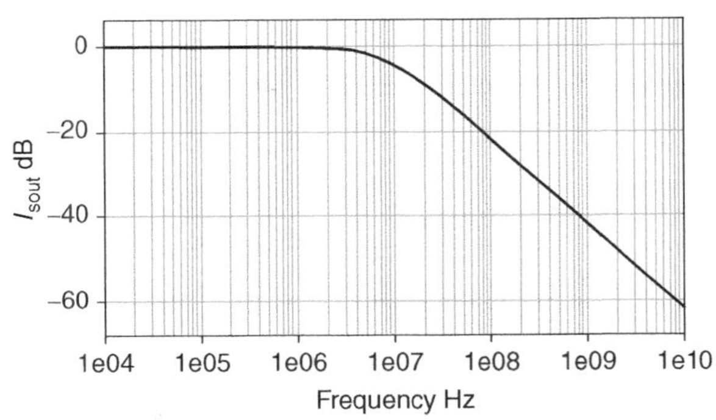

Figure 6.107 Parallel RL simulation of voltage magnitude frequency response.

Thinking of this parallel RL circuit in Figure 6.106 as a current divider, the source current, labeled I_S in the simulation schematic, will split between the resistor and the inductor according to their impedances. At low frequencies of the source current I_S, the inductor looks like a short circuit and almost all the source current goes through the inductor (and not through the resistor). The inductor current is also the output current I_{out}, and the transfer function shown in Figure 6.107 is $\boldsymbol{I}_{out}/\boldsymbol{I}_S$ in dB. Since low frequencies maximize I_{out}, this is a low-pass filter.

As the frequency of the input current I_S goes down, the inductor impedance $Z = j\omega L$ decreases toward zero. As shown in Figure 6.108, the output current I_{out} approaches the value of the source current I_S because the voltage across the inductor decreases toward zero. At zero frequency (DC), the current in the resistor is zero and the voltage across all three elements is zero.

The impedance (pure reactance) of the inductor changes with frequency, while the impedance (pure resistance) of the resistor does not. As the signal frequency goes up, the impedance of the inductor goes up, and more of the source current I_S goes through the resistor instead of the inductor. In the high-frequency limit, all of I_S goes through the resistor and none through the output inductor. The voltage across all the elements approaches I_S times resistance R.

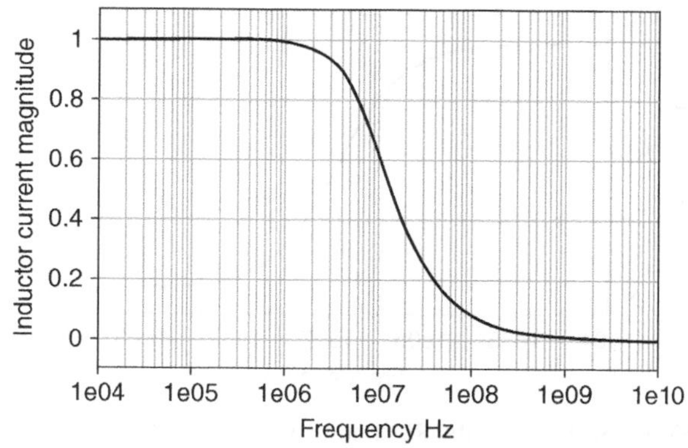

Figure 6.108 Parallel RL simulation of inductor output current frequency response.

Figure 6.109 shows the phase of the current transfer function $\boldsymbol{I}_{out}/\boldsymbol{I}_S$. At low frequencies, where the inductive reactance is very small, the phase is near zero. As frequency increases, the phase grows to −90°. The phase is −45° at the corner frequency ω_0 equals R over L.

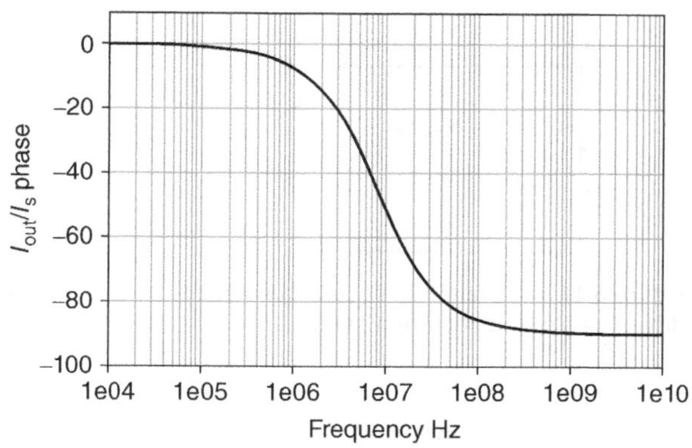

Figure 6.109 Parallel RL simulation of inductor output current phase response.

6.6.2.1 Lag Network I_{out}/I_S

Because the phase of the output current $\boldsymbol{I}_{out}$ lags the phase of the current source I_S, the parallel RL circuit in Figure 6.103 is a lag network.

Figure 6.109 shows the phase of the current transfer function $\boldsymbol{I}_{out}/\boldsymbol{I}_S$. At low frequencies, where the inductive reactance is very small, the phase is near zero. As frequency increases, the phase grows to −90°. The phase is −45° at the corner frequency ω_0.

6.6.2.2 Transfer Impedance V_{out}/I_S

Figure 6.110 shows the impedance transfer function $\boldsymbol{V}_{out}/\boldsymbol{I}_S$, where V_{out} is the inductor voltage, the only voltage in the circuit, and $\boldsymbol{I}_S$ is 1. At low frequencies, where current flows through the inductor, the voltage is small. As frequency increases, the inductive reactance increases and the voltage across the inductor increases. The result is a high-pass filter behavior for the output voltage.

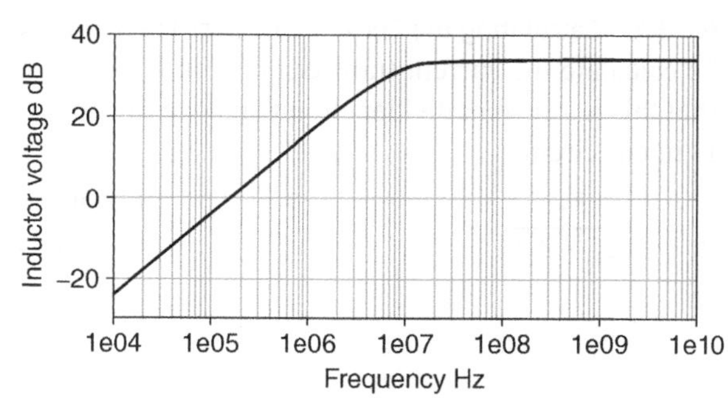

Figure 6.110 Parallel RL circuit simulation of transfer impedance magnitude response.

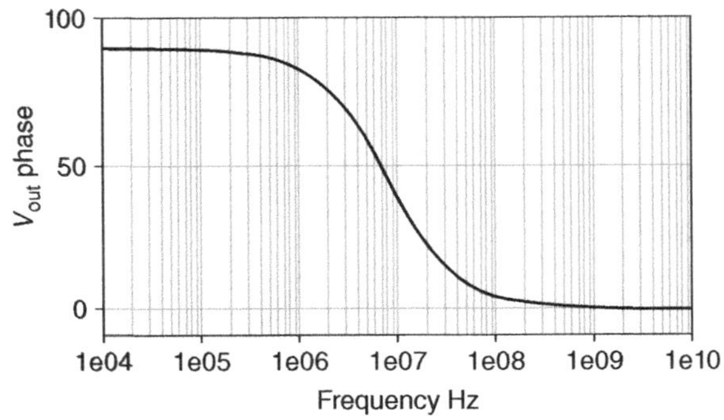

Figure 6.111 Parallel RL circuit simulation of transfer impedance phase response.

Figure 6.111 shows the phase of the transfer function $\boldsymbol{V}_{out}/\boldsymbol{I}_S$, which is the opposite of the phase response of the current transfer function in Figure 6.109.

6.6.3 Parallel RL Current Transfer Function Calculation

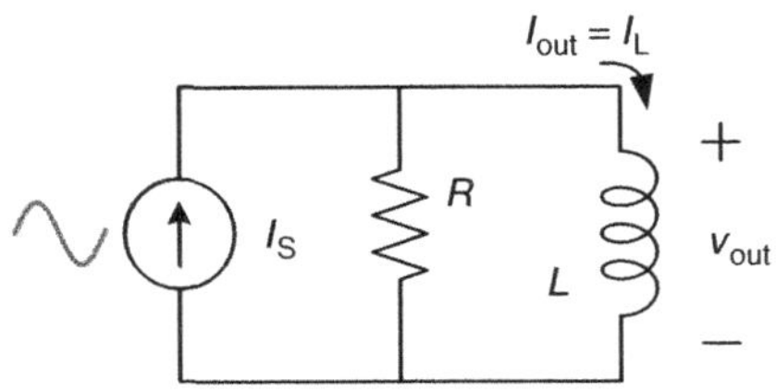

Figure 6.112 Parallel RL circuit schematic driven by sinusoidal current source.

$$\frac{\boldsymbol{I}_{out}}{\boldsymbol{I}_S} = \frac{1}{(j\omega L/R + 1)} = H(j\omega) \tag{6.363}$$

Equation 6.363 is the steady-state frequency domain current transfer function in phasor notation.

Figure 6.112 shows a parallel RL filter driven by a current source.

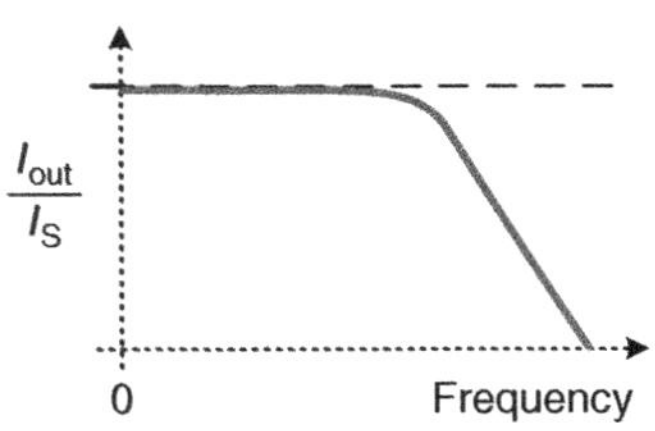

Figure 6.113 Parallel RL circuit Bode magnitude plot of current transfer function.

Figure 6.113 shows first-order low-pass filter frequency domain response from the current transfer function in Equation 6.363. I_S is one.

6.6.3.1 Calculating the Current Low-pass Filter Transfer Function

To calculate the current transfer function $\boldsymbol{I}_{out}/\boldsymbol{I}_S$, of the parallel RL circuit in Figure 6.112, start with the notion that this is a current divider circuit, and we can apply the current divider formula. $\boldsymbol{I}_{out}$ equals the inductor current $\mathbf{I}_L$ in our circuit.

$$i_1 = i_S \frac{Z_2}{(Z_1 + Z_2)} \tag{6.364}$$

Equation 6.364 states the current divider formula in terms of impedance Z, where i_S is a source current.

Plug into Equation 6.364 the quantities for the parallel LR circuit in Figure 6.112. Z_2 becomes resistance R and Z_1 becomes $j\omega L$, the impedance of the inductor, as shown in Equation 6.365.

$$\boldsymbol{I}_{out} = \boldsymbol{I}_S \frac{R}{(j\omega L + R)} \tag{6.365}$$

Divide both sides of Equation 6.365 by $\boldsymbol{I}_S$ to obtain Equation 6.366.

$$\frac{I_{out}}{I_S} = \frac{R}{(j\omega L + R)} \quad (6.366)$$

Equation 6.366 is the transfer function I_{out}/I_S for the parallel RL schematic in Figure 6.112.

To get another convenient form for the transfer function, divide both the numerator and denominator of Equation 6.366 by L to generate Equation 6.367.

$$\frac{I_{out}}{I_S} = \frac{R/L}{(j\omega + R/L)} = \frac{\omega_0}{(j\omega + \omega_0)} \quad (6.367)$$

$$\omega_0 = \frac{R}{L} \quad (6.368)$$

$$\frac{I_{out}}{I_S} = \frac{1}{(j\omega L/R + 1)} = \frac{1}{(j\omega/\omega_0 + 1)} \quad (6.369)$$

Equation 6.369 is the transfer function I_{out}/I_S for the parallel RL schematic in Figure 6.112.

6.6.3.2 Alternate Calculation of the Current Transfer Function I_{out}/I_S

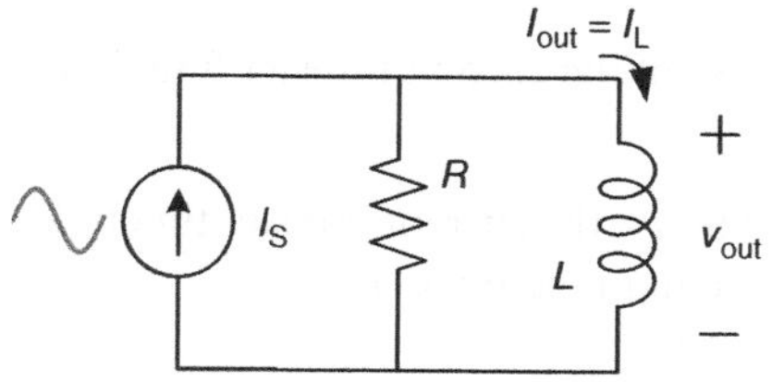

Figure 6.114 Parallel RL circuit schematic driven by sinusoidal current source.

$$\frac{I_{out}}{I_S} = \frac{R}{(j\omega L + R)} \quad (6.370)$$

Equation 6.370 is the steady-state frequency domain **current** transfer function, in phasor notation, for the parallel RL circuit in Figure 6.114.

Here is a second way to calculate the current transfer function I_{out}/I_S, using the parallel impedance formula rather than the current divider formula.

$$\frac{I_{out}}{I_S} = \frac{I_L}{I_S} = \frac{V_{out}/Z_L}{I_S} \quad (6.371)$$

In this circuit, in Figure 6.114, I_{out} is the inductor current I_L. The inductor voltage is also the output voltage V_{out}, the only voltage in this parallel circuit. I_{out} is V_{out} divided by the impedance of the inductor Z_L.

We know that the output voltage V_{out} equals the source current I_S times the parallel impedance Z_P of the resistor and the inductor together. Substitute $I_S Z_P$ into Equation 6.371, replacing V_{out}, to obtain Equation 6.372.

$$\frac{V_{out}/Z_L}{I_S} = \frac{I_S Z_P/Z_L}{I_S} \quad (6.372)$$

In Equation 6.372, cancel the source current I_S term on the top and bottom (numerator and denominator) to get Equation 6.373.

$$\frac{I_{out}}{I_S} = \frac{I_L}{I_S} = \frac{I_S Z_P/Z_L}{I_S} = \frac{Z_P}{Z_L} \quad (6.373)$$

In Equation 6.373, I_{out} over I_S equals the impedance of the parallel combination Z_P over the inductor impedance Z_L.

$$Z_P = \frac{Z_R Z_L}{Z_R + Z_L} \quad (6.374)$$

Equation 6.374 is the formula for the impedance of the parallel combination of the resistor R and the inductor L. Substitute Equation 6.374 into Equation 6.373 for I_{out} over I_S, replacing Z_P, to get Equation 6.375.

$$\frac{I_{out}}{I_S} = \frac{Z_P}{Z_L} = \frac{Z_R Z_L}{Z_R + Z_L} \cdot \frac{1}{Z_L} \quad (6.375)$$

In Equation 6.375 cancel Z_L terms in the numerator and the denominator to get Equation 6.376.

$$\frac{I_{out}}{I_S} = \frac{Z_P}{Z_L} = \frac{Z_R Z_L}{Z_R + Z_L} \cdot \frac{1}{Z_L} = \frac{Z_R Z_L}{Z_R + Z_L}\frac{1}{Z_L} = \frac{Z_R}{Z_R + Z_L} \quad (6.376)$$

$$\frac{I_{out}}{I_S} = \frac{R}{R + j\omega L} \quad (6.377)$$

or

$$\frac{I_{out}}{I_S} = \frac{1}{1 + {}^{j\omega L}/_{R}} = \frac{1}{1 + j\omega(L/R)} \quad (6.378)$$

Equations 6.377 and 6.378 are the steady-state frequency domain **current** transfer function, in phasor notation, for the parallel RL circuit in Figure 6.114.

6.6.4 Parallel RL Current Transfer Function Magnitude Calculation

To plot the frequency response we need to calculate the magnitude and phase of the current transfer function of the first-order parallel RL circuit driven by a current source in Figure 6.114.

$$\frac{I_{out}}{I_S} = \frac{1}{(j\omega L/R + 1)} = H(j\omega) \quad (6.379)$$

Start with Equation 6.379, the current transfer function of the parallel RL circuit in Figure 6.114.

$$C = a + jb \quad (6.380)$$

$$|C| = \sqrt{a^2 + b^2} \quad (6.381)$$

The magnitude of a complex number C, shown in Equation 6.380, or a complex function $G(\omega)$, shown in Equation 6.382, is the square root of the sum of the squares of the real and imaginary parts, shown in Equations 6.381 and 6.383.

$$G(\omega) = R(\omega) + jX(\omega) \quad (6.382)$$

$$|G(\omega)| = \sqrt{[R(\omega)]^2 + [X(\omega)]^2} \quad (6.383)$$

The current transfer function in Equation 6.379 is the quotient of a real valued number 1 divided by a complex function.

$$\left|\frac{A}{B}\right| = \frac{|A|}{|B|} \quad (6.384)$$

$$\left|\frac{A(\omega)}{B(\omega)}\right| = \frac{|A(\omega)|}{|B(\omega)|} \quad (6.385)$$

The magnitude of the ratio of two complex numbers, A and B in Equation 6.384, or functions $A(\omega)$ and $B(\omega)$ in Equation 6.385, equals the ratio of the magnitudes of those same two complex numbers or functions.

Apply Equations 6.383 and 6.385 to Equation 6.379 to calculate the magnitude of the RL low-pass filter current transfer function using the square root of the sum of the squares. Notice in the last term of Equation 6.383 that the complex operator j does not get squared in calculating the magnitude.

$$\left|\frac{I_{out}}{I_S}\right| = \frac{I_{out}}{I_S} = \frac{1}{|j\omega L/R + 1|} = \frac{1}{\sqrt{(\omega L/R)^2 + 1}} \quad (6.386)$$

$$|H(j\omega)| = \left|\frac{I_{out}}{I_S}\right| = \frac{1}{\sqrt{(\omega L/R)^2 + 1}} \quad (6.387)$$

Equation 6.387 is the magnitude of the RL low-pass filter current transfer function $H(j\omega)$.

6.6.5 Parallel RL Current Transfer Function Phase Calculation

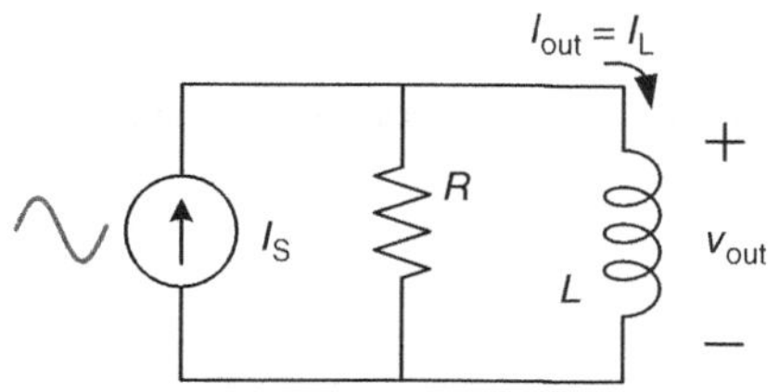

Figure 6.115 Parallel RL circuit schematic driven by sinusoidal current source.

$$\frac{\boldsymbol{I}_{out}}{\boldsymbol{I}_S} = \frac{1}{(j\omega L/R + 1)} = H(j\omega) \tag{6.388}$$

Equation 6.388 is the steady-state frequency domain current transfer function in phasor notation of the circuit in Figure 6.115.

6.6.5.1 Calculating the Phase Response of a Complex Transfer Function

The angle theta of a complex number $Z = R + jX$ (in the case of complex impedance Z), and the phase phi-of-omega of a complex function H, shown in Equation 6.389, are the arctangent of the imaginary part X over the real part R as shown in Equation 6.390.

$$H(\omega) = R(\omega) + jX(\omega) \tag{6.389}$$

$$\phi(\omega) = \tan^{-1}\frac{X(\omega)}{R(\omega)} \tag{6.390}$$

Equation 6.391 shows that the phase angle of a quotient of complex functions $F(j\omega)$ and $G(j\omega)$ is the phase angle of the numerator minus the phase angle of the denominator.

$$\angle H(j\omega) = \angle\frac{F(j\omega)}{G(j\omega)} = \frac{\angle\theta_F}{\angle\theta_G} = \angle(\theta_F) - \angle(\theta_G) \tag{6.391}$$

Apply Equations 6.390 and 6.391 to the current transfer function in Equation 6.388 to calculate the phase angle of the current transfer function in Equation 6.392.

$$\theta(\omega) = \angle H(j\omega) = \angle\left(\frac{1}{j\omega\frac{L}{R} + 1}\right) = \tan^{-1}(1) - \tan^{-1}(\omega L/R) = 0 - \tan^{-1}(\omega L/R) \tag{6.392}$$

$$\theta(\omega) = \angle H(j\omega) = -\tan^{-1}(\omega L/R) \tag{6.393}$$

Equation 6.393 is the **phase** response of the **current** transfer function in Equation 6.388 for the parallel RL low-pass filter in Figure 6.115.

$$\theta(\omega) = \angle H(j\omega) = -\tan^{-1}(\omega/\omega_0) \tag{6.394}$$

Equation 6.394 is the general form of the phase response in terms of ω_0.

$$\omega_0 = R/L \tag{6.395}$$

In Equation 6.395, ω_0 is the corner or break frequency of the series RL low-pass filter.

Figure 6.116 plots the phase response, in Equation 6.393, for the parallel RL circuit in Figure 6.115, a **lag network**.

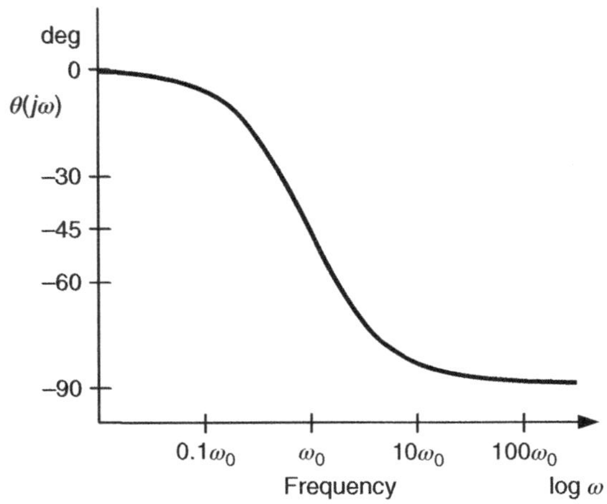

Figure 6.116 Parallel RL circuit current transfer function output phase.

6.6.6 Parallel RL Impedance Transfer Function Calculation

$$\frac{\boldsymbol{V}_{\text{out}}}{\boldsymbol{I}_{\text{S}}} = \frac{j\omega L}{1 + j\omega L/R} \tag{6.396}$$

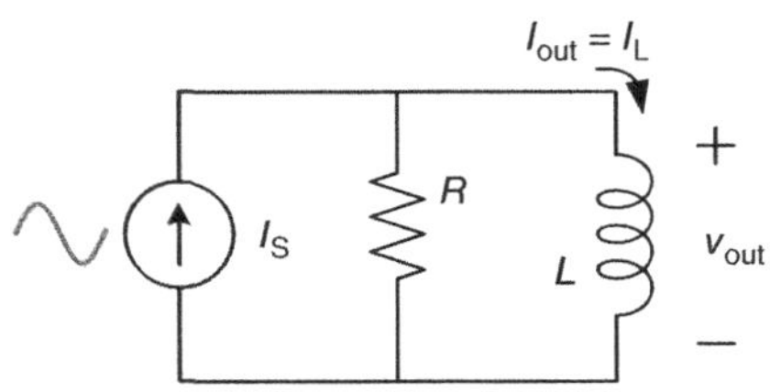

Figure 6.117 Parallel RL circuit schematic driven by sinusoidal current source.

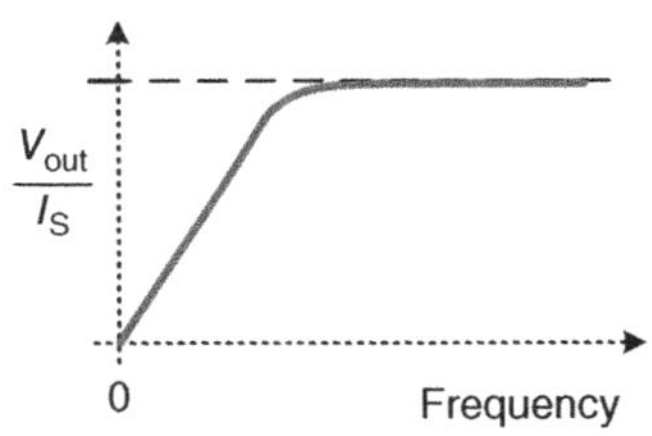

Figure 6.118 Parallel RL circuit transfer impedance magnitude frequency response.

Equation 6.396 is the impedance transfer function, also called the transimpedance, for the parallel RL circuit in Figure 6.117. Figure 6.118 plots the high-pass filter impedance transfer function magnitude.

At high frequencies, more current goes through the resistor, making $\boldsymbol{V}_{\text{out}}$ larger, making the impedance transfer function in Equation 6.396 high pass.

6.6.6.1 Calculate the RL Circuit Impedance Transfer Function $\frac{\boldsymbol{V}_{\text{out}}}{\boldsymbol{V}_{\text{S}}}$

Use a current divider approach to calculate the steady-state frequency domain transfer function, of the output voltage $\boldsymbol{V}_{\text{out}}$ over the input current $\mathbf{I}_{\text{S}}$, for the parallel RL circuit in Figure 6.117. Notice the use of j as the imaginary operator rather than i used in other mathematics and physics literature.

$$Z_{\text{L}} = j\omega L \tag{6.397}$$

Equation 6.397 gives the impedance Z_{L} of an ideal inductor.

$$Z_{\text{R}} = R \tag{6.398}$$

Equation 6.398 gives the impedance Z_{R} of an ideal resistor.

$$\boldsymbol{V}_{\text{out}} = \boldsymbol{I}_{\text{S}} Z_{\text{P}} \qquad (6.399)$$

Equation 6.399 expresses the output voltage $\boldsymbol{V}_{\text{out}}$, of the parallel *RL* schematic in Figure 6.117, using Ohm's law, where Z_{P} is the impedance of the parallel *RL* combination.

$$Z_{\text{P}} = \frac{Z_1 Z_2}{Z_1 + Z_2} \qquad (6.400)$$

Equation 6.400 shows the formula for the impedance Z_{P} of a parallel combination of two components with impedances Z_1 and Z_2.

Substitute the expression for Z_{P}, from Equation 6.400, into Equation 6.399, for Z_{P}, to get Equation 6.401 for the output voltage $\boldsymbol{V}_{\text{out}}$, changing Z_1 to Z_{R} and Z_2 to Z_{L}.

$$\boldsymbol{V}_{\text{out}} = \boldsymbol{I}_{\text{S}} Z_{\text{P}} = \frac{Z_{\text{R}} Z_{\text{L}}}{Z_{\text{R}} + Z_{\text{L}}} \qquad (6.401)$$

Divide both sides of equation by $\mathbf{I}_{\text{S}}$ to solve for the desired impedance transfer function $\boldsymbol{V}_{\text{out}}/\mathbf{I}_{\text{S}}$ in Equation 6.402.

$$\boldsymbol{V}_{\text{out}} = \boldsymbol{I}_{\text{S}} Z_{\text{P}} = I_{\text{S}} = \frac{Z_{\text{R}} Z_{\text{L}}}{Z_{\text{R}} + Z_{\text{L}}} \qquad (6.402)$$

Substitute Z_{R} and Z_{L}, from Equations 6.397 and 6.398, into Equation 6.402 to express the impedance formula in terms of *R* and *L* in Equation 6.403.

$$\frac{\boldsymbol{V}_{\text{out}}}{\boldsymbol{I}_{\text{S}}} = \frac{Z_{\text{R}} Z_{\text{L}}}{Z_{\text{R}} + Z_{\text{L}}} = \frac{Rj\omega L}{R + j\omega L} \qquad (6.403)$$

Divide Equation 6.403 by *R* to get the final expression in Equation 6.404.

$$\frac{\boldsymbol{V}_{\text{out}}}{\boldsymbol{I}_{\text{S}}} = \frac{j\omega L}{1 + j\omega L/R} \qquad (6.404)$$

Equation 6.404 is the **impedance transfer function** $\boldsymbol{V}_{\text{out}}/\mathbf{I}_{\text{S}}$ in for the parallel *RL* circuit in Figure 6.117.

$$\omega_0 = \frac{R}{L} \qquad (6.405)$$

From Equation 6.405, substitute *R* over *L* into Equation 6.404 to express Equation 6.406 in terms of ω_0.

$$\frac{\boldsymbol{V}_{\text{out}}}{\boldsymbol{V}_{\text{S}}} = \frac{j\omega L}{1 + j\omega L/R} = \frac{j\omega L}{1 + j\omega/\omega_0} = Z_{\text{L}} \frac{\boldsymbol{I}_{\text{out}}}{\boldsymbol{I}_{\text{S}}} \qquad (6.406)$$

Notice that this impedance transfer function in Equation 6.406 is also equal to the $\boldsymbol{I}_{\text{out}}/\boldsymbol{I}_{\text{S}}$ transfer function multiplied by the impedance of the inductor $j\omega L$.

6.6.7 Parallel *RL* Current Phasor Diagrams – Lag Network

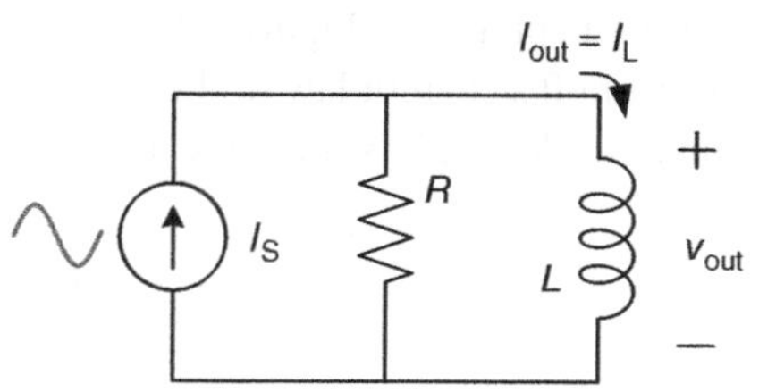

Figure 6.119 Parallel RL circuit schematic driven by sinusoidal current source.

$$\theta(\omega) = \angle H(j\omega) = -\tan^{-1}(\omega L/R) \qquad (6.407)$$

Equation 6.407, plotted in Figure 6.120, is the **phase** response of the **current** transfer function $\mathbf{I}_{\text{out}}/\mathbf{I}_{\text{S}}$ of the parallel *RL* low-pass filter current in Figure 6.119.

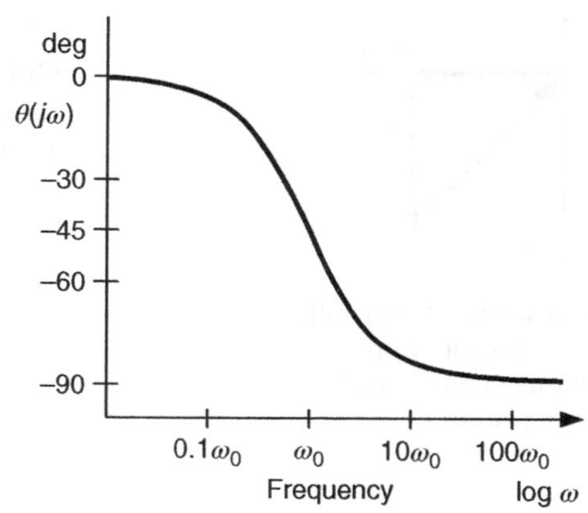

Figure 6.120 Parallel RL circuit current transfer function output phase.

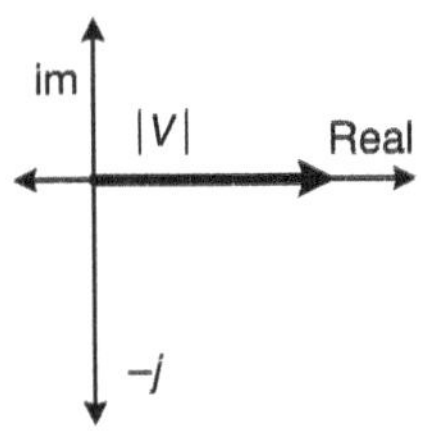

Figure 6.121 Parallel RL phasor diagram of voltage on real axis as a reference.

Use the voltage phasor **V** as the reference for zero phase, as shown in Figure 6.121, because in a parallel circuit, the voltage is the same for all the circuit elements. Plot the voltage phasor magnitude along the positive real axis. The unit for voltage is volts.

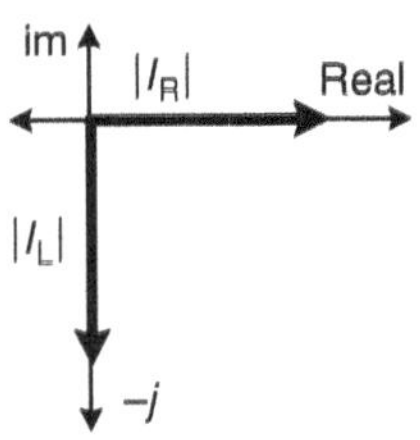

Figure 6.122 Parallel RL phasor diagram of inductor and resistor currents.

Plot the magnitude of the resistor current, $\mathbf{I}_R$, along the positive real axis in Figure 6.122 because the phase of the voltage across a resistor, relative to the current through the resistor, is zero. The currents are on a different magnitude scale from the voltage.

Plot the magnitude of the inductor current, $\mathbf{I}_L$, along the negative imaginary axis in Figure 6.122 because the current through an inductor lags the voltage across the inductor by 90°. Counterclockwise rotation is the positive phase direction.

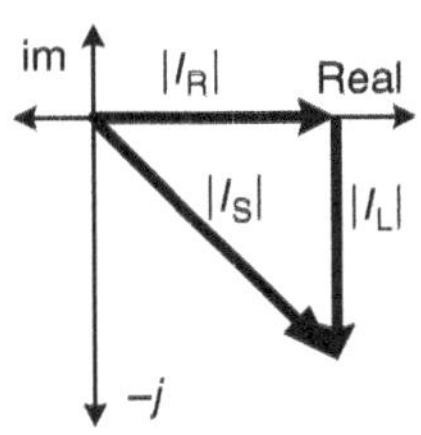

Figure 6.123 Parallel RL phasor diagram calculating source current phasor.

Set the inductor current phasor head-to-tail with the resistor current phasor for vector addition, as shown in 6.123. Connect the two phasors to get the magnitude and phase of the current phasor $\mathbf{I}_S$ through the current source.

The direction of the $\mathbf{I}_S$ phasor in Figure 6.123 expresses Kirchhoff's current law in the form of Equation 6.408.

$$|\mathbf{I}_S| = |\mathbf{I}_R| + |\mathbf{I}_L| \tag{6.408}$$

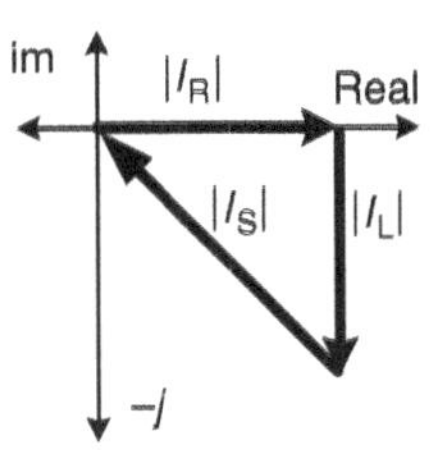

Figure 6.124 Parallel RL phasor diagram with opposite orientation of source phasor.

The opposite direction orientation of the $\mathbf{I}_S$ phasor in Figure 6.124, compared to Figure 6.123, reflects Kirchhoff's current law in the form of Equation 6.409, but the angle and magnitude result is the same as Figure 6.123.

$$|\mathbf{I}_R| + |\mathbf{I}_L| - |\mathbf{I}_S| = 0 \tag{6.409}$$

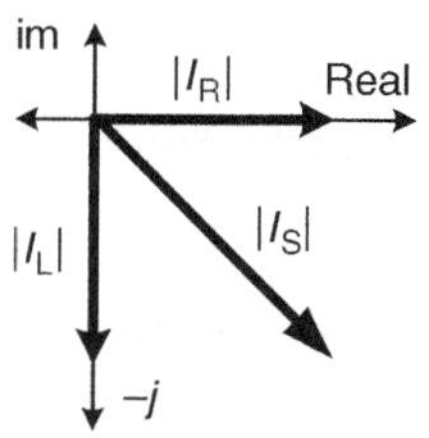

Figure 6.125 Parallel RL phasor diagram with current source phasor at −45°.

Figure 6.125 shows that the $\mathbf{I}_S$ phasor lags the resistor current phasor, in the resistor, by 45°.

In Figure 6.125, the current through the inductor is $\mathbf{I}_{out}$, which lags the $\mathbf{I}_S$ by 45°. This parallel circuit is a lag network.

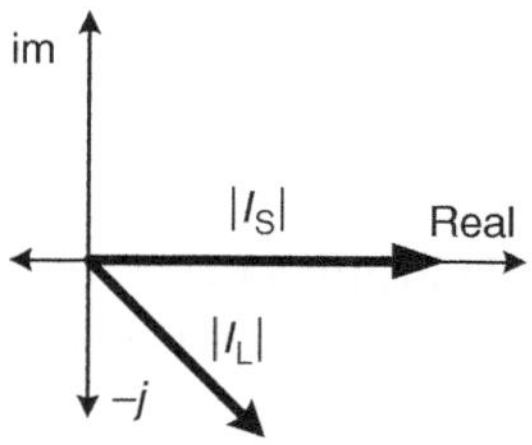

Figure 6.126 Parallel RL phasor diagram with current source phasor on real axis.

To complete the calculation of the phase of the output current $\mathbf{I}_L$ at the corner frequency, rotate the phasors in the counterclockwise (positive phase) direction by 45°, as shown in Figure 6.126. This rotation puts the $\mathbf{I}_S$ phasor on the real axis, with zero phase, making $\mathbf{I}_S$ the reference. Relative to $\mathbf{I}_S$, the phase of the output current $\mathbf{I}_{out} = \mathbf{I}_L$ is −45°.

6.7 Parallel RC High-pass Filter Frequency Response

6.7.1 Parallel RC Frequency Domain Results Overview

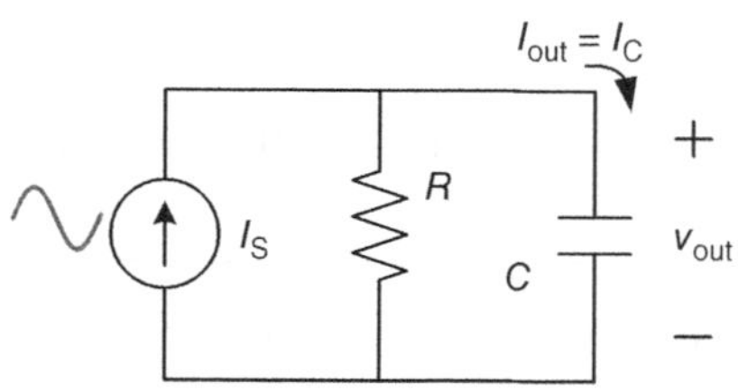

Figure 6.127 Parallel RC circuit schematic driven by sinusoidal current source.

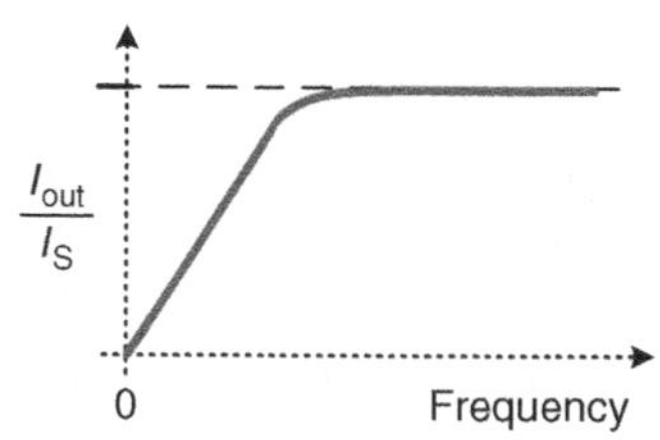

Figure 6.128 Parallel RC circuit Bode magnitude plot of current transfer function.

At high frequencies, the shunt capacitor in the schematic, in Figure 6.127, is low impedance and allows significant current to flow through the capacitor. At low frequencies, the capacitor is high impedance and doesn't allow much current to flow as output current I_{out}. Thus we know that for the current transfer function $\mathbf{I}_{out}/\mathbf{I}_S$, where the capacitor current I_C is $\mathbf{I}_{out}$, Figure 6.127 is a high-pass filter, as illustrated in Figure 6.128.

$$H(j\omega) = \frac{j\omega/\omega_0}{1 + j\omega/\omega_0} \tag{6.410}$$

Equation 6.410 is in the general format for high-pass filter transfer functions. This form is useful for easily calculating specific transfer functions.

$$\omega_0 = 1/RC \tag{6.411}$$

From Equation 6.411 for ω_0, substitute 1 over RC into the transfer function in Equation 6.410 in place of ω_0, to obtain Equation 6.412.

$$\frac{I_{out}}{I_S} = \frac{j\frac{\omega}{1/RC}}{1 + j\frac{\omega}{1/RC}} \qquad (6.412)$$

In Equation 6.412, rearrange the *RC* terms to obtain the result in Equation 6.413.

$$\frac{I_{out}}{I_S} = \frac{j\omega RC}{(1 + j\omega RC)} \qquad (6.413)$$

Equation 6.413 shows the steady-state frequency domain current transfer function, in phasor notation, for the parallel RC high-pass filter circuit in Figure 6.127.

$$|H(j\omega)| = \left|\frac{I_{out}}{I_S}\right| = \frac{\omega/\omega_0}{\sqrt{1 + (\omega/\omega_0)^2}} \qquad (6.414)$$

Equation 6.414 is in the general format for a first-order high-pass filter transfer function magnitude in terms of angular frequency ω and break frequency ω_0.

From Equation 6.411 for ω_0, substitute 1 over *RC* into the transfer function magnitude in Equation 6.414 in place of ω_0, to obtain Equation 6.415.

$$\left|\frac{I_{out}}{I_S}\right| = \frac{\omega RC}{\sqrt{1 + (\omega RC)^2}} = |H(j\omega)| \qquad (6.415)$$

For the parallel RC high-pass filter in Figure 6.127, Equation 6.415 is the magnitude of the transfer function $H(j\omega)$ in phasor notation.

$$\theta = \angle H(j\omega) = \frac{\pi}{2} - \tan^{-1}\left(\frac{\omega}{\omega_0}\right) \qquad (6.416)$$

Equation 6.416 is the general formula for a first-order high-pass filter transfer function phase response in terms of angular frequency ω and break frequency ω_0.

From Equation 6.411 for ω_0, substitute 1 over *RC* into the transfer function phase in Equation 6.416 in place of ω_0, to obtain Equation 6.417.

$$\theta = \angle H(j\omega) = \frac{\pi}{2} - \tan^{-1}\left(\frac{\omega}{1/RC}\right) = 90^\circ - \tan^{-1}(\omega RC) \qquad (6.417)$$

Equation 6.417 shows the **current** transfer function **phase** response, in phasor notation, for the series RC high-pass filter circuit in Figure 6.127.

Figure 6.129 plots the phase response, in Equation 6.417, of the current transfer function in Equation 6.413 and the schematic in Figure 6.127.

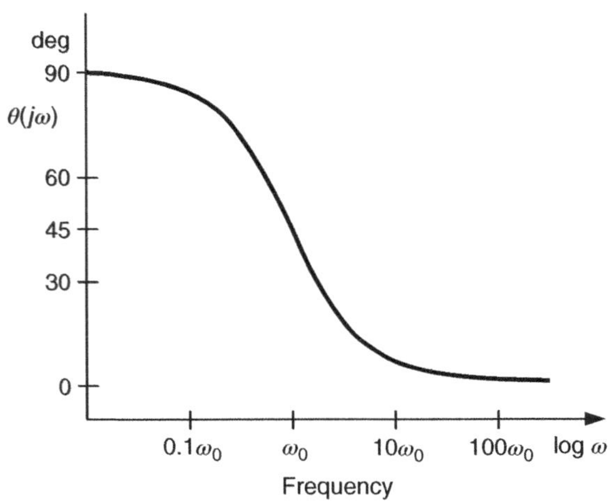

Figure 6.129 Parallel RC circuit current transfer function output phase.

6.7.2 Parallel RC Frequency Domain Example Intuitive Analysis

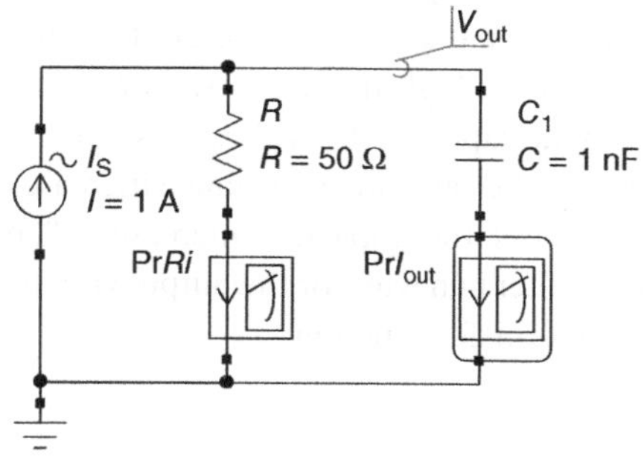

Figure 6.130 Parallel RC circuit simulation schematic driven by current source.

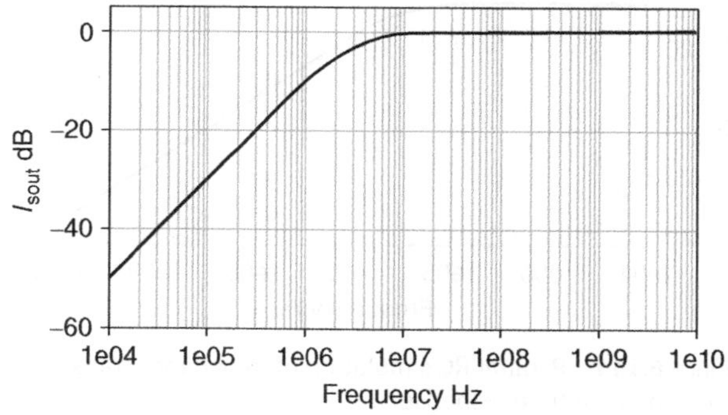

Figure 6.131 Parallel RC circuit Bode magnitude plot of current transfer function.

Thinking of the parallel RC circuit in Figure 6.130 as a current divider, the source current, labeled I_S in the simulation schematic, will split between the resistor and the capacitor according to their impedances. At high frequencies of the current source, I_S, the capacitor looks like a short circuit and almost all the source current goes through the capacitor (and not through the resistor). The capacitor current is I_{out} and the current transfer function shown in Figure 6.131 is $\mathbf{I}_{out}/\mathbf{I}_S$ in dB. Since high frequencies maximize I_{out}, this is a high-pass filter.

As the frequency of the input current I_S goes down, the capacitor impedance Z equals 1 over $j\omega C$, and increases towards infinity (open circuit). As shown in Figure 6.132, at low frequencies the output current I_{out} approaches zero, because the current through the capacitor decreases toward zero. At zero frequency (DC), the current in the resistor is I_S and the voltage across all three elements is I_S times resistance R.

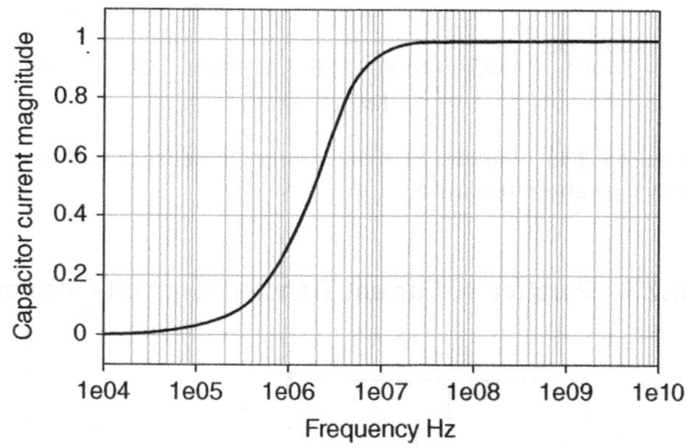

Figure 6.132 Parallel RC simulation of capacitor output current frequency response.

Figure 6.133 shows the phase of the current transfer function $\boldsymbol{I}_{out}/I_S$. At high frequencies, where the capacitive reactance is very small, the phase is near 90 degrees. As frequency decreases, the phase decreases to zero. The phase is 45° at the corner frequency ω_0.

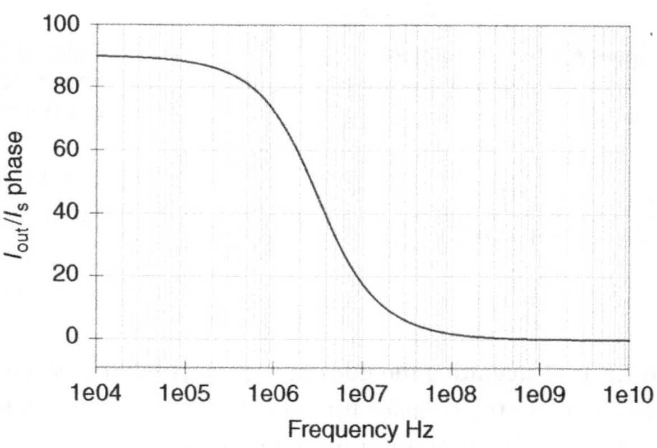

Figure 6.133 Parallel RC simulation of current transfer function phase response.

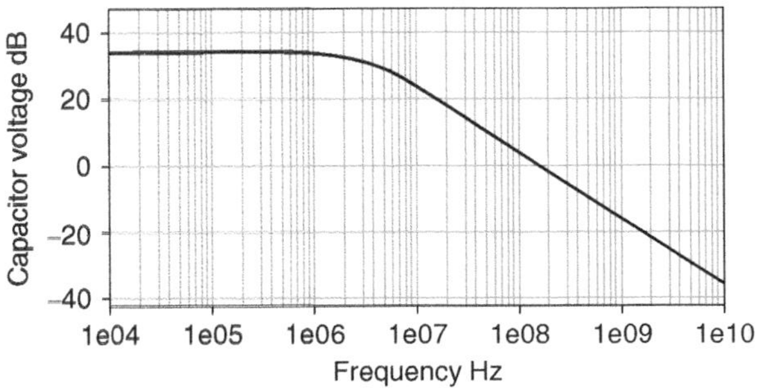

Figure 6.134 Parallel RC simulation of impedance transfer function magnitude response.

Figure 6.134 shows the impedance transfer function $\boldsymbol{V}_{\text{out}}/\boldsymbol{I}_{\text{S}}$, where V_{out} is the capacitor voltage, labeled V_{out} in the simulation schematic in Figure 6.130, the only voltage in the circuit. At high frequencies, where current flows through the capacitor, the voltage is small. As frequency decreases, the capacitive reactance increases and the voltage across the capacitor increases. The result is a low-pass filter behavior for the output voltage and the impedance transfer function.

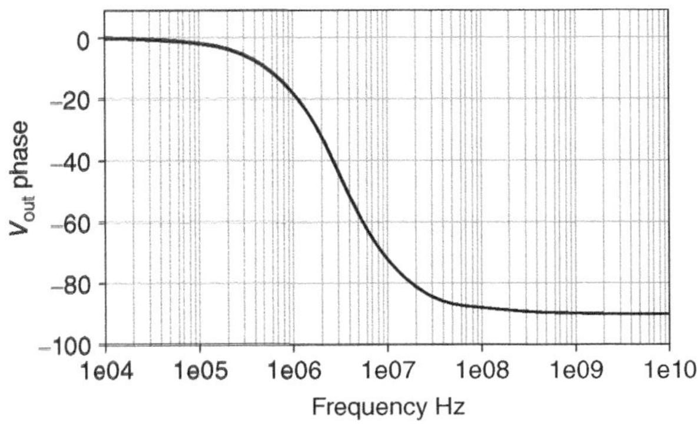

Figure 6.135 Parallel RC simulation of impedance transfer function phase response.

Figure 6.135 shows the phase of the impedance transfer function $\boldsymbol{V}_{\text{out}}/\boldsymbol{I}_{\text{S}}$, which is the opposite of the phase response of the current transfer function in Figure 6.131.

6.7.3 Parallel RC Current Transfer Function Calculation

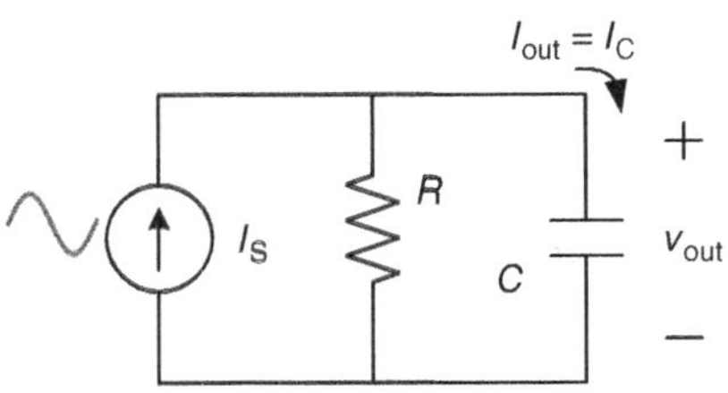

Figure 6.136 Parallel RC circuit schematic driven by sinusoidal current source.

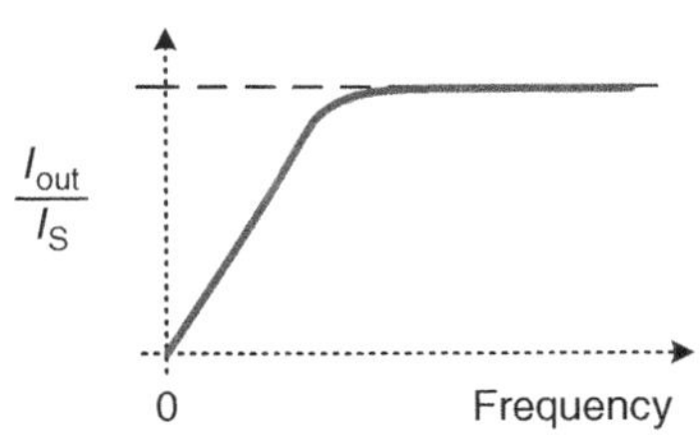

Figure 6.137 Parallel RC circuit Bode magnitude plot of current transfer function.

$$\frac{\boldsymbol{I}_{\text{out}}}{\boldsymbol{I}_{\text{S}}} = \frac{j\omega RC}{(1 + j\omega RC)} \tag{6.418}$$

For the parallel RC circuit in Figure 6.136, the current transfer function, in Equation 6.418, is plotted in Figure 6.137. Because the source I_{S} is set to 1, $\mathbf{I}_{\text{out}}$ in Figure 6.137 represents the whole current transfer function $\mathbf{I}_{\text{out}}$ over $\mathbf{I}_{\text{S}}$.

6.7.3.1 Calculating the Current High-pass Filter Transfer Function

To calculate the transfer function $\mathbf{I}_{\text{out}}/\mathbf{I}_{\text{S}}$, in Equation 6.418, start with a current divider approach. $\mathbf{I}_{\text{out}}$ equals $\mathbf{I}_{\text{C}}$ in this parallel RC circuit in Figure 6.136.

Equation 6.419 states the current divider formula in terms of impedance Z, where I_S is a source current.

$$i_1 = i_S \frac{Z_2}{(Z_1 + Z_2)} \tag{6.419}$$

Substitute into Equation 6.419 the impedance quantities for the parallel R circuit in Figure 6.136. Z_2 becomes resistance R and Z_1 becomes 1 over $j\omega C$, the impedance of the capacitor in Equation 6.420.

$$\boldsymbol{I}_{out} = \boldsymbol{I}_S \frac{R}{(1/j\omega C + R)} \tag{6.420}$$

Multiply the numerator and denominator of Equation 6.420 by $j\omega C$ to obtain Equation 6.421.

$$\boldsymbol{I}_{out} = \boldsymbol{I}_S \frac{j\omega RC}{(1 + j\omega RC)} \tag{6.421}$$

Divide both sides of Equation 6.421 by $\mathbf{I}_S$ to obtain the transfer function for $\mathbf{I}_{out}/\mathbf{I}_S$ in Equation 6.422.

$$\frac{\boldsymbol{I}_{out}}{\boldsymbol{I}_S} = \frac{j\omega RC}{(1 + j\omega RC)} \tag{6.422}$$

Equation 6.422 is the transfer function for $\mathbf{I}_{out}/\mathbf{I}_S$ for the parallel RC circuit in Figure 6.136.

6.7.4 Parallel RC Input Impedance Calculation

The impedance, Z_{in} in Equation 6.423, looking into the parallel RC circuit in Figure 6.138, is the parallel combination of the resistor impedance R and the capacitor impedance 1 over $j\omega C$.

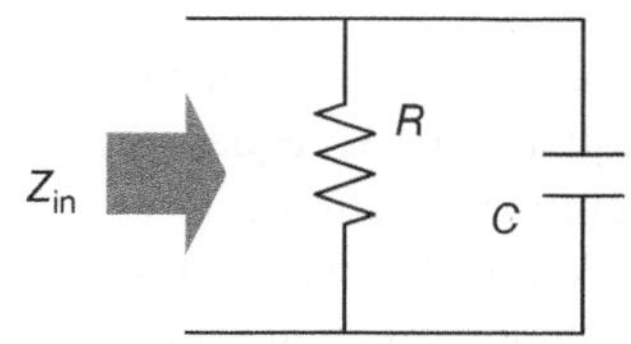

Figure 6.138 Parallel RC input impedance schematic diagram.

$$Z_{in} = Z_R \| Z_C \tag{6.423}$$

The impedance, Z_{in} in Equation 6.423, is the resistor impedance Z_R in parallel with the capacitor impedance Z_C.

$$Z_P = \frac{Z_1 Z_2}{Z_1 + Z_2} \tag{6.424}$$

Equation 6.424, from Equation 2.79, is the formula for two impedances in parallel.

Apply Equation 6.424 to Equation 6.423 to obtain Equation 6.425.

$$Z_{in} = \frac{Z_R Z_C}{Z_R + Z_C} \tag{6.425}$$

In Equation 6.425, substitute in the formulae for impedance R, of the resistor, and the impedance 1 over $j\omega C$ of the capacitor, to obtain Equation 6.235.

$$Z_{in} = \frac{R\dfrac{1}{j\omega C}}{R + \dfrac{1}{j\omega C}} \tag{6.426}$$

Multiply the top and bottom of Equation 6.235 by $j\omega C$ to obtain the result for Z_{in} in Equation 6.236.

$$Z_{in} = \frac{R}{j\omega RC + 1} \tag{6.427}$$

Equation 6.427 is the input impedance of the parallel RC circuit in Figure 6.138.

6.7.5 Parallel RC Current Transfer Function Magnitude Calculation

$$\frac{\boldsymbol{I}_{\text{out}}}{\boldsymbol{I}_S} = \frac{j\omega RC}{(1 + j\omega RC)} = H(j\omega) \tag{6.428}$$

Equation 6.428 is the current transfer function, in phasor notation, for the parallel RC circuit in Figure 6.139.

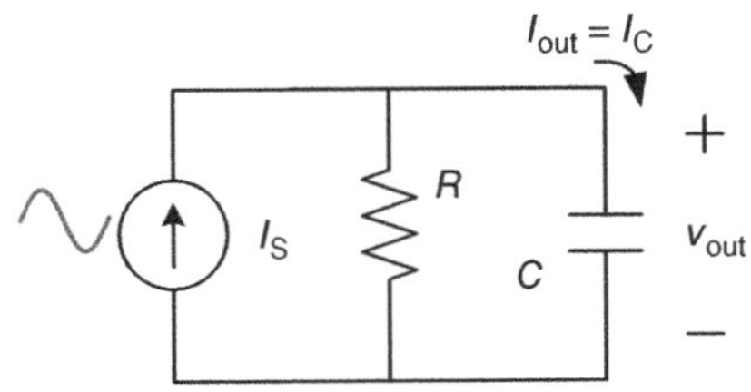

Figure 6.139 Parallel RC circuit schematic driven by sinusoidal current source.

To plot the frequency response, of the parallel RC high-pass filter in Figure 6.139, we need to calculate the magnitude and phase of the transfer function in Equation 6.428.

The magnitude of a complex function, $G(\omega)$ in Equation 6.429, is the square root of the sum of the squares of the real part $R(\omega)$ and the imaginary part $X(\omega)$, as shown in Equation 6.430.

$$G(\omega) = R(\omega) + jX(\omega) \tag{6.429}$$

$$|G(\omega)| = \sqrt{[R(\omega)]^2 + [X(\omega)]^2} \tag{6.430}$$

The parallel high-pass filter current transfer function, in Equation 6.428, is the quotient of two complex functions.

The magnitude of the ratio of two complex functions $A(\omega)$ and $B(\omega)$, in Equation 6.431, equals the ratio of the magnitudes of those same two complex functions.

$$\left|\frac{A(\omega)}{B(\omega)}\right| = \frac{|A(\omega)|}{|B(\omega)|} \tag{6.431}$$

Apply the quotient of magnitudes Equation 6.431, to the high-pass filter current transfer function in Equation 6.428, to get Equation 6.432.

$$\left|\frac{\boldsymbol{I}_{\text{out}}}{\boldsymbol{I}_S}\right| = \frac{|\boldsymbol{I}_{\text{out}}|}{|\boldsymbol{I}_S|} = \frac{|j\omega CR|}{|1 + j\omega CR|} \tag{6.432}$$

Equation 6.432 shows the magnitude of the transfer function equal to the quotient of the magnitudes of the numerator and denominators of the transfer function.

To Equation 6.432, apply the square root of the sum of the squares Equation 6.430, to get Equation 6.433. Notice that the j does not get squared.

$$\left|\frac{\boldsymbol{I}_{\text{out}}}{\boldsymbol{I}_S}\right| = \frac{\sqrt{(\omega CR)^2}}{\sqrt{1 + (\omega CR)^2}} = \frac{\omega RC}{\sqrt{1 + (\omega RC)^2}} \tag{6.433}$$

$$\left|\frac{\boldsymbol{I}_{\text{out}}}{\boldsymbol{I}_S}\right| = \frac{\omega RC}{\sqrt{1 + (\omega RC)^2}} = |H(j\omega)| \tag{6.434}$$

Equation 6.434 is the magnitude of the RC high-pass filter current transfer function $H(j\omega)$ in phasor notation.

6.7.6 Parallel RC Current Transfer Function Phase Calculation

$$\frac{\boldsymbol{I}_{\text{out}}}{\boldsymbol{I}_S} = H(j\omega) = \frac{j\omega RC}{(1 + j\omega RC)} = \frac{F(\omega)}{G(\omega)} \tag{6.435}$$

$$\angle H(j\omega) = \frac{\angle F(j\omega)}{\angle G(j\omega)} = \frac{\angle\theta_F}{\angle\theta_G} = \angle\theta_F - \angle\theta_G \tag{6.436}$$

A first-order filter transfer function, shown in Equation 6.435, can be a quotient of two complex expressions $F(\omega)$ and $G(\omega)$.

Equation 6.436 shows how the phase angle of a quotient of complex functions is the phase angle of the numerator minus the phase angle of the denominator.

$$\angle H(j\omega) = \angle F(j\omega) - \angle G(j\omega) \quad (6.437)$$

$$F = j\omega RC \quad (6.438)$$

$$G = 1 + j\omega RC \quad (6.439)$$

Equation 6.437 uses different notation than Equation 6.436 to state that the phase angle of a quotient of two complex functions, $F(\omega)$ and $G(\omega)$, equals the phase of the numerator minus the phase of the denominator.

Substitute values for the numerator, from Equation 6.438, and the denominator, Equation 6.439, into Equation 6.436 for the phase of a quotient of complex functions. The phase of the transfer function will be the phase of the numerator, $j\omega RC$, minus the phase of the denominator, the quantity 1 plus $j\omega RC$ as shown in Equation 6.440.

$$\angle H(j\omega) = \angle(j\omega RC) - \angle(1 + j\omega RC) \quad (6.440)$$

$$\theta(j\omega) = \angle(H(\omega) = -\tan^{-1}\frac{x(\omega)}{R(\omega)} \quad (6.441)$$

The phase angle theta of a complex function, like the transfer function $H(j\omega)$ in Equation 6.434, is the arctangent of the imaginary part $X(\omega)$ over the real part $R(\omega)$, as shown in Equation 6.441.

Apply the formula, in Equation 6.441, for the phase of a complex function, to both terms separately in Equation 6.440, the numerator and the denominator, as shown in Equations 6.442 and 6.443.

Numerator F

$$\angle(j\omega) = \tan^{-1}[X/R] = \tan^{-1}[j\omega RC/0] = \tan^{-1}(\infty) = \frac{\pi}{2} = 90^{\circ} \quad (6.442)$$

The phase angle of the numerator is 90° because the numerator is a purely imaginary number.

Denominator G

$$\angle(1 + j\omega RC) = \tan^{-1}[\omega RC/(1)] = \tan^{-1}(\omega RC) \quad (6.443)$$

The phase angle of the denominator is arctangent of omega times RC.

Substitute results from Equations 6.442 and 6.443 back into Equation 6.440 to obtain the phase angle of $H(j\omega)$ in Equation 6.444.

$$\angle H(j\omega) = \angle(j\omega RC) - \angle(1 + j\omega RC) = 90^{O} - \tan^{-1}(\omega RC) \quad (6.444)$$

$$\angle H(j\omega) = \frac{\pi}{2} - \tan^{-1}(\omega RC) \quad (6.445)$$

Equation 6.445, plotted in Figure 6.140, is the **phase** response of the RC high-pass filter circuit in Figure 6.139.

$$\omega_0 = 1/RC \quad (6.446)$$

Substitute the value 1 over RC from Equation 6.446 into Equation 6.446 to get the phase of the transfer function in terms of ω_0 in Equation 6.447.

$$\theta = \angle H(j\omega) = \frac{\pi}{2} - \tan^{-1}\left(\frac{\omega}{1/RC}\right) = \frac{\pi}{2} - \tan^{-1}\left(\frac{\omega}{\omega_0}\right) \quad (6.447)$$

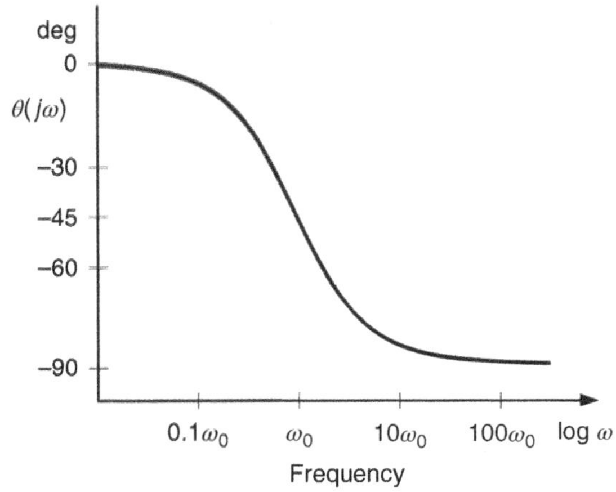

Figure 6.140 Parallel RC circuit transfer impedance phase response.

Figure 6.140 shows the phase response, from Equation 6.447, in terms of ω_0, of the RC high-pass filter circuit in Figure 6.139.

6.7.7 Parallel RC Phasor Diagrams – Lead Network

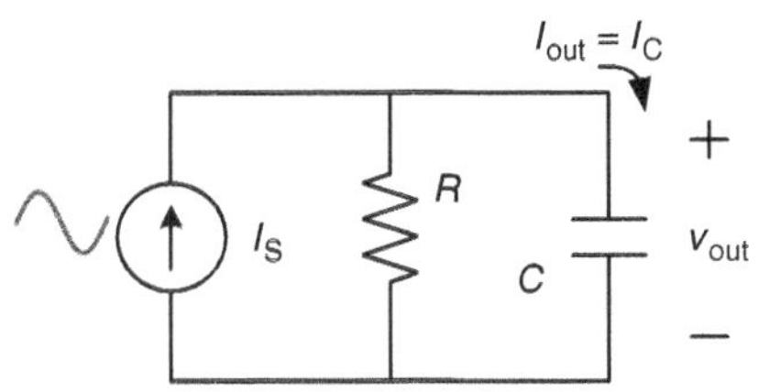

Figure 6.141 Parallel RC circuit schematic driven by sinusoidal current source.

$$\angle H(j\omega) = \frac{\pi}{2} - \tan^{-1}(\omega RC) \tag{6.448}$$

Equation 6.448 is the **phase** response of the **current** transfer function $\mathbf{I}_{out}/\mathbf{I}_S$ of the parallel RC high-pass filter circuit in Figure 6.141.

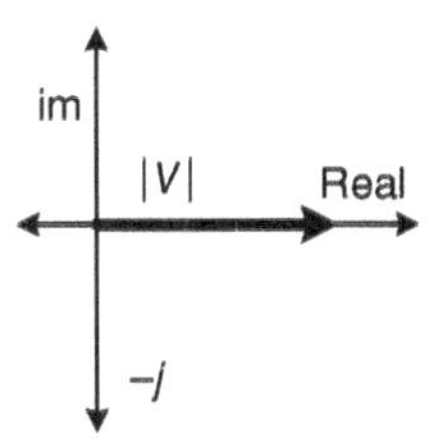

Figure 6.142 Parallel RC phasor diagram of source voltage phasor on real axis.

Use the voltage phasor **V** as the reference for zero phase, as shown in Figure 6.142, because in a parallel circuit, the voltage is the same for all the circuit elements. Plot the voltage phasor magnitude along the positive real axis. The unit for voltage is volts.

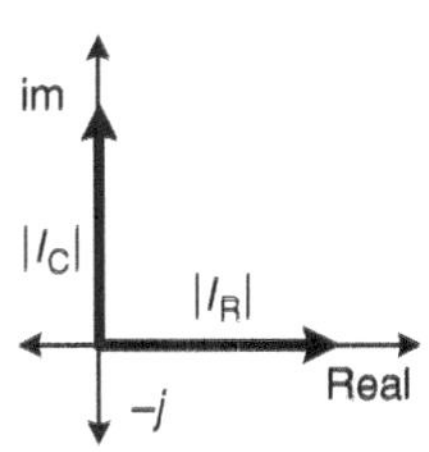

Figure 6.143 Parallel RC phasor diagram of capacitor and resistor current phasors.

Plot the magnitude of the resistor current phasor, $\mathbf{I}_R$, along the positive real axis in Figure 6.143 because the phase of the voltage across a resistor, relative to the current through the resistor, is zero. The currents are on a different magnitude scale from the voltage.

Plot the magnitude of the capacitor current phasor, $\mathbf{I}_C$, along the positive imaginary axis, as shown in Figure 6.143, because the current through a capacitor leads the voltage across the capacitor by 90°. Counterclockwise rotation is the positive phase direction.

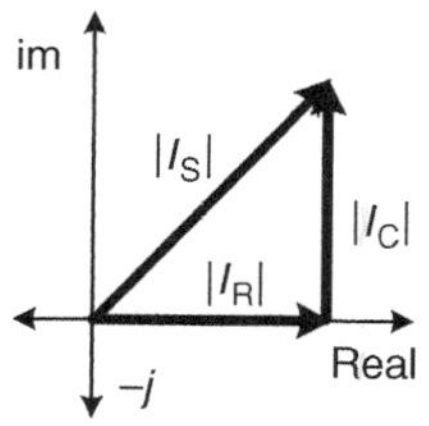

Figure 6.144 Parallel RC phasor diagram calculating source current phasor.

Set the capacitor current phasor head-to-tail with the resistor current phasor for vector addition. Connect the two phasors to get the magnitude and phase of the current phasor $\mathbf{I}_S$ through the current source, as shown in Figure 6.144.

The direction of the $\mathbf{I}_S$ phasor in Figure 6.144 expresses Kirchhoff's current law in the form of Equation 6.449.

$$|\mathbf{I}_S| = |\mathbf{I}_R| + |\mathbf{I}_C| \tag{6.449}$$

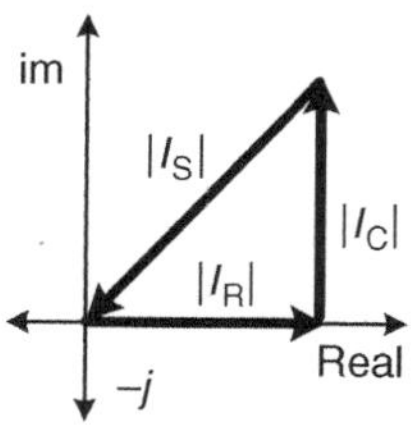

Figure 6.145 Parallel RL phasor diagram with opposite orientation of source phasor.

The opposite direction of the $\mathbf{I}_S$ phasor, in Figure 6.145 compared to Figure 6.144, reflects Kirchhoff's current law in the form of Equation 6.450, but the angle and magnitude result is the same as Figure 6.144.

$$|\mathbf{I}_R| + |\mathbf{I}_C| - |\mathbf{I}_S| = 0 \tag{6.450}$$

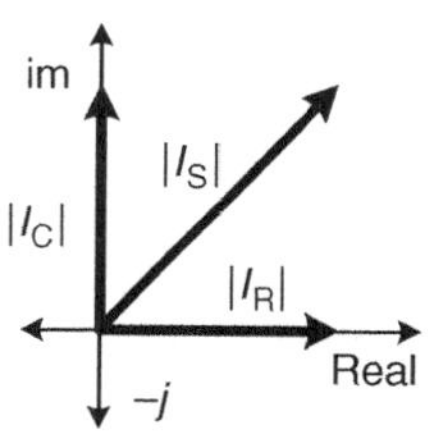

Figure 6.146 Parallel RL phasor diagram with current source phasor at 45°.

Figure 6.146 shows that the $\mathbf{I}_S$ phasor leads the resistor current phasor, in the resistor, by 45°.

The current through the capacitor is $\mathbf{I}_{out}$, which leads the source current $\mathbf{I}_S$ by 45°. This parallel circuit is a lead network.

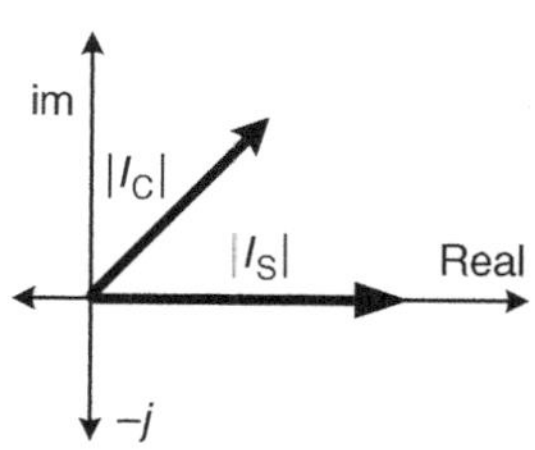

Figure 6.147 Parallel RL phasor diagram with current source phasor on real axis.

To complete the calculation of the phase of the output current $\mathbf{I}_C$ at the corner frequency, rotate the phasors in the clockwise (negative phase) direction by 45°, as shown in Figure 6.147. This rotation puts the $\mathbf{I}_S$ phasor on the real axis, with zero phase, making $\mathbf{I}_S$ the reference. Relative to $\mathbf{I}_S$, the phase of the output current $\mathbf{I}_{out}$ equals $\mathbf{I}_C$ is 45°.

7

Discharging and Charging First-Order RC and RL Circuits

Charging refers to the establishment of currents and voltages in a circuit, much like charging a battery. Charging can be thought of as how initial conditions in circuits occur, how initial conditions get put in place. Discharging refers to initial currents and voltages dissipating through or in the resistors of a circuit. For circuit analysis, it is convenient to separately consider natural response, related to discharging, and forced response, related to charging. Complete response then considers both zero-input response and zero-state response together. This chapter on charging and discharging focuses more on the switches and how they operate, results equations, and practical examples. Later chapters discuss more mathematical derivation of natural and step responses.

7.1 Discharging RC and RL Circuits – Natural Response

This discussion of discharging RC and RL circuits introduces the natural response of a circuit, also called the zero-input response (ZIR). Voltage sources and ideal switches are used to charge the circuits with voltage and currents, which will become the initial conditions of the natural response. The moment that the switch changes represents the beginning of the natural response analysis. The decaying exponential equations are derived in Section 8.2.2.

7.1.1 Discharging Circuits with Switches

The two circuits, in Figures 7.1a and 7.1b, exhibit exponentially decaying natural response behavior after the voltage sources, which charged the circuits, are removed from the circuit by the switches.

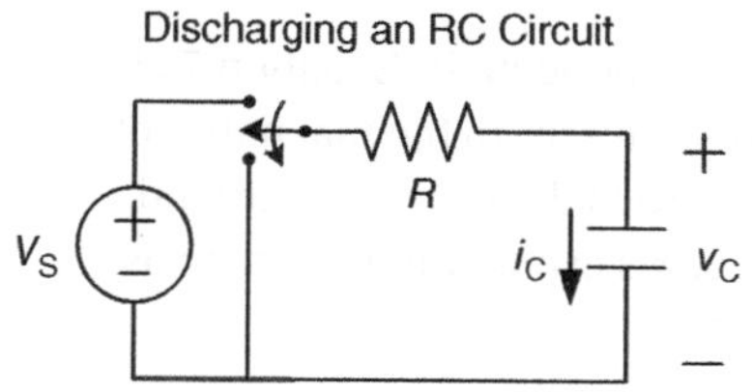

Figure 7.1a RC circuit switching to show natural response discharging.

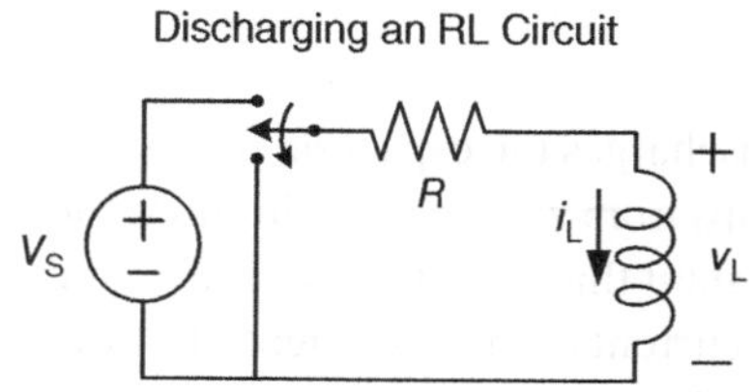

Figure 7.1b RL circuit switching to show natural response discharging.

First and Second Order Circuits and Equations: Technical Background and Insights, First Edition. Robert O'Rourke.

In Figures 7.1a and 7.1b, the voltage sources have charged the circuit for a very long time while the switches were in the upper position. Then the ideal switches move instantaneously, at time equals zero, to the lower positions, where they connect the resistors in parallel with the reactive elements. Thus the switches remove the voltage sources from the circuits, and the resulting natural response can be called zero input.

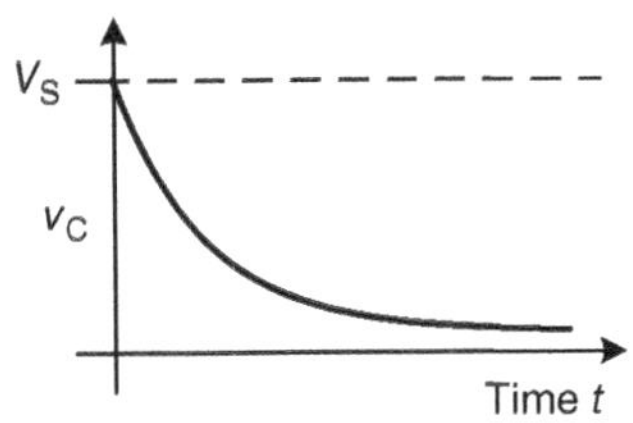

Figure 7.2a RC natural response capacitor voltage.

$$v_C = V_S e^{-t/RC} \tag{7.1a}$$

At the instant that the switch, in Figure 7.1a, connects the resistor to ground, the capacitor charge drains exponentially, as current through the resistor, as shown in Figure 7.2a. Equation 7.1a is the decaying capacitor voltage, starting from the source voltage V_S.

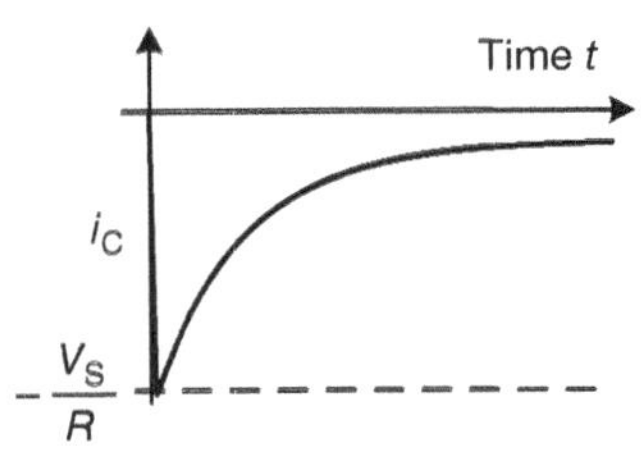

Figure 7.3a RC natural response capacitor current.

$$i_C = \frac{-V_S}{R} e^{-\frac{t}{RC}} \tag{7.2a}$$

The current that discharges the capacitor, in Equation 7.2a, jumps from zero to a maximum value of $-V_S/R$, at the instant that the switch changes, and then the capacitor current decays exponentially down to zero, as shown in Figure 7.3a.

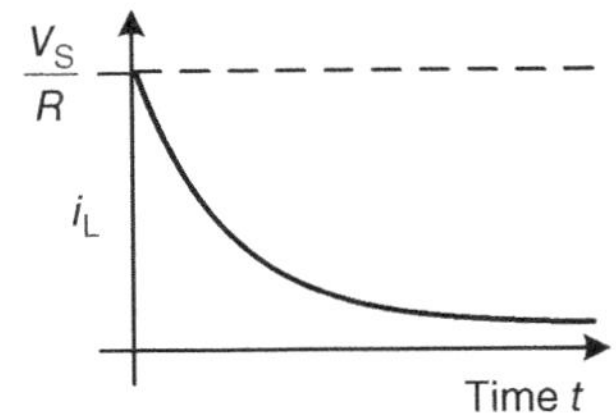

Figure 7.2b RL natural response inductor current.

$$i_L = \frac{V_S}{R} e^{-tR/L} \tag{7.1b}$$

At the instant that the switch, in Figure 7.1b, connects the resistor to ground, the inductor current dissipates in the resistor exponentially as shown in Figure 7.2b. Equation 7.1b is the decaying inductor current.

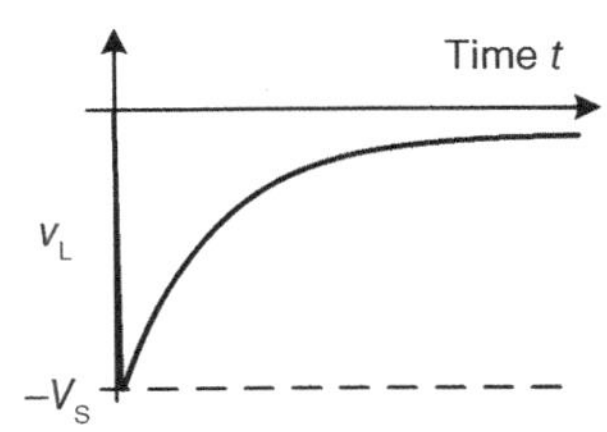

Figure 7.3b RL natural response inductor voltage.

$$v_L = -V_S e^{-\frac{R}{L}} \tag{7.2b}$$

The voltage across the inductor, in Equation 7.2b, jumps immediately from zero to the maximum of $-V_S$ when the switch connects the resistor to ground. Then this voltage decays exponentially to zero as shown in Figure 7.3b.

The rising and falling exponential curves are described by a time constant τ. For the RC circuit, the time constant is RC. For the LR circuit, the time constant is L/R.

Notice how the capacitor current makes a sudden change in value and the inductor voltage makes a sudden change in value. But the capacitor voltage does not make a sudden change and the inductor current does not make a sudden change.

7.1.2 Circuit Topologies Before Switching

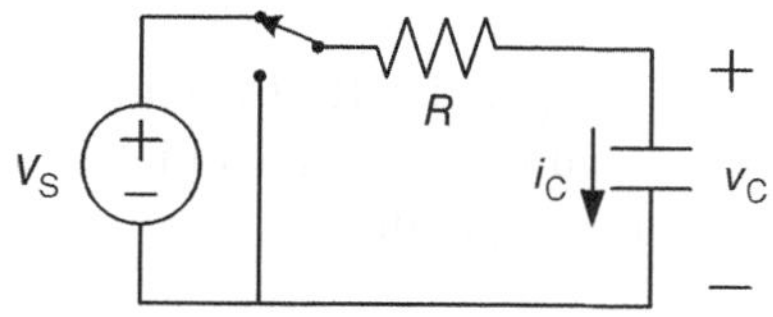

Figure 7.4a RC schematic switched to charge the capacitor voltage initial condition.

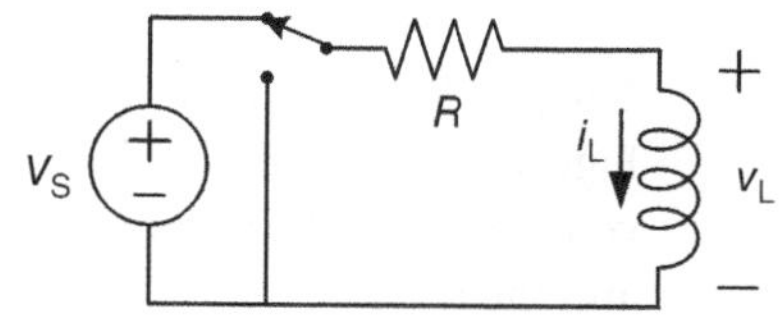

Figure 7.4b RL schematic switched to set up the inductor current initial condition.

In the schematics in Figures 7.4a and 7.4b the switches are in the upper position.

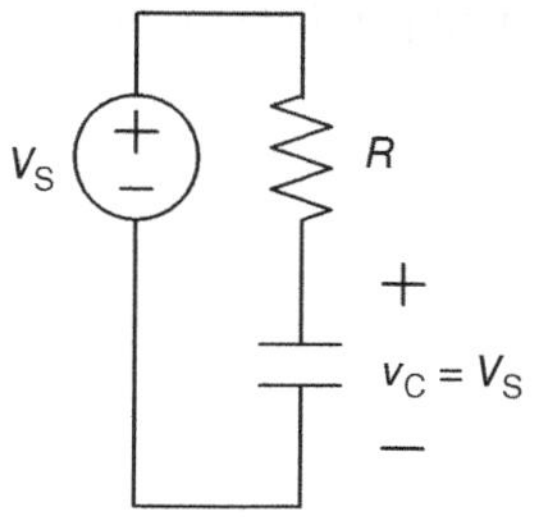

Figure 7.5a RC schematic, without switches, showing charging by voltage source.

Figures 7.5a and 7.5b show a different orientation of the circuits in Figures 7.4a and 7.4b, leaving out the switches.

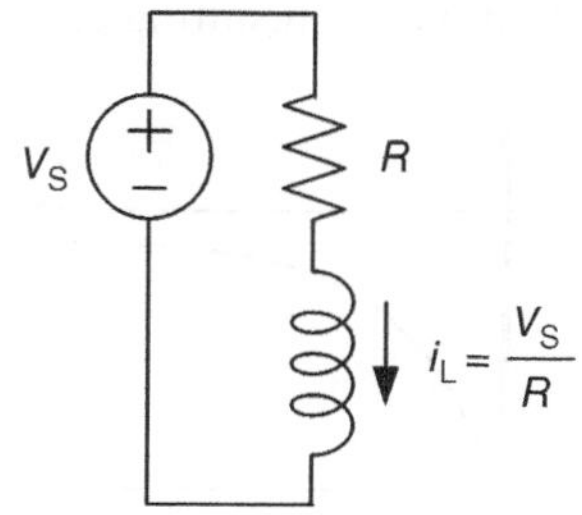

Figure 7.5b RL schematic, without switches, showing charging by voltage source.

In steady state, after a long time, in the circuit in Figure 7.5a, there is no current. All the source voltage V_S is across the capacitor.

In steady state, after a long time, in the circuit in Figure 7.5b, there is no voltage across the inductor. All the source voltage V_S is across the resistor.

7.1.3 Circuit Topologies After Switching – Natural Response

The circuits in Figures 7.6a and 7.6b have the switches in the lower positions, after being switched. As soon as the switches are thrown, Figure 7.6a becomes Figure 7.7a, and Figure 7.6b becomes Figure 7.7b.

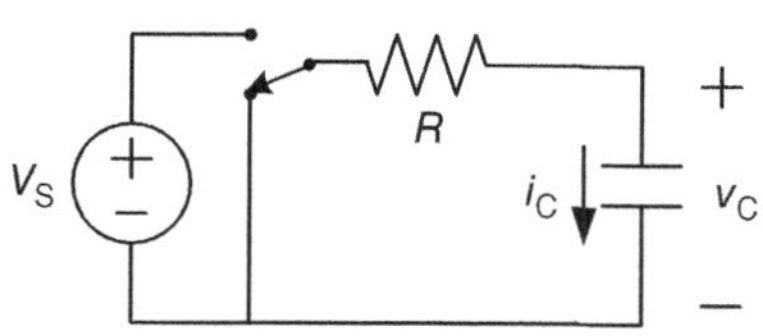

Figure 7.6a RC schematic switched for natural response discharging

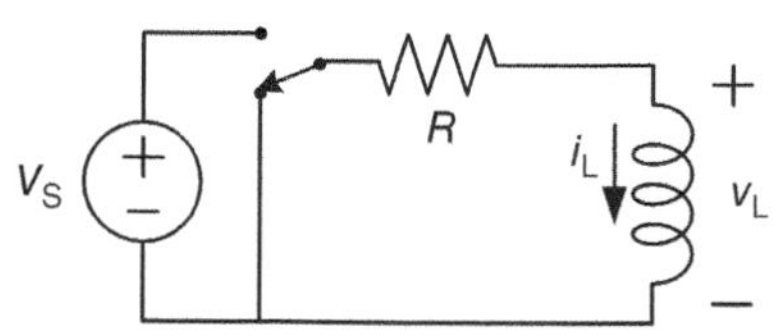

Figure 7.6b RL schematic switched for natural response discharging.

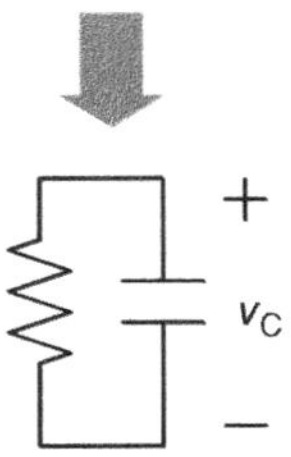

The parallel RC circuit in Figure 7.7a has an initial voltage across the capacitor and the resistor.

Figure 7.7a RC natural response circuit with switch not included.

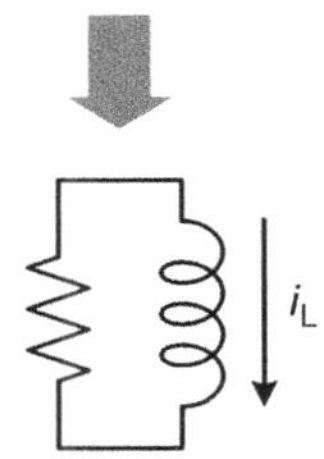

The parallel RL circuit in Figure 7.7b has an initial current in the inductor.

Figure 7.7b RL natural response circuit with switch not included.

After the switches move, from the upper position to the lower position, and the voltage sources are removed from the circuits, the resulting equivalent circuits in Figures 7.7a and 7.7b are simple two-element parallel circuits.

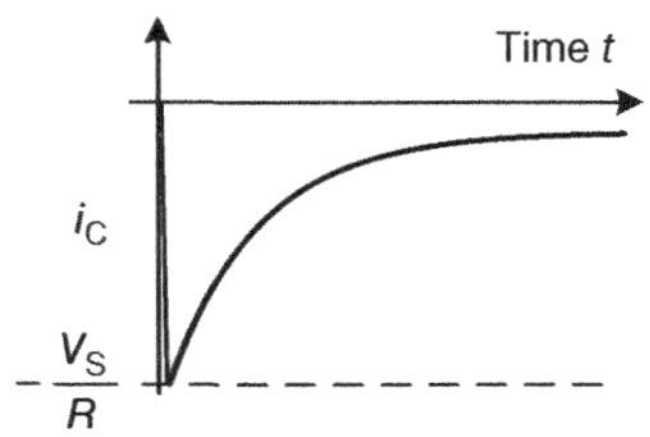

Figure 7.8a RC natural response capacitor current discharging.

$$i_C = \frac{-V_S}{R} e^{-\frac{t}{RC}} \tag{7.3a}$$

The capacitor voltage, in Figure 7.7a, causes current to flow counter-clockwise. The current flows down through the resistor and up through the capacitor. In Equation 7.3a, this current is negative, relative to v_C and neutralizes the charge accumulated on the capacitor (Figure 7.8a).

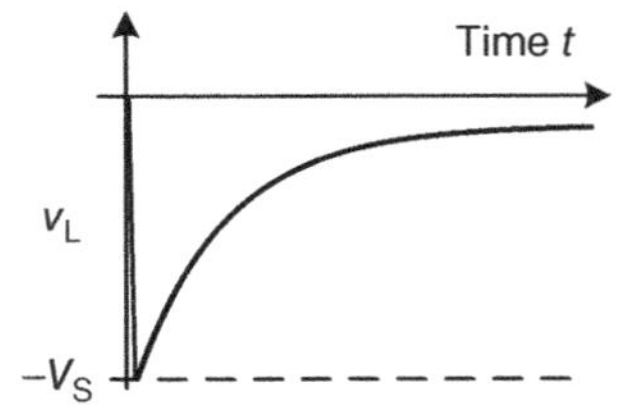

Figure 7.8b RL natural response inductor voltage decaying.

$$v_L = -V_S e^{-\frac{tR}{L}} \tag{7.3b}$$

The inductor current, in Figure 7.7b, flows clockwise, down through the inductor and then up through the resistor. The voltage drop across the resistor, shown in Equation 7.3b, also drops across the inductor and is negative relative to the marked v_L polarity (Figure 7.8b).

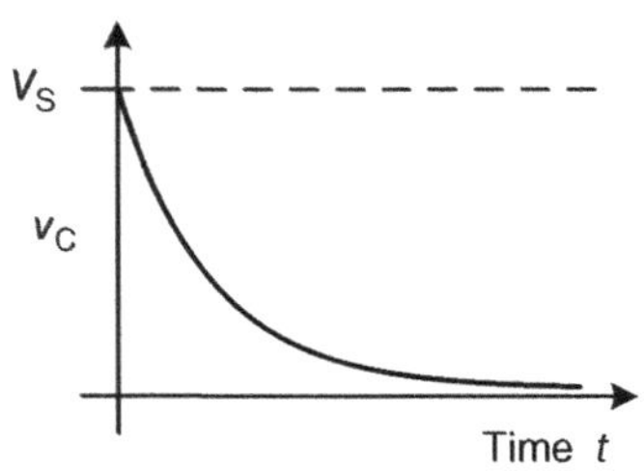

Figure 7.9a RC natural response decaying capacitor voltage.

$$v_C = V_S e^{-t/RC} \tag{7.4a}$$

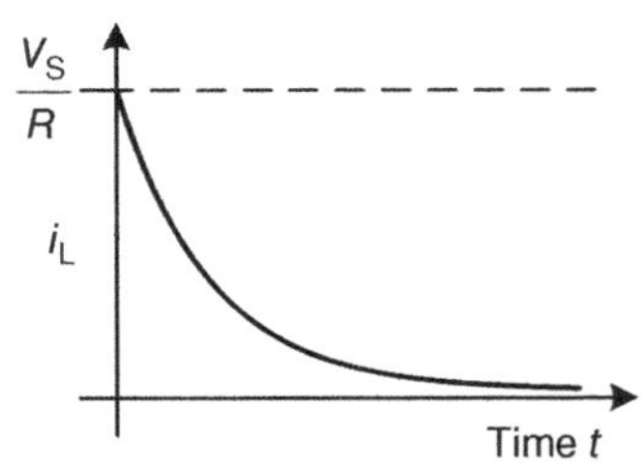

Figure 7.9b RL natural response decaying inductor current.

$$i_L = \frac{V_S}{R} e^{-tR/L} \tag{7.4b}$$

The capacitor natural response voltage, shown in Figure 7.9a and described by Equation 7.4a, decays exponentially down to zero.

The inductor natural response current, shown in Figure 7.9b and described by Equation 7.4b, decays exponentially down to zero.

7.2 Charging RC and RL Circuits – Step Response

Where discharging a circuit starts with reactive elements, capacitors and inductors, that have are already charged with an initial voltage or an initial current, charging circuits starts with the opposite assumption; currents or voltages are being applied to the circuits.

Where discharging a circuit relates to natural response, charging circuit relates to zero-state response and step response. Zero-state response (ZSR) starts with a circuit in zero state, that is, no currents nor voltages present in the circuit when the input is applied. Charging is an informal term; a circuit with voltages or currents present can be further charged or driven with an external voltage or current signal source.

7.2.1 Charging Circuits with Switched Sources

The exponential response equations shown here are derived in Chapter 9 on step response.

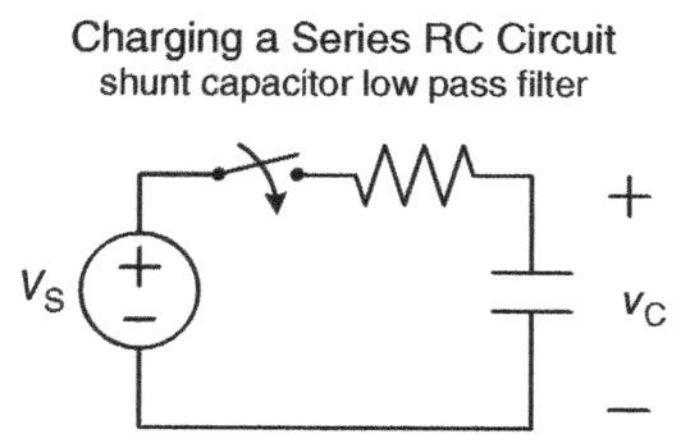

Figure 7.10a RC circuit with step response switch for charging.

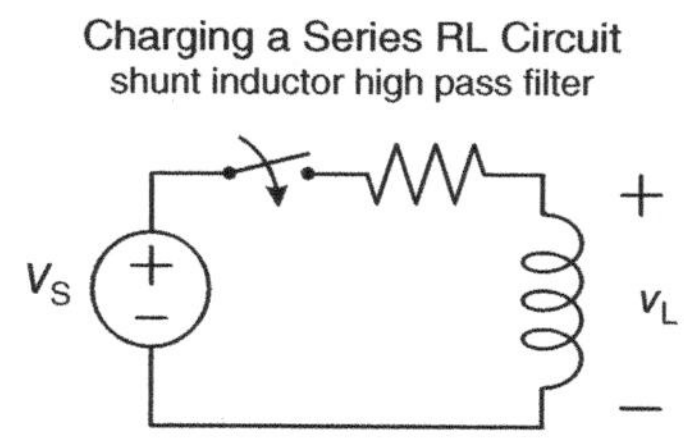

Figure 7.10b RL circuit with step response switch for charging.

Before the switches in Figures 7.10a and 7.10b close, there are no currents flowing in the resistors, capacitor, and inductor. Neither are there voltages across the resistors, capacitor, or inductor. (In Figure 7.10a and 7.11a, it is possible that an open circuit RC could have charge on the capacitor; we are not considering that case here.)

The closing switches connect the voltage sources to each circuit, as shown in Figures 7.11a and 7.11b.

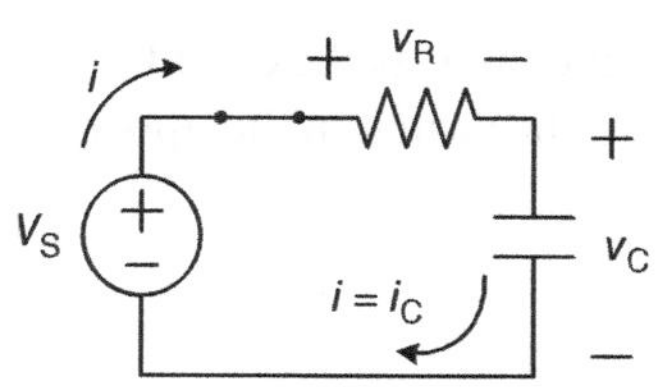

Figure 7.11a RC circuit with closed switch for step response charging.

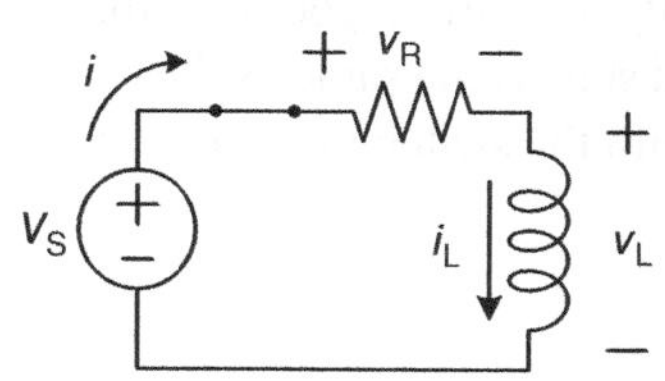

Figure 7.11b RL circuit with closed switch for step response charging.

7.2.2 Step Response Curves and Equations

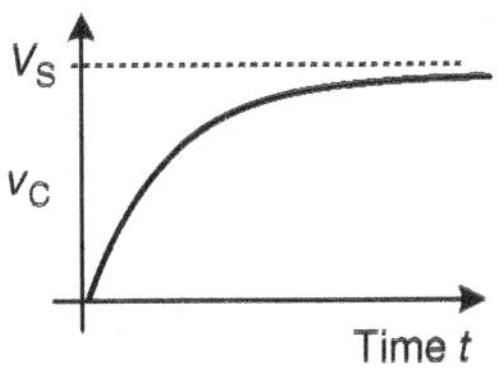

Figure 7.12a RC circuit step response capacitor voltage charging.

$$v_C = V_S\left(1 - e^{-t/RC}\right) \tag{7.5a}$$

At the instant that the switch closes, the capacitor voltage begins to charge up exponentially, until it reaches the same voltage as the source voltage V_S. The capacitor voltage is shown in Figure 7.12a and expressed in Equation 7.5a.

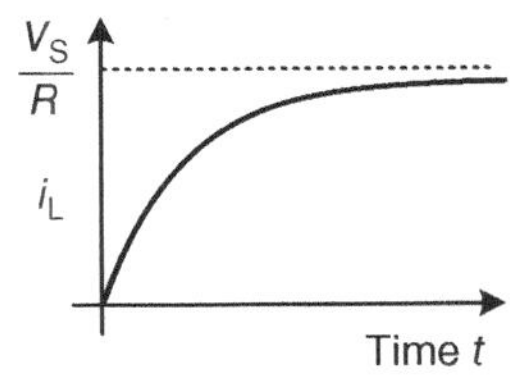

Figure 7.12b RL circuit step response inductor voltage charging.

$$i_L = \frac{V_S}{R}\left(1 - e^{-tR/L}\right) \tag{7.5b}$$

At the instant that the switch closes, the inductor current begins to build up exponentially, until it reaches the level V_S/R. The inductor current is shown in Figure 7.12b and expressed in Equation 7.5b.

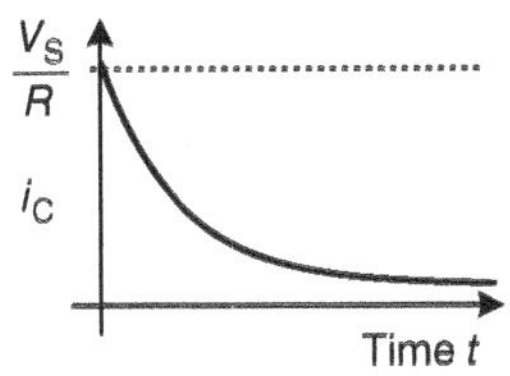

Figure 7.13a RC circuit step response capacitor current decaying.

$$i_C = \frac{V_S}{R}e^{-t/RC} \tag{7.6a}$$

As the capacitor voltage increases to V_S, the current decreases, as shown in Figure 7.13a and expressed in Equation 7.6a. Eventually the current goes to zero. Notice how this charged state is the same as the circuit in Figure 7.4a, with the switch in the upper position.

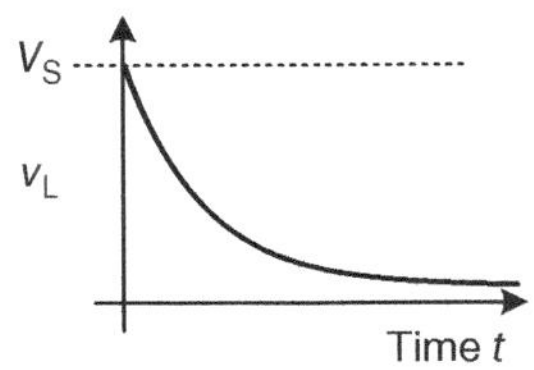

Figure 7.13b RL circuit step response inductor voltage decaying.

$$v_L = V_S e^{-tR/L} \tag{7.6b}$$

As the inductor current increases to V_S/R, the voltage across the inductor decreases, as shown in Figure 7.13b and Equation 7.6b.
Notice how this circuit is the same as the circuit shown in Figure 7.4b, with the switch in the upper position.

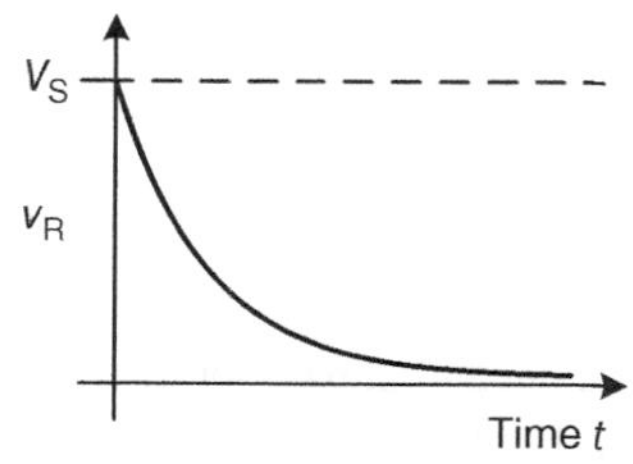

Figure 7.14a RC circuit step response resistor voltage decaying.

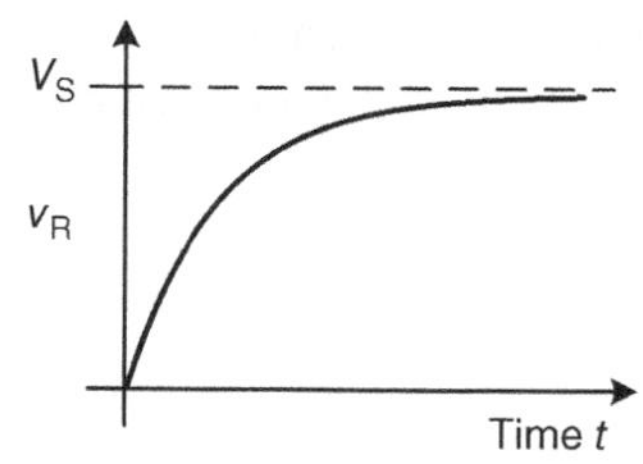

Figure 7.14b RL circuit step response resistor current increasing.

As the voltage across the capacitor increases, the voltage across the resistor decreases, as shown in Figure 7.14a.

As the current through the inductor increases, the voltage across the resistor also increases as shown in Figure 7.14b.

7.3 The Exponential Time Constant τ (Tau)

By introducing the time constants tau (τ), the first-order decaying exponentials can be described more generally than R, C, and L terms, as shown in Equations 7.7a and 7.7b. The corner frequency ω_0 is the inverse of τ, as shown in Equations 7.8a and 7.8b.

$$\tau = RC \tag{7.7a}$$

Equation 7.7a shows tau for the RC circuit.

$$\tau = R/L \tag{7.7b}$$

Equation 7.7b shows tau for the RL circuit.

$$\omega_0 = \frac{1}{\tau} = \frac{1}{RC} \tag{7.8a}$$

Equation 7.8a shows ω_0 for the RC circuit.

$$\omega_0 = \frac{1}{\tau} = \frac{R}{L} \tag{7.8b}$$

Equation 7.8b shows ω_0 for the RL circuit.

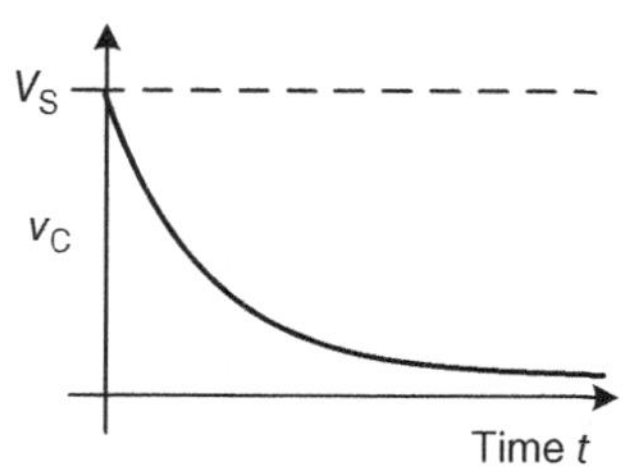

Figure 7.15a RC natural response decaying capacitor voltage.

$$v_C = V_S e^{-t/RC} = V_S e^{-t/\tau} \tag{7.9a}$$

Figure 7.15a, the same as Figure 7.2a, shows the decaying exponential response of the RC circuit, with the switch closed, in Figure 7.11a. Equation 7.9a expresses the decaying voltage in terms of t over tau.

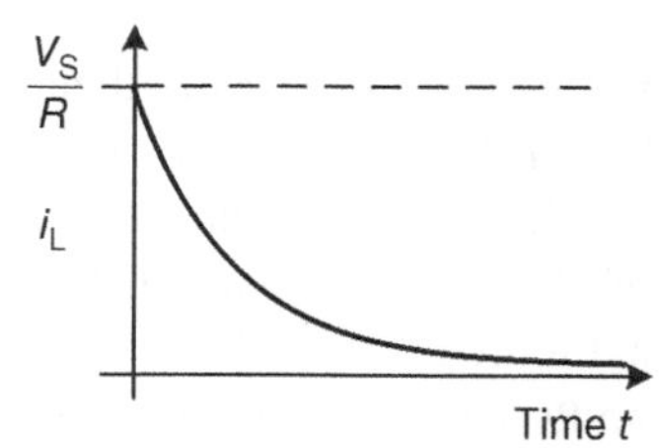

Figure 7.15b RL natural response decaying inductor current.

$$i_L = \frac{V_S}{R} e^{-tR/L} = \frac{V_S}{R} e^{-t/\tau} \tag{7.9b}$$

Figure 7.15b, the same as Figure 7.2b, shows the decaying exponential response of the RL circuit, with the switch closed, in Figure 7.11b. Equation 7.9b expresses the decaying current in terms of t over tau.

When t = tau, the exponent of e is negative one, and e-to-the-minus-one (e^{-1}) equals about 0.368. At time $t = \tau$ the function (e.g., current or voltage) has decayed about 63%.

$$e^{-t/RC} = e^{-t/\tau} \tag{7.10}$$

$e^{-1} \cong 0.368$ $\qquad e^{-2} \cong 0.135$ $\qquad e^{-3} \cong 0.05$

$$f(t) = Ae^{-t/\tau} \tag{7.11}$$

Figure 7.16 shows a generic decaying exponential function, from Equation 7.11, with the horizontal time axis marked in terms of tau (τ).

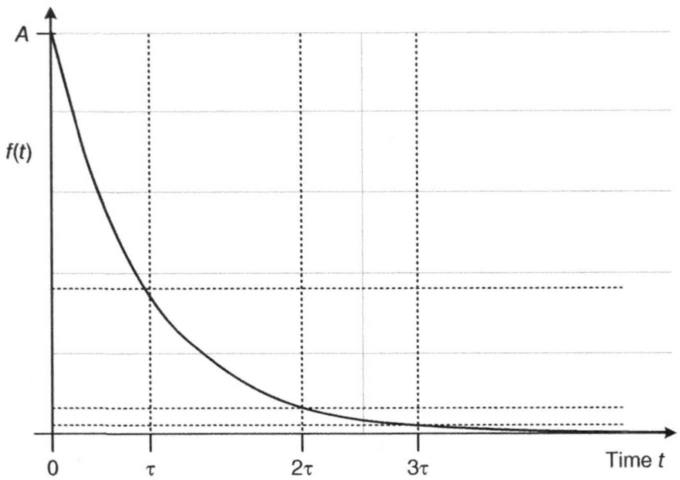

Figure 7.16 Graph of exponential function decay marked in units of time constant tau.

7.4 Pulse Train Time Constants Simulation Example

A voltage pulse train, driving an RC circuit, demonstrates charging, discharging, and the effect of the time constant τ, which equals RC for this circuit. The higher the value of the capacitor, the longer the time constant tau, and the more slowly the circuit responds to the rising and falling edges of the pulse train signal source.

Figure 7.16 shows an RC circuit driven by a 1-V pulse train with a pulse width of 1 ms and a period of 2 ms. The circuit topology is equivalent to Figure 7.10a, where v_C across the capacitor is comparable to V_{out} here in Figure 7.17.

$$v_C = V_S \left(1 - e^{-t/RC}\right) \tag{7.12}$$

Equation (7.12) (the same as Equation 7.5a) describes the capacitor voltage rising after the rising edge of the source voltage.

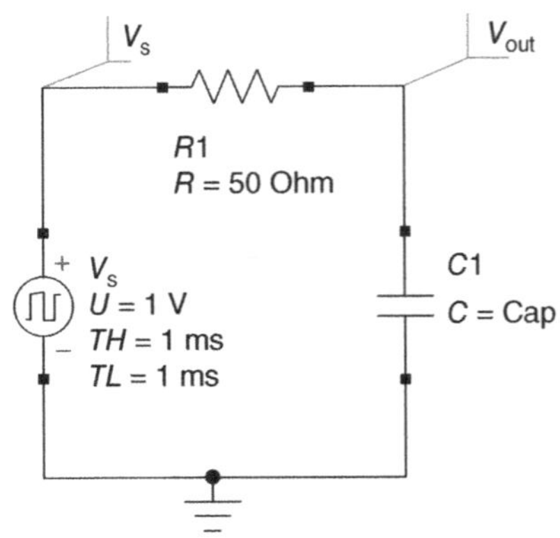

Figure 7.17 RC circuit QUCS simulation schematic with pulse train voltage input.

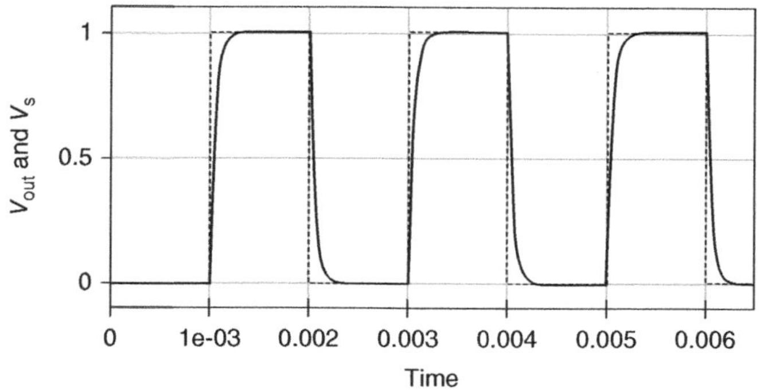

Figure 7.18 RC circuit pulse train simulation voltage response with 1 micro-Farad capacitance.

In Figures 7.18, 7.19, and 7.20, the dashed-line square wave signal is the source voltage V_S. The solid lines are the capacitor voltage rising after the rising edge of the source voltage.

When the capacitance value "Cap" equals 1 μF, RC equals 50 Ω times 1 μF, which equals 50 μs or 0.05 ms, shown in Figure 7.18.

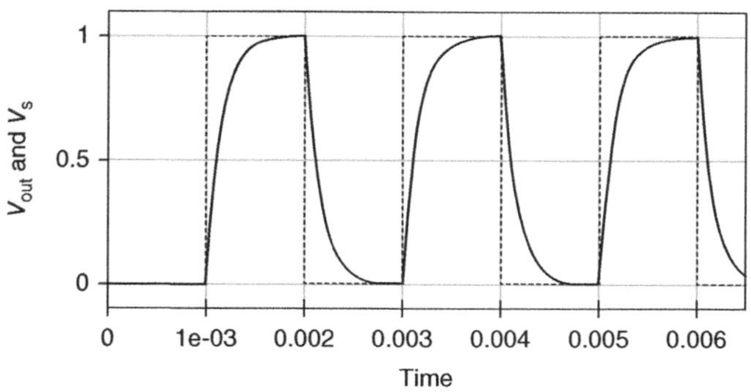

Figure 7.19 RC circuit pulse train simulation voltage response with 3 micro-Farad capacitance.

When the capacitance value "Cap" equals 3 μF, *RC* equals 50 Ω times 1 μF, which equals 150 μs = or 0.15 ms, as shown in Figure 7.19. 3τ is 450 μs or 0.45 ms. The V_{out} signal has moved vertically most of the distance within half the pulse width of 500 μs.

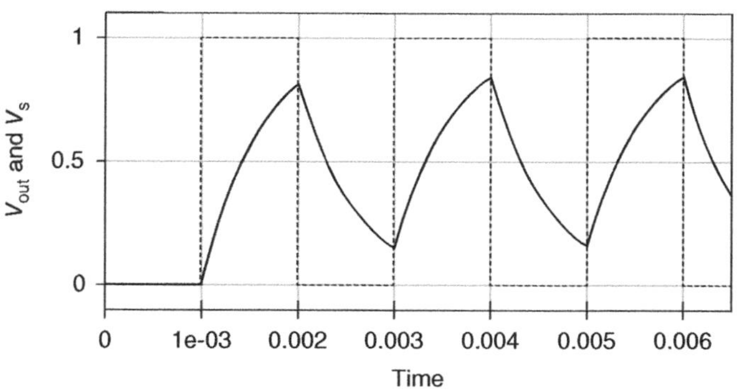

Figure 7.20 RC circuit pulse train simulation voltage response with 12 micro-Farad capacitance.

When the capacitance equals 12 μF, 2τ equals 1.2 ms, which is greater than the pulse width of 1 ms, as shown in Figure 7.19. The solid line is V_{out} and the dashed line is V_S.

8

Natural Response of RC and RL Circuits

8.1 RC and RL Circuits Natural Response Summary

Natural response is the circuit response behavior that reflects only the nature of the circuit, and not the nature of any input. Natural response is also called zero-input response (ZIR). Natural response comes from initial conditions in the circuit, like initial currents in inductors or initial voltages or charge on capacitors.

RC natural response

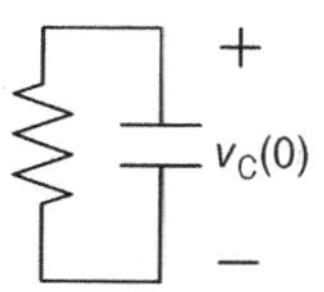

$$C\frac{dv(t)}{dt} + \frac{v(t)}{R} = 0 \quad (8.1a)$$

Figure 8.1a Parallel RC circuit with initial capacitor voltage.

RL natural response

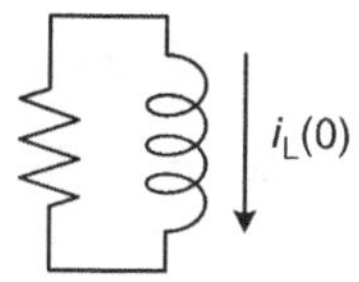

$$L\frac{di(t)}{dt} + i(t)R = 0 \quad (8.1b)$$

Figure 8.1b Parallel RL circuit with initial inductor current.

The natural response of an RL or RC circuit with no sources is described by a homogeneous linear differential equation. The initial conditions $v_C(0)$ and $i_L(0)$ appear in the schematics in Figures 8.1a and 8.1b as well as in Equations 8.2a and 8.2b.

$$v(t) = v_C(0)\,e^{\frac{-t}{RC}} \quad (8.2a)$$

$$i(t) = i_L(0)\,e^{\frac{-tR}{L}} \quad (8.2b)$$

The natural response of the RC circuit in Figure 8.1a is an initial capacitor voltage $v_C(0)$ draining current away exponentially through the resistor, discharging the capacitor to zero, described by Equation 8.2a.

The natural response of this RL circuit in Figure 8.1b is an initial inductor current $i_L(0)$ dissipating exponentially to zero through the resistor, described in Equation 8.2b.

First and Second Order Circuits and Equations: Technical Background and Insights, First Edition. Robert O'Rourke.

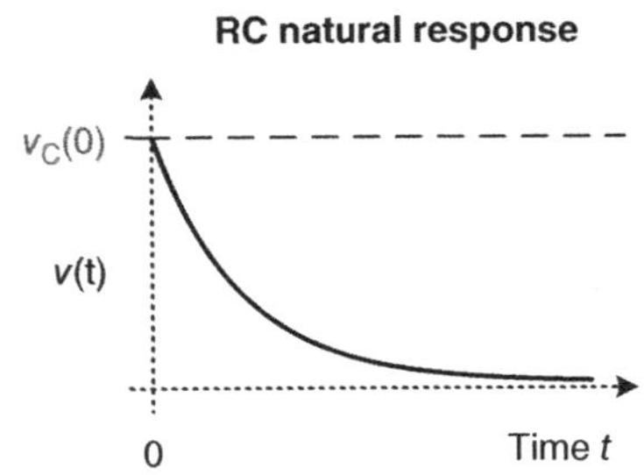

Figure 8.2a RC natural response capacitor voltage decaying from initial condition value.

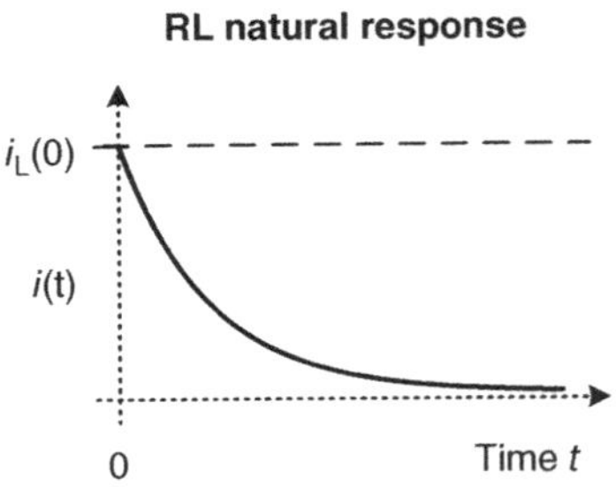

Figure 8.2b RL natural response inductor current decaying from initial condition value.

Figure 8.2a shows the natural response voltage $v(t)$, across the capacitor and resistor in Figure 8.1a, decaying exponentially from the initial condition amount $v_C(0)$ to zero, according to Equation 8.2a.

Figure 8.2b shows the natural response current $i(t)$, through the inductor and resistor in Figure 8.1b, decaying exponentially from the initial condition amount $i_L(0)$ to zero, according to Equation 8.2b.

8.2 RC and RL Natural Response Derivation

8.2.1 Two Single-Pole Single-Throw Switches for Natural Response

The circuits in Figures 8.3a and 8.3b generate the initial conditions for analyzing first-order natural response. Before time t equals 0, the two switches on the left side of each circuit are closed, connecting the sources to the reactive elements. Before time t equals 0, the two switches on the right are open, disconnecting the resistors from the reactive elements.

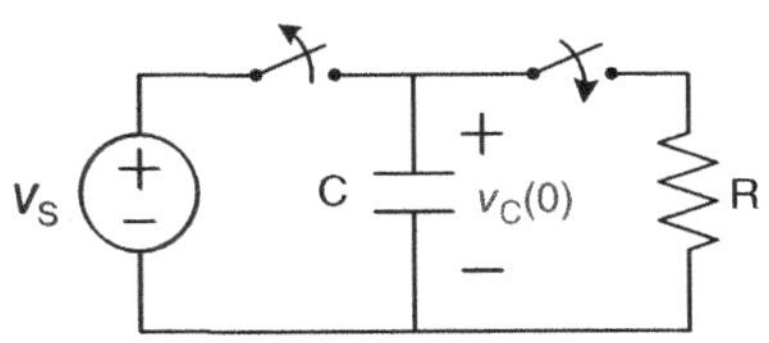

Figure 8.3a Parallel RC circuit schematic with switches to model natural response.

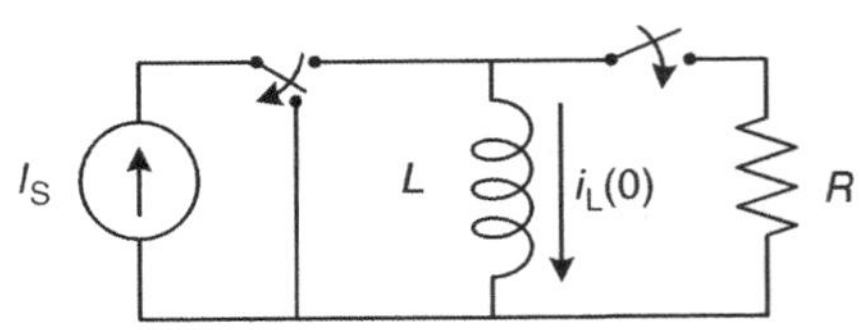

Figure 8.3b Parallel RL circuit schematic with switches to model natural response.

In Figure 8.3a the voltage source V_s charges the capacitor C to an initial voltage $V_C(0) = V_s$. At time t equals 0, the switch on the left side opens and removes the voltage source from the circuit. Also at t equals 0 the right-hand switch connects the capacitor to the resistor, and we see the natural response of the RC circuit based on the initial capacitor voltage $V_C(0)$.

In Figure 8.3b the current source circulates a current I_s through the inductor L establishing the initial condition $i_L(0)$. At time t equals 0, the switch on the left side removes the current source from the circuit. At the same time t equals 0, the right-hand switch closes and connects the inductor L to the resistor R, and we see the natural response of the RL circuit based on the initial inductor current $i_L(0)$.

After the switches in Figures 8.3a and 8.3b are thrown at time $t = 0$, the circuits in Figures 8.3a and 8.3b change to the parallel RC and RL topologies shown in Figures 8.4a and 8.4b.

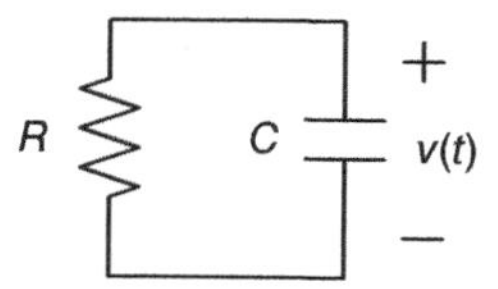

Figure 8.4a Parallel RC circuit marked with $v(t)$ corresponding to natural response in Figure 8.2a.

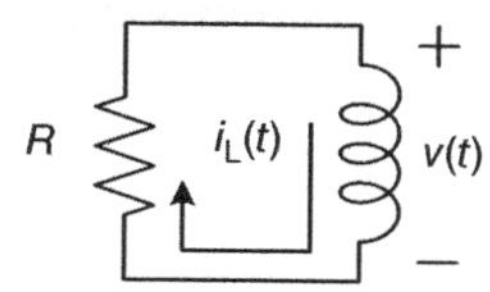

Figure 8.4b Parallel RL circuit marked with $i_L(t)$ corresponding to $i(t)$ in Figure 8.2b.

Starting with initial condition voltage $v_C(0)$ across the capacitor in Figure 8.4a, the parallel RC circuit exhibits natural response, and the voltage $v_C(0)$ becomes $v(t)$ as it decays exponentially over time.

Starting with the initial condition current $i_L(0)$ through the inductor in Figure 8.4b, the parallel RL circuit exhibits natural response and the current $i_L(0)$ becomes $i(t)$ as it decays exponentially over time.

8.2.2 RC and RL ZIR Differential Equation Derivation

Parallel RC natural response

Write Kirchhoff's current law (KCL) for the RC circuit in Figure 8.4a, summing the currents going out of the top node, as shown in Equation 8.3a.

$$i_C(t) + i_R(t) = 0 \tag{8.3a}$$

Equation 8.4a shows the current through a capacitor as a function of the voltage across the capacitor.

$$i_C = C\frac{dv}{dt} \tag{8.4a}$$

Substitute Equation 8.4a, for the capacitor current i_C, and Ohm's law, for the resistor current, into KCL Equation 8.3a to get Equation 8.5a.

$$C\frac{dv(t)}{dt} + \frac{v(t)}{R} = 0 \tag{8.5a}$$

In Equation 8.5a, move the resistor current term to the right-hand side to obtain Equation 8.6a.

$$C\frac{dv(t)}{dt} = -\frac{v(t)}{R} \tag{8.6a}$$

Put all the voltage terms from Equation 8.6a on the left side to get Equation 8.7a.

$$\frac{dv(t)}{v(t)} = -\frac{1}{RC}dt \tag{8.7a}$$

Parallel RL natural response

Write Kirchhoff's voltage law (KVL) for the RL circuit in Figure 8.4b, summing the voltages clockwise around the loop, as shown in Equation 8.3b.

$$v_L(t) + v_R(t) = 0 \tag{8.3b}$$

Equation 8.4b shows the voltage across an inductor as a function of the current through the inductor.

$$v_L = L\frac{di}{dt} \tag{8.4b}$$

Substitute Equation 8.4b for the inductor voltage v_L, and Ohm's law, for the resistor voltage, into KVL Equation 8.3b to get Equation 8.5b.

$$L\frac{di(t)}{dt} + i(t)R = 0 \tag{8.5b}$$

In Equation 8.5b, move the resistor voltage term to the right-hand side to obtain Equation 8.6b.

$$L\frac{di(t)}{dt} = -i(t)R \tag{8.6b}$$

Put all the current terms from Equation 8.6b on the left side to get Equation 8.7b.

$$\frac{di(t)}{i(t)} = -\frac{R}{L}dt \tag{8.7b}$$

In mathematical terms, the differential equation solution process shown here is called separation of variables.

8.2.3 RC and RL ZIR Differential Equation Solutions

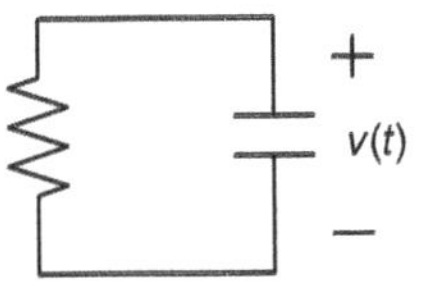

Figure 8.5a Parallel RC circuit marked with $v(t)$ corresponding to natural response in Figure 8.2a.

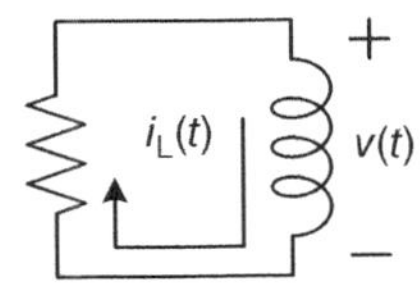

Figure 8.5b Parallel RL circuit marked with $i_L(t)$ corresponding to $i(t)$ in Figure 8.2b.

Integrate Equation 8.7a

$$\int \frac{dv(t)}{v(t)} = -\frac{1}{RC}\int dt \tag{8.8a}$$

Integrate Equation 8.7b

$$\int \frac{di(t)}{i(t)} = -\frac{R}{L}\int dt \tag{8.8b}$$

In order to perform the integration on these expressions in Equations 8.8a and 8.8b we first note that the expression on the left is a form for which we can find a formula in an integral table.

$$\int \frac{dx}{x} = \ln x + C \tag{8.9}$$

In Equation 8.9, the integral of 1 over x dx is the natural logarithm of x plus a constant C.

Apply Equation 8.9 to integrate Equation 8.8a and obtain Equation 8.10a for the natural logarithm of the voltage v in the RC circuit.

$$\ln v(t) = -\frac{1}{RC}t + K \tag{8.10a}$$

Apply Equation 8.9 to integrate Equation 8.8b and obtain Equation 8.10b for the natural logarithm of the current i in the RL circuit.

$$\ln i(t) = -\frac{R}{L}t + K \tag{8.10b}$$

In Equation 8.10a, K is an undetermined constant. Express K as the natural logarithm of k, also an undetermined constant, as shown in Equation 8.11a.

$$\ln v(t) = -\frac{1}{RC}t + \ln k \tag{8.11a}$$

In Equation 8.10b, K is an undetermined constant. Express K as the natural logarithm of k, also an undetermined constant, as shown in Equation 8.11b.

$$\ln i(t) = -\frac{R}{L}t + \ln k \tag{8.11b}$$

$$\ln e^{x} = x \tag{8.12}$$

Equation 8.12, from the definition of the natural logarithm, expresses a variable x as the natural log of an exponential.

Apply Equation 8.12 to the middle term of Equation 8.11a, making all the terms in Equation 8.11a natural log terms, as shown in Equation 8.13a.

$$\ln v(t) = \ln e^{\frac{-t}{RC}} + \ln k \tag{8.13a}$$

Apply Equation 8.12 to the middle term of Equation 8.11b, making all the terms in Equation 8.11b natural log terms, as shown in Equation 8.13b.

$$\ln i(t) = \ln e^{\frac{-tR}{L}} + \ln k \tag{8.13b}$$

The sum of two logarithms of functions a and b equals the log of the product ab, as shown in Equation 8.14.

$$\ln(a) + \ln(b) = \ln(ab) \tag{8.14}$$

Apply Equation 8.14 to Equation 8.13a to consolidate to one log term in Equation 8.15a.

$$\ln V(t) = \ln\left(ke^{\frac{-t}{RC}}\right) \quad (8.15a)$$

Exponentiate Equation 8.15a to the base e to get Equation 8.16a. (Take the antilogarithm of Equation 8.15a.)

$$v(t) = ke^{\frac{-t}{RC}} \quad (8.16a)$$

To evaluate the constant k in Equation 8.16a, consider the time t equals 0 when we know the voltage from the initial condition, given in Equation 8.17a.

$$v(0) = v_C(0) = ke^0 \quad (8.17a)$$

Taking e to the power 0 equals 1. As shown in Equation 8.18a, the constant k equals the initial voltage $v_C(0)$ in Equation 8.17a.

$$v_C(0) = k \quad (8.18a)$$

Substitute Equation 8.18a into Equation 8.16a in place of k to obtain Equation 8.19a.

Apply Equation 8.14 to Equation 8.13b to consolidate to one log term in Equation 8.15b.

$$\ln i(t) = \ln\left(ke^{\frac{-tR}{L}}\right) \quad (8.15b)$$

Exponentiate Equation 8.15b to the base e to get Equation 8.16b. (Take the antilogarithm of Equation 8.15b.)

$$i(t) = ke^{\frac{-tR}{L}} \quad (8.16b)$$

To evaluate the constant k in Equation 8.16b, consider the time t equals 0 when we know the current from the initial condition, given in Equation 8.17b

$$i(0) = i_L(0) = ke^0 \quad (8.17b)$$

Taking e to the power 0 equals 1. As shown in Equation 8.18b, the constant k equals the initial current $i_L(0)$ in Equation 8.17b.

$$i_L(0) = k \quad (8.18b)$$

Substitute Equation 8.18b into Equation 8.16b in place of k to obtain Equation 8.19b.

RC natural response

$$v(t) = v_C(0)e^{\frac{-t}{RC}} \quad (8.19a)$$

Equation 8.19a is the formula for the natural response of the RC circuit in Figure 8.5a.

RL natural response

$$i(t) = i_L(0)e^{\frac{-tR}{L}} \quad (8.19b)$$

Equation 8.19b is the formula for the natural response of the RL circuit in Figure 8.5b.

Equations 8.16a and 8.16b can be general for many different initial conditions. For this example, with the schematics and sources in Figures 8.5a and 8.5b to generate the initial conditions, before the natural response starts, values of initial conditions can be calculated using the sources.

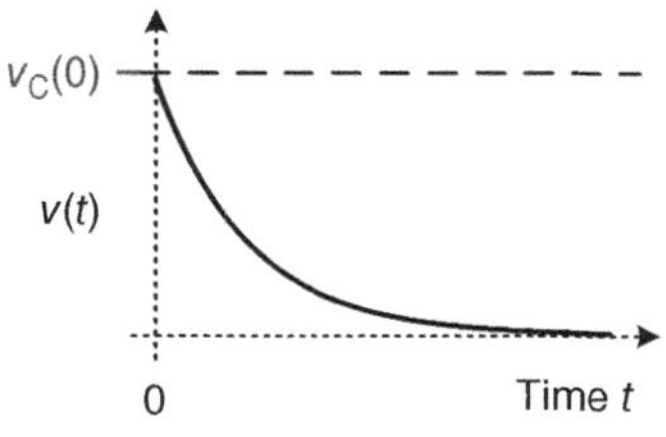

Figure 8.6a RC natural response capacitor voltage decaying from initial condition value.

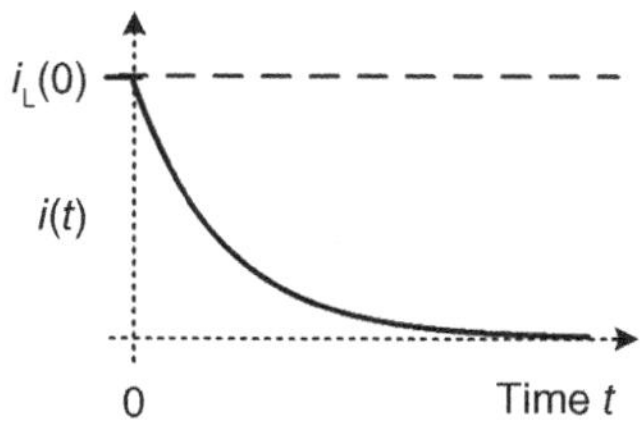

Figure 8.6b RL natural response inductor current decaying from initial condition value.

For the RC circuit in Figure 8.5a, Figure 8.6a shows the decaying exponential natural response voltage of Equation 8.19a.

The voltage graph in Figure 8.6a is labeled $v(t)$, and not $v_C(t)$ because $v(t)$ is the voltage for both the resistor and the capacitor, the only voltage in the circuit.

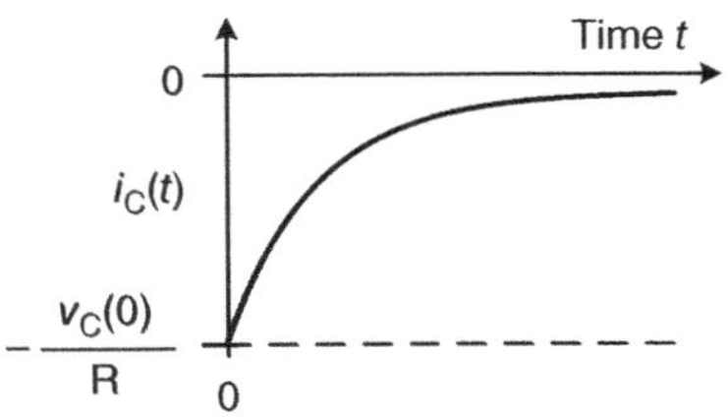

Figure 8.7a RC natural response capacitor current.

Figure 8.7a shows RC natural response current through the capacitor in Figure 8.5a. Current flows down through the resistor and the up through the capacitor, against the polarity, making the capacitor current negative. By comparison, the resistor current would be positive, equal to the voltage in Figure 8.6a divided by R.

RC natural response

$$i(t) = \frac{-V_C}{R} e^{\frac{-t}{RC}} \tag{8.20a}$$

Equation 8.20a is the formula for the natural response current of the RC circuit in Figure 8.5a.

For the RL circuit in Figure 8.5b, Figure 8.6b shows the decaying exponential natural response current of Equation 8.19b.

The current graph in Figure 8.6a is labeled $i(t)$, and not $i_L(t)$ because $i(t)$ is the current for both the resistor and the inductor, the only current in the circuit.

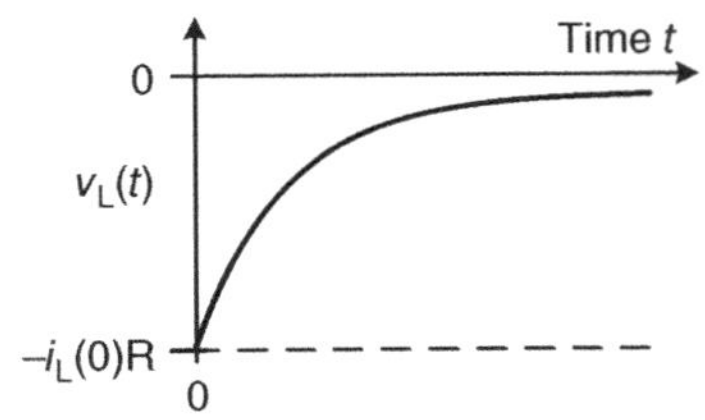

Figure 8.7b RL natural response inductor voltage.

Figure 8.7b shows the natural response voltage across the inductor in Figure 8.5b. Current flows down through the inductor and then up through the resistor. Relative to the voltage polarity shown in the schematic, the voltage across the inductor is negative.

RL natural response

$$v(t) = -Ri_L(0)e^{\frac{-tR}{L}} \tag{8.20b}$$

Equation 8.20b is the formula for the natural response voltage of the RL circuit in Figure 8.5b.

8.2.4 Alternative Derivation of RC and RL ZIR Differential Equations

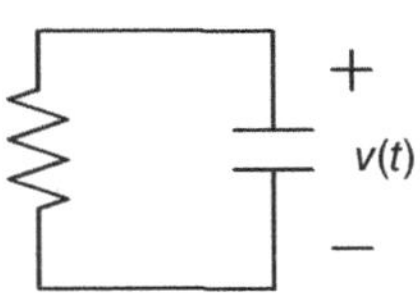

Figure 8.8a Parallel RC circuit marked with $v(t)$ corresponding to natural response in Figure 8.2a.

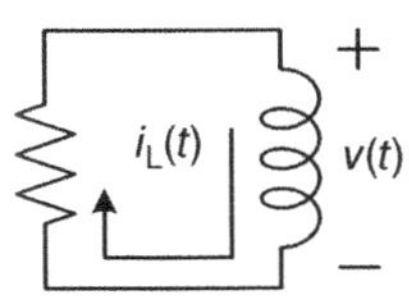

Figure 8.8b Parallel RL circuit marked with $i_L(t)$ corresponding to $i(t)$ in Figure 8.2b.

Write KVL for the RC circuit in Figure 8.8a, as shown in Equation 8.21a.

$$v_C(t) + v_R(t) = 0 \tag{8.21a}$$

Equation 8.22a shows the voltage across a capacitor as a function of the current through the capacitor.

$$v_C(t) = \frac{1}{C}\int_{t_0}^{t} i(x)\,dx + v_C(t_0) \tag{8.22a}$$

Substitute Equation 8.22a, for the capacitor voltage v_C, and Ohm's law for the resistor voltage, into Equation 8.21a to get Equation 8.23a.

$$\frac{1}{C}\int_{t_0}^{t} i(x)\,dx + iR = 0 \tag{8.23a}$$

Differentiate Equation 8.23a to get Equation 8.24a.

$$\frac{1}{C}i(t) + \frac{di(t)}{dt}R = 0 \tag{8.24a}$$

Rearrange terms in Equation 8.24a to obtain Equation 8.25a.

$$\frac{di(t)}{dt}R = -\frac{1}{C}i(t) \tag{8.25a}$$

Divide Equation 8.25a by R and $i(t)$, and multiply Equation 8.25a by dt, to obtain Equation 8.26a.

$$\frac{di(t)}{i(t)} = -\frac{1}{RC}dt \tag{8.26a}$$

Integrate both sides of Equation 8.26a to get Equation 8.27a.

$$\int \frac{di}{i} = -\frac{1}{RC}\int dt \tag{8.27a}$$

$$\int \frac{dx}{x} = \ln x + C \tag{8.28}$$

Using Equation 8.28, perform the integration on Equation 8.27a to obtain Equation 8.29a.

$$\ln i(t) = -\frac{1}{RC}t + \ln K \tag{8.29a}$$

Write KCL for the RL circuit in Figure 8.8b, summing the currents going out of the top node, as shown in Equation 8.21b.

$$i_L(t) + i_R(t) = 0 \tag{8.21b}$$

Equation 8.22b shows the current through an inductor as a function of the voltage across the inductor.

$$i_L(t) = \frac{1}{L}\int_{t_0}^{t} v(x)\,dx + i(t_0) \tag{8.22b}$$

Substitute Equation 8.22b, for the inductor current i_L, and Ohm's law for the resistor current, into Equation 8.21b to get Equation 8.23b.

$$\frac{1}{L}\int_{t_0}^{t} v(x)\,dx + \frac{v}{R} = 0 \tag{8.23b}$$

Differentiate Equation 8.23b to get Equation 8.24b.

$$\frac{1}{L}v(t) + \frac{dv(t)}{dt}\frac{1}{R} = 0 \tag{8.24b}$$

Rearrange terms Equation 8.24b to obtain Equation 8.25b.

$$\frac{dv}{dt}\frac{1}{R} = -\frac{1}{L}v(t) \tag{8.25b}$$

Divide Equation 8.25b by $v(t)$ and multiply Equation 8.25b by R and dt to obtain Equation 8.26b.

$$\frac{dv}{v(t)} = -\frac{R}{L}dt \tag{8.26b}$$

Integrate both sides of Equation 8.26b to get Equation 8.27b.

$$\int \frac{dv}{v} = -\frac{R}{L}\int dt \tag{8.27b}$$

The integral of $1/x\,dx$ is the natural logarithm of x plus a constant.

Using Equation 8.28, perform the integration on Equation 8.27b to obtain Equation 8.29b.

$$\ln v(t) = -\frac{R}{L}t + \ln K \tag{8.29b}$$

From this point in the derivation, the remaining steps can follow the example in the first derivation.

8.3 RC Natural Response (ZIR) Time Constants and Initial Current

This section calculates the slope of the natural response exponential curve in Equation 8.30 at time t equals zero. This provides the derivative of the $v_C(t)$ natural response curve evaluated at t equals zero.

The natural response voltage of an RC circuit starts with the initial condition capacitor voltage V_0 as shown in Figures 8.9 and 8.10 and Equation 8.30.

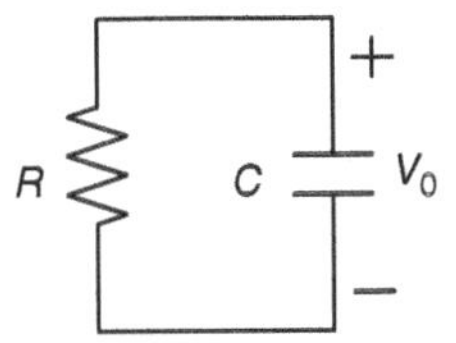

Figure 8.9 Parallel RC circuit marked with an initial condition capacitor voltage V_0.

$$v_C(t) = V_0 e^{(-t/RC)} \qquad (8.30)$$

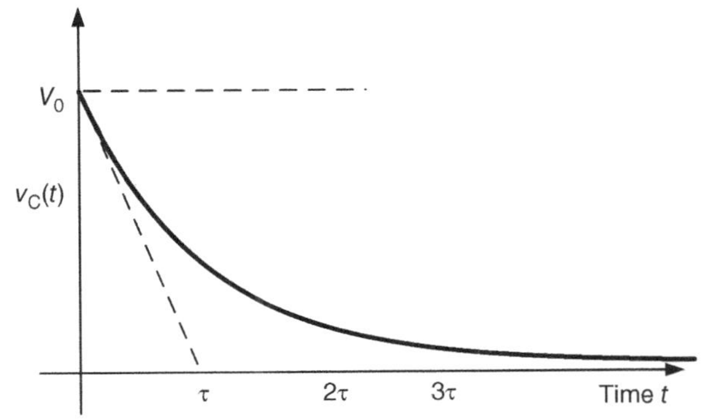

Figure 8.10 RC natural response decaying exponential curve with the horizontal time axis marked in units of the time constant tau τ.

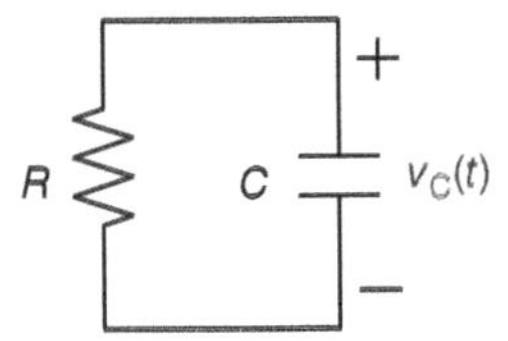

Figure 8.11 Parallel RC circuit marked with v-sub-c of t corresponding to the curve in Figure 8.10.

After time t equals 0, the natural response function of time $v_C(t)$, shown in Figure 8.10 and depicted in Figure 8.11, decays exponentially as a function of time according to Equation 8.30. A graph of Equation 8.30 is shown in Figure 8.10.

What is the slope of this decaying exponential function at time t equals 0?

8.3.1 Slope of Exponential Function at time 0

$$i_C = C\frac{dv_C(t)}{dt} \qquad (8.31)$$

Equation 8.31 describes the current through a capacitor as a function of the voltage across the capacitor.

$$\frac{i_C}{C} = \frac{dv_C(t)}{dt} \qquad (8.32)$$

Divide both sides of Equation 8.31 by capacitance C to get Equation 8.32.

Rearrange terms in Equation 8.32 to get Equation 8.33.

$$\frac{dv_C(t)}{dt} = v_C' = \frac{i_C}{C} \qquad (8.33)$$

Equation 8.33 states that the slope of exponential curve, of capacitor voltage in RC natural response, is the capacitor current i_C divided by C.

$$\frac{dv_C(0)}{dt} = v_C'(0) = \frac{i_C(0)}{C} = \frac{-V_0}{RC} \qquad (8.34)$$

At time t equals 0, as shown in Equation 8.34, the initial capacitor current equals $-V_0$ over R because the loop current goes through the capacitor against the positive polarity of voltage on the capacitor.

$$v'_C(0) = \frac{i_C(0)}{C} = \frac{-V_0}{RC} = -\frac{V_0}{\tau} \quad (8.35)$$

The time constant τ equals RC, giving Equation 8.35 for the slope of the RC natural response exponential curve at $t = 0$.

Checking this result in Equation 8.35 with the mnemonic of slope equal to rise over run, Figure 8.10 shows the rise of $-V_0$ and the run of τ. As the high value of V_0 is to the left, the slope has negative value.

The slope, in Equation 8.35, can be used as "m" in the straight line equation formula y equals mx plus b, in Equation 8.36, where y is the voltage function plotted on the vertical axis and m is the slope of the line. The independent variable in the horizontal axis is x and b is the y-intercept, the value on the y-axis at x equals zero.

$$y = \mathrm{m}x + \mathrm{b} \quad (8.36)$$

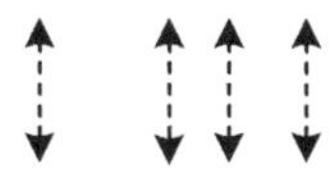

$$v(t) = \frac{-V_0}{\tau}t + V_0 \quad (8.37)$$

Figure 8.12 Relation between the common slope intercept formula and the specific voltage formula here.

Compare the straight line formula, in Equation 8.36, to a voltage straight line formula incorporating the known slope $-V_0/\tau$ as shown in Figure 8.12. In Equation 8.37, V_0 is the y-intercept corresponding to b. This line in Equation 8.37 is the dashed straight line shown in Figure 8.10.

8.3.2 Solve for the X-Intercept

Given the slope $-V_0/\tau$, and the y-intercept b, one can solve for the x-intercept, which is labeled τ on the horizontal axis of the graph in Figure 8.10.

$$v(t) = \frac{-V_0}{\tau}t + V_0 \quad (8.38)$$

To solve for the x intercept of the dashed straight line in Figure 8.10, set $v(t)$ in Equation 8.38 equal to zero to get Equation 8.39.

$$0 = \frac{-V_0}{\tau}t + V_0 \quad (8.39)$$

Rearrange terms in Equation 8.39.

$$\frac{V_0}{\tau}t = +V_0 \quad (8.40)$$

Cancel V_0 from both sides of Equation 8.40 to obtain Equation 8.41 for the x intercept τ, where the line crosses the x axis which is time t.

$$t = \tau \quad (8.41)$$

8.4 Natural Response of Series RL with Voltage Source

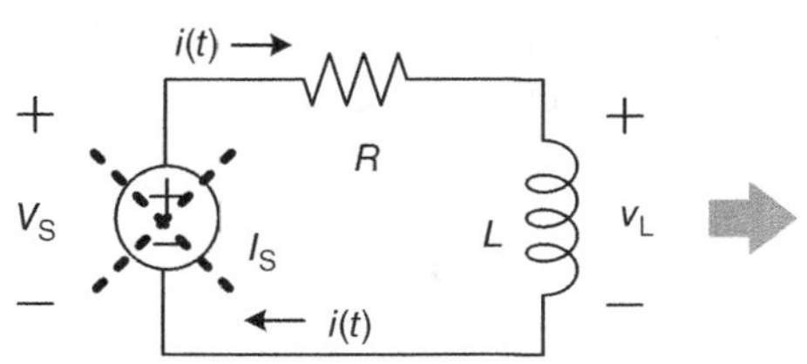

Figure 8.13a Series RL circuit with voltage source being removed to visualize natural response.

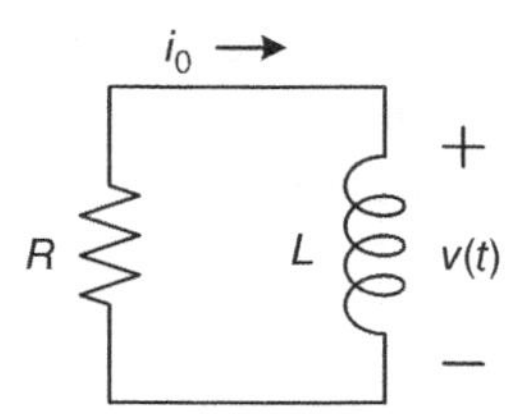

Figure 8.13b Parallel RL circuit with initial condition current i_0 from the voltage source.

The starting point for this natural response analysis is the resistor–inductor (RL) circuit, in Figure 8.13a, driven by a voltage source. The natural response of a circuit comes only from initial conditions, not from any input voltage or current source. Therefore we remove the voltage source from the circuit for natural response analysis.

Recall that removing a voltage source from a circuit leaves a short circuit, as if the voltage in the source were turned down to zero volts. Removing the voltage source from Figure 8.13a leads to the parallel RL circuit in Figure 8.13b.

The original series RL circuit on the left in Figure 8.13a, with the voltage source V_S, becomes the parallel RL circuit on the right in Figure 8.13b for calculating natural response. Natural response comes from the initial inductor current i_0.

The natural response of a circuit is the solution to a homogeneous differential equation. There is no source term in the differential equation just as there is no voltage source in the natural response circuit in Figure 8.13b. The initial current in this RL circuit is analogous to an initial condition to a differential equation.

8.4.1 Generating Series RL Initial Conditions with Switches

How does the initial condition get there? How should we think about initial condition on the inductor current intuitively? Looking at switch circuits, that can put initial inductor current into place, gives us some intuition. Other analyses, like step response, use different switches in circuits, and the comparison to natural response switches can be insightful.

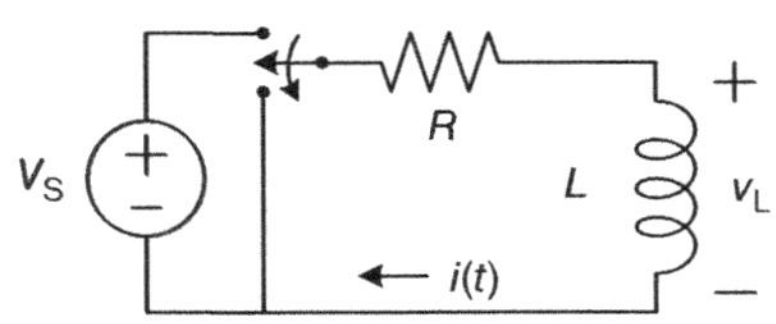

Figure 8.14 Series RL circuit schematic with switch to model initial condition inductor current.

In the upper position, the switch in Figure 8.14 connects the voltage source into the circuit, causing a current $i(t)$ to flow. Having the switch in the upper position corresponds to the starting circuit in Figure 8.13a with the voltage source. In steady state, with DC current, there is no voltage across the inductor and the DC current in the inductor becomes the initial condition current i_0.

At time t equals 0, when the switch in Figure 8.14 moves to the lower position, the voltage source is removed from the circuit in Figure 8.13a, making it equivalent to the parallel RL circuit in Figure 8.13b.

At time t equals 0, the initial inductor current $i_0(t)$ in Figure 8.13b immediately becomes the decaying exponential current $i(t)$. Also note that after t equals 0, the inductor voltage $v_L(t)$ in Figure 8.13a is now across the resistor in parallel with the inductor. The subscript on inductor voltage changes from $v_L(t)$ to $v(t)$ in Figure 8.13b to reflect this change from inductor-specific voltage to general voltage for the entire circuit.

8.4.2 Series RL Natural Response Differential Equation Solution

To solve for the time domain natural response of the parallel RL circuit, write KVL. Starting from the top and going around the loop clockwise, as shown in Figure 8.15, sum all the (two) voltage drops as functions of current. The voltage across the inductor plus the voltage across the resistor equals zero, as shown in Equation 8.42.

$$L\frac{di(t)}{dt} + Ri(t) = 0 \tag{8.42}$$

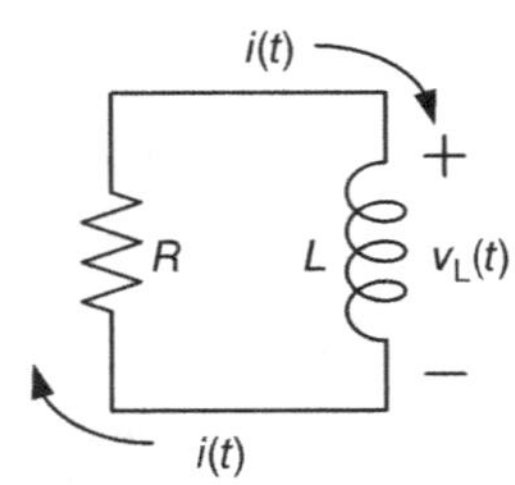

Figure 8.15 Parallel RL circuit showing reference polarity current direction.

The voltage across the resistor, from Ohm's law, is $i(t)$ times the resistance R. The voltage across the inductor is the inductance L times the derivative of the inductor current with respect to time, shown in Equation 8.43.

$$v_L = L\frac{di\,(t)}{dt} \tag{8.43}$$

What happened to i_0?
Why isn't i_0 called $i_0(t)$?
Initial conditions, such as i_0, are single, instantaneous constant values at the beginning of a natural response behavior involving variables like $i(t)$. As a constant value, i_0 is not a function of time.

In Equation 8.42, move the Ri term to the right side of the equal sign, as shown in Equation 8.44.

$$L\frac{di\,(t)}{dt} = -Ri\,(t) \tag{8.44}$$

Multiply both sides of Equation 8.44 by dt to obtain Equation 8.45.

$$Ldi = -Ri\,(t)\,dt \tag{8.45}$$

Divide both sides of Equation 8.45 by $i(t)$ to obtain Equation 8.46.

$$L\frac{di}{i\,(t)} = -Rdt \tag{8.46}$$

Divide both sides of Equation 8.46 by L to obtain Equation 8.47.

$$\frac{di}{i\,(t)} = -\frac{R}{L}dt \tag{8.47}$$

Integrate both sides of Equation 8.47 using the formula in Equation 8.48 for the integral of 1 over x for the left side of Equation 8.47.

$$\int \frac{1}{x}\,dx = \ln x + C \tag{8.48}$$

$$\int K\,dx = Kx + C \tag{8.49}$$

Use the formula in Equation 8.49 for the integral of a constant for the right side of Equation 8.47.

$$\ln i\,(t) + C_1 = -\frac{R}{L}t + C_2 \tag{8.50}$$

Equation 8.50 is the result of the integration of Equation 8.47. C_1 and C_2 are arbitrary constants where the 1 and 2 simply distinguish them.

In Equation 8.50 combine C_1 and C_2 into one constant on the left side of the equation. For convenience, choose this arbitrary constant to be a negatively valued natural logarithm of a constant i_0. This leads to the ln i_0 term in Equation 8.51.

$$\ln\,i\,(t) - \ln\,i_0 = -\frac{R}{L}t \tag{8.51}$$

$$\ln a - \ln b = \ln\,(a/b) \tag{8.52}$$

Use the law of logarithms in Equation 8.52 on the left side of Equation 8.51 to create a fraction term, as shown in Equation 8.53.

$$\ln\left(\frac{i\,(t)}{i_0}\right) = -\frac{R}{L}t \tag{8.53}$$

$$e^{\ln x} = x \tag{8.54}$$

Equation 8.54 shows that the exponent, to the base e, of the natural log, is the quantity itself.

Utilizing Equation 8.54, exponentiate both sides of Equation 8.53 to the base e to obtain Equation 8.55.

$$\frac{i(t)}{i_0} = e^{(-R/L)t} \quad (8.55)$$

Multiply both sides of Equation 8.55 by i_0, resulting in Equation 8.56.

$$i(t) = i_0 e^{(-R/L)t} \quad (8.56)$$

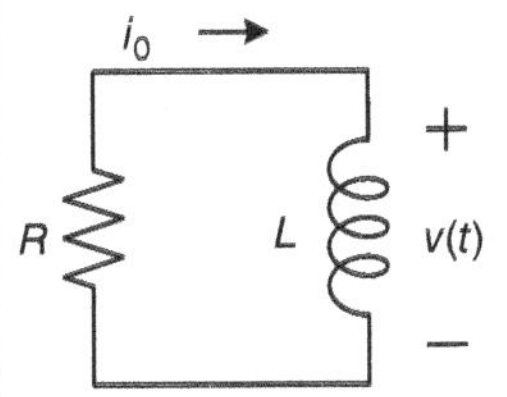

Figure 8.16 Parallel RL circuit with initial condition inductor current i_0.

i_0 is the initial condition of the mathematical integral, corresponding to the initial current in the inductor. i_0 is the initial amount of current that decays away to zero in the natural response.

In Equation 8.56 and in Figure 8.16, i_0 is the initial condition of the mathematical integral, corresponding to the initial current in the inductor i_0.

Equation 8.56 and Figure 8.16 present a general **natural response** for an unknown initial current i_0.

This example started out with a circuit, shown in Figure 8.14, with a voltage source that charges the circuit with the initial current. Having established the initial inductor current i_0, the voltage source gets magically removed from the circuit, and replaced by a short circuit, for the natural response. The value of the initial inductor current inductor i_0 can be calculated using that source voltage V_S.

$$i(0) = i_0 e^0 \quad (8.57)$$

To solve for i_0, based on the circuit with the voltage source in Figure 8.13a, calculate $i(t)$ for t equals zero in Equation 8.57. e-to-the-0 equals 1.

$$\frac{V_S}{R} = i_0 \quad (8.58)$$

From the circuit in Figure 8.13a, we can see that the initial current is V_S over R.

Substitute the value for i_0, from Equation 8.58, into Equation 8.56, to obtain Equation 8.59.

$$i(t) = \frac{V_S}{R} e^{(-R/L)t} \quad (8.59)$$

Equation 8.59 is the **natural response** and zero-input response (**ZIR**) current of the RL circuit in Figures 8.13b and 8.16.

From Equation 8.59 for $i(t)$, the value for $v(t)$ can be calculated by multiplying by R as shown in Equation 8.60.

$$v(t) = -V_S e^{(-R/L)t} \quad (8.60)$$

Equation 8.60 is the natural response and zero-input response (ZIR) voltage across the parallel RL circuit in Figures 8.13b and 8.16.

Note that the current flows up through the resistor in Figures 8.13b and 8.16, making the voltage, across both the resistor and the inductor, negative relative to the marked voltage polarity across the inductor, as shown in Figure 8.17b.

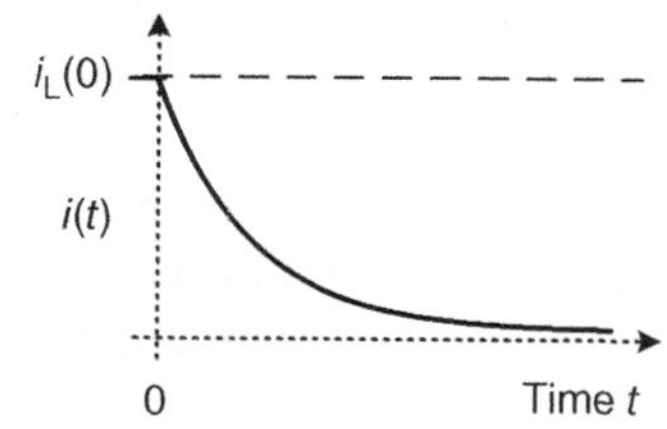

Figure 8.17a RL natural response inductor current $i(t)$ decaying from initial condition inductor current value of $i_L(0)$.

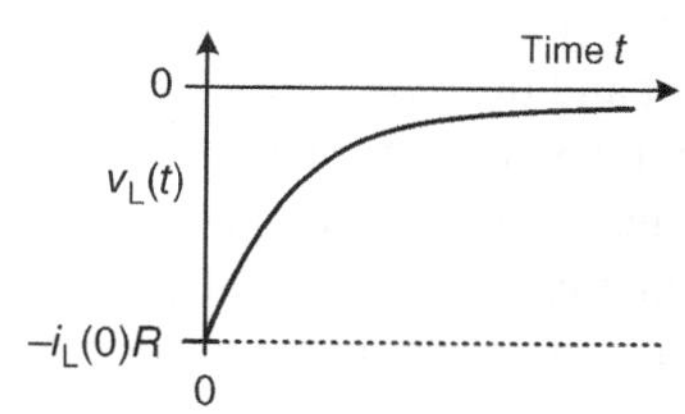

Figure 8.17b RL natural response inductor voltage $v(t)$ decaying.

Figure 8.17a shows the natural response current $i_L(t)$, through the inductor and resistor in Figures 8.13b and 8.16, decaying exponentially from the initial condition amount $i_L(0)$ to zero.

Figure 8.17b shows the natural response voltage $v(t)$, across the inductor and resistor in Figures 8.13b and 8.16, decaying exponentially from the initial condition amount $i_L(0)$ times R to zero.

8.5 First-Order RC and RL Natural Response Summary

In first-order electrical circuits, natural response comes from stored energy, in the form of initial inductor currents and capacitor voltages, which decay over time. Natural response is also called zero-input response (ZIR) in contrast to zero-state response (ZSR).

Natural response reveals the characteristics of the circuit, not the characteristics of any input.

The analysis of natural response intentionally excludes input signals. Natural response can be modeled using switch configurations. In circuit simulation, such as SPICE, most ideal lumped element inductor and capacitor models have the ability to set an initial condition that can be used for simulating natural response.

For mathematical rigor, we imagine circuit inputs having an infinite amount of time to settle, allowing transients to go to zero, to create the initial conditions for the start of the natural response time period.

First-order natural responses are exponential functions described by homogeneous differential equations.

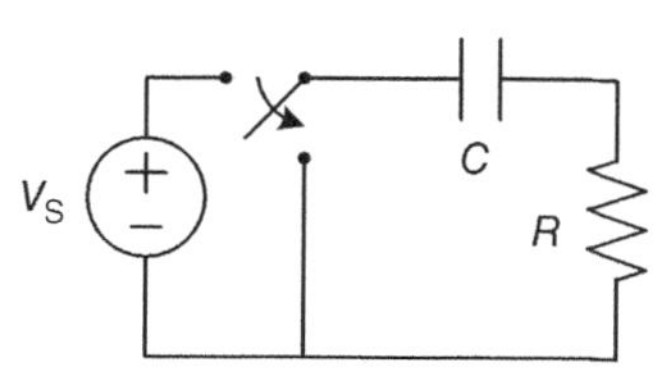

Figure 8.18 Series RC circuit with one switch to model the initial condition capacitor voltage for natural response.

The switch in Figure 8.18 can be used to model the generation of initial condition voltage on the capacitor, and then natural response when the switch changes from the upper position to the lower position.

In the upper position, the switch in Figure 8.18 connects the capacitor and resistor in series with the voltage source. This charges the capacitor with a voltage equal to the source voltage V_S. After the capacitor is charged, no current flows in the circuit.

When the switch moves instantaneously from the upper position to the lower position, the voltage source is removed from the circuit and the resistor and capacitor are in parallel. The voltage across the capacitor generates current in the resistor, which drains the charge from the capacitor eventually down to zero.

Note that operating the switch in the opposite direction models step response. In the lower position before switching, the resistor and capacitor in parallel drain away any charge stored in the capacitor. This creates the zero state. When the switch moves from the lower position to the upper position, a step function voltage is applied to the series RC circuit.

9

First-Order Step Response of RC and RL Circuits

9.1 First-Order Step Response Summary Overview

9.1.1 Step Response Introduction

Step response is the time domain response of a circuit to an instantaneous step (ideal – no rise time) input. Step response is one form of zero-state response (ZSR). Step response relates closely to the charging discussion in Chapter 7.

Zero state means that there aren't initial conditions in the circuit, no currents nor voltages. Any previous currents or voltages would have completely dissipated or decayed away to zero. Mathematically and rigorously, for all previous initial conditions to decay away, this means that an infinite amount of time must precede the step input. Thus, Figure 9.1 is marked as extending out to infinity in negative time before zero.

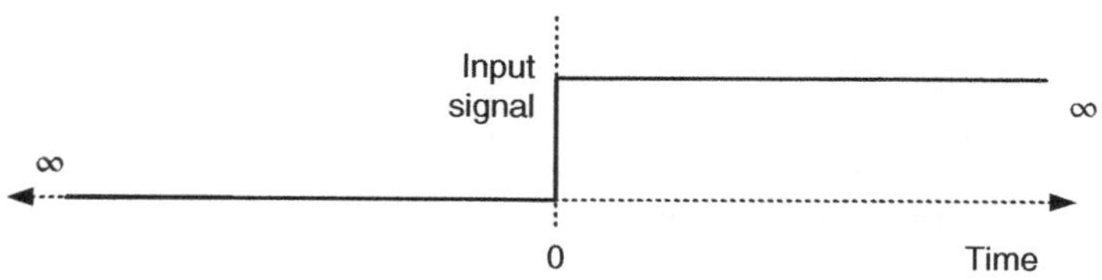

Figure 9.1 Unit step function signal versus time.

9.1.2 Differential Equation Solution Using a Particular Solution

In mathematical terms, step response corresponds to a non-homogeneous differential equation, which has a function term corresponding to the signal source in the circuit. A common approach to calculating the general solution, to a non-homogeneous differential equation, is to add one particular solution of the non-homogeneous solution to the general solution of the homogeneous solution. In circuit terms, one can find the step response of a circuit by adding one specific solution, of the non-homogeneous step response differential equation, to the natural response of that circuit.

Equation 9.1 shows the general solution v_{out}, to a circuit driven by a voltage source.

$$v_{\text{out}} = v_{\text{h}} + v_{\text{p}} \tag{9.1}$$

v_{h} = Solution to the homogeneous differential equation (natural response of the circuit).

v_{p} = Particular solution to the non-homogeneous differential equation (a specific solution of the step response).

First and Second Order Circuits and Equations: Technical Background and Insights, First Edition. Robert O'Rourke.

9.1.3 Step Response Topology Summary

Table 9.1 summarizes several different first-order circuit topologies and their voltage and current response to a step input in the time domain.

Table 9.1 The shapes and orientations of voltage and current time domain responses to several first order circuit topologies driven by step function inputs.

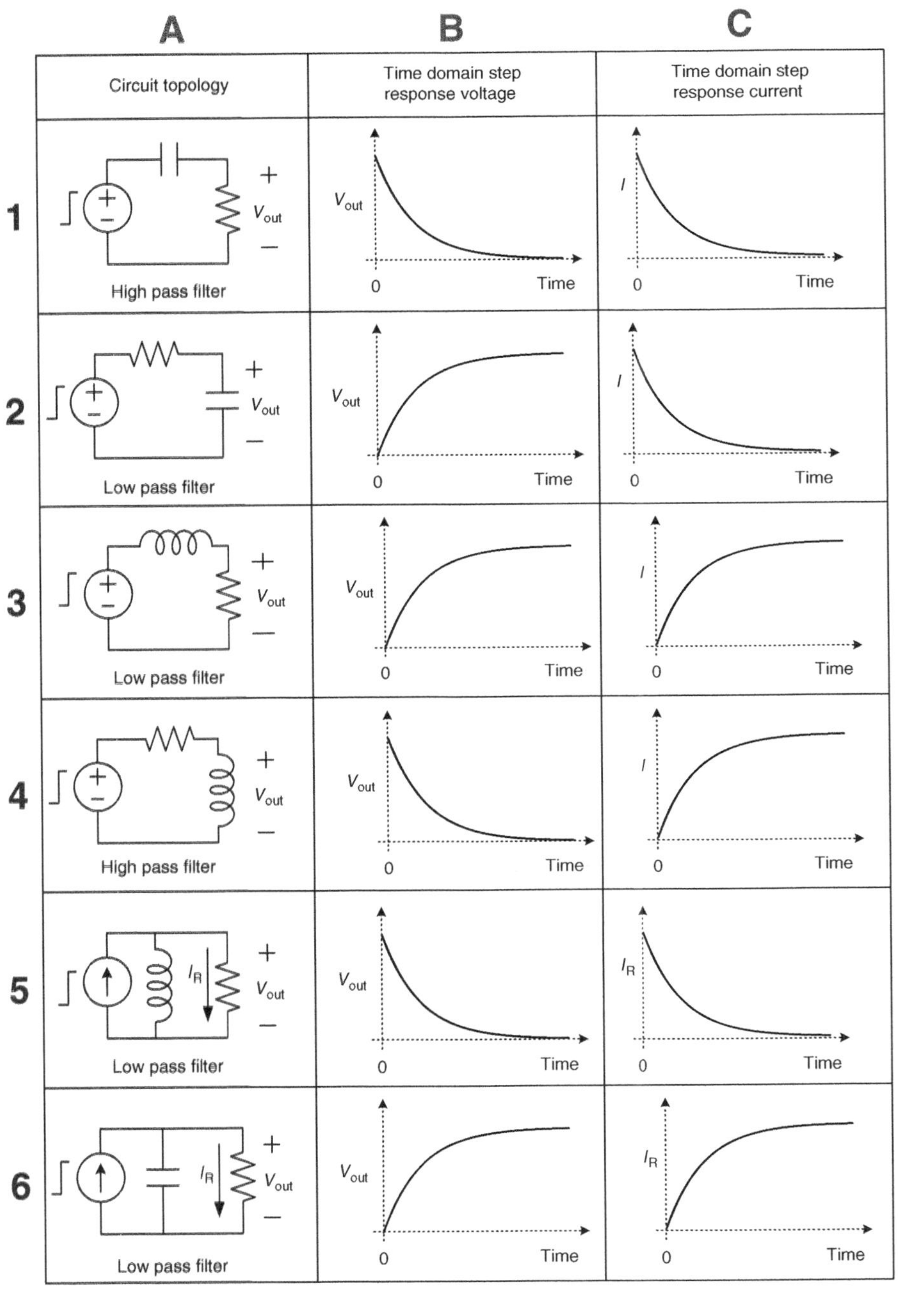

Note that Table 9.1 only presents well-behaved responses with exponential responses. There are other step responses that don't have exponential responses. If a current source drives a capacitor in series, there can be a voltage ramp, as charge accumulates on the capacitor, going up indefinitely. A voltage step input across a capacitor produces an impulse response. A current step input in series with an inductor also produces an impulse response.

9.1.4 First-Order Step Response Equation Summary

Here are step response voltages and currents, for $t > 0$, for two first-order circuits. Figure 9.2a shows a series RC circuit driven by a voltage step function $v_S(t)$ (Figure 9.3a), which equals the DC voltage amplitude V_S multiplied by the unit step function $u_S(t)$. Figure 9.2b shows a parallel RL circuit driven by a current step function $i_S(t)$ (Figure 9.3b), which equals the DC current amplitude I_S multiplied by the unit step function $u_S(t)$.

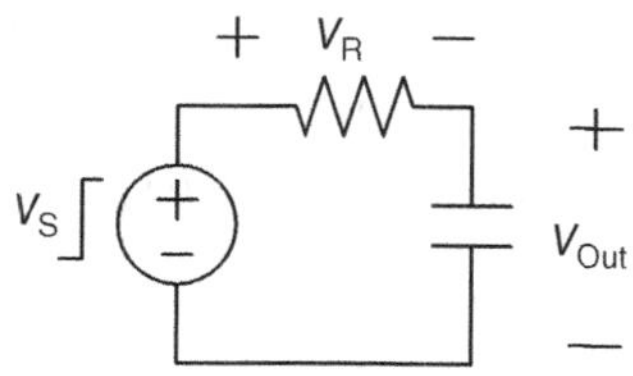

Figure 9.2a Series RC circuit driven by voltage step function.

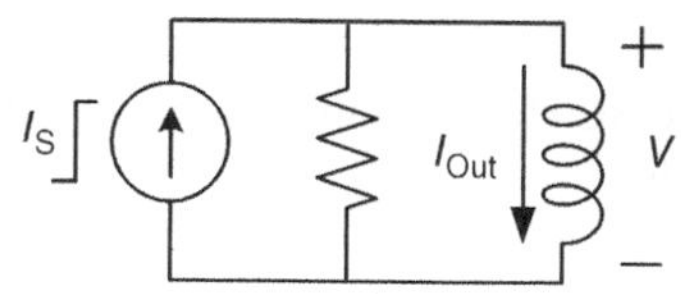

Figure 9.2b Parallel RL circuit driven by current step function.

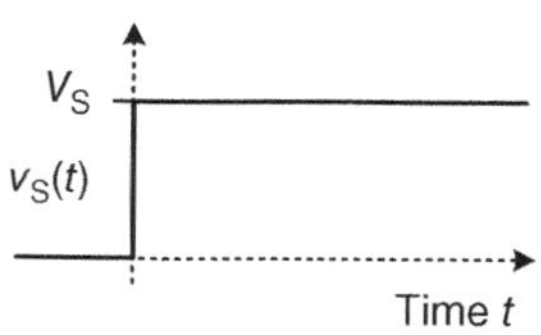

Figure 9.3a Graph of voltage step function $v_S(t)$ with amplitude V_S.

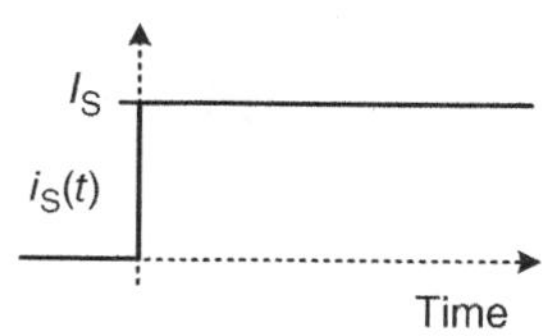

Figure 9.3b Graph of current step function $i_S(t)$ with amplitude I_S.

For the RC circuit in Figure 9.2a, Figure 9.4a shows the voltage step responses in Equations 9.2a and 9.3a. For the RL circuit in Figure 9.2b, Figure 9.4b shows the current step responses in Equations 9.2b and 9.3b

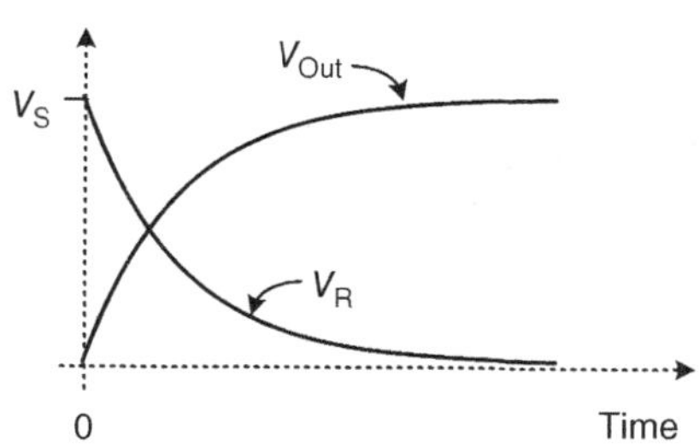

Figure 9.4a Series RC circuit voltage responses to voltage step function.

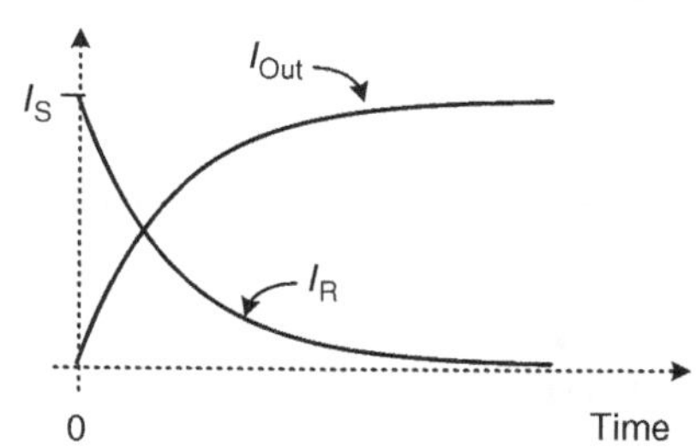

Figure 9.4b Parallel RL circuit current responses to current step function.

$$V_{out} = V_S\left(1 - e^{\frac{-t}{RC}}\right) \tag{9.2a}$$

$$V_R = V_S e^{\frac{-t}{RC}} \tag{9.3a}$$

$$I_{out} = I_S\left(1 - e^{\frac{-tR}{L}}\right) \tag{9.2b}$$

$$I_R = I_S e^{-\frac{tR}{L}} \tag{9.3b}$$

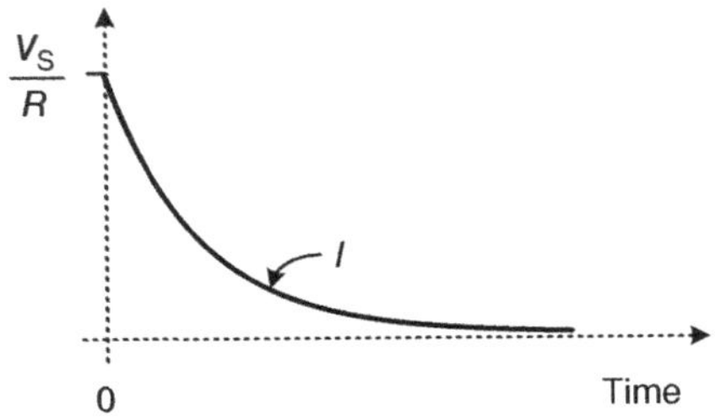

Figure 9.5a Series RC circuit current response to voltage step function.

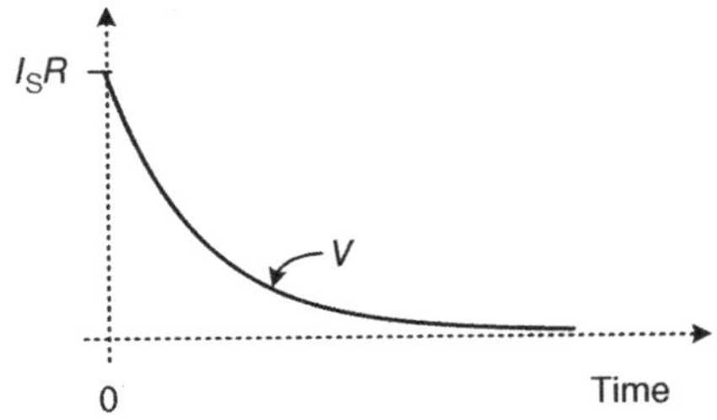

Figure 9.5b Parallel RL circuit voltage response to current step function.

$$I = \frac{V_S}{R} e^{\frac{-t}{RC}} \tag{9.4a}$$

$$V = I_S R e^{-\frac{tR}{L}} \tag{9.4b}$$

Figure 9.5a shows the RC step response loop current in Equation 9.4a. Figure 9.5b shows the RL step response voltage in Equation 9.4b.

9.1.5 Overview of Step Response of Series RL with Voltage Source

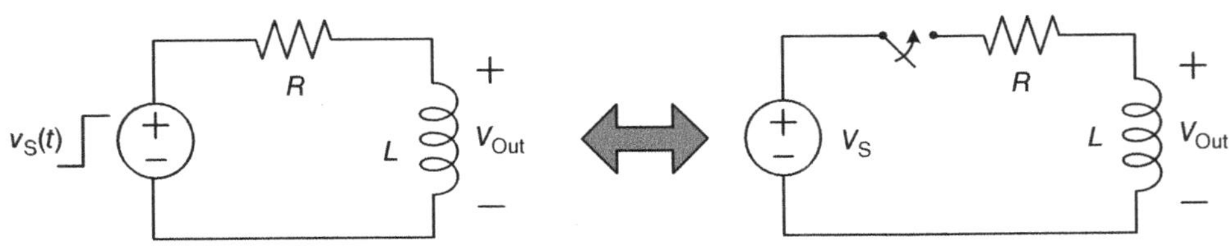

Figure 9.6a Series RL circuit driven by voltage step function $v_S(t)$.

Figure 9.6b Series RL circuit with DC voltage V_S and switch modeling step function.

$$v_S(t) = V_S\, u(t) \tag{9.5}$$

$$u(t) = \begin{cases} 0 \text{ for } t < 0 \\ 0 \text{ for } t \geq 0 \end{cases} \tag{9.6}$$

The circuit in Figure 9.6a, with the step symbol by the input voltage source, is equivalent to the circuit in Figure 9.6b with a switch. The switch closes at time t equals 0, implementing a step input in Equation 9.6, where $u(t)$ is the well-known step function described in Equation 9.6.

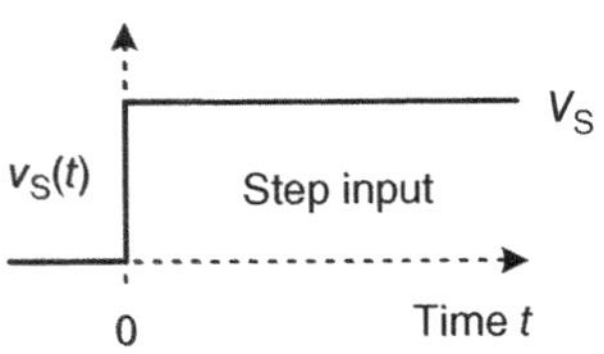

Figure 9.7 Graph of input step function $v_S(t)$ with amplitude V_S.

V_S is a constant DC value that, when multiplied by the step function $u(t)$, creates the voltage source input waveform $v_S(t)$ shown in Figure 9.7.

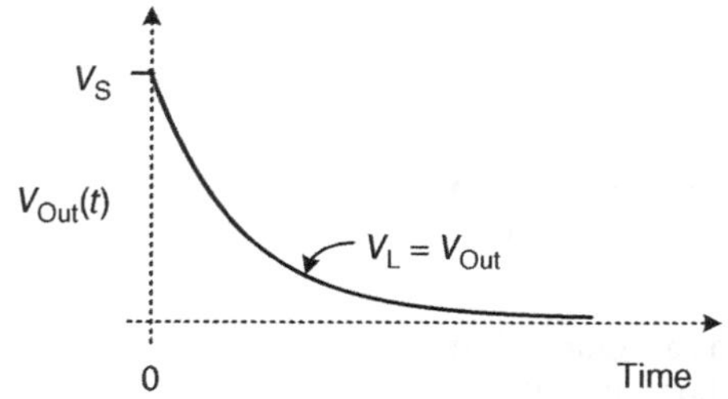

Figure 9.8 Series RL inductor voltage response to voltage step function input.

At time t equals 0, the stepped source voltage V_S initially drops entirely across the inductor because no instantaneous change in current is allowed in an ideal inductor. An instantaneous voltage across the inductor would imply an instantaneous current in the inductor.

As inductor current increases from zero, the voltage across the resistor increases. Eventually V_{out} across the inductor decays to zero as shown in Figure 9.8 and described by Equation 9.7.

$$V_{out}(t) = V_S e^{-\frac{tR}{L}} \qquad (9.7)$$

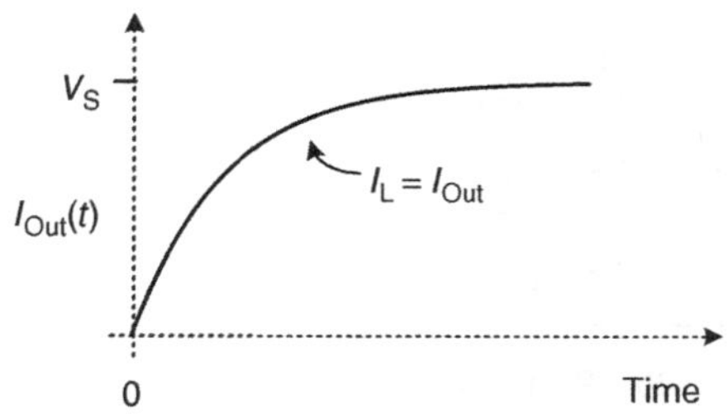

Figure 9.9 Series RL current response to voltage step function input.

The step response of the RL circuit, driven by a voltage source, causes an inductor current increasing from zero, as shown in Figure 9.9 and described in Equation 9.8.

$$I_{out}(t) = \frac{V_S}{R}\left(1 - e^{-\left(\frac{Rt}{L}\right)}\right) \qquad (9.8)$$

9.2 Intuitive Analysis of RC and RL Step Response

9.2.1 Step Response Intuitive Analysis Setup

This intuitive analysis shows how to visually inspect a first-order circuit and determine whether the current and voltage waveforms are ***rising*** or ***decaying*** exponentials, given that the response is already known to be exponential and the time constants are already known and understood. The series RC circuit in Figure 9.10a is driven by the voltage step function $v_S(t)$, with DC magnitude V_S, in Figure 9.11a. The parallel RL circuit in Figure 9.10b is driven by the current step function $i_S(t)$, with DC magnitude I_S, in Figure 9.11b.

Series Capacitor Low Pass Filter

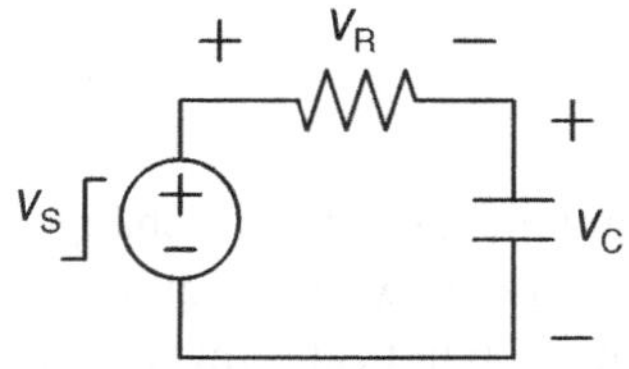

Figure 9.10a Series RC circuit driven by voltage step function.

Shunt Inductor High Pass Filter

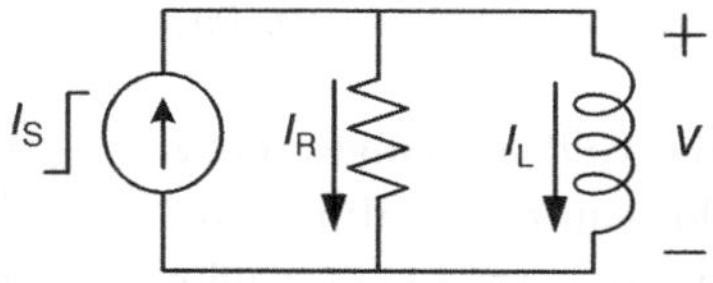

Figure 9.10b Parallel RL circuit driven by current step function.

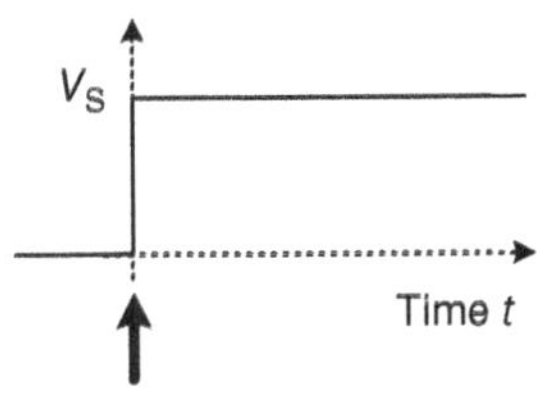

Figure 9.11a Graph of voltage step function $v_S(t)$ with amplitude V_S.

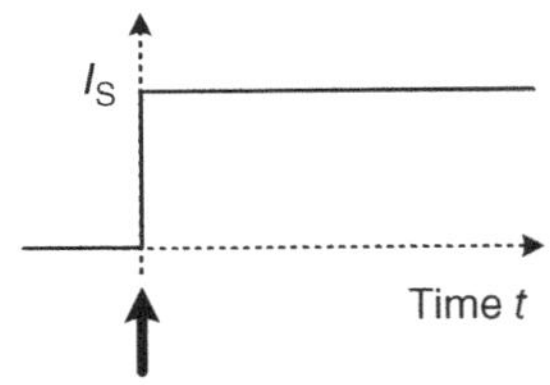

Figure 9.11b Graph of current step function $i_S(t)$ with amplitude I_S.

At the instant that the step function changes from zero to one, what happens to the voltages and currents in each circuit? In step response, all initial conditions, capacitor voltages and inductor currents, are zero for $t < 0$.

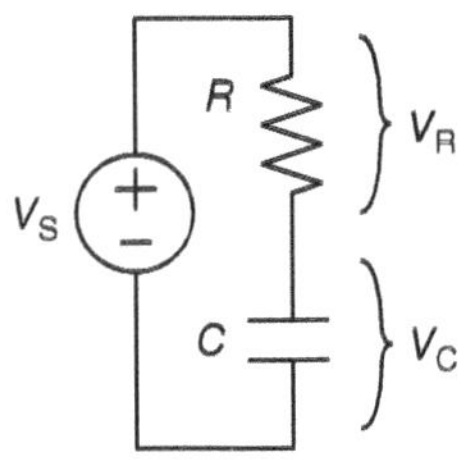

Figure 9.12a Series RC circuit driven by DC voltage V_S.

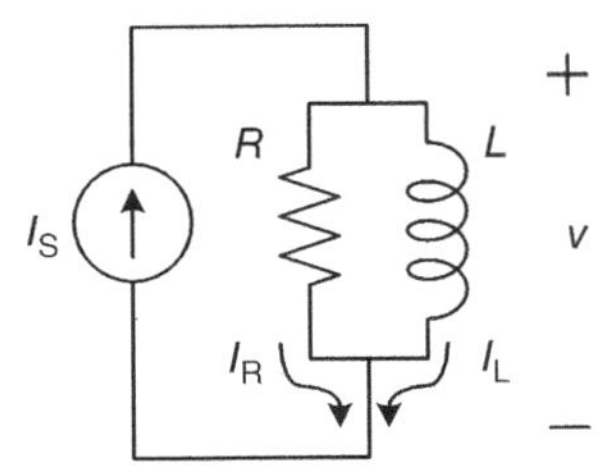

Figure 9.12b Parallel RL circuit driven by DC current I_S.

In the RC circuit in Figures 9.10a and 9.12a, V_s is dropped across the series combination of the resistor and capacitor. How much voltage is dropped across the resistor and how much voltage is dropped across the capacitor at exactly t equals 0, right after the voltage step rises?

In the RL circuit in Figures 9.10b and 9.12b, I_s is split between the parallel resistor and inductor. How much current goes through the resistor and how much current goes through the inductor at exactly t equals 0, right after the current step rises?

9.2.2 Step Response Branch Relationship Intuition

$$i_C(t) = C\frac{dv_C(t)}{dt} \tag{9.9a}$$

What intuition can the capacitor branch relationship Equation 9.9a provide?

$$i_C \neq C\frac{\Delta v_C}{dt} \tag{9.10a}$$

There can be no instantaneous voltage changes across a capacitor because it would require an infinite current. Equation 9.10a indicates this impossibility with an inequality for a finite change in capacitor voltage (Δv_C) divided by an infinitesimally small change in time dt.

$$v_L(t) = L\frac{di_L(t)}{dt} \tag{9.9b}$$

What intuition can the inductor branch relationship in Equation 9.9b provide?

$$v_L \neq L\frac{\Delta i_L}{dt} \tag{9.10b}$$

There can be no instantaneous current changes through an inductor because it would require an infinite voltage. Equation 9.10b indicates this impossibility with an inequality for a finite change in inductor current (Δi_L) divided by an infinitesimally small change in time dt.

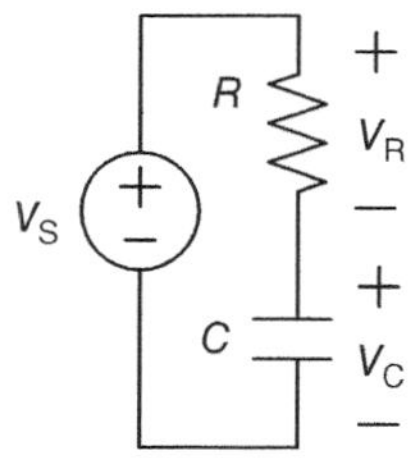

Figure 9.13a Series RC circuit driven by DC voltage V_S.

At the instant t equals 0, in the RC circuit in Figure 9.13a, the source voltage V_S, dropped across the resistor, causes an initial current V_S/R that starts charging the capacitor.

$$i_R(t)\big|_{t=0} = \frac{V_S}{R} = i_C(t)\big|_{t=0} \text{ At } t = 0 \tag{9.11a}$$

9.2.3 Step Response Intuitive Analysis Results

From that initial value of loop current, V_S over R, described in Equation 9.11a, the current decays exponentially, as described in Equation 9.12a and plotted in Figure 9.14a.

$$i(t) = \frac{V_S}{R} e^{-t/RC} \tag{9.12a}$$

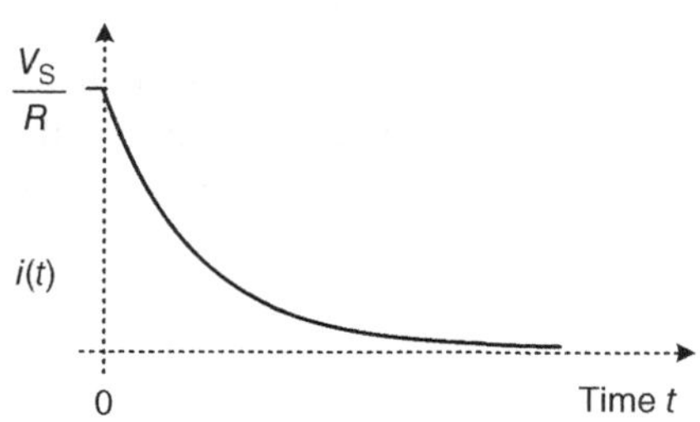

Figure 9.14a Series RC circuit current response to voltage step function.

At t equals 0 in the RC circuit in Figure 9.13a, all the source voltage V_S in the drops across the resistor, so v_R starts high, as shown in Figure 9.15a. The current $i(t)$ also starts high, as shown in Figure 9.14a. Both decay exponentially over time to zero.

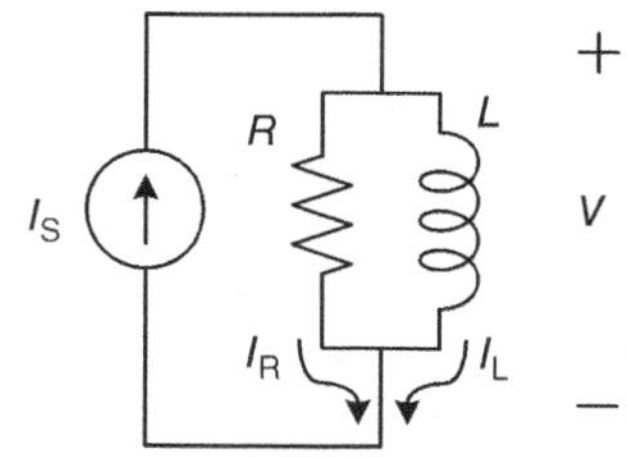

Figure 9.13b Parallel RL circuit driven by DC current I_S.

At the instant t equals 0, in the RL circuit in Figure 9.13b, the source current I_S, through the resistor, causes voltage $I_S R$ across the inductor that leads to current through the inductor.

$$v_L(t)\big|_{t=0} = I_S R = v(t)\big|_{t=0} \qquad \text{At } t = 0 \tag{9.11b}$$

From that initial value of voltage, $I_S R$, across all the circuit elements, described in Equation 9.11b, the voltage decays exponentially, as described in Equation 9.12b and plotted in Figure 9.14b.

$$v(t) = I_S e^{-tR/L} \tag{9.12b}$$

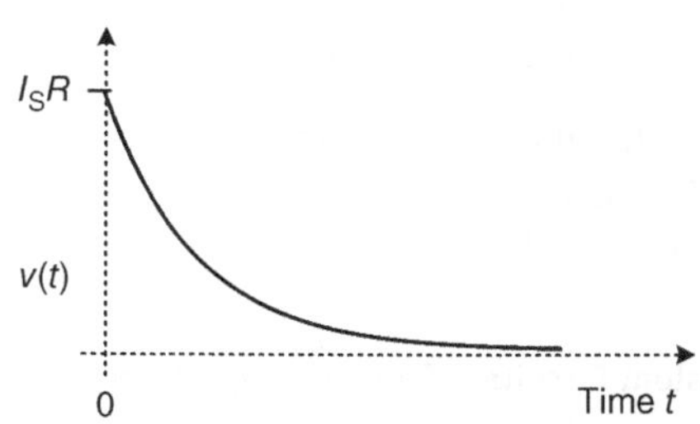

Figure 9.14b Parallel RL circuit voltage response to current step function.

At t equals 0 in the RL circuit in Figure 9.13b, all the current initially goes through the resistor, so i_R starts also high, as shown in Figure 9.15b. The voltage $v(t)$ also starts high, as shown in Figure 9.14b. Both decay exponentially over time to zero.

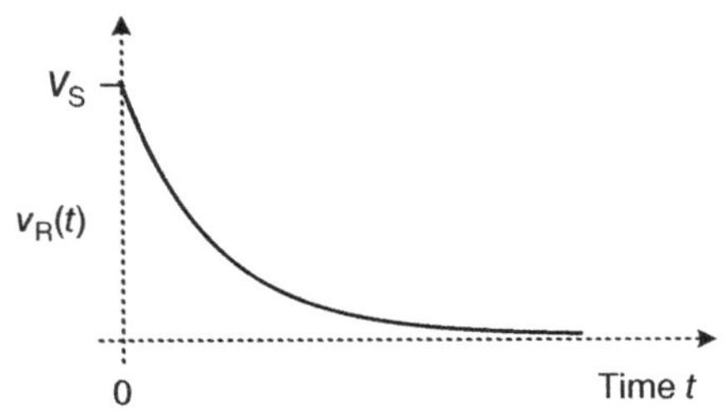

Figure 9.15a Series RC circuit resistor voltage response to voltage step function.

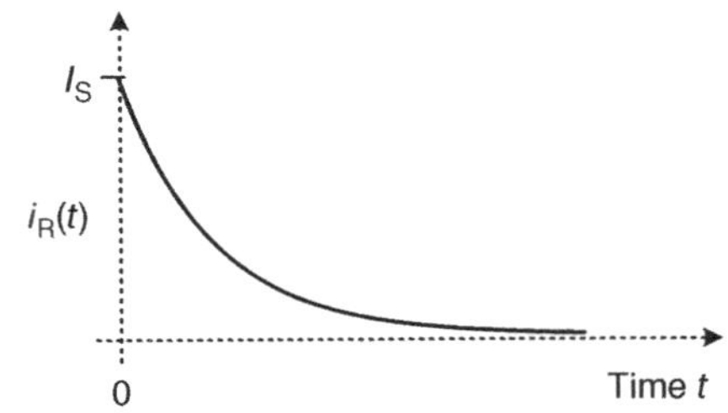

Figure 9.15b Parallel RL circuit resistor current response to current step function.

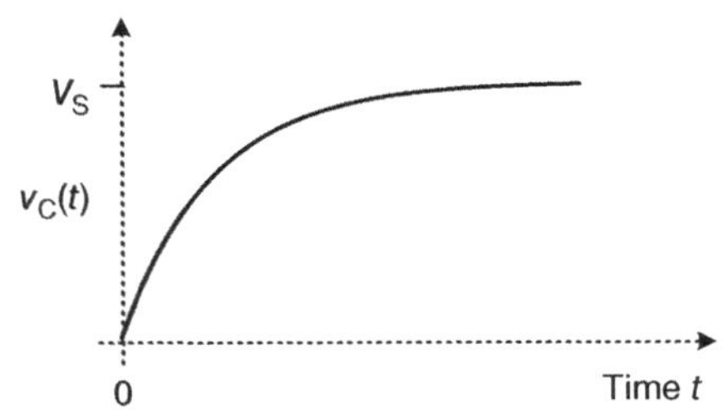

Figure 9.16a Series RC circuit capacitor voltage response to voltage step function.

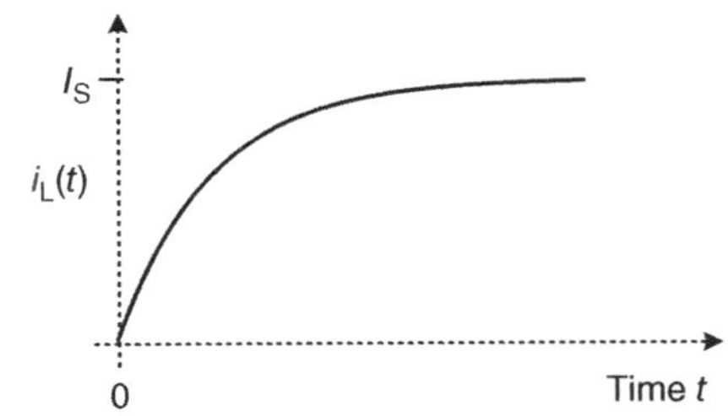

Figure 9.16b Parallel RL circuit inductor current response to current step function.

At t equals 0, there is no voltage across the capacitor, so v_C is low and grows exponentially over time, as shown in Figure 9.16a.

Comparing Figures 9.15a and 9.16a, as the current charges the capacitor for $t > 0$, the voltage across the capacitor $v_C(t)$, increases, decreasing the voltage across the resistor $v_R(t)$. (Why? KVL says that $v_R(t)$ plus $v_C(t)$ always add up to V_s, which is a constant.)

In the steady state, after a long time, the capacitor charges up to V_S. There is no voltage across the resistor and current flow stops altogether.

At t equals 0, there is no current through the inductor, so $i_L(t)$ is low and grows exponentially over time, as shown in Figure 9.16b.

Comparing Figures 9.15b and 9.16b, as more current flows through the inductor, less current flows through the resistor. (Why? KCL says that all the currents entering or leaving a node must add to zero.)

In the steady state, after a long time, the inductor builds up current to I_S. There is no current through the resistor and the output voltage across the resistor and the inductor goes to zero.

9.2.3.1 Equivalent Circuits – Same Voltage Response

Thévenin and Norton's theorems can show that the two circuits, in Figures 9.17a and 9.17b, are equivalent. They have the same output voltage across the inductor.

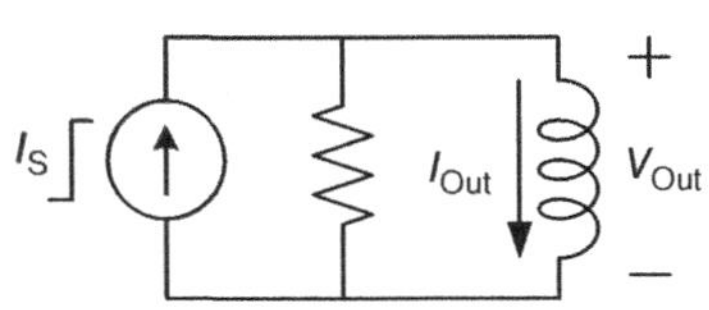

Figure 9.17a Parallel RL circuit driven by current step function.

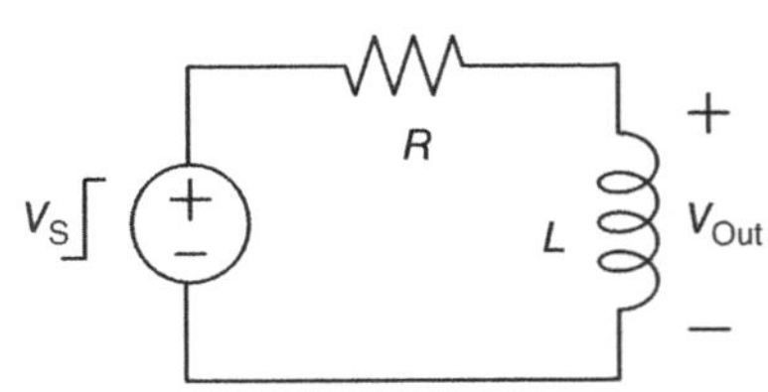

Figure 9.17b Series RL circuit driven by voltage step function.

9.3 Series RC Step Response Solution Using a Particular Solution

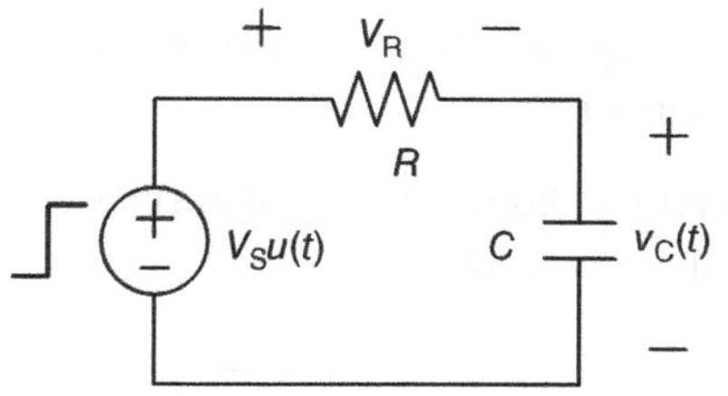

Figure 9.18 Series RC circuit driven by voltage step function.

This section calculates the step response, of the series RC circuit in Figure 9.18, driven by a voltage source step function. The solution approach calculates a particular solution to the non-homogeneous differential equation, that can be added to the natural response homogeneous differential equation solution, to generate the general solution to the step response non-homogeneous differential equation.

$$v_{out} = v_h + v_p \tag{9.13}$$

v_h in Equation 9.13 is the voltage solution to the homogeneous differential equation, the natural response of the undriven RC circuit.

v_p in Equation 9.13 is one specific voltage solution to the non-homogeneous differential equation, the forced response of the driven RC circuit.

The general solution to the homogeneous equation, and one particular solution to the non-homogeneous differential equation, provide the most general solution to the non-homogeneous differential equation corresponding to driven circuits. Figure 9.19 shows the unit step function, with an amplitude of 1.

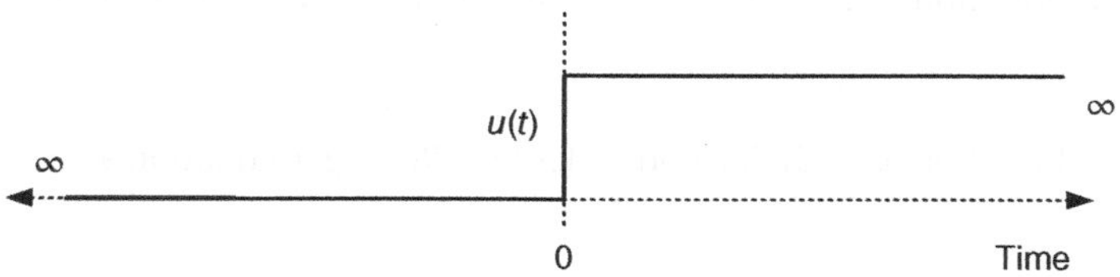

Figure 9.19 Unit step function signal versus time.

$$u(t) = 0 \quad \text{for } t < 0 \tag{9.14}$$

$$u(t) = 1 \quad \text{for } t \geq 0 \tag{9.15}$$

Equations 9.14 and 9.15 define the unit step function $u(t)$ shown in Figure 9.19.

$u(t)$ is the unit step function used in mathematics including with Laplace transforms.

$$v_S(t) = 0 \quad \text{for } t < 0 \tag{9.16}$$

$$v_S(t) = V_S \quad \text{for } t \geq 0 \tag{9.17}$$

V_S is a DC voltage. $v_S(t)$ is a function of time. $v_S(t)$ equals V_S times the unit step function $u(t)$.

One can often solve step response differential equations for time $t > 0$, eliminating the $u(t)$ term from the equations.

Equations 9.16 and 9.17 define the voltage input signal used to analyze step response in the RC circuit in Figure 9.18.

For the driven RC circuit in Figure 9.18, write Kirchhoff's voltage law (KVL), as shown in Equation 9.18.

$$v_C(t) + v_R(t) = v_S \tag{9.18}$$

Equation 9.18 says that the source voltage v_S equals the voltage v_C across the capacitor plus the voltage v_R across the resistor.

$$v_C(t) + Ri(t) = v_S \tag{9.19}$$

In Equation 9.18, apply Ohm's law to the resistor voltage v_R to get Equation 9.19.

$$i_C(t) = C\frac{dv_C(t)}{dt} \tag{9.20}$$

Equation 9.20, the branch relationship for an ideal capacitor, states the current through a capacitor in terms of the voltage across the capacitor.

In this series RC, the current is the same in all three components. Therefore, the capacitor current in Equation 9.20 can be used as the current $i(t)$ throughout the circuit.

Substitute the capacitor current Equation 9.20 into the Ohm's law resistor voltage term in Equation 9.19.

$$v_C(t) + RC\frac{dv_C(t)}{dt} = V_S \tag{9.21}$$

The non-homogeneous differential Equation 9.21 describes the forced response in terms of capacitor voltage.

9.3.1 Natural Response of the Homogeneous RC Circuit

Taking the non-homogeneous differential in Equation 9.21, and removing the source term V_s, leaves a homogeneous differential equation, in Equation 9.22 that corresponds to Figure 9.20.

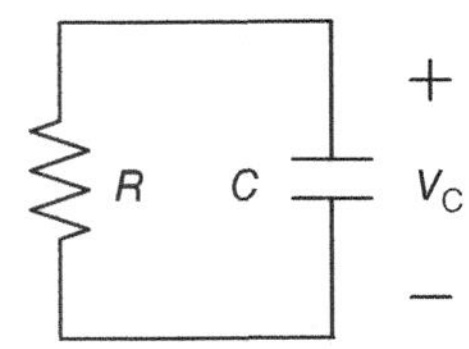

Figure 9.20 Parallel RC circuit for natural response.

$$v_C(t) + RC\frac{dv_C(t)}{dt} = 0 \tag{9.22}$$

In Equation 9.22, move the RC derivative term to the other side as shown in Equation 9.23.

$$v_C(t) = -RC\frac{dv_C(t)}{dt} \tag{9.23}$$

In Equation 9.23, multiply both sides by dt to get Equation 9.24.

$$v_C(t)dt = -RCdv_C(t) \tag{9.24}$$

In Equation 9.24, divide both sides by $-RC$ to get Equation 9.25.

$$\frac{-1}{RC}v_C(t)dt = dv_C(t) \tag{9.25}$$

In Equation 9.25, divide both sides by $v_C(t)$ to get Equation 9.26.

$$\frac{-1}{RC}dt = \frac{1}{v_C(t)}dv_C(t) \tag{9.26}$$

Integrate both side of Equation 9.26 to get Equation 9.27.

$$\int \frac{-1}{RC}dt = \int \frac{1}{v_C(t)}dv_C(t) \tag{9.27}$$

Perform the integrations in Equation 9.27 to get Equation 9.28. The integral of 1 over x is ln$x$. The constant of integration from each side can be combined into a single constant K_1.

$$\frac{-t}{RC} + K_1 = \ln v_C(t) \tag{9.28}$$

Apply Equation 9.29 to Equation 9.28 to obtain Equation 9.30, the equation for the capacitor voltage of the undriven RC circuit.

$$e^{x+y} = e^x e^y \tag{9.29}$$

In Equation 9.29, the exponential of a sum is the product of exponentials.

In Equation 9.28, K_1 is an unknown constant. Exponentiating K_1 doesn't really change it, as indicated in Equation 9.30.

$$v_C(t) = K_1 e^{\frac{-t}{RC}} \tag{9.30}$$

Equation 9.30 is v_h in Equation 9.13 for v_{out} equals v_h plus v_p. In Equation 9.30, v_{out} equals $v_C(t)$, which is also $v(t)$, as restated in Equation 9.31.

$$v(t) = K_1 e^{\frac{-t}{RC}} \tag{9.31}$$

Equation 9.31 is the general solution to the homogeneous equation, with an unknown arbitrary constant K_1, and the next step toward the general solution is to find one particular solution to the non-homogeneous differential equation.

9.3.2 Generate a Particular Solution from a Constant Function

Equation 9.32 is the non-homogeneous differential equation for the RC circuit in Figure 9.18, with the forcing function V_S on the right-hand side of the equation.

$$v(t) + RC\frac{dv(t)}{dt} = V_S \quad (9.32)$$

Equation 9.32 is the non-homogeneous differential equation for step response, generated from KVL where $v(t)$ is the capacitor voltage.

Because V_S is a constant for t greater than 0, try an unknown constant A as the particular solution to the non-homogeneous differential Equation 9.32.

$$A + (RC)0 = V_S \quad (9.33)$$

Substitute an unknown constant A into Equation 9.32 as a proposed particular solution function for $v(t)$. The derivative of a constant is zero, as shown in Equation 9.33.

$$A = V_S \quad (9.34)$$

From Equation 9.33, eliminate the 0 middle term to obtain Equation 9.34.

$$v(t) = K_1 e^{\frac{-t}{RC}} + V_S \quad (9.35)$$

Use Equations 9.31 and 9.34 to construct the general solution Equation 9.35 according to Equation 9.13 where $v(t)$ is the capacitor voltage.

Use initial conditions to calculate the value of the unknown constant K_1. At time t equals 0, before the rising edge, $v_C(0)$ equals 0.

$$v(0) = 0 = K_1 e^{\frac{0}{RC}} + V_S \quad (9.36)$$

Substitute t equals 0 into Equation 9.35 to get Equation 9.36.

$$0 = K_1 + V_S \quad (9.37)$$

In Equation 9.36, the exponential e-to-the-0 equals 1, yielding Equation 9.37.

$$K_1 = -V_S \quad (9.38)$$

The previously unknown constant K_1 equals minus V_S, as shown in Equation 9.38.

$$v(t) = -V_S e^{\frac{-t}{RC}} + V_S \quad (9.39)$$

From Equation 9.38, substitute $-V_S$ into Equation 9.35 for K_1, to obtain Equation 9.39.

$$v(t) = V_S - V_S e^{\frac{-t}{RC}} \quad (9.40)$$

Rearrange the terms in Equation 9.39 to obtain Equation 9.40.

Bring out the V_S term in Equation 9.40 to obtain Equation 9.41.

$$v(t) = v_{out} = V_S\left(1 - e^{\frac{-t}{RC}}\right) \quad (9.41)$$

Equation 9.41 is the general solution for the step response of the low-pass filter RC circuit in Figure 9.18 where $v(t)$ is the capacitor voltage.

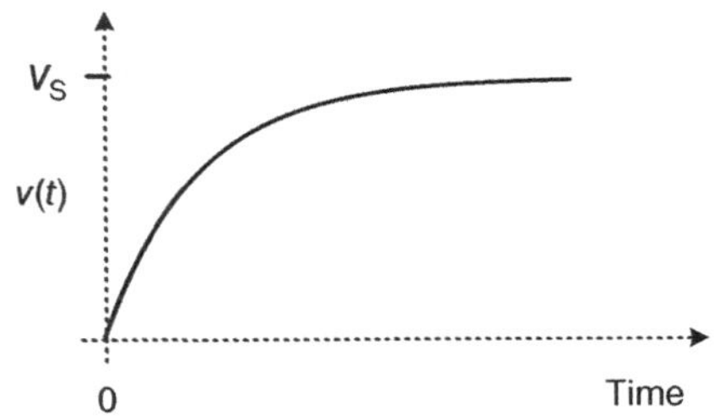

Figure 9.21 Series RC circuit capacitor output voltage response to voltage step function.

Figure 9.21 plots Equation 9.41, the general solution for the step response of the low-pass filter RC circuit in Figure 9.18. The capacitor voltage $v_C(t)$ equals $v(t)$.

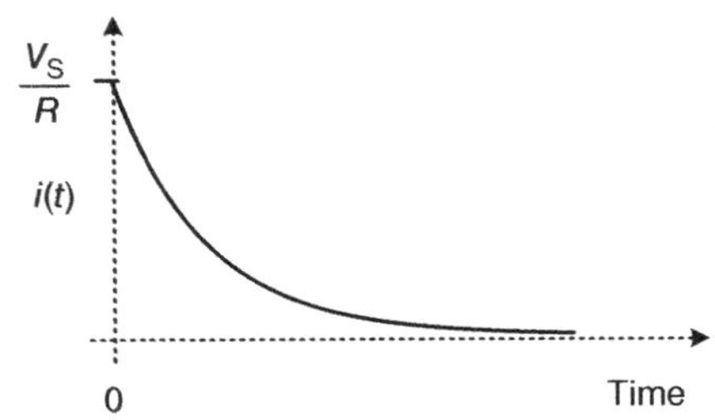

Figure 9.22 Series RC circuit current response to voltage step function.

Figure 9.22 plots the loop current $i(t)$, from Equation 9.42, in the general solution for the step response of the low-pass filter RC circuit in Figure 9.18. This current is related to the resistor voltage v_R by Ohm's law.

$$i(t) = \frac{V_S}{R} e^{\frac{-t}{RC}} \tag{9.42}$$

9.4 Series RL Step Response Solution Using a Particular Solution

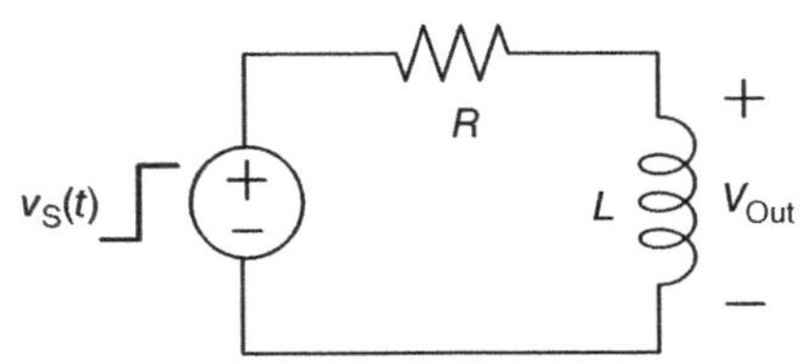

Figure 9.23 Series RL circuit driven by voltage step function $v_S(t)$.

This section calculates the step response, of the series RL circuit in Figure 9.23, driven by a voltage source step function. The solution approach calculates a particular solution that can be added to the natural response homogeneous differential equation solution to generate the general solution to the step response non-homogeneous differential equation.

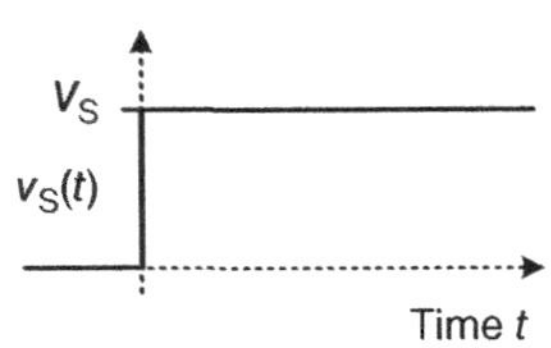

Figure 9.24 Graph of voltage step function $v_S(t)$ with amplitude V_S.

$$v_S(t) = V_S\, u(t) \tag{9.43}$$

$$v_S(t) = 0 \qquad \text{for } t < 0 \tag{9.44}$$

$$v_S(t) = V_S \qquad \text{for } t \geq 0 \tag{9.45}$$

V_S is a DC voltage. $v_S(t)$ is a function of time, as shown in Figure 9.24. $v_S(t)$ equals V_S times the unit step function $u(t)$. Often we solve for time $t > 0$, eliminating the $u(t)$ term from the equation.

Equation 9.46 shows the general solution v_{out}, to the RL circuit driven by a voltage source in Figure 9.23.

$$v_{\text{out}} = v_{\text{h}} + v_{\text{p}} \tag{9.46}$$

v_{h} = Solution to the homogeneous differential equation.

v_{p} = Particular solution to the non-homogeneous differential equation.

v_{h} in Equation 9.46 is the voltage solution to the homogeneous differential equation, the natural response of the undriven RL circuit.

v_{p} in Equation 9.46 is one specific voltage solution to the non-homogeneous differential equation, the forced response of the driven RL circuit.

This solution process follows a standard differential equation solution approach The general solution to the homogeneous equation, and one particular solution to the non-homogeneous differential equation, provide the most general solution to the non-homogeneous differential equation corresponding to driven circuits.

To begin the step response analysis of the driven RL circuit in Figure 9.23, write Kirchhoff's voltage law (KVL), as shown in Equation 9.47.

$$v_{\text{S}} = v_{\text{R}}(t) + v_{\text{L}}(t) \tag{9.47}$$

Equation 9.47 says that the source voltage v_{S} equals the voltage v_{L} across the inductor plus the voltage v_{R} across the resistor.

$$v_{\text{L}} = L\frac{di_{\text{L}}(t)}{dt} \tag{9.48}$$

The voltage across an inductor equals the inductance times the time derivative of the inductor current with respect to time, as shown in Equation 9.48.

$$V_{\text{S}} = Ri(t) + L\frac{di(t)}{dt} \tag{9.49}$$

In Equation 9.47, apply Ohm's law to the resistor voltage v_{R}, and apply Equation 9.48 to the inductor voltage to get Equation 9.49. The non-homogeneous differential Equation 9.49 describes the forced response in terms of inductor current.

9.4.1 Natural Response of the Homogeneous RL Circuit

Taking the non-homogeneous differential in Equation 9.49, and removing the source term V_{s}, leaves a homogeneous differential equation, in Equation 9.50, that corresponds to Figure 9.25.

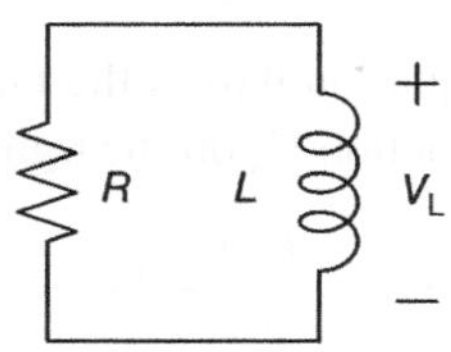

Figure 9.25 Parallel RL circuit for natural response.

$$Ri(t) + L\frac{di(t)}{dt} = 0 \tag{9.50}$$

In Equation 9.50, move the *RL* derivative term to the other side as shown in Equation 9.51.

$$Ri(t) = -L\frac{di(t)}{dt} \tag{9.51}$$

In Equation 9.51, multiply both sides by *dt* to get Equation 9.52.

$$Ri(t)dt = -Ldi(t) \tag{9.52}$$

In Equation 9.52, divide both sides by $-L$ to get Equation 9.53.

$$-\frac{R}{L}i(t)dt = di(t) \tag{9.53}$$

In Equation 9.53, divide both sides by $i(t)$ to get Equation 9.54.

$$-\frac{R}{L}dt = \frac{di(t)}{i(t)} \quad (9.54)$$

Integrate both sides of Equation 9.54 to get Equation 9.55.

$$\int -\frac{R}{L}dt = \int \frac{di(t)}{i(t)} \quad (9.55)$$

Perform the integrations in Equation 9.55, using the integral formula in Equation 9.56, to get Equation 9.57. The constant of integration from each side can be combined into a single constant K_1.

$$\int \frac{1}{x}dx = \ln x \quad (9.56)$$

Exponentiate both sides of Equation 9.57, using Equation 9.58, to get Equation 9.59 for the inductor current of the undriven RL circuit.

$$K_1 - \frac{R}{L}t = \ln i(t) \quad (9.57)$$

Equation 9.58 says that the exponential of a sum is the product of exponentials.

$$e^{x+y} = e^x e^y \quad (9.58)$$

In Equation 9.57, K_1 is an unknown constant. Exponentiating K_1 doesn't really change that, so it appears the same K_1 in Equation 9.60.

$$K_1 e^{-tR/L} = i(t) \quad (9.59)$$

Considering $i(t)$ in terms of the entire loop current, rather than the inductor current in Equation 9.48, we have dropped the L subscript on current.

$$i(t) = K_1 e^{-tR/L} \quad (9.60)$$

Equations 9.59 and 9.60, the natural response equation for the inductor current of the undriven RL circuit, is v_h in Equation 9.46 v_{out} equals v_h plus v_p.

Equation 9.60 is the general solution to the homogeneous equation, with an unknown arbitrary constant K_1, and the next step toward the general solution is to find one particular solution to the non-homogeneous differential equation.

9.4.2 Generate a Particular Solution from a Constant Function

Equation 9.61 is the non-homogeneous differential equation for the RL circuit in Figure 9.23, with the forcing function V_S on the right-hand side of the equation.

$$Ri(t) + L\frac{di(t)}{dt} = V_S \quad (9.61)$$

Equation 9.61 is the non-homogeneous differential equation for step response, generated from KVL.

Because V_S is a constant for t greater than 0, try an unknown constant A as the particular solution to the non-homogeneous differential Equation 9.61.

$$RA + L\frac{dA}{dt} = V_S \quad (9.62)$$

Substitute an unknown constant A into Equation 9.61 as a proposed particular solution for the current function, obtaining Equation 9.62.

$$RA + L \cdot 0 = V_S \quad (9.63)$$

In Equation 9.62, the derivative of the constant A is zero, resulting in L multiplied by zero in Equation 9.63.

$$A = \frac{V_S}{R} \quad (9.64)$$

Remove the middle L term from Equation 9.63 to obtain Equation 9.64.

$$i_L(t) = K_1 e^{-tR/L} + \frac{V_S}{R} \quad (9.65)$$

Use Equations 9.60 and 9.64 to construct the general solution Equation 9.65 according to Equation 9.13.

V_S over R in Equation 9.65 is the particular solution needed for constructing the general solution.

$$i_L(t) = 0 = K_1 e^{-0(R/L)} + \frac{V_S}{R} \quad (9.66)$$

Use initial conditions to calculate the value of the unknown constant K_1. At time t equals 0, before the rising edge of the current step function input, the current $i(t)$ equals 0, illustrated in Equation 9.66.

$$0 = K_1 + \frac{V_S}{R} \quad (9.67)$$

The exponential e-to-the-0 equals 1. Simplifying Equation 9.66 leads to Equation 9.67.

$$K_1 = -\frac{V_S}{R} \quad (9.68)$$

The previously unknown constant K_1 equals minus V_S over R, as shown in Equation 9.68.

$$i(t) = -\frac{V_S}{R} e^{-tR/L} + \frac{V_S}{R} \quad (9.69)$$

Plug $-V_S$ into Equation 9.65 for K_1 to obtain Equation 9.69.

$$i(t) = \frac{V_S}{R} - \frac{V_S}{R} e^{-tR/L} \quad (9.70)$$

Rearrange terms in Equation 9.69 to obtain Equation 9.70.

Factor out the V_S term in Equation 9.70 to obtain the result in Equation 9.71

$$i(t) = \frac{V_S}{R}(1 - e^{-tR/L}) \quad (9.71)$$

Equation 9.71 is the general solution for the **step response** of the RL circuit in Figure 9.23.

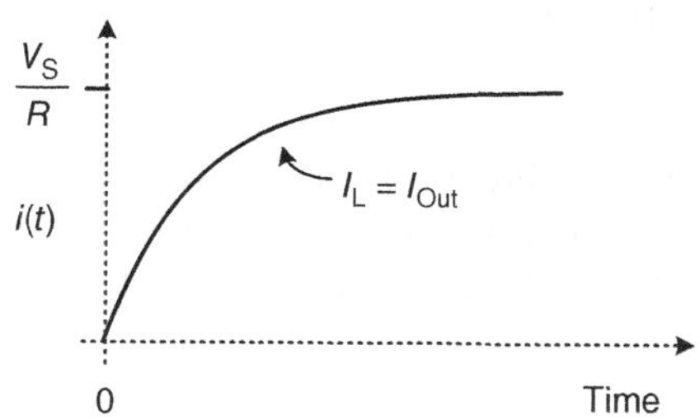

Figure 9.26 Series RL circuit current response to voltage step function.

Figure 9.26 plots Equation 9.71, the general solution for the step response loop current of the RL circuit in Figure 9.23. The loop current $i(t)$ equals I_{out}.

$$v_R(t) = i(t)R = V_S(1 - e^{-tR/L}) \quad (9.72)$$

For the series RL circuit in Figure 9.23, the **resistor voltage** can be calculated by multiplying both sides of Equation 9.71 by R, as shown in Equation 9.72.

To calculate the inductor voltage v_{out} for the series RL circuit in Figure 9.23, subtract the resistor voltage v_R from the source voltage V_S, as shown in Equation 9.73.

$$V_S - v_R(t) = v_L(t) \quad (9.73)$$

Rearrange Equation 9.73 and substitute Equation 9.72 into Equation 9.73 for v_R, as shown in Equation 9.74.

$$v_L(t) = v_{out}(t) = V_S - V_S(1 - e^{-tR/L}) \quad (9.74)$$

Multiply out Equation 9.74 to obtain Equation 9.75 for v_L which equals v_{out}.

$$v_{out}(t) = V_S - V_S + V_S e^{-tR/L} \quad (9.75)$$

Simplify Equation 9.75, where the two V_S terms cancel, leading to Equation 9.76.

$$V_{out}(t) = V_S e^{-\frac{tR}{L}} \quad (9.76)$$

Equation 9.76 provides the inductor step response voltage $v_L(t)$ equals $v_{out}(t)$, for the series RL circuit in Figure 9.23.

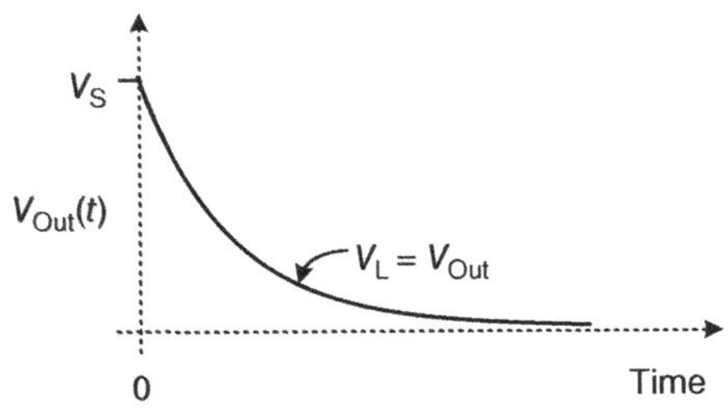

Figure 9.27 Series RL circuit inductor output voltage response to voltage step function.

Figure 9.27 plots Equation 9.76, the inductor voltage V_{out}, for the RL circuit in Figure 9.23.

9.5 Series RL Step Response with Voltage Source

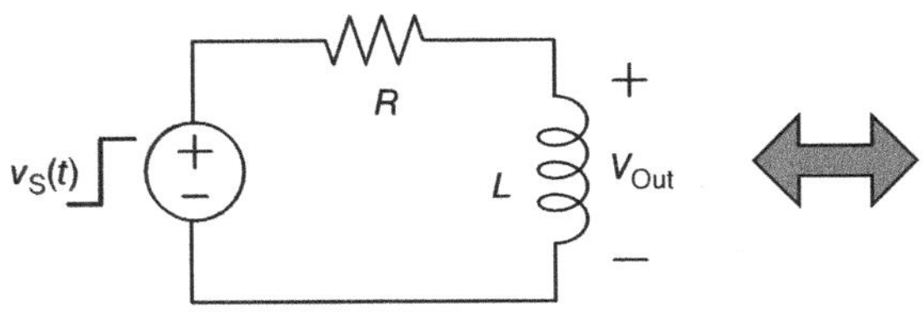

Figure 9.28a Series RL circuit driven by voltage step function $v_S(t)$.

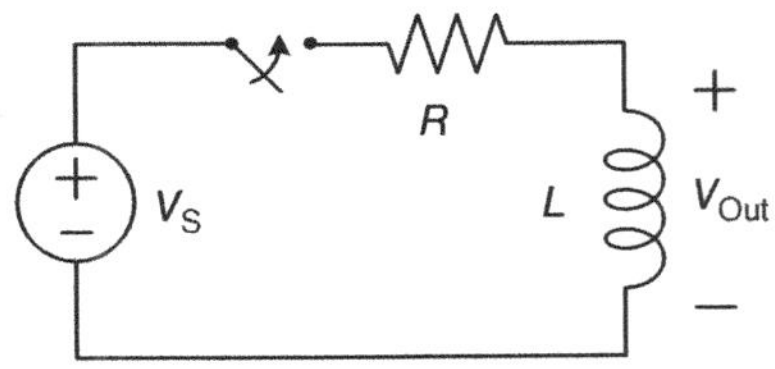

Figure 9.28b Series RL circuit with DC voltage V_S and switch modeling step function.

This section derives and solves the step response of a series RL circuit using a different approach than the previous section. The switch circuit in Figure 9.28b provides the same step response as step function in Figure 9.24, as reflected in the step symbol on the left side of the voltage source in Figure 9.28a. This analysis carries the forcing function through the whole differential equation derivation and solution.

$$v_R + v_L = V_S u(t) \tag{9.77}$$

Apply Kirchhoff's voltage law (KVL), to the series RL circuit in Figure 9.28a,b, which includes the step function $u(t)$ describing the switching behavior at t equals 0.

More details on the KVL equations and the step function are presented at the beginning of a previous section, starting at Equation 9.43.

$$Ri(t) + L\frac{di(t)}{dt} = V_S u(t) \tag{9.78}$$

In Equation 9.77, apply Ohm's law to the resistor voltage v_R, and apply Equation 9.48 to the inductor voltage to get Equation 9.78.

$$Ri(t) + L\frac{di(t)}{dt} = V_S \tag{9.79}$$

For time $t > 0$ the driving function is a DC voltage V_S. Drop $u(t)$ from Equation 9.78.

$$L\frac{di(t)}{dt} = V_S - Ri(t) \tag{9.80}$$

Rearrange Equation 9.79 algebraically, to separate the variables, putting the current term $Ri(t)$ on the right side in Equation 9.80.

$$Ldi(t) = [V_S - Ri(t)]dt \tag{9.81}$$

Multiply both sides of Equation 9.80 by dt to get Equation 9.81.

$$\frac{Ldi(t)}{V_S - Ri(t)} = dt \tag{9.82}$$

Divide both sides of the Equation 9.81 by the quantity V minus Ri to obtain Equation 9.82.

$$\int \frac{Ldi}{-Ri(t) + V_S} = \int dt \tag{9.83}$$

Integrate both sides of Equation 9.82, as shown in Equation 9.83.

$$L\int \frac{di}{-Ri(t) + V_S} = \int dt \tag{9.84}$$

Inductance is a constant that can be taken outside of the integral, as shown in Equation 9.84.

$$\int \frac{1}{ax+b}\,dx = \frac{1}{a}\ln(ax+b) \tag{9.85}$$

For the left side of the Equation 9.84, use the integral formula in Equation 9.85. For "a" substitute in minus R and for "b" use the source voltage V_S to get Equation 9.87.

$$\int K\,dx = Kx + C \quad K = 1 \tag{9.86}$$

For the right side of Equation 9.84, the integral of one, use the formula in Equation 9.86 for the integral of a constant.

$$\frac{-L}{R}\ln(V_S - Ri(t)) = t + C \tag{9.87}$$

In Equation 9.87, C is a constant of integration corresponding to an initial condition of the circuit.

Solve for the constant of integration, C, using initial conditions.

At time t equals 0, the current $i(t)$ also equals 0 because current in the inductor cannot instantaneously change when a finite voltage is applied is applied across the inductor.

$$\frac{-L}{R}\ln(V_S - Ri(t)) = t + C \tag{9.88}$$

In Equation 9.87, substitute in 0 for t and 0 for i(0), as shown in Equation 9.88, to obtain Equation 9.89.

$$\frac{-L}{R}\ln V_S = C \tag{9.89}$$

$$\frac{-L}{R}\ln(V_S - Ri(t)) = t + \frac{-L}{R}\ln V_S \tag{9.90}$$

From Equation 9.89, substitute the initial condition formula for C into Equation 9.87 to get Equation 9.90.

$$\frac{-L}{R}\ln(V_S - Ri(t)) - \frac{L}{R}\ln V_S = t \tag{9.91}$$

In Equation 9.90, move the ln V_S term to the left side of the equation to get Equation 9.91.

$$\frac{-L}{R}[\ln(V_S - Ri(t)) + \ln V_S] = t \tag{9.92}$$

In Equation 9.91, gather the terms multiplied by $-L$ over R, as shown in Equation 9.92.

$$[\ln(V_S - Ri(t)) + \ln V_S] = -(R/L)t \tag{9.93}$$

Multiply both sides of Equation 9.92 by $-R$ over L to get Equation 9.93.

$$\ln(a) - \ln(b) = \ln(a/b) \tag{9.94}$$

Use the law of logarithms in Equation 9.94 to combine terms on the left side of Equation 9.93, resulting in Equation 9.95.

$$\ln\left[\frac{V_S - Ri(t)}{V_S}\right] = -(R/L)t \tag{9.95}$$

$$\frac{V_S - Ri(t)}{V_S} = e^{-(R/L)t} \tag{9.96}$$

Exponentiate both sides of Equation 9.95 to obtain Equation 9.96.

$$\frac{V_S}{V_S} - \frac{Ri(t)}{V_S} = e^{-(R/L)t} \tag{9.97}$$

Separate the left side of Equation 9.96 into two terms over V_S, as shown in Equation 9.97.

$$\frac{-Ri(t)}{V_S} = -1 + e^{-(R/L)t} \tag{9.98}$$

Subtract the V_S over V_S term from the left-hand side of Equation 9.97 and subtract 1 from both sides, resulting in Equation 9.98.

$$i(t) = \frac{V_S}{R} - \frac{V_S}{R}e^{-(R/L)t} \tag{9.99}$$

Multiply both sides of Equation 9.98 by $-V_S$ over R to solve for $i(t)$ in Equation 9.99.

Factor out the V_S over R term from the right side of Equation 9.99 to get the result in Equation 9.100.

$$i(t) = \frac{V_S}{R}(1 - e^{-(R/L)t}) \tag{9.100}$$

Equation 9.100 is the forced response of the driven RL circuit in Figures 9.28a and 9.28b.

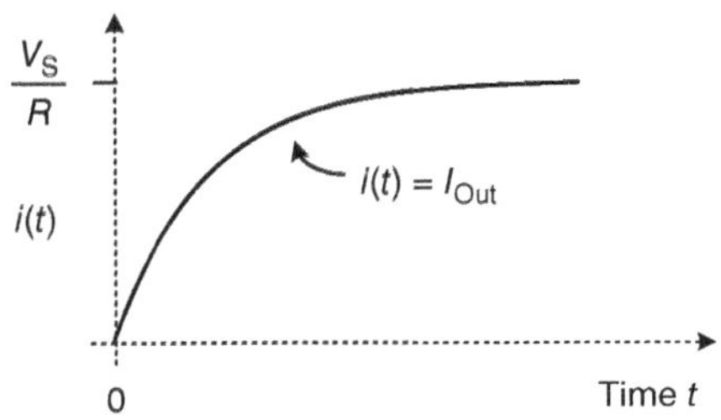

Figure 9.29 Series RL circuit current response to voltage step function.

Equation 9.100 is plotted in Figure 9.29.

Equation 9.100 and Figure 9.29 match Equation 9.71 and Figure 9.26.

9.6 First-Order Step Response Summary

In first-order electrical circuits, step response comes from an applied step function, an ideal instantaneous change with zero rise time.

The ideal step function is mathematically significant and it can be handled in the Laplace transform *s*-domain.

Step response is a forced response described by non-homogeneous differential equations. Step response is a zero-state response (ZSR), but not every ZSR is a step response.

The analysis of step response intentionally excludes initial conditions. The start point of the step function at minus infinity allows time for any existing energy to dissipate from the circuit to put it into a zero state before the step response step function is applied.

Step response can be modeled using switch configurations. In circuit simulation, such as SPICE, the step function needs to have a finite rise time slow enough to avoid numerical difficulties for the simulator.

For mathematical rigor, we imagine circuit inputs having an infinite amount of time to settle, allowing transients to go to zero, to create the initial conditions for the natural response.

10

Complete Response of First-Order RC and RL Circuits

10.1 First-Order Complete Response Summary Overview

10.1.1 Complete Response Definition Equations

The complete response of a circuit is the response to both initial conditions and input signals. The complete response of a circuit is the sum of the natural response and the step or other driven response.

$$v_C = v_N + v_S \quad (10.1a) \qquad i_C = i_N + i_S \quad (10.1b)$$

Equations 10.1a and 10.1b show complete response v_C is the sum of natural response v_N and the step response v_S. More generally, complete response is the sum of ZIR (zero-input response) and ZSR (zero-state response) as shown in Equation 10.2.

$$v_{out(complete)} = v_{out(ZIR)} + v_{out(ZSR)} \quad (10.2)$$

Natural response and step response analyses are constructed to exclude one another. Natural response is zero-input response, excluding input signals like step functions. Step response is a zero-state response, where zero state specifically excludes initial conditions.

Complete response captures all the circuit response behavior, the combination of both natural response and step response. After all, a step function can be applied to a circuit that already has some stored energy, some initial inductor currents, or initial capacitor voltages.

10.1.2 Complete Response RC and RL Results

This section presents examples of complete response for series RC and RL circuits, generated by adding the natural response to the step response. The two solutions are then rearranged to correspond to transient response plus steady-state response.

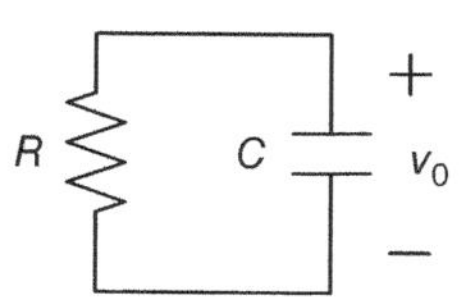

Figure 10.1a Parallel RC circuit with initial condition capacitor voltage v_0.

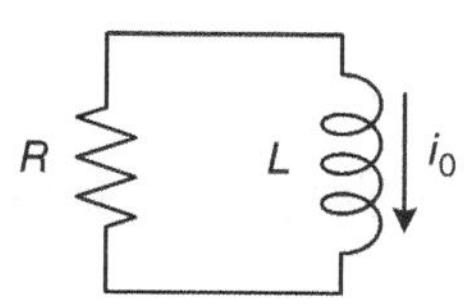

Figure 10.1b Parallel RL circuit with initial condition inductor current i_0.

$$v(t) = v_0 e^{(-t/RC)} \quad (10.3a) \qquad i(t) = i_0 e^{(-R/L)t} \quad (10.3b)$$

First and Second Order Circuits and Equations: Technical Background and Insights, First Edition. Robert O'Rourke.

Equation 10.3a is the natural response ZIR voltage for the RC circuit in Figures 10.1a and 8.2a, where v_0 and $v_C(0)$ are the initial condition voltage on the capacitor.

Equation 10.3b is the natural response ZIR current for the RL circuit in Figures 10.1b and 8.2b, where i_0 and $i_L(0)$ are the initial condition current in the inductor.

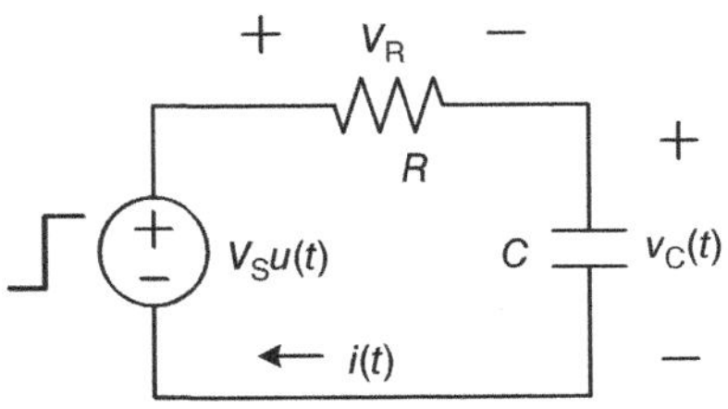

Figure 10.2a Series RC circuit driven by voltage step function.

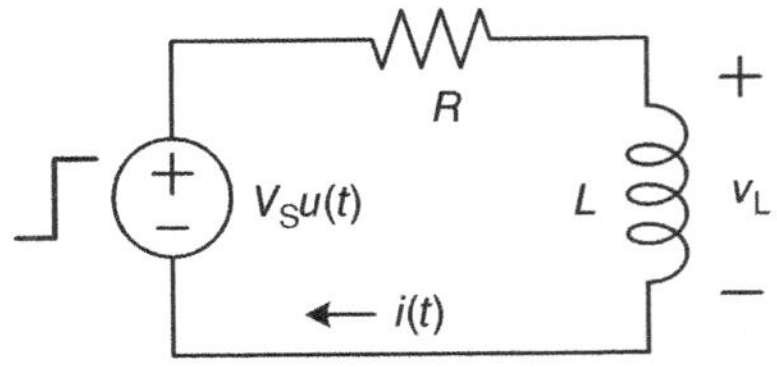

Figure 10.2b Series RL circuit driven by voltage step function.

$$v_C(t) = V_S(1 - e^{(-t/RC)}) \quad (10.4a)$$

Equation 10.4a is the step response voltage for the RC circuit in Figure 10.2a, where at time equals zero the voltage source V_S changes instantaneously from 0 V to V_S.

$$i(t) = \frac{V_S}{R}(1 - e^{(-R/L)t}) \quad (10.4b)$$

Equation 10.4b is the step response voltage for the RL circuit in Figure 10.2b, where at time equals zero the voltage source V_S changes instantaneously from 0 V to V_S.

$$v_C(t) = v_0 e^{(-t/RC)} + V_S(1 - e^{(-t/RC)}) \quad (10.5a)$$

Equation 10.5a is the complete response for the RC circuit in Figure 10.2a with a step function input and also the initial condition of voltage v_0 on the capacitor. The first term on the right-hand side of Equation 10.5a is the natural response and the second term is the step response.

$$i(t) = i_0 e^{(-R/L)t} + \frac{V_S}{R}(1 - e^{(-R/L)t}) \quad (10.5b)$$

Equation 10.5b is the complete response for the RL circuit in Figure 10.2b with a step function input and also the initial condition of current i_0 in the inductor. The first term on the right-hand side of Equation 10.5b is the natural response and the second term is the step response.

$$v_C(t) = V_S + (v_0 - V_S)e^{(-t/RC)} \quad (10.6a)$$

Equation 10.6a is the complete response for the RC circuit in Figure 10.2a. The terms in Equation 10.5a have been rearranged. V_S is the steady-state response and the right-hand term of Equation 10.6a is the transient response, which decays exponentially to zero.

$$i(t) = \frac{V_S}{R} + \left(i_0 - \frac{V_S}{R}\right) e^{(-R/L)t} \quad (10.6b)$$

Equation 10.6b is the complete response for the RL circuit in Figure 10.2b. The terms in Equation 10.5b have been rearranged. V_S over R is the steady-state response and the right-hand term of Equation 10.6b is the transient response.

10.2 Series RC Complete Response Examples

For a series RC circuit, this section explores two examples of initial capacitor voltage. The first example is 0.5 V $v_C(0)$. The second example is −0.5 V $v_C(0)$.

10.2.1 First RC Example – No Initial Capacitor Voltage

Figure 10.3 shows a series RC circuit driven by a 1 V step function starting at time equals 20 ns. The 25 nF capacitor has the initial voltage $v_C(0)$ set to 0. The resistor voltage is defined as V_S minus *VC*, establishing a polarity of positive current flowing clockwise around the loop.

With no initial conditions in the schematic in Figure 10.3, this circuit shows step response, for comparison and contrast to the following two examples of complete response.

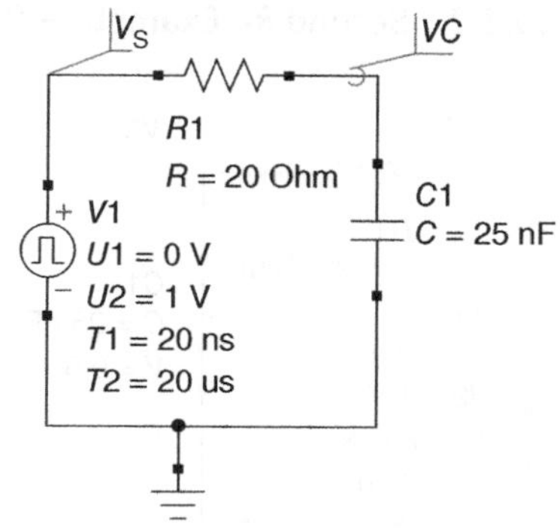

Figure 10.3 Series RC circuit simulation schematic with no initial condition.

Figure 10.4 shows the 1 V step function as a dashed line. The starting time of 20 ns is chosen to be able to see the signal not overlapped with the vertical axis. The capacitor voltage, *VC* in the schematic, builds up toward the source voltage V_S.

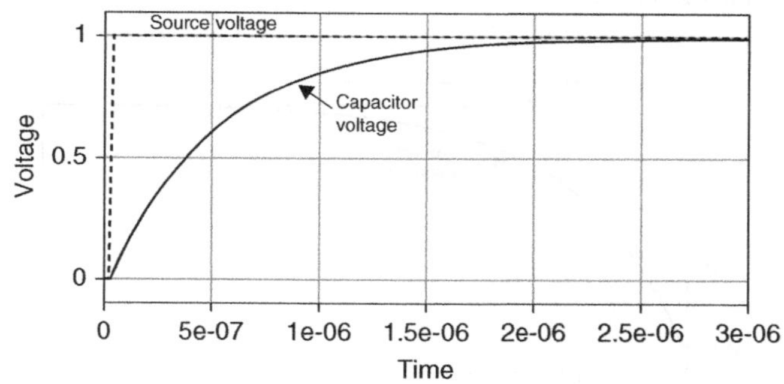

Figure 10.4 Series RC step response simulation plot of capacitor voltage.

Figure 10.5 shows the resistor current, defined by V_1 (V_S) minus *VC*, which is positive in the clockwise direction. It jumps up and then decays to zero exponentially over time. The negative source current indicates that the current polarity for the voltage source is positive coming out the bottom of the source and flowing in the counterclockwise direction.

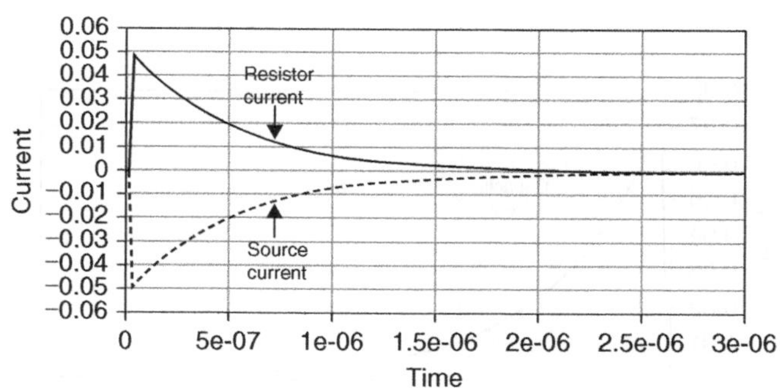

Figure 10.5 Series RC step response simulation plot of currents.

10.2.2 Second RC Example – Plus 0.5 V Initial Capacitor Voltage

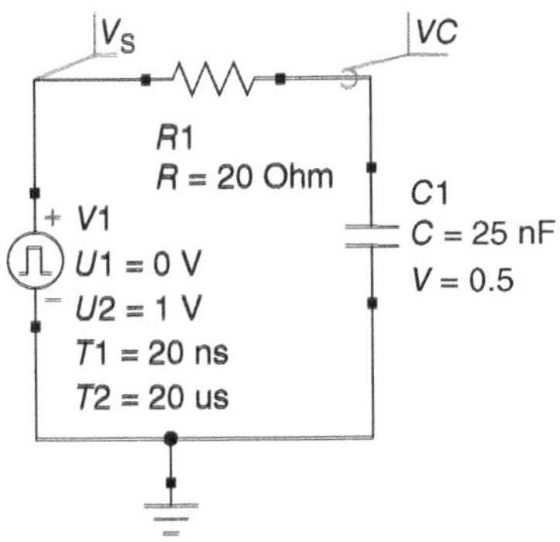

Figure 10.6 Series RC circuit simulation schematic with 0.5 V initial condition.

Figure 10.6 shows a series RC circuit driven by a 1 V step function starting at time equals 20 ns. The 25 nF capacitor has the initial voltage $v_C(0)$ set to 0.5 V. The resistor voltage is defined as V_S minus VC, establishing a polarity of positive current flowing clockwise around the loop. The polarity for the initial condition voltage is opposite of the resistor voltage polarity. Before the step function, VC is at negative 0.5 V and 25 mA current flows in the positive clockwise direction around the loop.

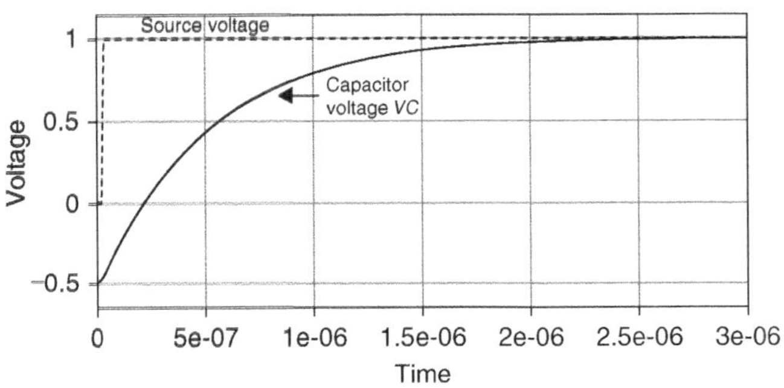

Figure 10.7 RC step response simulation plot of capacitor voltage with 0.5 V initial condition.

Figure 10.7 shows the 1 V step function as a dashed line. The solid line is the capacitor voltage, VC in the schematic. Because of a 0.5 V initial condition opposite in polarity to the clockwise positive current polarity in these graphs, the capacitor voltage starts at −0.5 V. After the step function, the capacitor voltage, VC in the schematic, increases 1.5 V over time, toward the source voltage of 1 V.

The 20 ns delay, from the initial conditions in place at time equals 0, to the step, has a small effect on the results.

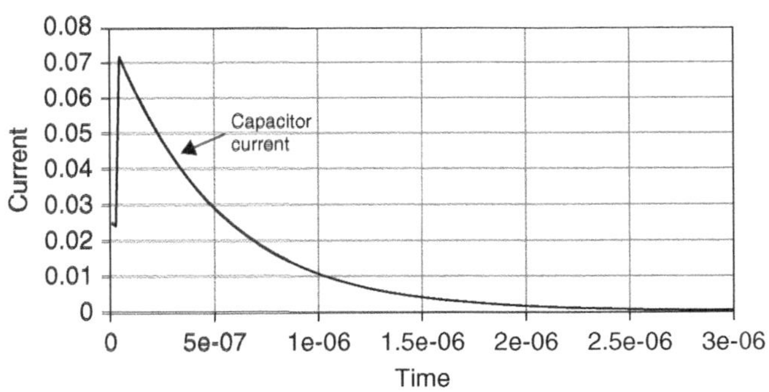

Figure 10.8 RC step response simulation plot of capacitor current with 0.5 V initial condition.

Figure 10.8 shows the loop current through the resistor and capacitor, positive in the clockwise direction. Because the capacitor voltage starts at −0.5 V, the initial voltage drop across the resistor is 1.5 V, resulting in 75 mA of initial current after the voltage input step. The capacitor current decreases exponentially down to zero over time.

10.2.3 Third RC Example – Minus 0.5 V Initial Capacitor Voltage

Figure 10.9 shows a series RC circuit driven by a 1 V step function starting at time equals 20 ns. The 25 nF capacitor has the initial voltage $v_C(0)$ set to −0.5. The resistor voltage is defined as V_S minus *VC*, establishing a polarity of positive current flowing clockwise around the loop. The polarity for the initial condition voltage is opposite of the resistor voltage polarity, making the *VC* node 0.5 V before the step function.

The 20 ns delay, from the initial conditions in place at time equals 0, to the step, has a small effect on the results.

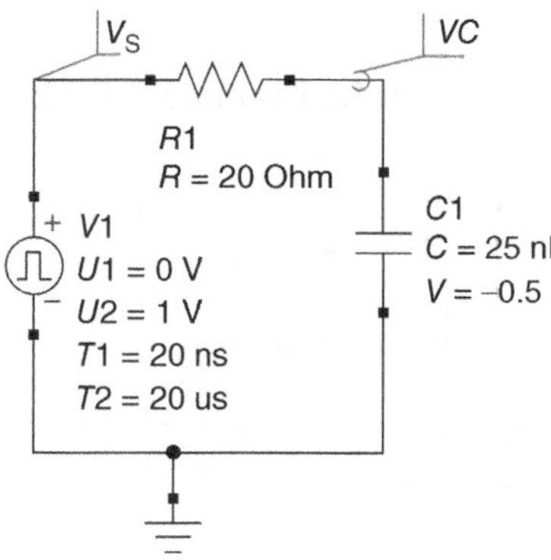

Figure 10.9 Series RC circuit simulation schematic with −0.5 V initial condition.

Figure 10.10 shows the 1 V step function as a dashed line. The starting time of 20 ns is chosen to be able to see the signal not overlapped with the vertical axis. The capacitor voltage, *VC* in the schematic, builds up toward the 1 V source voltage, starting from 0.5 V.

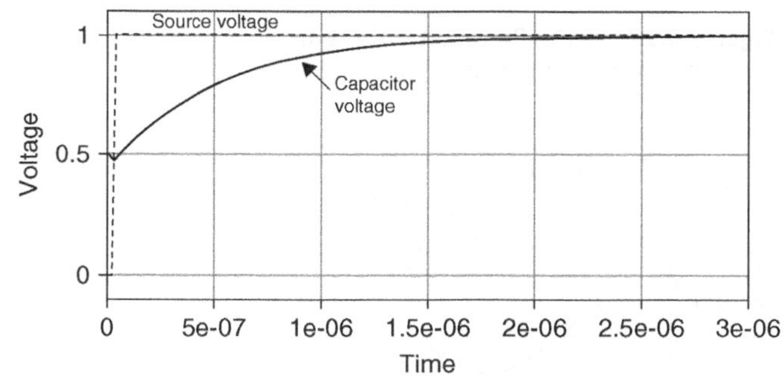

Figure 10.10 Series RC step response simulation plot of capacitor voltage.

Figure 10.11 shows the resistor current, defined by V_S minus *VC*, which is positive in the clockwise direction. It is the loop current also through the capacitor. Before the 1 V step input, node *VC* is at 0.5 V, generating current toward V_S. After the 1 V step input, the loop current jumps up from −25 to +25 mA, and then decays to zero exponentially over time.

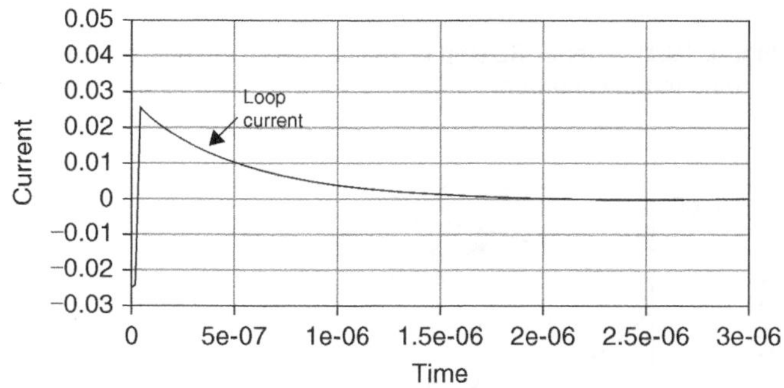

Figure 10.11 Series RC step response simulation plot of capacitor current.

10.3 RL Complete Response Example and Intuitive Analysis

For a series RL circuit, this section explores three examples of initial inductor current. The first example is zero inductor current, making it a step response example. The second example has the initial inductor flowing in the same direction as the loop current generated by the step function. In the third and final example, the initial condition inductor current flows in the opposite direction as the current driven by the input voltage source.

10.3.1 First RL Example – No Initial Inductor Current

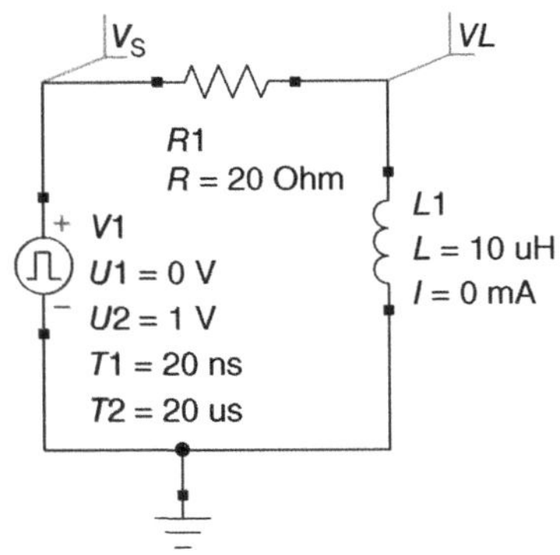

Figure 10.12 Series RL circuit simulation schematic with no initial condition.

Figure 10.12 shows a series RL circuit driven by a 1 V step function starting at time equals 20 ns. The 10 μH inductor has the initial current set to 0. The resistor voltage is defined as V_S minus VL, establishing a polarity of positive current flowing clockwise around the loop.

Figure 10.13 shows the 1 V step function as a dashed line. The starting time of the step function, at 20 ns, is chosen to be able to see the signal not overlapped with the vertical axis.

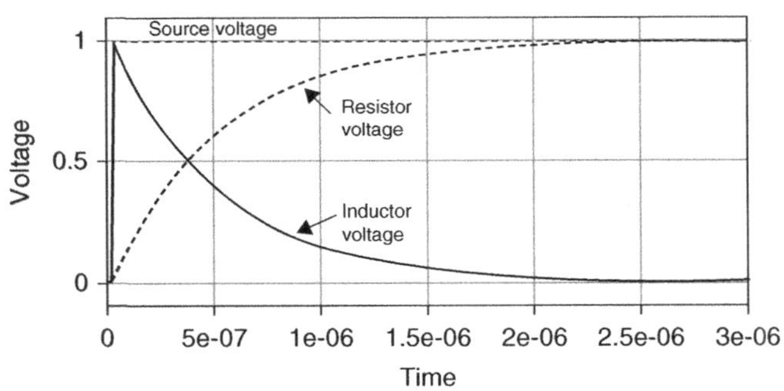

Figure 10.13 Series RL step response simulation plot of voltages with no initial condition.

Figure 10.13 shows the inductor voltage (solid line), initially jumping up from zero to the source voltage of 1 V, and then decaying exponentially with time. The dashed curved line is the voltage across the resistor, which builds up from zero over time to 1 V.

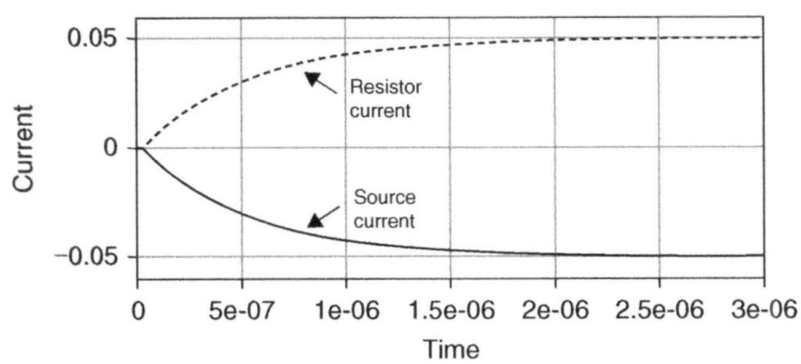

Figure 10.14 Series RL step response simulation plot of currents with no initial condition.

Figure 10.14 shows the resistor current, with the dashed line, as positive, flowing clockwise around the loop, building up over time. By Ohm's law, the resistor current and voltage, in Figure 10.13, have the same shape and polarity. The current starts at zero and ends up at 50 mA. The negative source current indicates a voltage source polarity of positive current coming down out of the voltage source, which means counterclockwise around the loop.

10.3.2 Second RL Example – Additive Initial Current

This series RL circuit example, shown in Figure 10.15, has an initial inductor current of 25 mA in the clockwise direction. The initial condition inductor current flows down out of the bottom of the inductor. The minus sign on the 25 mA current specification means that the positive polarity for the initial current refers to current coming out the top. Even though the initial inductor current is labeled as minus 25 mA, it is additive to the loop current chosen polarity of clockwise, established by the definition of resistor voltage as V_S minus VL, making i_0 equal to 25 mA in the equations and plots.

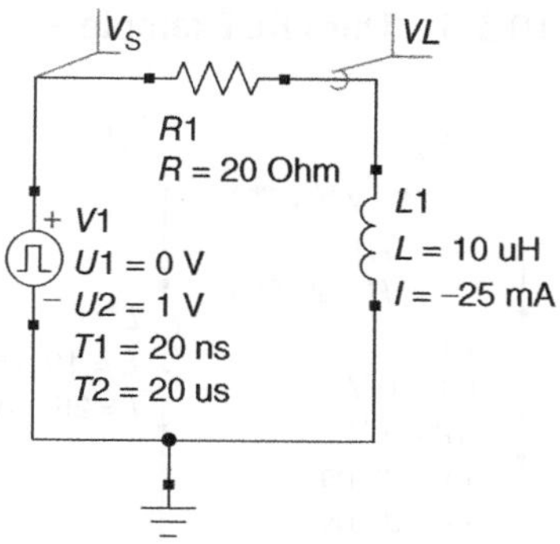

Figure 10.15 Series RL circuit simulation schematic with −25 mA initial condition.

The 25 mA initial condition current in the inductor creates a positive voltage across the resistor, making the inductor voltage negative to satisfy Kirchhoff's voltage law (KVL) before the step function is applied. The solid line in Figure 10.16 is inductor voltage, which jumps up 1 V, from −0.5 V up to +0.5 V. Because of the initial condition inductor current, the resistor voltage builds up from 0.5 to 1 V and the inductor voltage decays exponentially by 0.5 V.

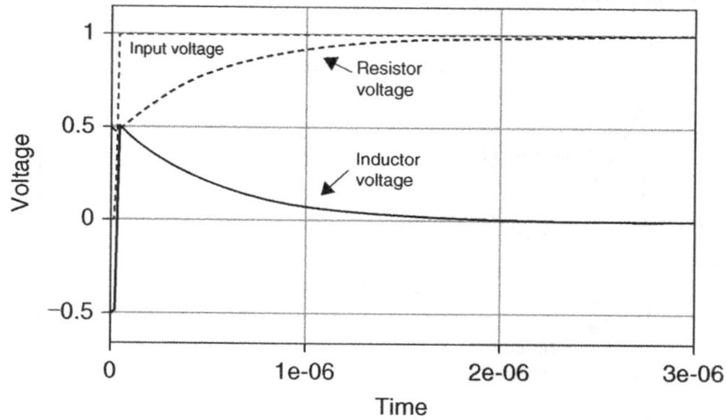

Figure 10.16 RL step response simulation plot of voltages with −25 mA initial condition.

The solid line in Figure 10.17 is the current building up to a final value of 50 mA, which is V_S over R.

Because of the 25 mA initial condition inductor current i_0, the bracketed quantity in Equation 10.7, i_0 minus V_S over R equals −25 mA.

Equation 10.7 shows the complete response for the series RL circuit in Figure 10.15.

$$i(t) = \frac{V_S}{R} + \left(i_0 - \frac{V_S}{R}\right) e^{(-R/L)t} \tag{10.7}$$

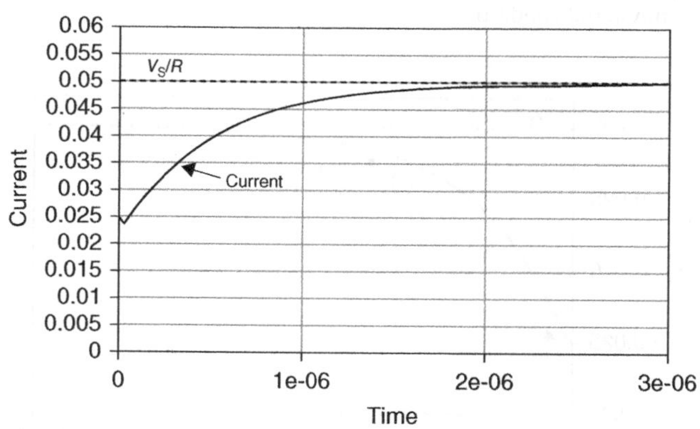

Figure 10.17 RL step response simulation plot of currents with −25 mA initial condition.

10.3.3 Third RL Example – Subtractive Initial Current

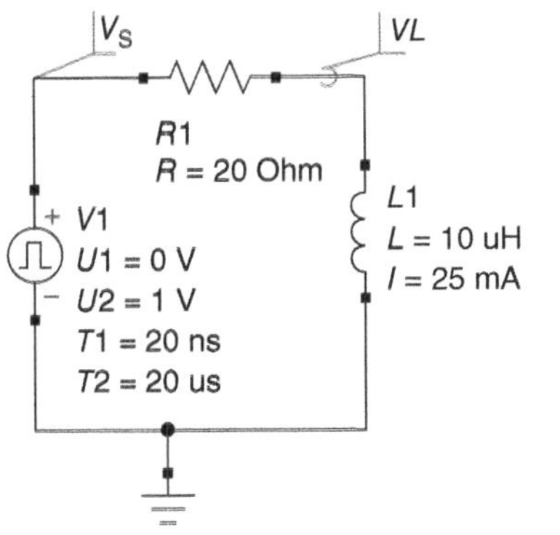

Figure 10.18 Series RL circuit simulation schematic with 25 mA initial condition.

This series RL circuit example, in Figure 10.18, has an initial inductor current of 25 mA in the counterclockwise direction. The initial condition inductor current flows up out of the inductor. Even though the initial inductor current is labeled as positive 25 mA it subtracts from the loop current chosen polarity of clockwise, established by the definition of resistor voltage as V_S minus VL, making i_0 equal to −25 mA in the equations and plots.

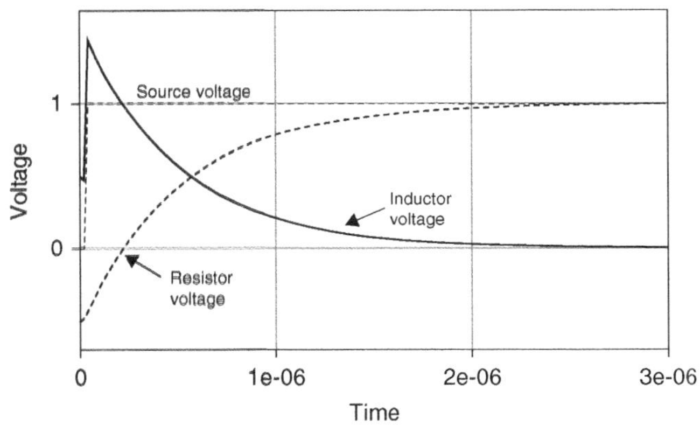

Figure 10.19 RL step response simulation plot of voltages with 25 mA initial condition.

The 25 mA initial condition current in the inductor creates a negative voltage across the resistor, making the inductor voltage positive to satisfy KVL before the step function is applied. The solid line in Figure 10.19 is inductor voltage, which jumps up 1 V, from +0.5 up to +1.5 V. Because of the initial condition inductor current, the resistor builds up from −0.5 to 1 V and the inductor voltage decays exponentially by 1.5 V.

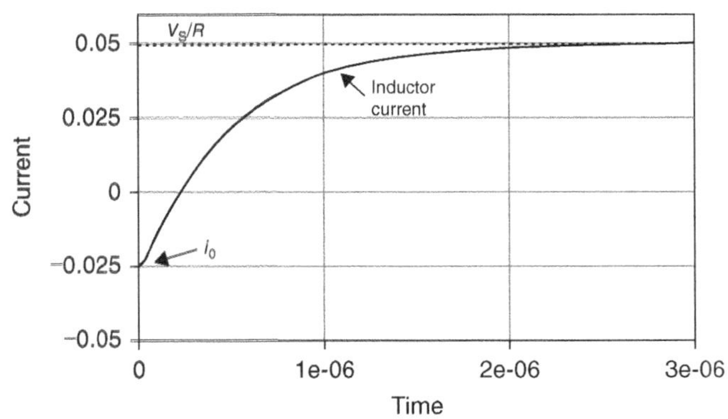

Figure 10.20 RL step response simulation plot of current with 25 mA initial condition.

The solid line in Figure 10.20 is the current building up to a final value of 50 mA, which is V_S over R.

Because of the −25 mA initial condition inductor current i_0, the bracketed quantity in Equation 10.8, i_0 minus V_S over R equals −75 mA.

Equation 10.8 shows the complete response for the series RL circuit in Figure 10.18.

$$i(t) = \frac{V_S}{R} + \left(i_0 - \frac{V_S}{R}\right) e^{(-R/L)t} \tag{10.8}$$

10.4 Complete Response with Switches

10.4.1 Switch Models for Complete Response

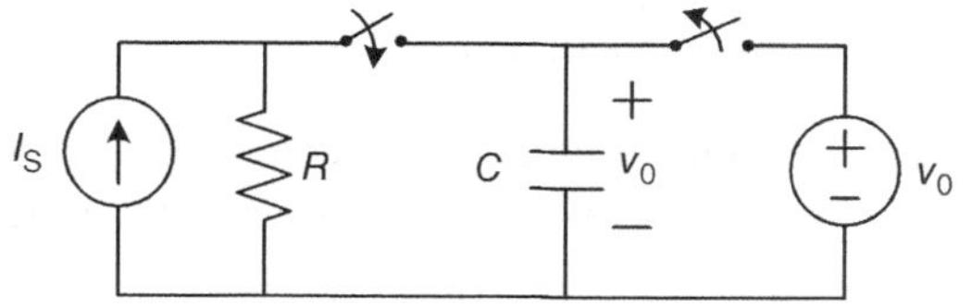

Figure 10.21a RC circuit schematic with switches to model complete response.

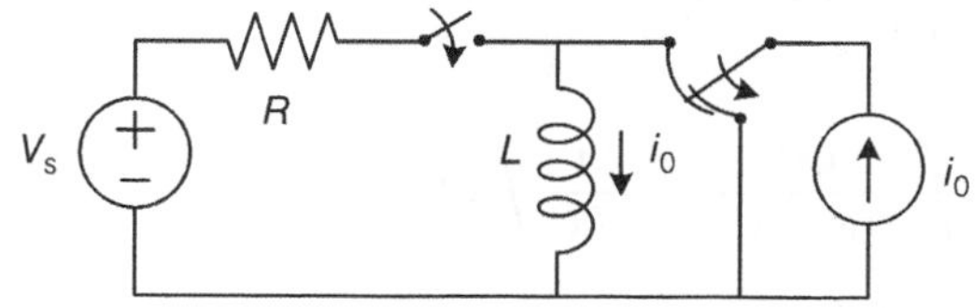

Figure 10.21b RL circuit schematic with switches to model complete response.

The switch models in Figures 10.21a and 10.21b[1] can generate complete response. The right-hand side sources, for v_0 in Figure 10.21a and i_0 in Figure 10.21b, set up the initial conditions, while the right-hand side switches are closed and the left-hand side switches are open. The left-hand side switches close to apply the step function signals, I_S in Figure 10.21a and V_S in Figure 10.21b. At the same time as the left-hand side switches close, the right-hand side switches open.

10.4.2 RC Complete Response Derivation with Direct Integration

This section calculates the complete response for the RC circuit in Figure 10.21a, where the switches move at time t equals zero. From the right-hand side voltage source v_0, the capacitor carries initial condition voltage of v_0. From the left-hand side current source I_S, a step function gets applied to the RC circuit at time t equals 0.

Right after the switches move, at time t equals 0, the RC circuit looks like Figure 10.22, where $v(t)$ refers to the voltage across the resistor and the capacitor, which has an initial value of v_0. After the step function at time t equals 0, the current source I_S is a constant value.

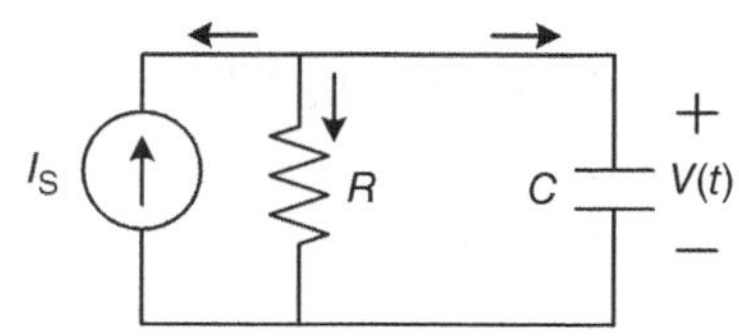

Figure 10.22 RC parallel circuit schematic driven by current source.

For the RC circuit in Figure 10.22, write Kirchhoff's current law (KCL), summing currents leaving the top node, as shown in Equation 10.9.

$$C\frac{dv(t)}{dt} + \frac{v(t)}{R} - I_S = 0 \quad (10.9)$$

Divide Equation 10.9 by C and move the I_0 term to the right-hand side as shown in Equation 10.10.

$$\frac{dv(t)}{dt} + \frac{v(t)}{RC} = \frac{I_S}{C} \quad (10.10)$$

In Equation 10.10, move the $v(t)$ over RC term to the right-hand side to obtain Equation 10.11.

$$\frac{dv(t)}{dt} = \frac{I_S}{C} - \frac{v(t)}{RC} \quad (10.11)$$

In Equation 10.11, multiply the numerator and the denominator of the I_0 term by RC (which is τ).

$$\frac{dv(t)}{dt} = \frac{RC \cdot I_S}{RC \cdot C} - \frac{v(t)}{RC} \quad (10.12)$$

In Equation 10.12, group the terms divided by RC and cancel C over C in the I_0 term, to obtain Equation 10.13.

1 The switch models in Figures 10.21a and 10.21b appear in the book "Introduction to Electronic Circuits" Third Edition, by Richard C. Dorf and James A. Svoboda, John Wiley & Sons. The RC and RL analyses are adapted from that same Section 9.3.

$$\frac{dv(t)}{dt} = \frac{RI_S - v(t)}{RC} \quad (10.13)$$

Cross multiply Equation 10.13 to separate the variables. Notice the sign changes in Equation 10.14.

$$\frac{dv(t)}{v(t) - RI_S} = \frac{-dt}{RC} \quad (10.14)$$

Integrate both sides of Equation 10.14. The constant minus 1 over RC term can move out of the integral sign. D is the constant of integration.

$$\int \frac{dv(t)}{v(t) - RI_S} = -\frac{1}{RC}\int dt + D \quad (10.15)$$

Using the integral formula in Equation 10.16, perform the integration of Equation 10.15, obtaining Equation 10.17.

$$\int \frac{1}{x} = \ln(x) \quad (10.16)$$

In Equation 10.17, D is the constant of integration.

$$\ln(v(t) - RI_0) = \frac{-t}{RC} + D \quad (10.17)$$

Using the exponential formula in Equation 10.18, rewrite Equation 10.17 as Equation 10.19.

$$e^{(x+y)} = e^{(x)}e^{(y)} \quad (10.18)$$

Since D is an unknown constant, exponentiating D in Equation 10.19 doesn't materially change it.

$$v(t) - RI_S = e^{-t/RC}e^{D} \quad (10.19)$$

In Equation 10.19, move the RI_0 term to the right-hand side and rewrite e^D as simply D in Equation 10.20.

$$v(t) = RI_S + De^{-t/RC} \quad (10.20)$$

Solve for the constant D in Equation 10.20, using initial conditions, by setting t equal to 0 in Equation 10.21.

$$v(0) = v_0 = RI_S + De^{-0/RC} \quad (10.21)$$

In Equation 10.21, e^0 equals 1, leading to Equation 10.22.

$$v_0 = RI_S + D \quad (10.22)$$

Solve Equation 10.22 for D, as shown in Equation 10.23.

$$D = v_0 - RI_S \quad (10.23)$$

From Equation 10.23, substitute the value for D into Equation 10.20 to obtain Equation 10.24.

$$v(t) = RI_S + (v_0 - RI_S)e^{-t/RC} \quad (10.24)$$

Equation 10.24 is the complete response for the RC circuit with two switches in Figure 10.21a.

Equation 10.24 is in the form of a steady-state term RI_0 plus a transient term multiplied times the exponential term.

Multiply out the terms in Equation 10.24 to get Equation 10.25.

$$v(t) = RI_S + v_0e^{-t/RC} - RI_Se^{-t/RC} \quad (10.25)$$

Group the RI_0 terms in Equation 10.25 to get Equation 10.26.

$$v(t) = v_0e^{-t/RC} + RI_S\left(1 - e^{-t/RC}\right) \quad (10.26)$$

Equation 10.26 is the complete response for the RC circuit with two switches in Figure 10.21a.

Equation 10.26 to is in the form of a natural response term with v_0 plus a step response term with RI_0.

10.4.3 RL Complete Response Derivation with Direct Integration

This section calculates the complete response for the RL circuit in Figures 10.21b and 10.23, where the switches move at time t equals zero.

From the right-hand side current source i_0, in Figure 10.23,[2] the inductor carries initial condition current of i_0. From the left-hand side voltage source V_S, a step function voltage gets applied to the RL circuit at time t equals 0.

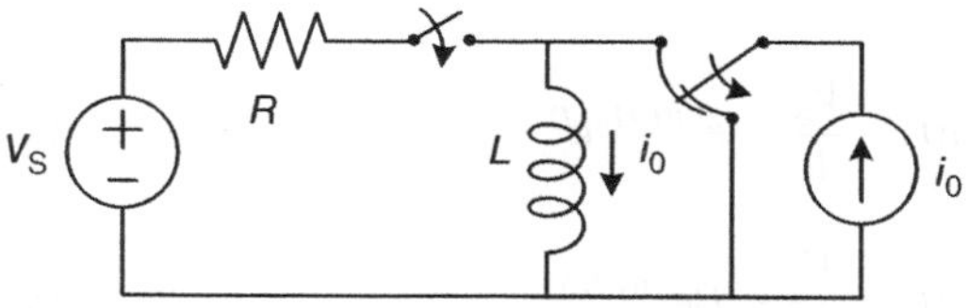

Figure 10.23 RL circuit schematic with switches to model complete response.

Right after the switches move, at time t equals 0, the RL circuit looks like Figure 10.24, where $i(t)$ refers to the loop current, which has an initial condition value of i_0. After the step function at time t equals 0, the voltage source V_S is a constant value.

For the RL circuit in Figure 10.24, write Kirchhoff's voltage law (KVL), summing voltages counter clockwise around the loop, starting at the inductor, as shown in Equation 10.27.

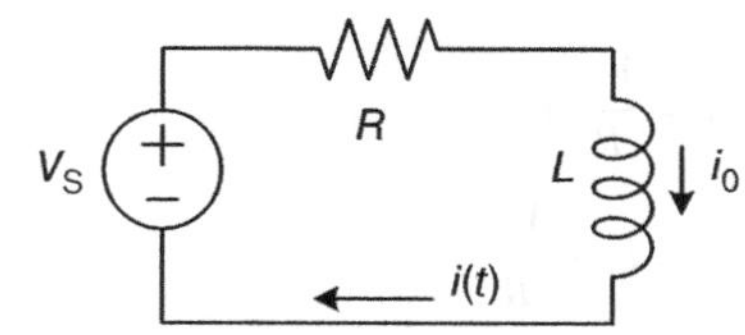

Figure 10.24 RL series circuit schematic driven by voltage source.

$$L\frac{di(t)}{dt} + Ri(t) - V_S = 0 \tag{10.27}$$

Divide Equation 10.27 by L and move the V_S term to the right-hand side as shown in Equation 10.28.

$$\frac{di(t)}{dt} + \frac{R}{L}i(t) = \frac{V_S}{L} \tag{10.28}$$

In Equation 10.28, multiply the numerator and the denominator of the V_S term by L over R (which is τ), as shown in Equation 10.29.

$$\frac{di(t)}{dt} + \frac{R}{L}i(t) = \frac{R}{L}\frac{L}{R}\frac{V_S}{L} \tag{10.29}$$

Simplify Equation 10.29 by cancelling two L terms on the right-hand side, resulting in Equation 10.30.

$$\frac{di(t)}{dt} + \frac{R}{L}i(t) = \frac{R}{L}\frac{V_S}{R} \tag{10.30}$$

In Equation 10.30, move the $i(t)$ term to the right-hand side.

$$\frac{di(t)}{dt} = \frac{R}{L}\frac{V_S}{R} - \frac{R}{L}i(t) \tag{10.31}$$

In Equation 10.31, group the R over L terms.

$$\frac{di(t)}{dt} = \frac{R}{L}\left(\frac{V_S}{R} - i(t)\right) \tag{10.32}$$

Cross multiply Equation 10.32 to separate the variables. Notice the sign changes that appear in Equation 10.33.

$$\frac{di(t)}{\left(i(t) - \frac{V_S}{R}\right)} = \frac{-R}{L}dt \tag{10.33}$$

Integrate both sides of Equation 10.33. The constant minus R over L term can move out of the integral sign. D is the constant of integration in Equation 10.34.

$$\int \frac{di(t)}{\left(i(t) - \frac{V_S}{R}\right)} = \frac{-R}{L}\int dt + D \tag{10.34}$$

Using the integral formula in Equation 10.35, perform the integration of Equation 10.34 obtaining Equation 10.36. D is the constant of integration.

$$\int \frac{1}{x} = \ln(x) \tag{10.35}$$

$$\ln\left(i(t) - \frac{V_S}{R}\right) = -t\frac{R}{L} + D \tag{10.36}$$

Using the exponential formula in Equation 10.37, rewrite Equation 10.36 as Equation 10.38.

---

2 The switch model in Figure 10.23 appears in the book "Introduction to Electronic Circuits" Third Edition, by Richard C. Dorf and James A. Svoboda, John Wiley & Sons. The RC and RL analyses are adapted from that same Section 9.3.

$$e^{(x+y)} = e^{(x)}e^{(y)} \quad (10.37)$$

Since D is an unknown constant, exponentiating D in Equation 10.38 doesn't materially change it.

$$i(t) - \frac{V_S}{R} = e^{-t(R/L)}e^{D} \quad (10.38)$$

In Equation 10.38, move the RI_0 term to the right-hand side and rewrite e^D as simply D, to obtain Equation 10.39.

$$i(t) = \frac{V_S}{R} + De^{-t(R/L)} \quad (10.39)$$

Use initial conditions to solve for the constant D in Equation 10.39, by setting t equal to 0.

$$i(0) = i_0 = \frac{V_S}{R} + De^{-0(R/L)} \quad (10.40)$$

In Equation 10.40, e^0 equals 1, leading to Equation 10.41.

$$i_0 = \frac{V_S}{R} + D \quad (10.41)$$

In Equation 10.41, D is the constant of integration. Solve Equation 10.41 for D, resulting in Equation 10.42.

$$D = i_0 - \frac{V_S}{R} \quad (10.42)$$

From Equation 10.42, substitute the value for D into Equation 10.39 to obtain Equation 10.43.

$$i(t) = \frac{V_S}{R} + \left(i_0 - \frac{V_S}{R}\right)e^{-t(R/L)} \quad (10.43)$$

Equation 10.43 is the complete response for the RL circuit with two switches in Figure 10.23.

Equation 10.43 is in the form of a steady-state term V_S over R plus a transient term multiplied times the exponential term.

Multiply out the terms in Equation 10.43 to get Equation 10.44.

$$i(t) = \frac{V_S}{R} + i_0e^{-t(R/L)} - \frac{V_S}{R}e^{-t(R/L)} \quad (10.44)$$

Group the V_S over R terms in Equation 10.44 to get Equation 10.45.

$$i(t) = i_0e^{-t(R/L)} + \frac{V_S}{R}(1 - e^{-t(R/L)}) \quad (10.45)$$

Equation 10.45 is the complete response for the RL circuit with two switches in Figures 10.23 and 10.21b.

Equation 10.45 to is in the form of a natural response term with i_0 plus a step response term with V_S over R.

10.5 Complete Response General Derivation and Formulae

The previous RC and RL complete response derivations, using direct integration, share a common structure in terms of the time constant τ (tau). This section shows a derivation of complete solution that is more general than the RC and RL derivations of complete solution done in the previous sections.

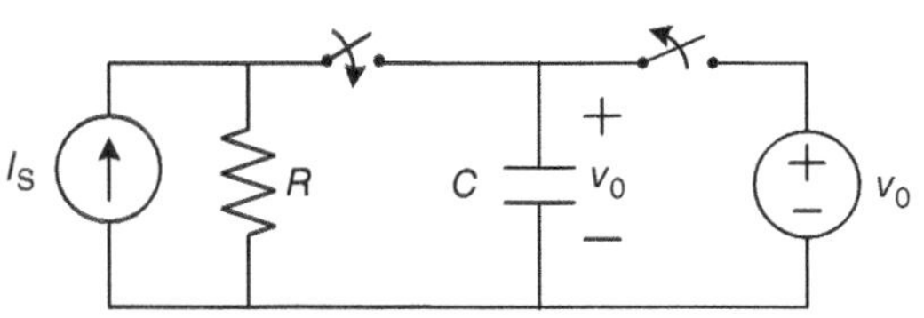

Figure 10.25a RC circuit schematic with switches to model complete response.

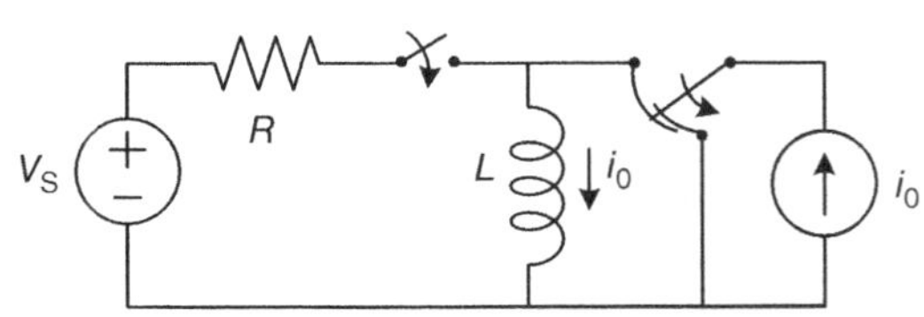

Figure 10.25b RL circuit schematic with switches to model complete response.

RC

For the RC circuit in Figure 10.25a, (originally Figure 10.21a), Equation 10.46a comes from the KCL Equation 10.9.

$$C\frac{dv(t)}{dt} + \frac{v(t)}{R} - I_S = 0 \tag{10.46a}$$

Divide Equation 10.46a by C and move the I_S term to the right-hand side as shown in Equation 10.47a.

$$\frac{dv(t)}{dt} + \frac{v(t)}{RC} = \frac{I_S}{C} \tag{10.47a}$$

Equation 10.48a defines τ for the RC circuit in Figure 10.25a.

$$\tau = RC \tag{10.48a}$$

$$K = \frac{I_S}{C} \tag{10.49a}$$

RL

For the RL circuit in Figure 10.25b (originally Figures 10.21b and 10.23) Equation 10.46b comes from the KVL Equation 10.27.

$$L\frac{di(t)}{dt} + Ri(t) - V_S = 0 \tag{10.46b}$$

Divide Equation 10.46b by L and move the V_S term to the right-hand side, as shown in Equation 10.47b.

$$\frac{di(t)}{dt} + \frac{R}{L}i(t) = \frac{V_S}{L} \tag{10.47b}$$

Equation 10.48b defines τ for the RL circuit in Figure 10.25b.

$$\tau = \frac{L}{R} \tag{10.48b}$$

$$K = \frac{V_S}{L} \tag{10.49b}$$

RC **RL**

$$\frac{dx(t)}{dt} + \frac{x(t)}{\tau} = K \tag{10.50}$$

Using K in Equations 10.49a and 10.49b, shift to the general form of the differential equation in terms of x and τ in Equation 10.50.

Equation 10.50 is the generalized differential equation for first-order complete response. K is not a constant of integration in this derivation.

In Equation 10.50, multiply the K term by τ to obtain Equation 10.51.

$$\frac{dx(t)}{dt} + \frac{x(t)}{\tau} = \frac{K\tau}{\tau} \tag{10.51}$$

In Equation 10.51, move the $x(t)$ to the right-hand side as shown in Equation 10.52.

$$\frac{dx(t)}{dt} = \frac{K\tau}{\tau} - \frac{x(t)}{\tau} \tag{10.52}$$

Rearrange Equation 10.52 to combine the τ terms, as shown in Equation 10.53.

$$\frac{dx(t)}{dt} = \frac{K\tau - x(t)}{\tau} \tag{10.53}$$

Divide both sides of Equation 10.53 by the $K\tau$ minus $x(t)$ term to get Equation 10.54.

$$\frac{dx(t)}{dt}\frac{1}{K\tau - x(t)} = \frac{1}{\tau} \tag{10.54}$$

Multiply both sides of Equation 10.54 by dt.

$$\frac{dx(t)}{K\tau - x(t)} = \frac{dt}{\tau} \tag{10.55}$$

Multiply both sides of Equation 10.55 by minus 1, rearranging the denominator of the left-hand side to get Equation 10.56.

$$\frac{dx(t)}{x(t) - K\tau} = \frac{-dt}{\tau} \tag{10.56}$$

Integrate both sides of Equation 10.56.

$$\int \frac{dx(t)}{x(t) - K\tau} = \int \frac{-dt}{\tau} \tag{10.57}$$

On the left-hand side of Equation 10.57, move the constant one over τ term outside of the integral sign, as shown in Equation 10.58.

$$\int \frac{dx(t)}{x(t) - K\tau} = \frac{-1}{\tau} \int dt \tag{10.58}$$

Using the integral formula in Equation 10.59, perform the integration of Equation 10.58, obtaining Equation 10.60.

$$\int \frac{1}{x} = \ln(x) \tag{10.59}$$

$$\ln(x(t) - K\tau) = \frac{-t}{\tau} + D \tag{10.60}$$

In Equation 10.60, D is the constant of integration.

$$e^{(x+y)} = e^{(x)}e^{(y)} \tag{10.61}$$

Using the exponential formula in Equation 10.61, rewrite Equation 10.60 as Equation 10.62.

$$x(t) - K\tau = e^{-t/\tau}e^{D} \tag{10.62}$$

Since D is an unknown constant, exponentiating D in Equation 10.62 doesn't materially change it. Rewrite e^D as D in Equation 10.63.

$$x(t) - K\tau = De^{-t/\tau} \tag{10.63}$$

Move the $K\tau$ term, in Equation 10.63, to the right-hand side to solve for the state variable $x(t)$, as shown in Equation 10.64.

$$x(t) = K\tau + De^{-t/\tau} \tag{10.64}$$

Shift back to the RC and RL details, using Equations 10.65a, 10.65b, 10.66a, and 10.66b, to obtain Equations 10.67a and 10.67b.

RC

$$\tau = RC \tag{10.65a}$$

$$K = \frac{I_S}{C} \tag{10.66a}$$

$$v(t) = RI_S + De^{-t/RC} \tag{10.67a}$$

Solve for the constant D in Equation 10.67a, by setting t equals zero to use initial conditions, as shown in Equation 10.68a.

$$v(0) = v_0 = RI_S + De^{-0/RC} \tag{10.68a}$$

In Equation 10.68a, e^0 equals 1, leading to Equation 10.69a.

$$v_0 = RI_S + D \tag{10.69a}$$

RL

$$\tau = \frac{L}{R} \tag{10.65b}$$

$$K = \frac{V_S}{L} \tag{10.66b}$$

$$i(t) = \frac{V_S}{R} + De^{-t(R/L)} \tag{10.67b}$$

Solve for the constant D in Equation 10.67b, by setting t equals zero to use initial conditions, as shown in Equation 10.68b.

$$i(0) = i_0 = \frac{V_S}{R} + De^{-0(R/L)} \tag{10.68b}$$

In Equation 10.68b, e^0 equals 1, leading to Equation 10.69b.

$$i_0 = \frac{V_S}{R} + D \tag{10.69b}$$

Solve Equations 10.69a and 10.69b for D, as shown in Equations 10.70a and 10.70b.

$$D = v_0 - RI_S \qquad (10.70a)$$

$$D = i_0 - \frac{V_S}{R} \qquad (10.70b)$$

From Equations 10.70a and 10.70b, substitute the value for D into Equations 10.68a and 10.68b.

$$v(t) = RI_S + (v_0 - RI_S)e^{-t/RC} \qquad (10.71a)$$

$$i(t) = \frac{V_S}{R} + \left(i_0 - \frac{V_S}{R}\right)e^{-t(R/L)} \qquad (10.71b)$$

Equations 10.71a and 10.71b are the complete response equations for the RC and RL circuits in Figures 10.25a and 10.25b. Equations 10.71a and 10.71b correspond to the steady-state form of complete response Equations 10.24 and 10.43.

11

First-Order Opamp Integrator and Differentiator Circuits

This chapter presents basic opamp integrator and differentiator circuits, which have a long history in analog circuits including analog computers. The chapter starts with an RC integrator circuit, which shows how a series capacitor can integrate a signal current source.

11.1 RC Integrator Circuit Step Response

This section calculates the step response of a series RC circuit, in Figure 11.1, driven by a current source. The voltage plot in Figure 11.2 demonstrates the integration behavior of the circuit. Constant current accumulates charge on the capacitor, and the integral of the constant current is the linearly increasing voltage shown in Figures 11.2 and 11.3.

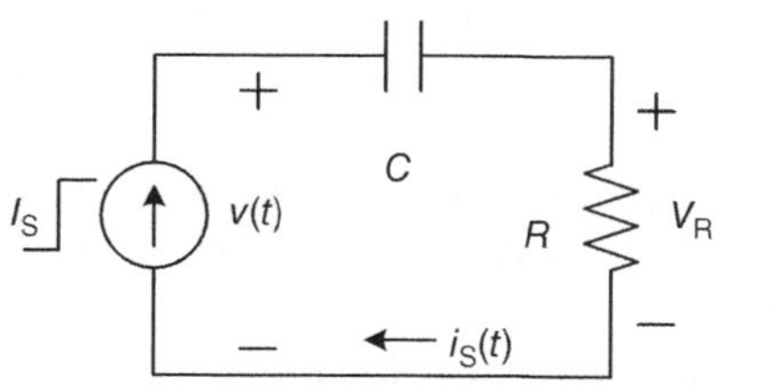

Figure 11.1 RC series circuit schematic driven by current step function $i_S(t)$.

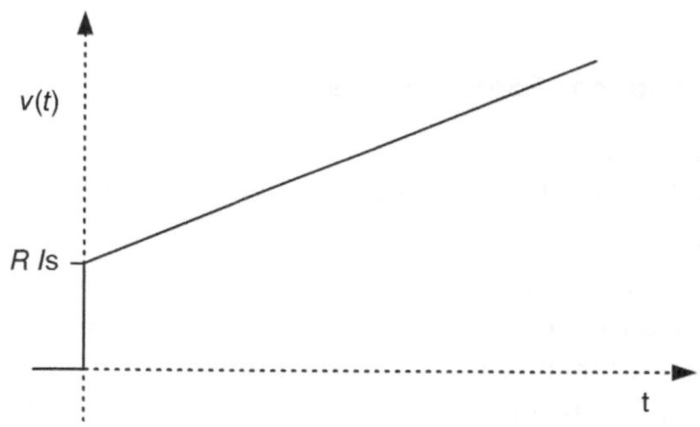

Figure 11.2 Graph of voltage $v(t)$ across the capacitor and the resistor.

$$i_S(t) = I_s\, u(t) \tag{11.1}$$

$$u(t) = \begin{cases} 0 \text{ for } t < 0 \\ 1 \text{ for } t \geq 0 \end{cases} \tag{11.2}$$

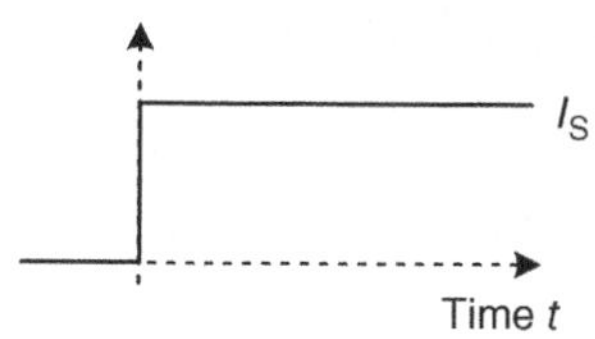

Figure 11.3 Current step function $i_S(t)$ with steady-state value I_S.

To derive the equation for the voltage $v(t)$, across the capacitor in series with the resistor, write Kirchhoff's voltage law (KVL) as shown in Equation 11.3.

$$v(t) = v_R(t) + v_C(t) \tag{11.3}$$

First and Second Order Circuits and Equations: Technical Background and Insights, First Edition. Robert O'Rourke.

$$v_C(t) = \frac{1}{C}\int_0^t i(\tau)d\tau \tag{11.4}$$

Equation 11.4 describes the voltage across an ideal lumped capacitor in terms of the current through the capacitor.

$$v(t) = Ri_S(t) + \frac{1}{C}\int_0^t i(\tau)d\tau \tag{11.5}$$

Substitute Equation 11.4 and Ohm's law into Equation 11.3 to obtain Equation 11.5.

$$v(t) = RI_S(t) + \frac{1}{C}\int_0^t i(\tau)d\tau \tag{11.6}$$

Consider only the time $t > 0$, so that $i_S(t)$ equals the constant value I_S as shown in Equation 11.6.

The RI_S term in Equation 11.6 corresponds to a rectangular area under the $v(t)$ curve in Figure 11.4. The integral term in Equation 11.6 corresponds to the triangular area in Figure 11.6 under the slanted line and above the RI_S area.

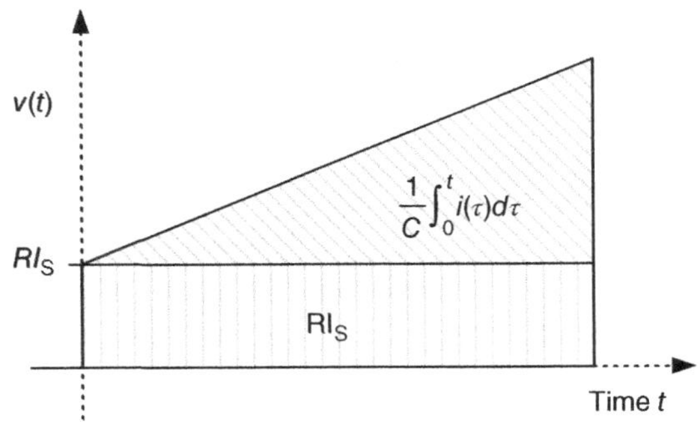

Figure 11.4 Graph of constant voltage across the resistor and growing capacitor voltage.

11.1.1 Description of Step Response

When the current step function is applied to the series RC circuit in Figure 11.1, there is an instantaneous constant current I_S in the entire loop. This instantaneous current causes an instantaneous voltage, of I_S times R, across the output resistor.

As time passes for t greater than 0, the capacitor integrates the current and builds the voltage across the capacitor, but does ***not*** reduce the output voltage V_R across the output resistor. Because the current source in Figure 11.1 is constant, the resistor current and therefore the output V_R voltage remain constant.

The voltage $v(t)$ across the current source, including the resistor voltage, increases linearly with time, since the capacitor is integrating a constant current value.

11.2 Opamp Integrator Circuit

An opamp integrator circuit integrates a current over time to get voltage, as expressed in Equation 11.7 for capacitor voltage in terms of current.

$$v_C(t) = \frac{1}{C}\int_0^t i(\tau)d\tau \tag{11.7}$$

11.2.1 Ideal Opamp Model Summary

These are the properties of an ideal opamp model:

- Input voltage ($v+$ to $v-$) is zero.
- Input current is zero.
- The output voltage v_{out} is whatever it needs to be to keep the input voltage equal to zero.

11.2.2 Ideal Opamp Integrator Circuit Intuitive Analysis

The schematic in Figure 11.5 is an inverting opamp integrator circuit. Applying the ideal opamp model to the circuit means that current running from the voltage source v_S, through R_S, goes past $v-$ and toward v_{out}. No current goes into the minus node nor plus node of the opamp. The current follows the arrows in Figure 11.5, flowing into the capacitor.

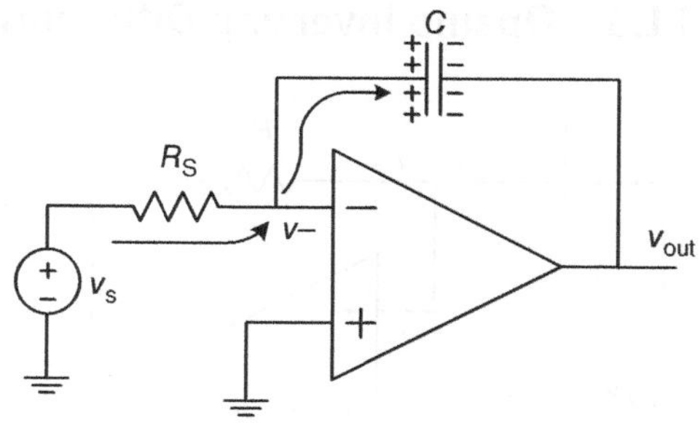

Figure 11.5 Opamp integrator circuit with accumulating charge marked on the capacitor.

Charge gets collected by the capacitor, creating voltage across the capacitor. This voltage, the integral of current in Equation 11.7, is what makes the integration behavior in the circuit in Figure 11.5.

Since the opamp input voltage difference is zero, using the ideal opamp model, the connection to the minus (−) terminal of the op-amp is at the same voltage as the plus (+) terminal, which is grounded. Holding the opamp input voltage at ground creates the voltage inversion, like the opamp inverting amplifier.

11.2.3 Ideal Opamp Integrator Circuit Derivation

Write Kirchhoff's current law (KCL) for the currents entering the $v-$ node, as shown in Figure 11.6 and Equation 11.8. (No current enters the minus terminal of the opamp.)

$$\frac{v_S(t) - 0}{R_S} + i_C(t) = 0 \tag{11.8}$$

$$i_C(t) = C\frac{dv(t)}{dt} \tag{11.9}$$

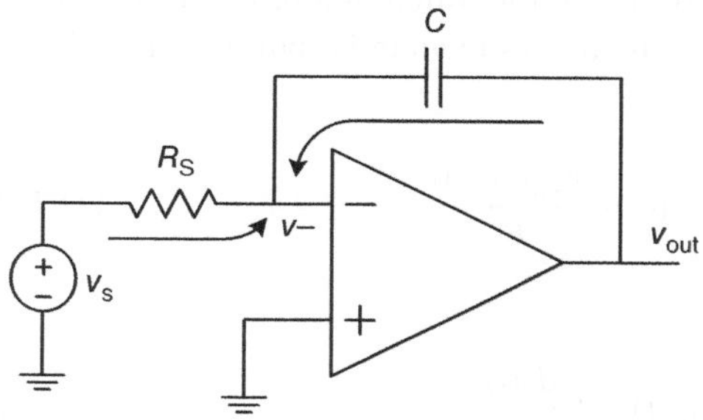

Figure 11.6 RC opamp integrator circuit schematic.

Substitute the capacitor branch relationship, in Equation 11.9, into the KCL Equation 11.8, to get Equation 11.10.

$$\frac{v_S(t)}{R_S} + C\frac{dv_{out}(t)}{dt} = 0 \tag{11.10}$$

Divide Equation 11.10 by C.

$$\frac{v_S(t)}{R_S C} + \frac{dv_{out}(t)}{dt} = 0 \tag{11.11}$$

In Equation 11.11 move the RC term to the right-hand side of the equals sign and multiply both sides by dt to obtain Equation 11.12.

$$dv_{out}(t) = -\frac{v_S(t)}{R_S C}dt \tag{11.12}$$

Integrate Equation 11.12, as shown in Equation 11.13.

$$\int dv_{out}(t) = -\int \frac{v_S(t)}{R_S C}dt \tag{11.13}$$

Perform the integration on Equation 11.13, using τ as a variable of integration. Take the constant 1 over RC term outside of the integral sign, as shown in Equation 11.14.

$$v_{out}(t) = \frac{1}{R_S C}\int_0^t v_s(\tau)d\tau + v_{out}(0) \tag{11.14}$$

The $v_{out}(t)$ term is the same as the capacitor voltage. If there is no initial condition capacitor voltage, then $v_{out}(0)$ equals zero.

The opamp circuit output $v_{out}(t)$ term is proportional to the integral of current, expressed by $v_S(t)$ over the R term outside the integral.

11.3 Opamp Inverting Differentiator Circuit

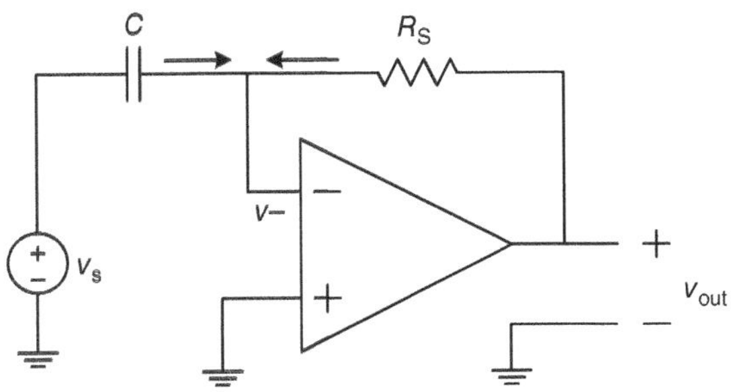

Figure 11.7 RC opamp differentiator circuit schematic.

The output voltage $v_{out}(t)$, of the opamp differentiator circuit, shown in Figure 11.7, is proportional to the derivative of the input voltage v_S, as expressed in Equation 11.15.

$$v_{out}(t) = -RC\frac{dv_S(t)}{dt} \tag{11.15}$$

11.3.1 Ideal Opamp Differentiator Circuit Derivation

To derive the differentiator circuit behavior, write KCL (Kirchhoff's current law) for the currents entering the $v-$ node, as shown in Equation 11.16. (No current enters the minus terminal of the ideal opamp.)

$$i_C(t) = \frac{v_{out}(t) - 0}{R_S} \tag{11.16}$$

Equation 11.16 states that the capacitor current entering the $v-$ node, shown in Figure 11.7, equals the resistor current coming from the output. (No current enters the minus terminal of the opamp.)

$$i_C(t) = C\frac{dv(t)}{dt} \tag{11.17}$$

Equation 11.17 shows the branch relationship for an ideal capacitor, the current through the capacitor as a function of voltage across the capacitor.

$$C\frac{dv_S(t)}{dt} = \frac{v_{out}(t)}{R_S} \tag{11.18}$$

From Equation 11.17, substitute the capacitor branch relationship into Equation 11.18 to get Equation 11.19. As the $v-$ node is at zero volts, the voltage across the capacitor is v_S.

$$R_S C\frac{dv_S(t)}{dt} = v_{out}(t) \tag{11.19}$$

Equation 11.19 shows how the differentiator output voltage $v_{out}(t)$ is proportional to the derivative of the input signal $v_S(t)$. When the $v-$ pin of the opamp is at zero volts, the voltage across the capacitor is v_S.

Part 3

Second-Order Circuits

12

Second-Order RLC Circuits Overview

12.1 What are Second-Order Circuits?

12.1.1 A Capacitor and an Inductor Together Make a Second Order Circuit

Second-order circuits have two reactive circuit elements, which store energy. Circuits with an inductor and capacitor together, where they can exchange energy, are second-order circuits.

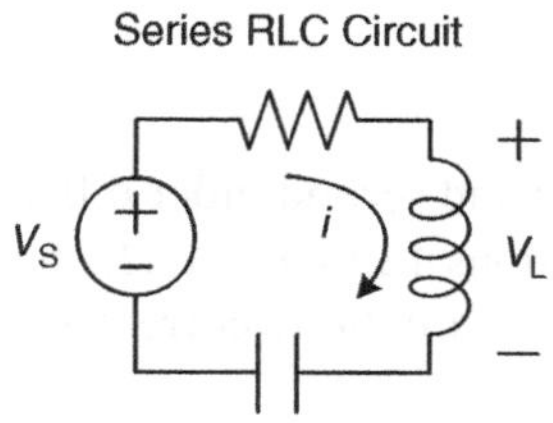

Figure 12.1a Series RLC circuit schematic driven by a voltage source.

Two reactive circuit elements, an inductor and a capacitor, make a circuit second order.

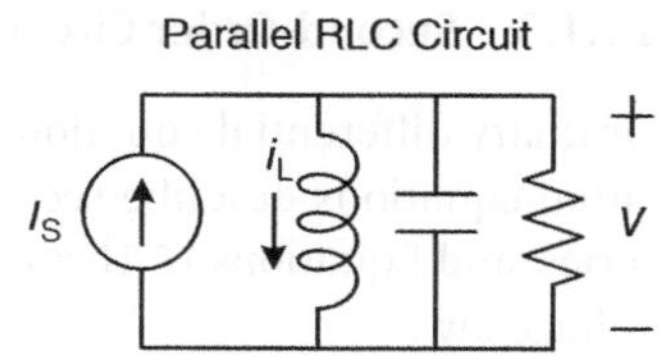

Figure 12.1b Parallel RLC circuit schematic driven by a voltage source.

These two circuits in Figures 12.1a and 12.1b, series and parallel RLC, are second-order circuits.

12.1.2 RLC Circuits Have a Resonant Frequency

Second-order RLC circuits have a resonant frequency where impedances and frequency responses exhibit peak values, maximum or minimum. In frequency domain, these resonant frequencies can be used to create filters, which are frequency selective circuits.

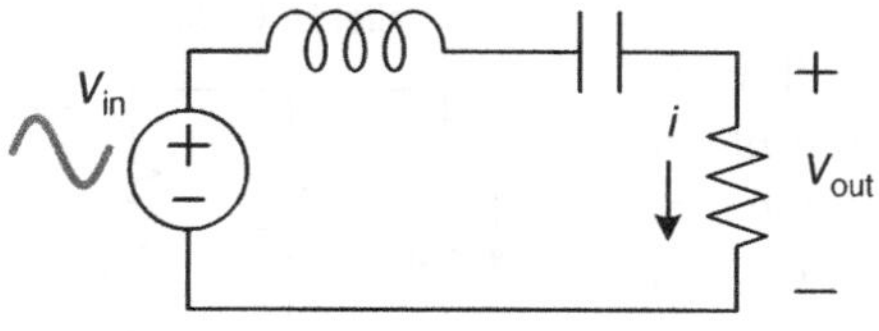

Figure 12.2a Series RLC circuit schematic driven by sinusoidal voltage.

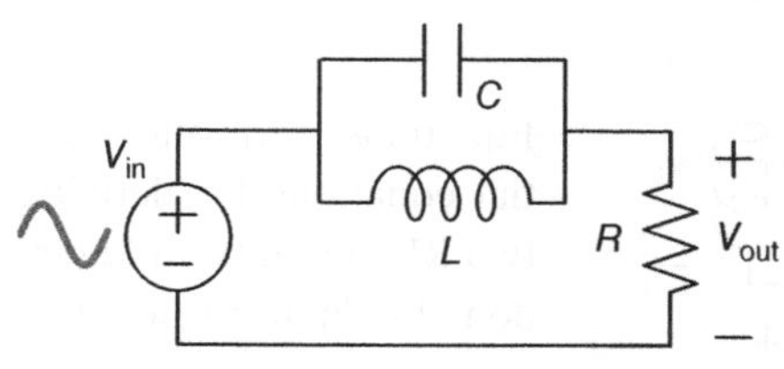

Figure 12.2b Parallel-series RLC circuit schematic driven by sinusoidal voltage.

First and Second Order Circuits and Equations: Technical Background and Insights, First Edition. Robert O'Rourke.

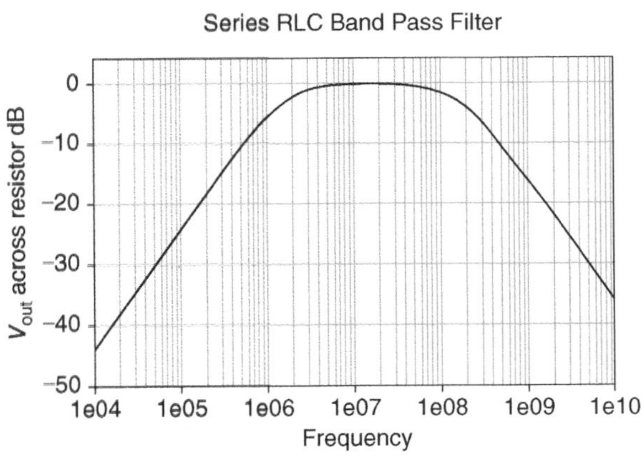

Figure 12.3a Series RLC band pass simulated frequency response voltage in dB.

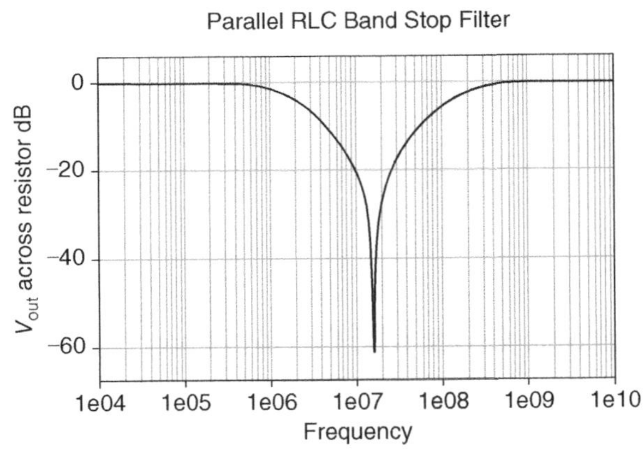

Figure 12.3b RLC band stop filter simulated frequency response voltage in dB.

Figure 12.2a shows a series RLC band pass filter with the output voltage plotted in Figure 12.3a as a function of frequency.

Figure 12.2b shows a parallel RLC band stop filter with the output voltage plotted in Figure 12.3b as a function of frequency.

12.1.3 Second-Order Circuit Differential Equations

Ordinary differential equations, with respect to time, describe lumped element circuits and second-order differential equations describe second-order RLC circuits in the time domain. These Equations 12.1a and 12.2a, for series, and Equations 12.1b and 12.2b, for parallel RLC circuits, come from applications of Kirchhoff's current and voltage laws.

Series RLC circuit

$$LC\frac{d^2v_C}{dt^2} + RC\frac{dv_C}{dt} + v_C = 0 \tag{12.1a}$$

or

$$LC\frac{d^2i}{dt^2} + RC\frac{di}{dt} + i = 0 \tag{12.2a}$$

Parallel RLC circuit

$$LC\frac{d^2v}{dt^2} + \frac{L}{R}\frac{dv}{dt} + v = 0 \tag{12.1b}$$

or

$$LC\frac{d^2i_L}{dt^2} + \frac{L}{R}\frac{di_L}{dt} + i_L = 0 \tag{12.2b}$$

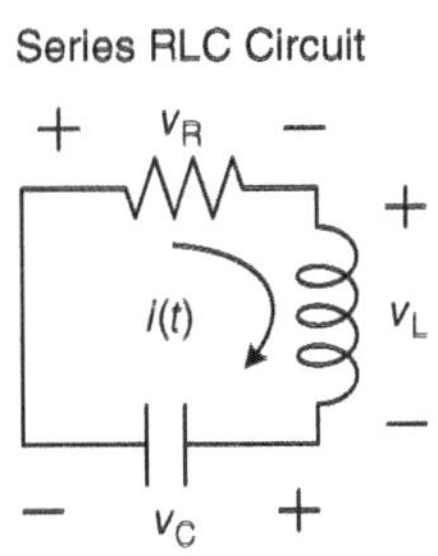

Figure 12.4a Series RLC undriven circuit schematic.

Equations 12.1 and 12.2 are homogeneous differential equations for deriving the natural response of the two RLC circuits, in Figures 12.4a and 12.4b, in the time domain. There are no signal sources driving these circuits.

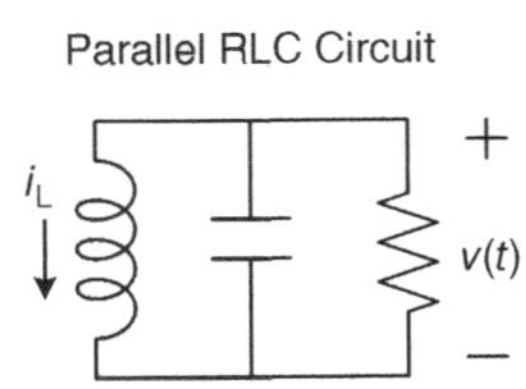

Figure 12.4b Parallel RLC undriven circuit schematic.

12.1.4 Second-Order Circuits Can Exhibit Overshoot and Ringing

With two complementary reactive circuit elements, capable of exchanging energy between each other, underdamped second-order RLC circuits exhibit overshoot and ringing in response to a step input.

The damping and shape of the time domain step response, shown in Figure 12.5, depends upon the resistance (loss) in the RLC circuit.

Second-order RLC circuits differ from first-order RC and RL circuits in the ability of RLC circuits to exhibit overshoot and ringing (in some situations but not all).

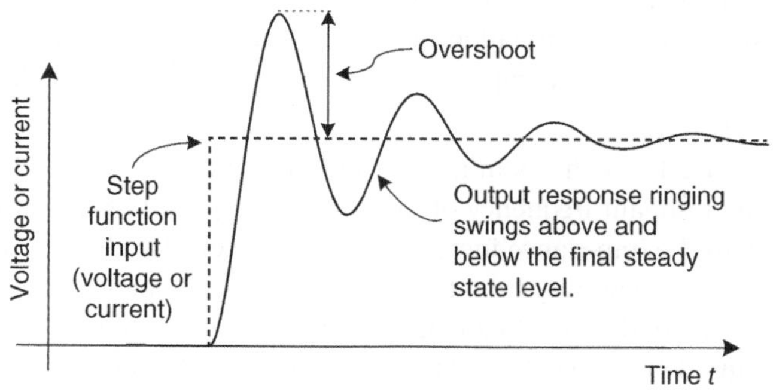

Figure 12.5 Generic second-order underdamped step response plot.

12.2 Resonance in the Frequency Domain

12.2.1 Series and Parallel Impedances

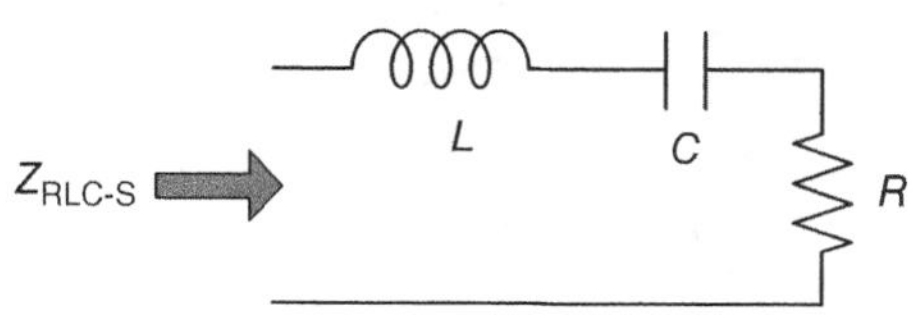

Figure 12.6a Series RLC circuit impedance schematic.

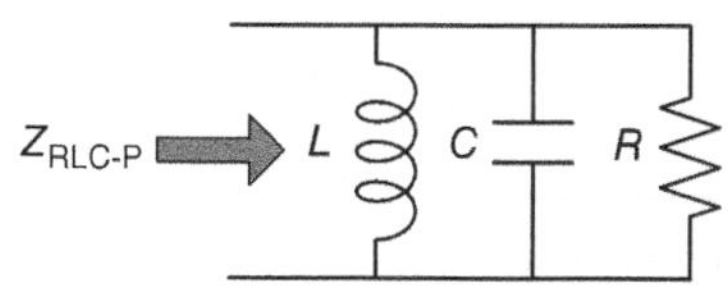

Figure 12.6b Parallel RLC circuit impedance schematic.

Series RLC impedance equation

$$Z_{\text{RCL-S}} = R + j\left(\omega L - \frac{1}{\omega C}\right) \tag{12.3a}$$

Parallel RLC impedance equation

$$Z_{\text{RLC-P}} = \frac{R}{1 + j\left(\omega C - \frac{1}{\omega L}\right)} \tag{12.3b}$$

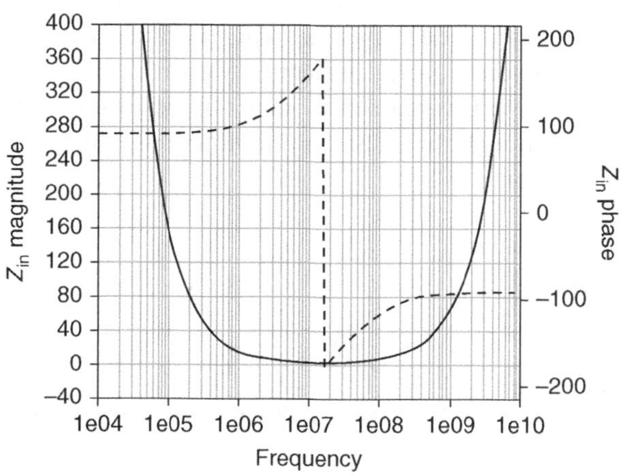

Figure 12.7a Series RLC simulated impedance plot magnitude and phase.

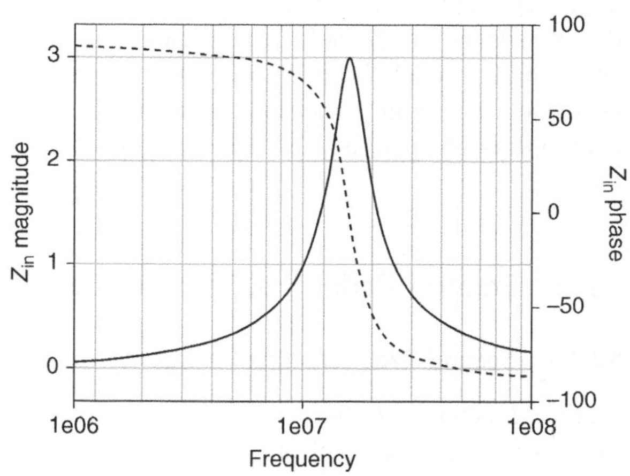

Figure 12.7b Parallel RLC simulated impedance plot magnitude and phase.

Equations 12.3a and 12.3b describe the impedances $Z_{RCL\text{-}S}$ and $Z_{RLC\text{-}P}$, looking into the series and parallel circuits as illustrated in Figures 12.6a and 12.6b. The frequency selectivity and resonance of these input impedances are shown in the simulation plots in Figures 12.7a and 12.7b. The solid lines in Figures 12.7a and 12.7b show the magnitude of the impedance and the dashed lines show the phase in degrees labeled on the right-hand vertical scales.

Figure 12.7a shows minimum in impedance, around the resonant frequency of the circuit in Figure 12.6a, with the impedance increasing sharply a few decades of frequency away from the minimum. The resonance corresponds to the frequency where the two ω terms in Equation 12.3a cancel. At series resonance, the input impedance looks like just the series resistor.

Figure 12.7b shows maximum in impedance, around the resonant frequency of the circuit in Figure 12.6b, with the impedance decreasing sharply away from the maximum. The resonance corresponds to the frequency where the two ω terms in Equation 12.3b cancel. At parallel resonance, the input impedance looks like just the resistor.

12.2.2 Circuit Elements Around Resonance

Series RLC circuit

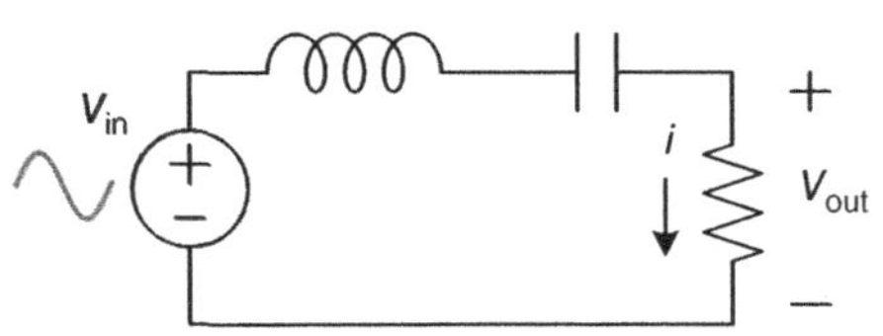

Figure 12.8a Series RLC circuit schematic driven by sinusoidal voltage.

Parallel RLC circuit

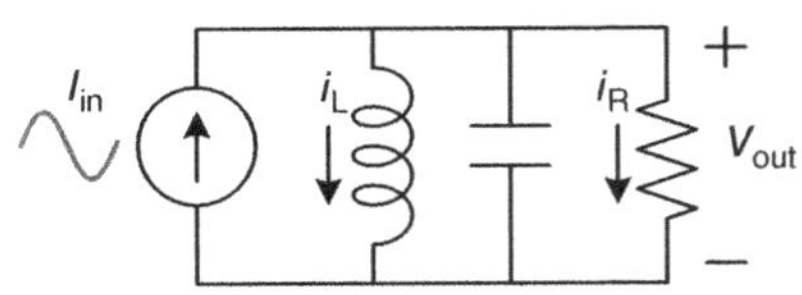

Figure 12.8b Parallel RLC circuit schematic driven by sinusoidal current.

In the series RLC circuit in Figure 12.8a, current flows best through the loop at resonance, when the impedance of the inductor and the capacitor cancel. At frequencies below resonance, the impedance of the capacitor increases, blocking current flow through the whole loop, including the resistor, as frequency decreases. At frequencies above resonance, the impedance of the inductor increases, blocking current flow through the inductor and the whole loop, as frequency goes up. Lower impedance means more current through the resistor and more output voltage.

In the parallel RLC circuit in Figure 12.8b, current flows best around the resistor at resonance, when the impedance of the inductor and the capacitor together is maximum. At frequencies below resonance, the impedance of the inductor decreases, allowing current flow through the inductor instead of through the resistor. At frequencies above resonance, the impedance of the capacitor decreases, allowing more current flow through the capacitor as frequency goes up. Maximum output voltage occurs at resonance because the current goes through the resistor rather than the inductor or capacitor.

12.3 Second-Order RLC Transfer Functions and Q

12.3.1 Second-Order Transfer Functions

Transfer functions are commonly used to quantify the input–output characteristics of circuits in the steady-state frequency domain, using phasor quantities, shown in bold face type.

$$\frac{\boldsymbol{V}_{\text{out}}}{\boldsymbol{V}_{\text{in}}} = \frac{1}{1 + j\left(\frac{\omega L}{R} - \frac{1}{\omega RC}\right)} \quad (12.4a)$$

Equation 12.4a is the voltage transfer function for the series RLC circuit in Figure 12.8a.

$$\left|\frac{\boldsymbol{V}_{\text{out}}}{\boldsymbol{V}_{\text{in}}}\right| = \frac{1}{\sqrt{1 + j\left(\frac{\omega L}{R} - \frac{1}{\omega RC}\right)^2}} \quad (12.5a)$$

Equation 12.5a is the magnitude of the series RLC transfer function in Equation 12.8a.

$$\frac{\boldsymbol{I}_{\text{out}}}{\boldsymbol{I}_{\text{in}}} = \frac{1}{1 + j\left(\omega RC - \frac{R}{\omega L}\right)} \quad (12.4b)$$

Equation 12.4b is the current transfer function for the parallel RLC circuit in Figure 12.8b.

$$\left|\frac{\boldsymbol{I}_{\text{out}}}{\boldsymbol{I}_{\text{in}}}\right| = \frac{1}{\sqrt{1 + \left(\omega RC - \frac{R}{\omega L}\right)^2}} \quad (12.5b)$$

Equation 12.5b is the magnitude of the parallel RLC transfer function in Equation 12.8b.

Transfer function magnitudes, plotted in graphs in dB and log frequency, are common for characterizing circuits and electronic products.

Equation 12.6a is the phase response for the series RLC circuit in Figure 12.8a.

$$\angle H(j\omega) = \frac{\pi}{2} - \tan^{-1}[\omega RC/(1 - \omega^2 LC)] \quad (12.6a)$$

Equation 12.6b is the phase response for the parallel RLC circuit in Figure 12.8b.

$$\angle H(j\omega) = -\tan^{-1}\left(\omega RC - \frac{R}{\omega L}\right) \quad (12.6b)$$

12.3.2 Quality Factor Q

The quality factor Q is a figure of merit related to loss, like energy lost by dissipation in a resistance. Q can be used for inductors and Q is commonly used with resonant circuits like filters.

$$Q = 2\pi\left(\frac{\text{energy stored}}{\text{energy dissipated per cycle}}\right) \quad (12.7)$$

$$Q = \frac{\omega_0}{\Delta\omega} = \frac{\omega_0}{\text{BW}} \quad (12.8)$$

As defined in Equation 12.7, Q is the amount of energy stored per cycle of sinusoidal signal, in comparison to the amount of energy lost or dissipated per cycle.

Quality factor is closely related to bandwidth, as defined in Equation 12.8, where BW stands for bandwidth, and $\Delta\omega$ represents a range of frequencies where a circuit response is above a certain level, often −3 dB.

In RLC circuits, both bandwidth and Q relate to energy dissipated in resistors. In series RLC circuits, lower resistance means lower loss, higher Q, and less bandwidth. In parallel circuits, higher resistance means lower loss, higher Q, and less bandwidth.

12.4 Two Time Domain Responses

Two main time domain response analyses are natural response and step response, where natural response is undriven and step response is driven. Natural response is also called zero-input response (ZIR) and step response is one type of zero-state response (ZSR). Natural response circuits, like Figure 12.4a,b, don't have voltage or current sources.

12.4.1 RLC Natural Response 3 Cases

For different values of R, L, and C, the natural response can fall into three different categories or cases, related to the mathematical form of the characteristic roots of the differential equation, including whether they are real, complex, or imaginary.

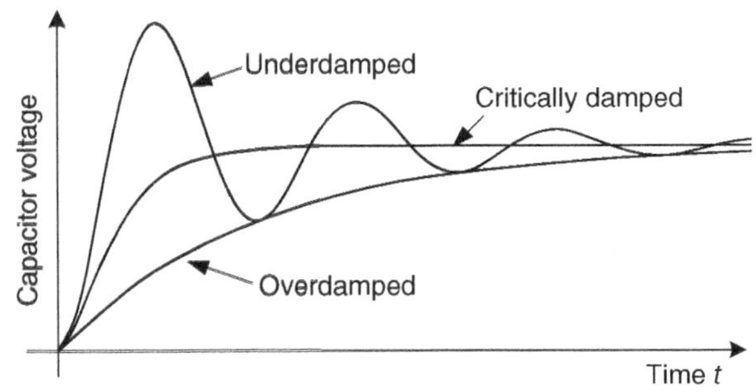

Figure 12.9 Plot of series RLC step response capacitor voltage for three cases.

For a series RLC circuit with an initial condition capacitor voltage, Figure 12.9 shows the natural response capacitor voltage for three values of the resistor to show three cases of natural response.

Table 12.1 shows three cases of natural response, including the mathematical nature of the characteristic roots, the corresponding value of quality factor Q, and a representative shape of loop current response in a series RLC circuit.

Table 12.1 Three response cases in terms of quality factor Q.

Overdamped Two real and unequal roots	$Q < \frac{1}{2}$	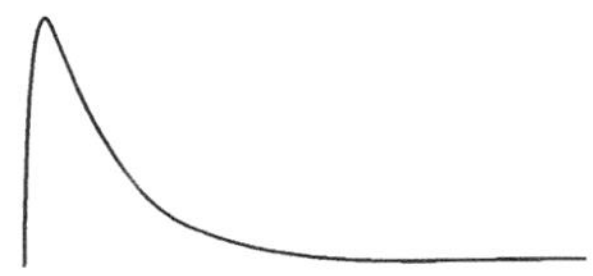
Critically Damped Two real and equal roots	$Q = \frac{1}{2}$	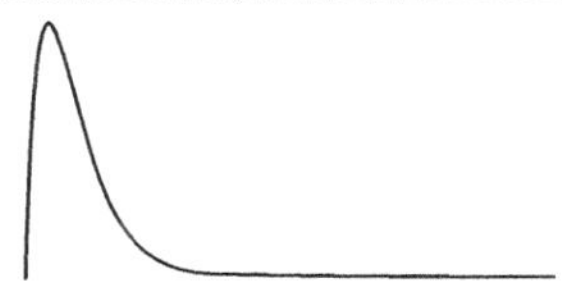
Underdamped Two complex conjugate roots	$Q > \frac{1}{2}$	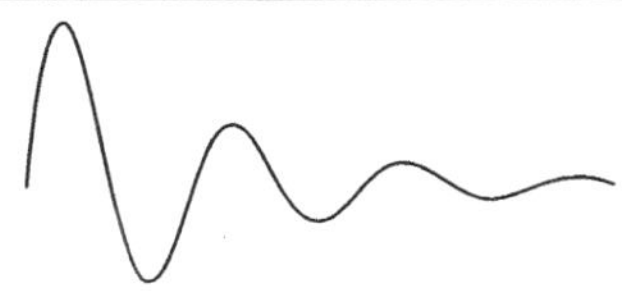

Notice how the quality factor Q appears in both time and frequency domain analyses. High Q means less loss. In the time domain, lower loss means more ringing in the response. In the frequency domain, higher Q means narrower bandwidth.

12.4.2 RLC Step Response 3 Cases

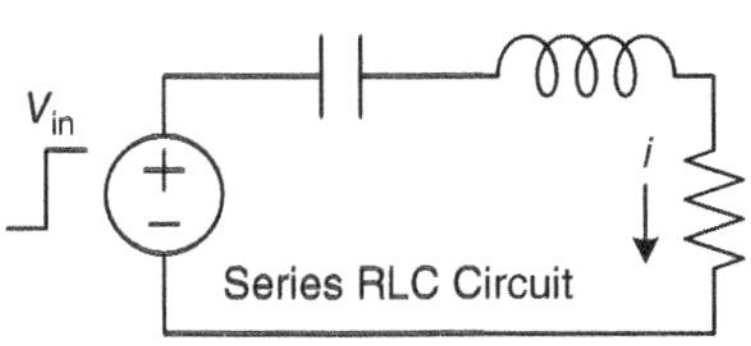

Figure 12.10 Series RLC circuit schematic driven by voltage step function.

Step response is the circuit response to an input voltage or current that is stepped instantaneously from 0 to a DC value. Figure 12.10 shows a step symbol, to the left of the voltage source, to indicate a step function input.

Like natural response, step response can have the same cases of response. Figure 12.5 shows an underdamped step response, where the step function is the dashed line, and the underdamped ringing signal is the solid line.

12.4.3 Complete Response is ZIR Plus ZSR

Complete response combines natural response and step response, ZIR and ZSR. Natural response looks at the response to initial conditions in the circuit, excluding voltage or current source driving signals, step response analyzes response to input signals and excludes initial conditions in the circuit. Complete response includes both initial conditions and external driving signals.

13

Second-Order RLC Frequency Response

13.1 Series and Parallel RLC Impedance

This section explores the steady-state frequency domain impedance of series and parallel RLC combinations, including intuition on the impedance, resonance, and circuit response minima and maxima.

13.1.1 Series and Parallel Impedance Summary

The impedances of series and parallel RLC combinations (Figures 13.1a and 13.1b) provide insight to the behavior of RLC circuits and resonance.

Series RLC schematic

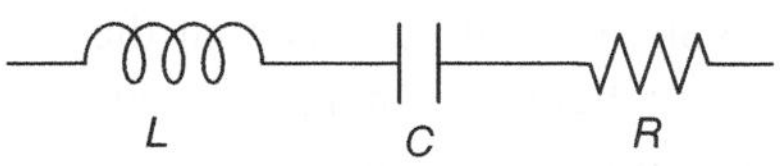

Figure 13.1a Series RLC schematic.

Parallel RLC schematic

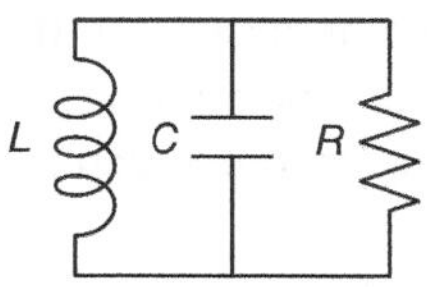

Figure 13.1b Parallel RLC schematic.

Series RLC impedance equation

$$Z_{\text{RLC-S}} = R + j\left(\omega L - \frac{1}{\omega C}\right) \tag{13.1a}$$

Parallel RLC impedance equation

$$Z_{\text{RLC-P}} = \frac{R}{1 + j\left(\omega C - \dfrac{1}{\omega L}\right)} \tag{13.1b}$$

First and Second Order Circuits and Equations: Technical Background and Insights, First Edition. Robert O'Rourke.

Series RLC impedance graph

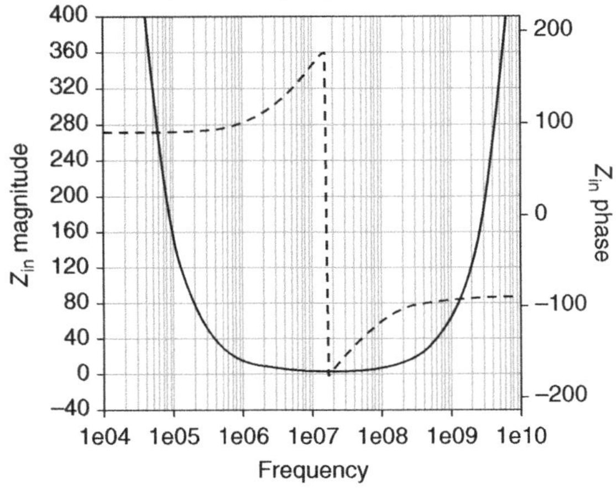

Figure 13.2a Series RLC simulation results graph of impedance.

Parallel RLC impedance graph

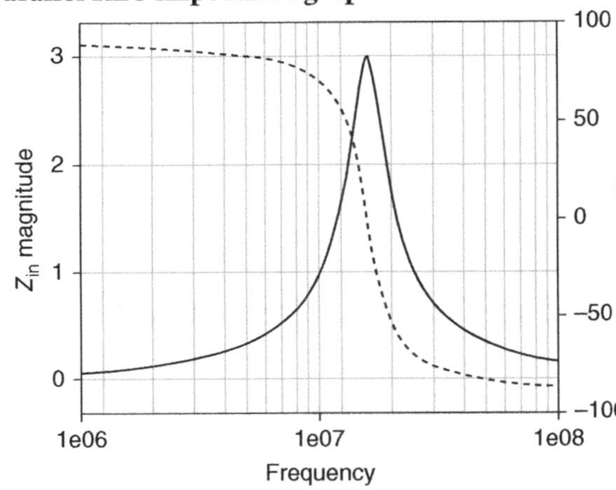

Figure 13.2b Parallel RLC simulation results graph of impedance.

Figure 13.2a shows a plot of series RLC input impedance magnitude, the solid line and the left axis scale, overlaid on the series RLC input impedance phase with the dashed line and the right axis scale.

Figure 13.2b shows a plot of parallel RLC input impedance magnitude, the solid line and the left axis scale, overlaid on the parallel RLC input impedance phase with the dashed line and the right axis scale.

The **minimum** value of the series RLC impedance, shown in Figure 13.3a, is R, and it occurs at the resonant frequency ω_0, where the imaginary term in the impedance Equation 13.1a becomes zero.

The **maximum** value of the parallel RLC impedance, shown in Figure 13.3b, is R, and it occurs at the resonant frequency ω_0, where the imaginary term in the denominator of Equation 13.1b becomes zero.

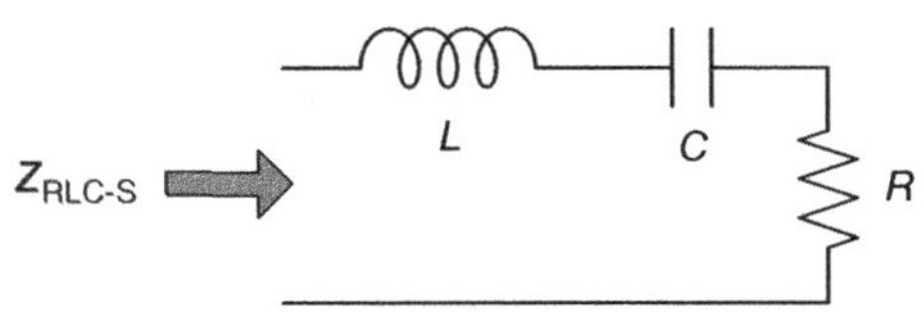

Figure 13.3a Series RLC impedance schematic.

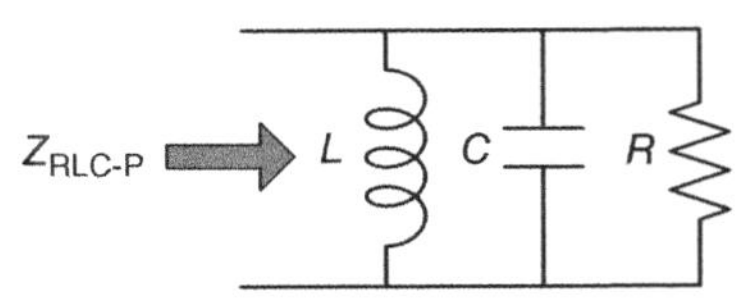

Figure 13.3b Parallel RLC impedance schematic.

13.1.1.1 Current Flow Above and Below Resonance

Series RLC circuit

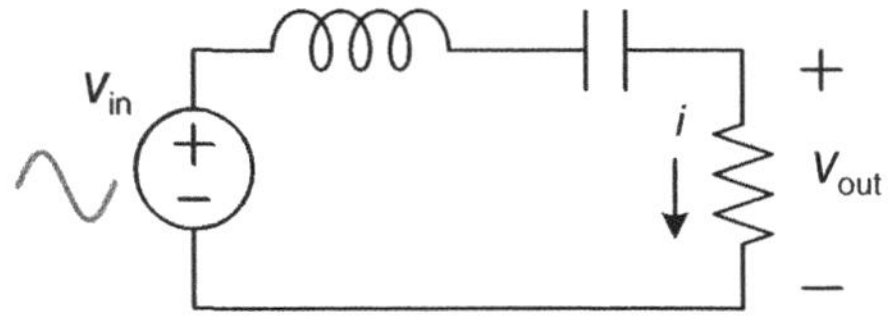

Figure 13.4a Series RLC circuit schematic driven by sinusoidal voltage.

Parallel RLC circuit

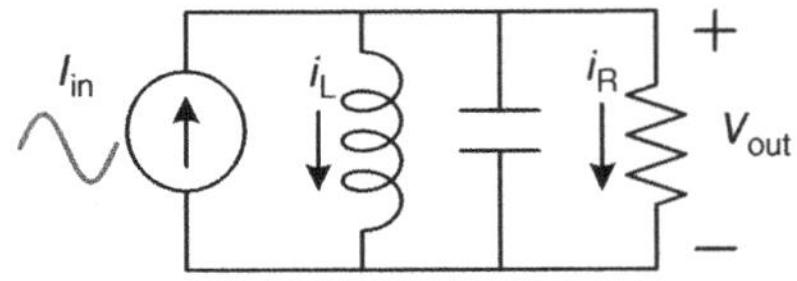

Figure 13.4b Parallel RLC circuit schematic driven by sinusoidal current.

In the series RLC circuit in Figure 13.4a, current flows best at the resonant frequency, where the impedance of the inductor and the capacitor cancels on another. At frequencies decreasing below the resonant frequency, the increasing capacitor impedance increasingly blocks current flow. At frequencies increasing above the resonant frequency, the increasing inductor impedance increasingly blocks current flow.

In the parallel RLC in Figure 13.4b, current flows through the output resistor best at resonance, where the impedance of the inductor and capacitor in parallel is high. At frequencies decreasing below the resonant frequency, the decreasing inductor impedance increasingly allows current flow through the inductor. At frequencies increasing above the resonant frequency, the decreasing capacitor impedance increasingly allows current flow through the capacitor.

13.1.2 Series RLC Impedance and Resonant Frequency

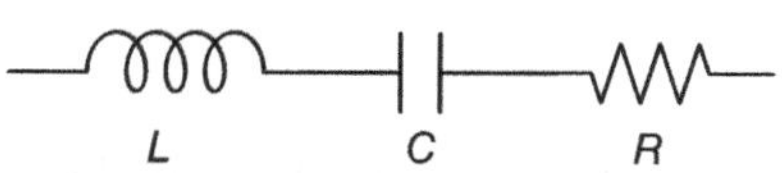

Figure 13.5 Series RLC schematic elements.

This section derives the formula for the impedance and resonant frequency ω_0 of the series RLC combination in Figures 13.5 and 13.6.

$$Z_L = j\omega L \quad (13.2)$$

Equation 13.2 is the impedance of an ideal inductor.

$$Z_C = \frac{1}{j\omega C} \quad (13.3)$$

Equation 13.3 is the impedance of an ideal capacitor.

$$Z_R = R \quad (13.4)$$

Equation 13.4 is the impedance of an ideal resistor.

For series circuits, impedances add.

$$Z_{RLC\text{-}S} = Z_L + Z_C + Z_R \quad (13.5)$$

For the series RLC in Figure 13.5, Equation 13.5 calculates the total impedance by summing the impedances of all three circuit elements.

$$Z_{RLC\text{-}S} = j\omega L + \frac{1}{j\omega C} + R \quad (13.6)$$

From Equations 13.2–13.4, substitute the impedance formulae for resistance, inductance, and capacitance into Equation 13.5 to get Equation 13.6.

$$Z_{RLC\text{-}S} = j\omega L + -j\frac{1}{\omega C} + R \quad (13.7)$$

In Equation 13.6, multiply the top and bottom of the middle capacitor impedance term by j to put minus j in the numerator to obtain Equation 13.7.

In Equation 13.7, gather terms to separate real and imaginary parts of the series RLC impedance equation in Equation 13.8.

$$Z_{RLC\text{-}S} = R + j\left(\omega L - \frac{1}{\omega C}\right) \quad (13.8)$$

Equation 13.8 is the impedance of the series RLC network in Figures 13.5 and 13.6.

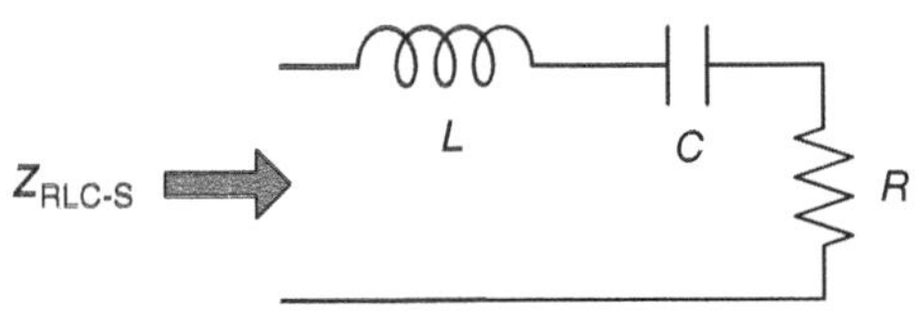

Figure 13.6 Series RLC impedance schematic.

In Figure 13.6, the arrow indicates perspective for input impedance, the series impedance looking into the RLC network.

13.1.2.1 Magnitude of Series RLC Impedance

The magnitude of a complex function, $G(\omega)$ in Equation 13.9, is the square root of the sum of the squares of the real part $R(\omega)$ and the imaginary part $X(\omega)$, as shown in Equation 13.10.

$$G(\omega) = R(\omega) + jX(\omega) \tag{13.9}$$

$$|G(\omega)| = \sqrt{[R(\omega)]^2 + [X(\omega)]^2} \tag{13.10}$$

To calculate the magnitude of the series RLC impedance, apply Equation 13.10 to Equation 13.8, to obtain Equation 13.11.

$$|Z_{RLC\text{-}S}| = \sqrt{R^2 + \left(\omega L - \frac{1}{\omega C}\right)^2} \tag{13.11}$$

Equation 13.11 is the formula for the magnitude of the series RLC networks in Figures 13.5 and 13.6.

13.1.2.2 Resonance of Series RLC Impedance

Resonance occurs at the angular frequency ω where the imaginary term in the impedance Equation 13.8 becomes zero. This resonant frequency is often designated with a zero subscript as ω_0.

To calculate the resonant frequency of series RLC in Figures 13.5 and 13.6, set the imaginary term in Equation 13.8 equal to zero, as shown in Equation 13.12.

$$\left(\omega L - \frac{1}{\omega C}\right) = 0 \tag{13.12}$$

Rearrange terms in Equation 13.12 to obtain Equation 13.13.

$$\omega L = \frac{1}{\omega C} \tag{13.13}$$

Multiply both sides of Equation 13.13 by ω, and divide both sides by L to obtain Equation 13.14.

$$\omega^2 = \frac{1}{LC} \tag{13.14}$$

Take the square root of both sides of Equation 13.14 to solve for ω_{in} terms of inductance L and capacitance C in Equation 13.15.

$$\omega = \frac{1}{\sqrt{LC}} \tag{13.15}$$

$$\omega_0 = \frac{1}{\sqrt{LC}} \tag{13.16}$$

Equation 13.16 is the resonant frequency ω_0 of an RLC network.

The resonant frequency in Equation 13.16 is the same for series RLC as it is for parallel RLC.

13.1.2.3 Series RLC Impedance Intuition

At resonance, the inductive and capacitive reactances are equal and opposite; they cancel one another.

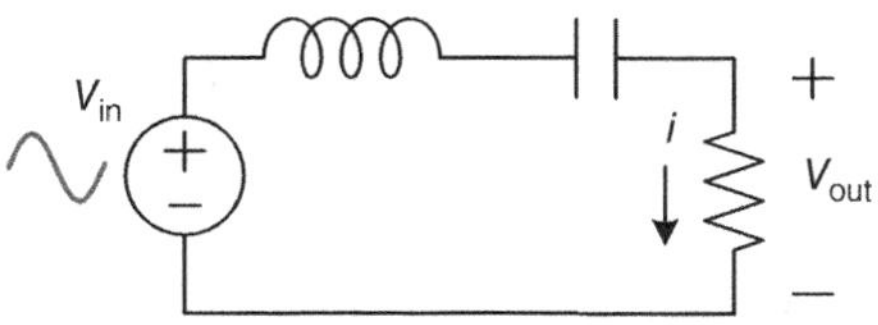

Figure 13.7a Series RLC circuit schematic driven by sinusoidal voltage.

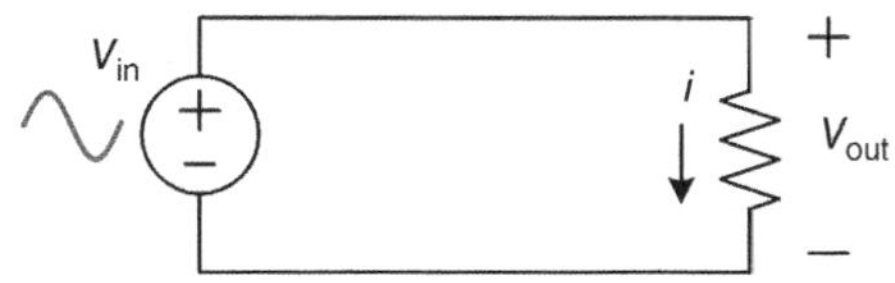

Figure 13.7b Sinusoidal voltage source across resistor modeling resonance.

Figure 13.7a shows a series RLC circuit driven in sinusoidal steady state.

At resonance, an RLC circuit becomes resistive, the L and C seemingly disappear, as shown in Figure 13.7b.

Inductors and capacitors are energy storage elements. Connected together, they can exchange energy back and forth between one another. Sinusoidal current is alternating current (AC). At resonance, the back-and-forth energy exchange between capacitor and inductor matches the back-and-forth alternating amplitude of the sinusoidal source. This match – this resonance – makes the series LC combination transparent to the sinusoidal source. This resonance is connected in series along one current path. The exchange of energy between the inductor and capacitor is in the same path as the current.

Energy in the capacitor comes from stored charge which creates a voltage. Energy in the inductor comes from magnetic flux and appears as current. At resonance, the rise and fall of voltage across the capacitor, due to the exchange of energy between the inductor and the capacitor, matches the rise and fall of the sinusoidal source voltage.

13.1.3 Series RLC Impedance Magnitude and Resonant Frequency Example

This section explores the magnitude of the input impedance in Equation 13.17 and calculates the resonant frequency for an RLC circuit in Figure 13.8.

$$|Z_{\text{RLC-S}}| = \sqrt{R^2 + \left(\omega L - \frac{1}{\omega C}\right)^2} \tag{13.17}$$

Equation 13.17 shows the magnitude of the impedance of a series RLC circuit, as shown in Figures 13.8 and 13.9.

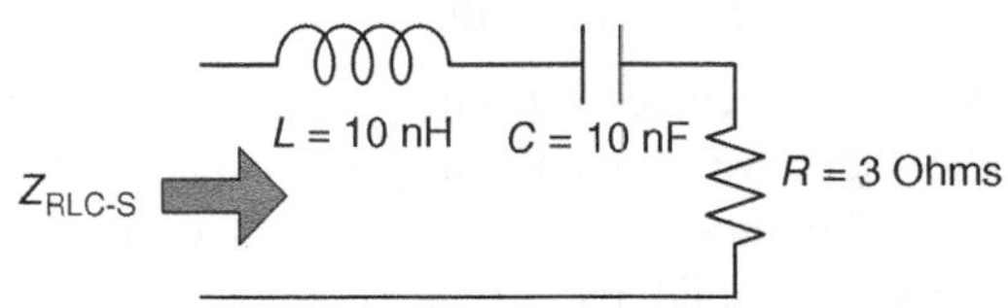

Figure 13.8 Series RLC impedance schematic.

Impedance is voltage divided by current. To analyze the input impedance $Z_{\text{RLC-S}}$, in the circuit schematic in Figure 13.8, look at the voltage, across the left-hand-side terminal, divided by the current going into the terminal.

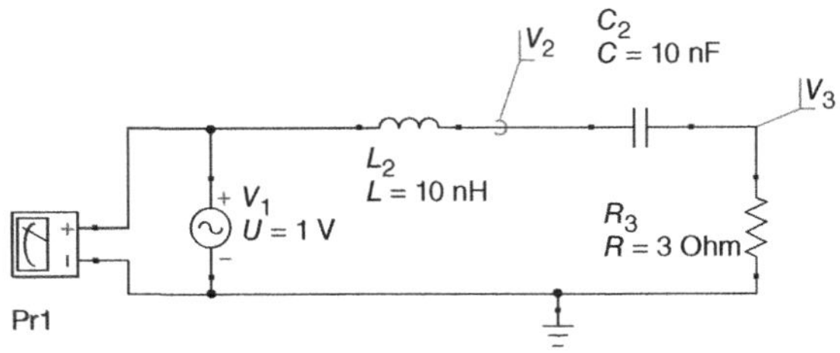

Figure 13.9 Series RLC simulation schematic with sinusoidal voltage source.

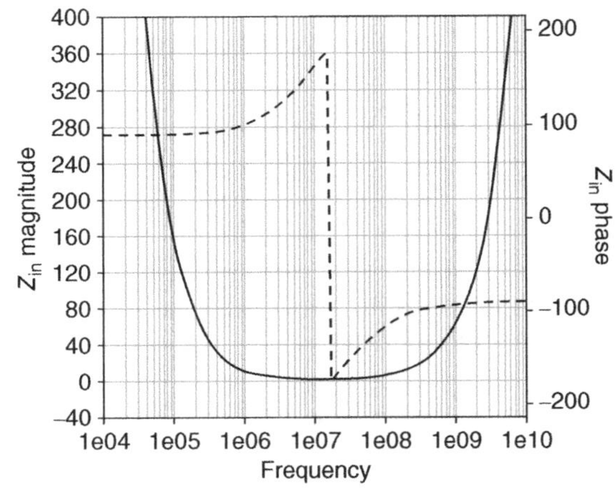

Figure 13.10 Series RLC simulation results graph of impedance.

Figure 13.9 shows a sinusoidal steady-state circuit simulation schematic for this series RLC input impedance analysis.

The solid line in Figure 13.10 is the magnitude of the input impedance in Equation 13.17. The plot is not in dB.

The minimum RLC series impedance (the solid line in Figure 13.10) occurs at resonance, where ω equals one over the square root of inductance L times capacitance C. At resonance, the entire ω term in Equation 13.17 goes to zero and the impedance reduces to the series resistance R. In this example, in the schematic in Figure 13.9, the resistance is 3 Ω. In Figure 13.10, the vertical section of the dashed phase line in Figure 13.10 shows the exact resonant frequency.

In Figure 13.10, the X-axis frequency scale is in Hertz. The equation for resonant frequency, in Equation 13.18, is in radians, where $\omega = 2\pi f$ where f = frequency.

$$\omega_0 = \frac{1}{\sqrt{LC}} = 2\pi f \tag{13.18}$$

$$f_0 = \frac{1}{2\pi\sqrt{LC}} = \frac{1}{2\pi\sqrt{(1e-8)(1e-8)}} = \frac{1e8}{2\pi} = 1.59e7 = 15.9\ \text{MHz} \tag{13.19}$$

Equation 13.19 shows the resonant frequency calculation in Hertz.

At low frequencies, capacitive reactance is larger than the inductive reactance and the capacitance dominates the impedance. In Figure 13.11, capacitive reactance is the solid line. At high frequencies, inductance dominates the impedance as the inductive reactance is larger than the capacitive reactance. Inductive reactance is the dashed line in Figure 13.11. Together, the two lines in Figure 13.11 trace the shape of the impedance plot (solid line) in Figure 13.10. In this series RLC, the impedances of the individual elements add to provide the overall impedance; this allows this overlay of two impedances to add up to the overall impedance.

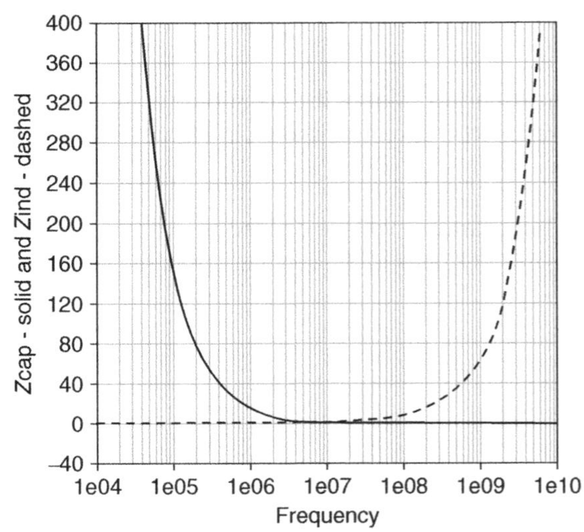

Figure 13.11 Series RLC simulation results graph of capacitor and inductor impedances.

13.1.4 Parallel RLC Impedance and Resonant Frequency

This section derives the formula for the impedance and resonant frequency ω_0 of the parallel RLC combination in Figures 13.1b and 13.12.

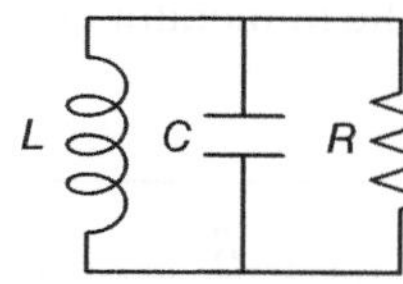

Figure 13.12 Parallel RLC schematic.

Equation 13.20 is the formula for the impedance Z_P of up to n parallel reactive components, as shown in Figure 13.13.

$$Z_P = \frac{1}{\frac{1}{Z_1} + \frac{1}{Z_2} + \cdots + \frac{1}{Z_n}} = \frac{1}{\sum_{m=1}^{n} \frac{1}{Z_m}} \tag{13.20}$$

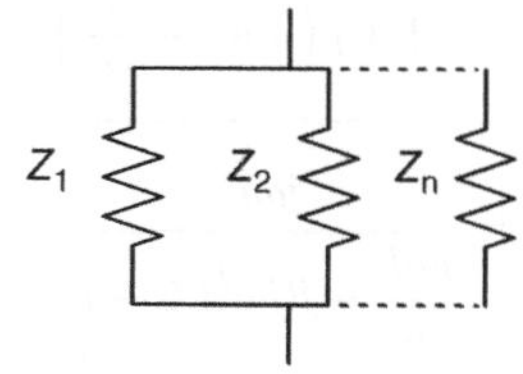

Figure 13.13 *N* resistive impedances in parallel.

Apply Equation 13.20 to the parallel RLC in Figure 13.12, to get $Z_{RLC\text{-}P}$ in Equation 13.21 in terms of Z_R, Z_L, and Z_C.

$$Z_{RLC\text{-}P} = \frac{1}{\frac{1}{Z_R} + \frac{1}{Z_L} + \frac{1}{Z_C}} \tag{13.21}$$

Substitute the individual impedance Equations 13.2–13.4 into Equation 13.21 to obtain Equation 13.22.

$$Z_{RLC\text{-}P} = \frac{1}{\frac{1}{R} + \frac{1}{j\omega L} + \frac{1}{\frac{1}{j\omega C}}} \tag{13.22}$$

In Equation 13.22, simplify the capacitive impedance term on the right side of the denominator, as shown in Equation 13.23.

$$Z_{RLC\text{-}P} = \frac{1}{\frac{1}{R} + \frac{1}{j\omega L} + j\omega C} \tag{13.23}$$

In the denominator of the impedance expression in Equation 13.23, multiply the top and bottom of the middle inductive term by *j*. This gives a minus sign in Equation 13.24 because *j* times *j* equals minus 1.

$$Z_{RLC\text{-}P} = \frac{1}{\frac{1}{R} + \frac{-1}{\omega L} + j\omega C} \tag{13.24}$$

Gather terms in the denominator of Equation 13.24 to group imaginary and real parts, as shown in Equation 13.25.

$$Z_{RLC\text{-}P} = \frac{1}{\frac{1}{R} + j\left(\omega C - \frac{1}{\omega L}\right)} \tag{13.25}$$

Equation 13.25 shows the impedance of a parallel RLC combination.

For another useful form of Equation 13.25, substitute G equals 1 over R as seen in Equation 13.26.

$$Z_{RLC\text{-}P} = \frac{1}{G + j(\omega C - 1/\omega L)} \tag{13.26}$$

Equation 13.26 shows the impedance of a parallel RLC combination.

Multiply the numerator and denominator of Equation 13.25 by R to get a different form in Equations 13.27 and 13.28.

$$Z_{\text{RLC-P}} = \frac{R}{1 + j\left(\omega C - \frac{1}{\omega L}\right)R} \quad (13.27)$$

$$Z_{\text{RLC-P}} = \frac{R}{1 + j\left(\omega RC - \frac{R}{\omega L}\right)} \quad (13.28)$$

Equations 13.27 and 13.28 show the impedance of a parallel RLC combination.

$$Z_{\text{RLC-P}} = \frac{j\omega L}{\frac{j\omega L}{R} + j(j\omega^2 LC - j)} \quad (13.29)$$

Multiply the numerator and denominator of the right-hand side of Equation 13.25 by $j\omega L$ to get a different form in Equation 13.29.

In Figure 13.14, the arrow indicates perspective for the impedance looking into the parallel RLC network.

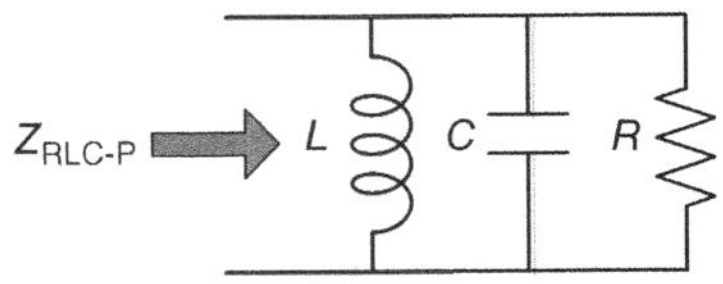

Figure 13.14 Parallel RLC impedance schematic.

13.1.4.1 Magnitude of Parallel RLC Impedance

The magnitude of a complex function, $G(\omega)$ in Equation 13.30, is the square root of the sum of the squares of the real part $R(\omega)$ and the imaginary part $X(\omega)$, as shown in Equation 13.31.

$$G(\omega) = R(\omega) + jX(\omega) \quad (13.30)$$

$$|G(\omega)| = \sqrt{[R(\omega)]^2 + [X(\omega)]^2} \quad (13.31)$$

The magnitude of the ratio of two complex functions $A(\omega)$ and $B(\omega)$ equals the ratio of the magnitudes of those same two complex functions, as shown in Equation 13.32.

$$\left|\frac{A(\omega)}{B(\omega)}\right| = \frac{|A(\omega)|}{|B(\omega)|} \quad (13.32)$$

To calculate the magnitude of the parallel RLC impedance in Equation 13.25, Equation 13.32 means that Equation 13.31 can be applied to the denominator of Equation 13.25 to get Equation 13.33.

$$|Z_{\text{RLC-P}}| = \frac{1}{\sqrt{\frac{1}{R^2} + \left(\omega C - \frac{1}{\omega L}\right)^2}} \quad (13.33)$$

Equation 13.33 is the magnitude of the impedance Equation 13.25 of the parallel RLC network in Figures 13.1b and 13.12.

In Equation 13.33, substitute 1 over G for R to obtain Equation 13.34.

$$|Z_{\text{RLC-P}}| = \frac{1}{\sqrt{G^2 + (\omega C - 1/\omega L)^2}} \quad (13.34)$$

Equation 13.34 is the magnitude of the impedance Equation 13.25 of the parallel RLC network in Figures 13.1b and 13.12.

13.1.5 Parallel RLC Admittance

This section derives the formula for the admittance of the parallel RLC combination in Figure 13.15.

$$Y = \frac{1}{Z} \qquad (13.35)$$

Admittance Y is the inverse of impedance Z, as shown in Equation 13.35.

$$Y_P = Y_1 + Y_2 + \cdots + Y_n = \sum_{m=1}^{n} Y_m \qquad (13.36)$$

Admittances in parallel add, as shown in Equation 13.36.

For the parallel RLC in Figure 13.15, the admittance is shown in Equation 13.37.

$$Y_{\text{RLC-P}} = Y_R + Y_L + Y_C = \frac{1}{Z_R} + \frac{1}{Z_L} + \frac{1}{Z_C} = \frac{1}{Z_{\text{RLC-P}}} \qquad (13.37)$$

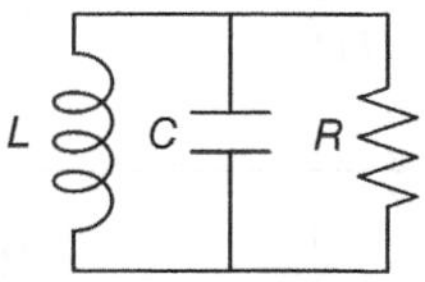

Figure 13.15 Parallel RLC schematic.

Substitute the individual impedance Equations 13.39, 13.41, and 13.43 into Equation 13.37 to obtain Equation 13.44.

$$Z_L = j\omega L \qquad (13.38)$$

Equation 13.38 is the impedance of an ideal inductor.

$$Y_L = \frac{-j}{\omega L} \qquad (13.39)$$

Equation 13.39 is the admittance of an ideal inductor.

$$Z_C = \frac{1}{j\omega C} \qquad (13.40)$$

Equation 13.40 is the impedance of an ideal capacitor.

$$Y_C = j\omega C \qquad (13.41)$$

Equation 13.41 is the admittance of an ideal capacitor.

$$Z_R = R = \frac{1}{G} \qquad (13.42)$$

Equation 13.42 is the impedance of an ideal resistor, where R is the resistance and G is conductance.

$$Y_R = \frac{1}{R} = G \qquad (13.43)$$

Equation 13.43 is the admittance of an ideal resistor, where R is the resistance and G is conductance.

$$Y_{\text{RLC-P}} = \frac{1}{R} - \frac{j}{\omega L} + j\omega C \qquad (13.44)$$

Equation 13.44 is the admittance of the parallel RLC in Figure 13.15.

13.1.5.1 Resonance of Parallel RLC

Resonance occurs at the angular frequency ω where the imaginary term in the impedance equation becomes zero. This resonant frequency is often designated with a zero subscript as ω_0.

$$\omega C - \frac{1}{\omega L} = 0 \qquad (13.45)$$

To calculate the resonant frequency of the parallel RLC in Figures 13.14 and 13.15, set the imaginary term in Equation 13.25 equal to zero, as shown in Equation 13.45.

$$\omega C = \frac{1}{\omega L} \qquad (13.46)$$

Rearrange terms in Equation 13.45 to obtain Equation 13.46.

$$\omega^2 = \frac{1}{LC} \qquad (13.47)$$

Multiply both sides of Equation 13.46 by ω and divide both sides by C to obtain Equation 13.47.

$$\omega = \frac{1}{\sqrt{LC}} \qquad (13.48)$$

Take the square root of both sides of Equation 13.47 to solve for ω in terms of inductance L and capacitance C in Equation 13.48.

$$\omega_0 = \frac{1}{\sqrt{LC}} \qquad (13.49)$$

Equations 13.16 and 13.49 are the resonant frequency ω_0 of an RLC network.

The resonant frequency Equation 13.49 is the same for series RLC as it is for parallel RLC.

The **maximum** value R, of the parallel RLC impedance, occurs at the resonant frequency ω_0, where the complex term in the denominators of Equations 13.25–13.28, 13.33, and 13.34 becomes zero.

The maximum impedance for a parallel RLC circuit, the solid line in Figure 13.16, occurs at resonance, where ω equals one over the square root of inductance L times capacitance C. At resonance, the impedance of the RLC combination equals the resistance R. In this example in Figure 13.16, the resistance is 10 Ω.

At low frequencies, capacitance reactance is larger than the inductive reactance and the capacitance dominates the phase. At high frequencies, inductance dominates the phase as the inductive reactance is larger than the capacitive reactance.

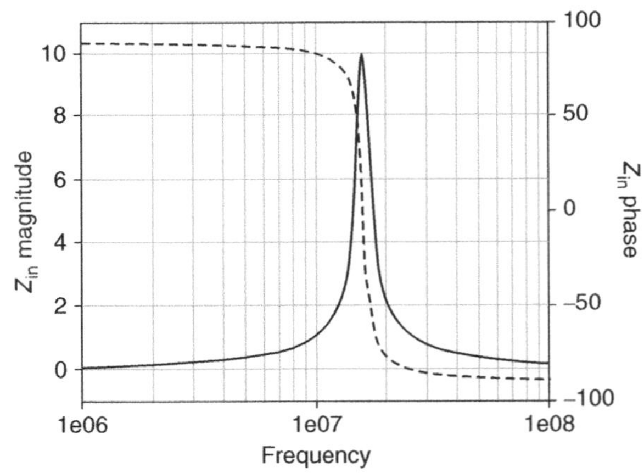

Figure 13.16 Parallel RLC impedance simulation results.

In Figure 13.16, the dashed line is the phase, scaled on the right-hand axis. Where the phase crosses zero is the exact resonance point.

In parallel circuits, admittances of individual elements add up to the overall admittance of the circuit. To see how the capacitance and inductance affect the impedance at low and high frequency, plot their admittances overlaid on the admittance of the parallel RLC. It ends up looking like the series RLC example of individual circuit element impedances overlaid on the impedance of the whole RLC.

13.1.5.2 Parallel RLC Impedance Intuition

At resonance, the inductive and capacitive reactances are equal and opposite and they cancel one another.

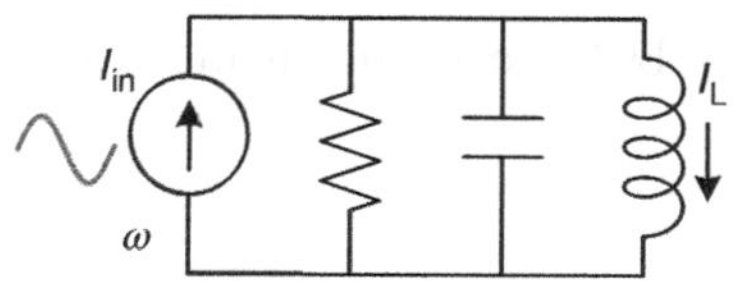

Figure 13.17a Parallel RLC circuit schematic driven by sinusoidal current.

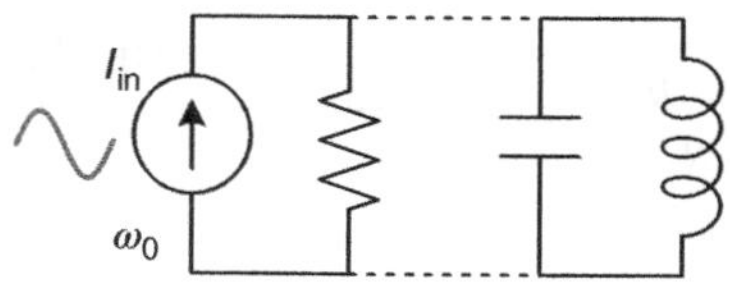

Figure 13.17b Parallel current source and resistor with LC seemingly disconnected.

Inductors and capacitors are energy storage elements. Connected together, they can exchange energy back and forth between one another. In parallel RLC, the capacitor and inductor are connected to each other at both ends. Both of their terminals are connected together, as shown in Figure 13.17a. The capacitor and inductor can exchange energy back and forth, without current from the rest of the circuit. At resonance, the back-and-forth exchange of energy between the capacitor and inductor matches the sinusoidal input source in a way that presents a high impedance to the source.

At resonance, indicated by the ω_0 by the source in Figure 13.17b, the parallel LC combination appears to be an open circuit, represented by dashed lines in the schematic in Figure 13.17b.

13.2 Second-Order RLC Frequency Response

This section explores second-order sinusoidal steady-state frequency response, including expressions for magnitude and phase. Bold-face type indicates phasor quantities. Notice the use of j as the imaginary operator rather than i used in other mathematics and physics literature.

13.2.1 Second-Order Series RLC Voltage Transfer Function

This sub-section calculates the steady-state frequency domain voltage transfer function of a series RLC circuit shown in Figure 13.18.

To calculate the transfer function of the series RLC output voltage, $\mathbf{V}_{out}$ over the input voltage $\mathbf{V}_{in}$, Equation 13.50 provides the series RLC impedance $Z_{RLC\text{-}S}$ from Equation 13.1a.

$$Z_{RLC\text{-}S} = j\omega L + \frac{1}{j\omega C} + R \tag{13.50}$$

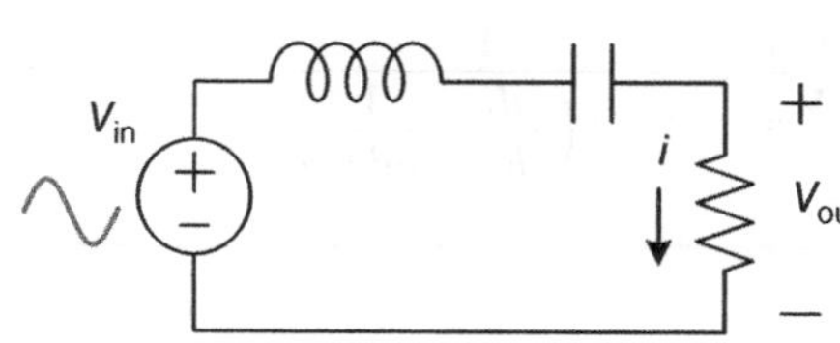

Figure 13.18 Series RLC schematic with sinusoidal voltage source.

Start with Ohm's law for $\mathbf{V}_{out}$ across the resistor in Equation 13.51. $\mathbf{V}_{in}$ over $Z_{RLC\text{-}S}$ is the loop current.

$$\boldsymbol{V}_{out} = \frac{\boldsymbol{V}_{in}}{Z_{RLC\text{-}S}} R \tag{13.51}$$

The output voltage across the resistor, $\mathbf{V}_{out}$ in Equation 13.51, equals the current in the loop times the impedance (resistance) of the resistor. The current in the loop equals the input voltage $\mathbf{V}_{in}$, divided by the series impedance of the inductor, capacitor, and resistor combined from Equation 13.50 and 13.1a.

Substitute the series RLC impedance formula, from Equation 13.50 into Equation 13.51 Ohm's law to obtain Equation 13.52.

$$\boldsymbol{V}_{\text{out}} = \frac{\boldsymbol{V}_{\text{in}}}{j\omega L + \frac{1}{j\omega C} + R} R \tag{13.52}$$

Divide both sides of Equation 13.52 by $\mathbf{V}_{\text{in}}$ to obtain Equation 13.53.

$$\frac{\boldsymbol{V}_{\text{out}}}{\boldsymbol{V}_{\text{in}}} = \frac{R}{j\omega L + \frac{1}{j\omega C} + R} \tag{13.53}$$

Equation 13.53, for $\mathbf{V}_{\text{out}}$ over $\mathbf{V}_{\text{in}}$, is the voltage transfer function for the series RLC in Figure 13.18.

Divide the numerator and denominator of the right side of Equation 13.53 by R to get Equation 13.54.

$$\frac{\boldsymbol{V}_{\text{out}}}{\boldsymbol{V}_{\text{in}}} = \frac{1}{\frac{j\omega L}{R} + \frac{1}{j\omega RC} + 1} \tag{13.54}$$

Equation 13.54 is the sinusoidal steady-state voltage transfer function for the series RLC circuit in Figure 13.18.

In Equation 13.54 multiply the denominator $j\omega RC$ term by j over j obtaining Equation 13.55.

$$\frac{\boldsymbol{V}_{\text{out}}}{\boldsymbol{V}_{\text{in}}} = \frac{1}{j\frac{\omega L}{R} - j\frac{1}{\omega RC} + 1} \tag{13.55}$$

In the denominator of Equation 13.55, factor out the j and rearrange terms to collect the imaginary term in Equation 13.56.

$$\frac{\boldsymbol{V}_{\text{out}}}{\boldsymbol{V}_{\text{in}}} = \frac{1}{1 + j\left(\frac{\omega L}{R} - \frac{1}{\omega RC}\right)} \tag{13.56}$$

Equation 13.56 is the sinusoidal steady-state voltage transfer function for the series RLC circuit in Figure 13.18.

13.2.1.1 Another Form of the Series RLC Transfer Function

For another form of the series RLC transfer function, go back to Equation 13.53. On the right-hand side of Equation 13.53, multiply the numerator and denominator by $j\omega C$ to obtain Equation 13.57.

$$\frac{\boldsymbol{V}_{\text{out}}}{\boldsymbol{V}_{\text{in}}} = \frac{j\omega RC}{-\omega^2 LC + 1 + j\omega RC} \tag{13.57}$$

Rearrange denominator terms in Equation 13.57 to obtain Equation 13.58.

$$\frac{\boldsymbol{V}_{\text{out}}}{\boldsymbol{V}_{\text{in}}} = \frac{j\omega RC}{1 - \omega^2 LC + j\omega RC} \tag{13.58}$$

Equation 13.58 is the sinusoidal steady-state voltage transfer function for the series RLC circuit in Figure 13.18.

13.2.1.2 Magnitude of Series RLC Transfer Function

The magnitude of a complex function, $G(\omega)$ in Equation 13.59, is the square root of the sum of the squares of the real part $R(\omega)$ and the imaginary part $X(\omega)$, as shown in Equation 13.60.

$$G(\omega) = R(\omega) + jX(\omega) \tag{13.59}$$

$$|G(\omega)| = \sqrt{[R(\omega)]^2 + [X(\omega)]^2} \tag{13.60}$$

The magnitude of the ratio of two complex functions $A(\omega)$ and $B(\omega)$ equals the ratio of the magnitudes of those same two complex functions, as shown in Equation 13.61.

$$\left|\frac{A(\omega)}{B(\omega)}\right| = \frac{|A(\omega)|}{|B(\omega)|} \tag{13.61}$$

To calculate the magnitude of the series RLC transfer function in Equation 13.56, apply Equation 13.61 to take the magnitude of the numerator and the denominator of Equation 13.56 to get Equation 13.62.

$$\frac{\boldsymbol{V}_{\text{out}}}{\boldsymbol{V}_{\text{in}}} = \frac{|1|}{\left|1 + j\left(\frac{\omega L}{R} - \frac{1}{\omega RC}\right)\right|} \tag{13.62}$$

In Equation 13.62, apply Equation 13.60, the square root of the sum of the squares, to get Equation 13.63.

$$\frac{\boldsymbol{V}_{\text{out}}}{\boldsymbol{V}_{\text{in}}} = \frac{1}{\sqrt{1 + j\left(\frac{\omega L}{R} - \frac{1}{\omega RC}\right)^2}} \tag{13.63}$$

Equation 13.63 is the **magnitude** of the series RLC transfer function from Equation 13.56 and Figure 13.18.

To calculate the magnitude of the series RLC transfer function in Equation 13.58, apply Equation 13.61 to take the magnitude of the numerator and the denominator of Equation 13.58 to get Equation 13.64.

$$\left|\frac{\boldsymbol{V}_{\text{out}}}{\boldsymbol{V}_{\text{in}}}\right| = \frac{|j\omega RC|}{|1 - \omega^2 LC + j\omega RC|} \tag{13.64}$$

Apply Equation 13.60 to the numerator and the denominator of Equation 13.64 to get Equation 13.65.

$$\left|\frac{\boldsymbol{V}_{\text{out}}}{\boldsymbol{V}_{\text{in}}}\right| = \frac{\sqrt{(\omega RC)^2}}{\sqrt{(1 - \omega^2 LC)^2 + (\omega RC)^2}} \tag{13.65}$$

Simplify the numerator of Equation 13.65 to get Equation 13.66.

$$\left|\frac{\boldsymbol{V}_{\text{out}}}{\boldsymbol{V}_{\text{in}}}\right| = \frac{\omega RC}{\sqrt{(1 - \omega^2 LC)^2 + (\omega RC)^2}} \tag{13.66}$$

Equation 13.66 is the **magnitude** of the series RLC transfer function from Equation 13.58 and Figure 13.18.

13.2.1.3 Phase of Series RLC Transfer Function

$$\theta(\omega) = \angle H(\omega) = \tan^{-1}\frac{X(\omega)}{R(\omega)} \tag{13.67}$$

Equation 13.67 shows in the general formula for the phase angle of a complex function $H(j\omega)$, where $X(\omega)$ is the imaginary part and $R(\omega)$ is the real part.

The series RLC transfer function, in Equation 13.56, is the quotient of two complex expressions $F(\omega)$ and $G(\omega)$, as shown in Equation 13.68.

$$H(\omega) = \frac{\boldsymbol{V}_{\text{out}}}{\boldsymbol{V}_{\text{in}}} = \frac{1}{1 + j\left(\frac{\omega L}{R} - \frac{1}{\omega RC}\right)} = \frac{F(\omega)}{G(\omega)} \tag{13.68}$$

$$F = 1 \quad (13.69)$$

$$G = 1 + j\left(\frac{\omega L}{R} - \frac{1}{\omega RC}\right) \quad (13.70)$$

In the case of the series RLC transfer function $H(j\omega)$ in Equation 13.68, the numerator is the purely real number 1.

Equation 13.71 shows how the phase angle of a quotient of complex functions is the phase angle of the numerator minus the phase angle of the denominator.

$$\angle H(j\omega) = \angle\frac{F(j\omega)}{G(j\omega)} = \frac{\angle\theta_F}{\angle\theta_G} = \angle(\theta_F) - \angle(\theta_G) \quad (13.71)$$

Into Equation 13.71, substitute the numerator and denominator from Equation 13.68, of the RC high-pass transfer function $H(j\omega)$, to get Equation 13.72.

$$\angle H(j\omega) = \frac{\angle 1}{\angle\left(1 + j\left(\frac{\omega L}{R} - \frac{1}{\omega RC}\right)\right)} = \angle(1) - \angle\left(1 + j\left(\frac{\omega L}{R} - \frac{1}{\omega RC}\right)\right) \quad (13.72)$$

Apply the formula in Equation 13.67, for the phase of a complex function, separately to both terms in Equation 13.72, the numerator and the denominator, as shown in Equations 13.73 and 13.74

Numerator F

$$\angle(1) = \tan^{-1}[X/R] = \tan^{-1}[0/1] = \tan^{-1}(0) = 0 \quad (13.73)$$

The phase angle of the numerator is 0 because the imaginary component is zero.

Denominator G

$$\angle\left(1 + j\left(\frac{\omega L}{R} - \frac{1}{\omega RC}\right)\right) = \tan^{-1}\left[\left(\frac{\omega L}{R} - \frac{1}{\omega RC}\right)/(1)\right] \quad (13.74)$$

Combine the results, from Equations 13.73 and 13.74, back into Equation 13.72 to obtain Equation 13.75.

$$\angle H(j\omega) = -\tan^{-1}\left(\frac{\omega L}{R} - \frac{1}{\omega RC}\right) \quad (13.75)$$

Equation 13.75 is the sinusoidal steady-state **phase** response of the series RLC circuit transfer function in Equation 13.56.

13.2.1.4 Phase of Another Form of Series RLC Transfer Function

$$\theta(\omega) = \angle H(\omega) = \tan^{-1}\frac{X(\omega)}{R(\omega)} \quad (13.76)$$

Equation 13.76 shows in the general formula for the phase angle of a complex function $H(j\omega)$.

$$H(\omega) = \frac{\mathbf{V}_{\text{out}}}{\mathbf{V}_{\text{in}}} = \frac{j\omega RC}{1 - \omega^2 LC + j\omega RC} = \frac{F(\omega)}{G(\omega)} \quad (13.77)$$

The series RLC transfer function, in Equation 13.57 is the quotient of two complex expressions $F(\omega)$ and $G(\omega)$, as shown in Equation 13.77.

$$F = j\omega RC \quad (13.78)$$

$$G = 1 - \omega^2 LC + j\omega RC \quad (13.79)$$

In the case of the series RLC transfer function $H(j\omega)$ in Equation 13.77, the numerator is the purely imaginary expression $j\omega RC$ in Equation 13.78.

Equation 6.80 shows how the phase angle of a quotient of complex functions is the phase angle of the numerator minus the phase angle of the denominator.

$$\angle H(j\omega) = \angle\frac{F(j\omega)}{G(j\omega)} = \frac{\angle\theta_F}{\angle\theta_G} = \angle(\theta_F) - \angle(\theta_G) \tag{13.80}$$

Into Equation 13.80, substitute the numerator, from Equation 13.78, and denominator from Equation 13.79, of the RC high-pass transfer function $H(j\omega)$ to get Equation 13.81.

$$\angle H(j\omega) = \frac{\angle j\omega RC}{\angle(1 - \omega^2 LC + j\omega RC)} = \angle(j\omega RC) - \angle(1 - \omega^2 LC + j\omega RC) \tag{13.81}$$

Apply the formula, in Equation 13.76, for the phase of a complex function, separately to both terms separately in Equation 13.81, the numerator and the denominator, as shown in Equations 13.82 and 13.83.

*Numerator **F***

$$\angle(j\omega RC) = \tan^{-1}[X/R] = \tan^{-1}[\omega RC/0] = \tan^{-1}(\infty) = 90 \tag{13.82}$$

The phase angle the numerator is π over 2 because the real component is zero.

*Denominator **G***

$$\angle(1 - \omega^2 LC + j\omega RC) = \tan^{-1}[\omega RC/(1 - \omega^2 LC)] \tag{13.83}$$

Substitute the results, in Equations 13.82 and 13.84, back into Equation 13.80 to obtain Equation 13.84.

$$\angle H(j\omega) = \frac{\angle j\omega RC}{\angle(1 - \omega^2 LC + j\omega RC)} = \frac{\pi}{2} - \tan^{-1}[\omega RC/(1 - \omega^2 LC)] \tag{13.84}$$

$\angle H(j\omega) = \frac{\pi}{2} - \tan^{-1}[\omega RC/(1 - \omega^2 LC)]$ (13.84)	Equation 13.84 is the sinusoidal steady-state phase response of the series RLC circuit transfer function Equation 13.57.

13.2.2 Series RLC Transfer Function Magnitude and Phase Simulation Example

This section simulates the magnitude and phase of the series RLC transfer function in Equation 13.85 and Figures 13.19 and 13.20.

$$\frac{\boldsymbol{V}_{\text{out}}}{\boldsymbol{V}_{\text{in}}} = \frac{1}{1 + j\left(\frac{\omega L}{R} - \frac{1}{\omega RC}\right)} \tag{13.85}$$

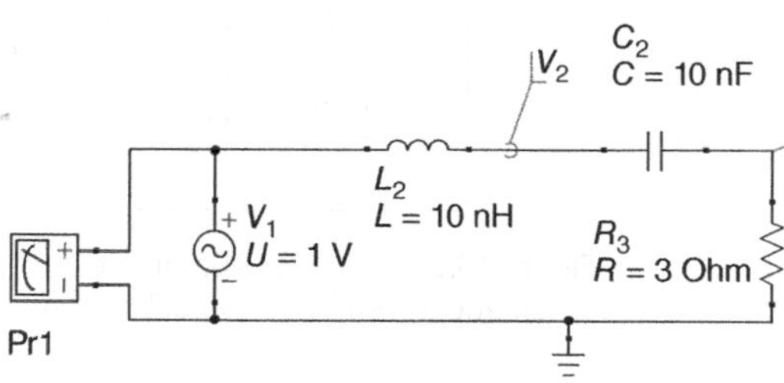

Figure 13.20 Series RLC simulation schematic with sinusoidal voltage source.

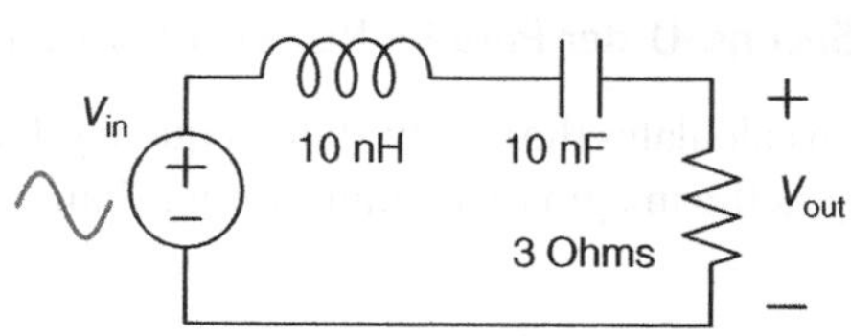

Figure 13.19 Series RLC schematic with sinusoidal voltage source.

V_{out} in Figure 13.19 corresponds to V_3 in the simulation schematic in Figure 13.20, where V_3 is the output voltage across the 3 Ω resistor. The transfer function for this series RLC circuit is Equation 13.85.

$$\frac{\mathbf{V}_{\text{out}}}{\mathbf{V}_{\text{in}}} = \frac{1}{\sqrt{1 + j\left(\frac{\omega L}{R} - \frac{1}{\omega RC}\right)^2}} \tag{13.86}$$

Equation 13.86 is the **magnitude** of the series RLC transfer function from Equation 13.85 and the transfer function magnitude is plotted in Figure 13.21.

In Figure 13.21 the peak of the magnitude is at resonance, where ωL equals 1 over ωC.

$$\angle H(j\omega) = -\tan^{-1}\left(\frac{\omega L}{R} - \frac{1}{\omega RC}\right) \tag{13.87}$$

Plotted in Figure 13.22, Equation 13.87 is the sinusoidal steady-state **phase** response of the series RLC circuit transfer function in Equation 13.85.

In Figure 13.22, at low frequencies, the capacitance dominates the response. At resonance the phase inverts from negative 180° to positive 180°, and the inductance dominates the high-frequency behavior.

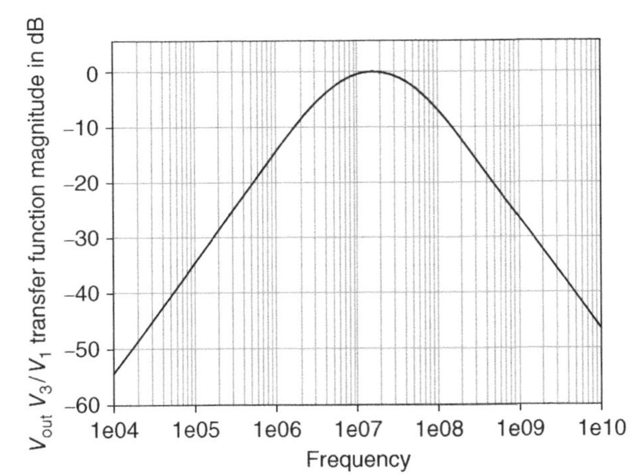

Figure 13.21 Series RLC simulation results graph of voltage transfer function magnitude.

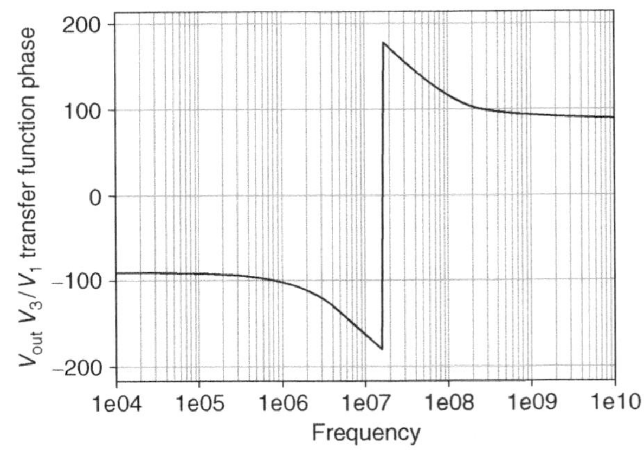

Figure 13.22 Series RLC simulation results graph of voltage transfer function phase.

13.2.3 Second-Order Parallel RLC Current Transfer Function

This section calculates the steady-state frequency domain current transfer function of a parallel RLC circuit. Notice the use of j as the imaginary operator rather than i used in other mathematics and physics literature.

$$Z_{\text{RLC-P}} = \frac{R}{1 + j\left(\omega C - \frac{1}{\omega L}\right)R} \tag{13.88}$$

Equation 13.88 provides the parallel RLC impedance $Z_{\text{RLC-P}}$ from Equation 13.1b.

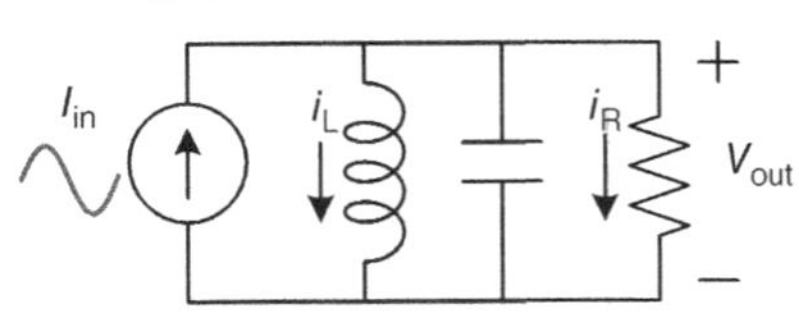

Figure 13.23 Parallel RLC circuit schematic driven by sinusoidal current.

Start with Ohm's law for $\mathbf{V}_{\text{out}}$ across all circuit elements.

$$\mathbf{V}_{\text{out}} = Z_{\text{RLC-P}}\mathbf{I}_{\text{in}} \tag{13.89}$$

The output voltage $\mathbf{V}_{out}$, in Equation 13.89, equals the source current $\mathbf{I}_{in}$ times the parallel impedance of the inductor, capacitor, and resistor combined. The output current $\mathbf{I}_{out}$ (the same as $\mathbf{I}_R$) then equals $\mathbf{V}_{out}$ divided by R.

Substitute the parallel RLC impedance formula, from Equation 13.88, into Equation 13.89 Ohm's law to obtain Equation 13.90.

$$\boldsymbol{V}_{out} = \frac{R}{1+j\left(\omega C - \frac{1}{\omega L}\right)R}\boldsymbol{I}_{in} \quad (13.90)$$

Divide both sides of Equation 13.90 by $\mathbf{I}_{in}$ and R.

$$\frac{\boldsymbol{V}_{out}}{R\boldsymbol{I}_{in}} = \frac{1}{1+j\left(\omega C - \frac{1}{\omega L}\right)R} \quad (13.91)$$

In Equation 13.91 $\mathbf{V}_{out}$ over R becomes $\mathbf{I}_{out}$ in Equation 13.92.

$$\frac{\boldsymbol{I}_{out}}{\boldsymbol{I}_{in}} = \frac{1}{1+j\left(\omega RC - \frac{R}{\omega L}\right)} \quad (13.92)$$

Equations 13.91 and 13.92 are the sinusoidal steady-state current transfer function for the parallel RLC circuit in Figure 13.23.

For another form of the current transfer function, on the right-hand side of Equation 13.92, multiply the numerator and denominator by $j\omega L$ to get Equation 13.93.

$$\frac{\boldsymbol{I}_{out}}{\boldsymbol{I}_{in}} = \frac{j\omega L}{j\omega L + j(j\omega^2 LC - j)R} \quad (13.93)$$

In Equation 13.93, multiply out the j in front of the bracketed term to obtain Equation 13.94. j times j equals minus 1.

$$\frac{\boldsymbol{I}_{out}}{\boldsymbol{I}_{in}} = \frac{j\omega L}{j\omega L + (-\omega^2 RLC + R)} \quad (13.94)$$

In Equation 13.94, rearrange the denominator to obtain Equation 13.95.

$$\frac{\boldsymbol{I}_{out}}{\boldsymbol{I}_{in}} = \frac{j\omega L}{(R - \omega^2 RLC) + j\omega L} \quad (13.95)$$

Equation 13.95 is the sinusoidal steady-state current transfer function for the parallel RLC circuit in Figure 13.23.

13.2.3.1 Magnitude of the Parallel RLC Current Transfer Function

To calculate the magnitude of the parallel RLC current transfer function in Equation 13.92, apply Equation 13.61 to take the magnitude of the numerator and the denominator of Equation 13.92 to get Equation 13.96.

$$\left|\frac{\boldsymbol{I}_{out}}{\boldsymbol{I}_{in}}\right| = \frac{|1|}{\left|1+j\left(\omega RC - \frac{R}{\omega L}\right)\right|} \quad (13.96)$$

Apply Equation 13.60 to Equation 13.96 to calculate the magnitude of the numerator and the denominator of Equation 13.96 to get Equation 13.97.

$$\left|\frac{\boldsymbol{I}_{out}}{\boldsymbol{I}_{in}}\right| = \frac{1}{\sqrt{1+\left(\omega RC - \frac{R}{\omega L}\right)^2}} \quad (13.97)$$

Equation 13.97 is the **magnitude** of the parallel RLC current transfer function from Equation 13.92 and Figure 13.23.

To calculate another form of the magnitude of the parallel RLC transfer function, in Equation 13.93 apply Equation 13.61 to take the magnitude of the numerator and the denominator of Equation 13.93 to get Equation 13.98.

$$\left|\frac{\boldsymbol{I}_{\text{out}}}{\boldsymbol{I}_{\text{in}}}\right| = \frac{|j\omega L|}{|(R-\omega^2RLC)+j\omega L|} \tag{13.98}$$

Apply Equation 13.60 to take the magnitude of the numerator and the denominator of Equation 13.98 to get Equation 13.99.

$$\left|\frac{\boldsymbol{I}_{\text{out}}}{\boldsymbol{I}_{\text{in}}}\right| = \frac{\sqrt{(j\omega L)^2}}{\sqrt{(R-\omega^2RLC)^2+(j\omega L)^2}} \tag{13.99}$$

Simplify the numerator of Equation 13.99 to get Equation 13.100.

$$\left|\frac{\boldsymbol{I}_{\text{out}}}{\boldsymbol{I}_{\text{in}}}\right| = \frac{j\omega L}{\sqrt{(R-\omega^2RLC)^2+(j\omega L)^2}} \tag{13.100}$$

Equation 13.100 is the **magnitude** of the parallel RLC current transfer function from Equation 13.93 and Figure 13.23.

13.2.3.2 Phase of Parallel RLC Current Transfer Function

$$\frac{\boldsymbol{I}_{\text{out}}}{\boldsymbol{I}_{\text{in}}} = \frac{1}{1+j\left(\omega RC-\frac{R}{\omega L}\right)} \tag{13.101}$$

Equation 13.101 is the sinusoidal steady-state current transfer function for the parallel RLC circuit in Figure 13.23.

$$\theta(\omega) = \angle H(\omega) = \tan^{-1}\frac{X(\omega)}{R(\omega)} \tag{13.102}$$

Equation 13.102 shows in the general formula for the phase angle of a complex function $H(j\omega)$, where $X(\omega)$ is the imaginary part and $R(\omega)$ is the real part.

$$H(\omega) = \frac{\boldsymbol{I}_{\text{out}}}{\boldsymbol{I}_{\text{in}}} = \frac{1}{1+j\left(\omega RC-\frac{R}{\omega L}\right)} = \frac{F(\omega)}{G(\omega)} \tag{13.103}$$

The parallel RLC current transfer function, in Equation 13.101, is the quotient of two complex expressions $F(\omega)$ and $G(\omega)$, as shown in Equation 13.103.

$$F = 1 \tag{13.104}$$

$$G = 1+j\left(\omega RC-\frac{R}{\omega L}\right) \tag{13.105}$$

In the case of this parallel RLC transfer function $H(j\omega)$ in Equation 13.101, the numerator is the purely real number 1, as shown in Equation 13.104.

Equation 13.106 shows how the phase angle of a quotient of complex functions is the phase angle of the numerator minus the phase angle of the denominator.

$$\angle H(j\omega) = \angle\frac{F(j\omega)}{G(j\omega)} = \frac{\angle\theta_{\text{F}}}{\angle\theta_{\text{G}}} = \angle(\theta_{\text{F}}) - \angle(\theta_{\text{G}}) \tag{13.106}$$

Into Equation 13.106, substitute the numerator and denominator from Equation 13.103, of the parallel RLC transfer function $H(j\omega)$, to get Equation 13.107.

$$\angle H(j\omega) = \frac{\angle 1}{\angle\left(1+j\left(\omega RC-\frac{R}{\omega L}\right)\right)} = \angle(1) - \angle\left(1+j\left(\omega RC-\frac{R}{\omega L}\right)\right) \tag{13.107}$$

Apply the formula, in Equation 13.102, for the phase of a complex function, separately to both terms in Equation 13.107, the numerator and the denominator, as shown in Equations 13.108 and 13.109.

Numerator F

$$\angle(1) = \tan^{-1}[X/R] = \tan^{-1}[0/1] = \tan^{-1}(0) = 0 \tag{13.108}$$

The phase angle of the numerator is zero because the numerator is a purely real number.

Denominator G

$$\angle\left(1 + j\left(\omega RC - \frac{R}{\omega L}\right)\right) = \tan^{-1}\left[\left(\omega RC - \frac{R}{\omega L}\right)/(1)\right] \tag{13.109}$$

$$\angle H(j\omega) = 0 - \tan^{-1}\left(\omega RC - \frac{R}{\omega L}\right) \tag{13.110}$$

Equation 13.110 is the sinusoidal steady-state phase response of the parallel RLC circuit in Figure 13.23 with the current transfer function in Equation 13.101.

13.2.4 Parallel RLC Impedance Magnitude and Phase Simulation Example

This section simulates the magnitude and phase of the parallel RLC transfer function in Equation 13.111 and Figure 13.24.

$$\frac{\mathbf{I}_{\text{out}}}{\mathbf{I}_{\text{in}}} = \frac{1}{1 + j\left(\omega RC - \frac{R}{\omega L}\right)} \tag{13.111}$$

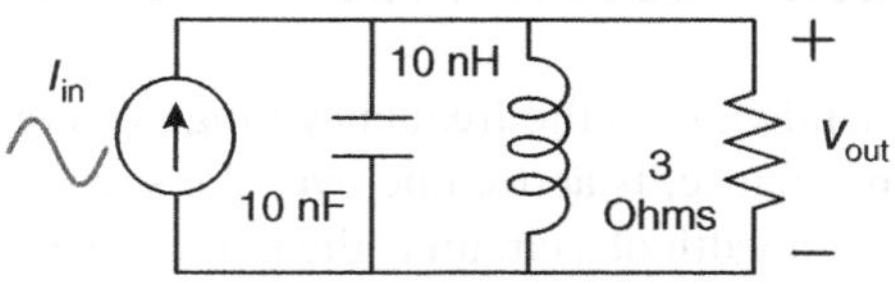

Figure 13.24 Parallel RLC circuit schematic driven by sinusoidal current.

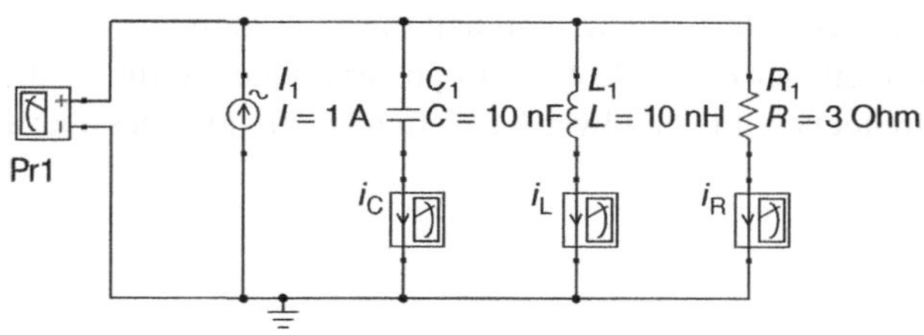

Figure 13.25 Parallel RLC simulation schematic driven by sinusoidal current.

$\mathbf{I}_{\text{out}}$ in Equation 13.111 corresponds to i_R in the simulation schematic in Figure 13.25. i_R is the output current through the 3 Ω resistor.

$$\left|\frac{\mathbf{I}_{\text{out}}}{\mathbf{I}_{\text{in}}}\right| = \frac{1}{\sqrt{1 + \left(\omega RC - \frac{R}{\omega L}\right)^2}} \tag{13.112}$$

Equation 13.112, the **magnitude** of the parallel RLC current transfer function, from Equation 13.111 and the schematics in Figures 13.24 and 13.25, is plotted in Figure 13.26.

In Figure 13.26 the peak of the magnitude is at resonance, where ωC equals 1 over ωL.

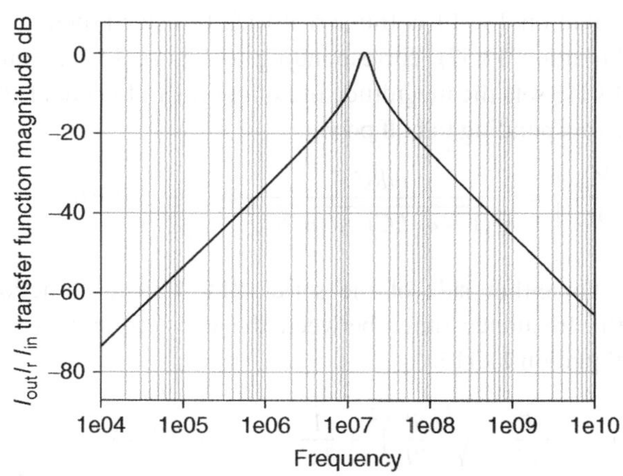

Figure 13.26 Parallel RLC simulation results plot of current transfer function magnitude.

$$\angle H(j\omega) = -\tan^{-1}\left(\omega RC - \frac{R}{\omega L}\right) \tag{13.113}$$

Plotted in Figure 13.27, Equation 13.113 is the sinusoidal steady-state **phase** response of the parallel RLC circuit transfer function in Equation 13.111.

In Figure 13.27, at high frequencies the capacitance dominates the response with lower impedance. At resonance the phase crosses through zero. At low frequencies, the inductance dominates the response behavior.

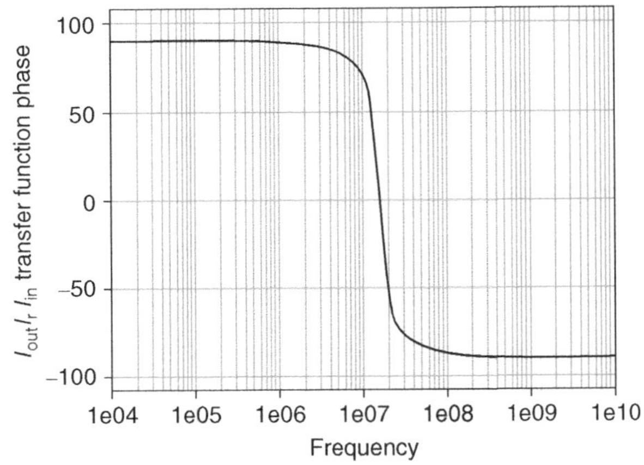

Figure 13.27 Parallel RLC simulation results plot of current transfer function phase.

13.3 Second-Order RLC Bandwidth and Quality Factor

Bandwidth is the frequency range where the circuit response magnitude, relative to the maximum response at resonance, is above one over the square root of two (0.707 or ~−3 dB), which corresponds to half-power. The bandwidth of a circuit is the frequency range where the circuit delivers more than half the maximum power.

In first-order RC and RL circuits, the corner or break frequency ω_0 marks the half-power frequency, the −3 dB point. In second-order circuits, ω_0 marks the resonant frequency, and bandwidth is defined by frequencies above and below resonance.

13.3.1 Series RLC Bandwidth

$$\frac{\boldsymbol{V}_{\text{out}}}{\boldsymbol{V}_{\text{in}}} = \frac{j\omega RC}{-\omega^2 LC + 1 + j\omega RC} \tag{13.114}$$

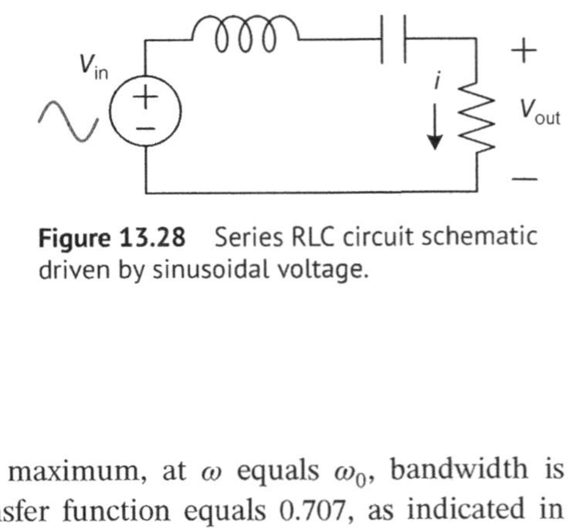

Figure 13.28 Series RLC circuit schematic driven by sinusoidal voltage.

Equation 13.114 is the series RLC transfer function (from Equation 13.57), corresponding to Figure 13.28. Equation 13.115 sets the magnitude of Equation 13.114 equal to 0.707, corresponding to half power.

$$\left|\frac{\boldsymbol{V}_{\text{out}}}{\boldsymbol{V}_{\text{in}}}\right| = \frac{\omega RC}{\sqrt{(1-\omega^2 LC)^2 + (\omega RC)^2}} = \frac{1}{\sqrt{2}} = 0.707 \tag{13.115}$$

On either side of the series RLC transfer function resonance maximum, at ω equals ω_0, bandwidth is the frequency range between the two frequencies where the transfer function equals 0.707, as indicated in Equation 13.115.

$$\omega_1 = -\frac{R}{2L} + \sqrt{\left(\frac{R}{2L}\right)^2 + \frac{1}{LC}} \tag{13.116}$$

In Equations 13.116 and 13.117, ω_1 and ω_2 are the two cutoff frequencies where the series RLC voltage transfer function equals 0.707.

$$\omega_2 = \frac{R}{2L} + \sqrt{\left(\frac{R}{2L}\right)^2 + \frac{1}{LC}} \tag{13.117}$$

The bandwidth (BW), shown in Equation 13.118, lies between these two frequencies ω_1 and ω_2 in Equations 13.116 and 13.117.

$$\mathrm{BW} = \omega_2 - \omega_1 \tag{13.118}$$

From Equations 13.116 and 13.117, Equation 13.119 calculates bandwidth equals R over L in Equation 13.120.

$$\omega_2 - \omega_1 = \left(\frac{R}{2L} + \sqrt{\left(\frac{R}{2L}\right)^2 + \frac{1}{LC}}\right) - \left(-\frac{R}{2L} + \sqrt{\left(\frac{R}{2L}\right)^2 + \frac{1}{LC}}\right) = \frac{R}{2L} - -\frac{R}{2L} = \frac{R}{L} \tag{13.119}$$

$$\mathrm{BW} = \frac{R}{L} \tag{13.120}$$

For a series RLC circuit, larger resistance means greater bandwidth. This is apparent from Equation 13.120, where bandwidth BW is proportional to the resistance R.

13.3.2 Series RLC Bandwidth Example

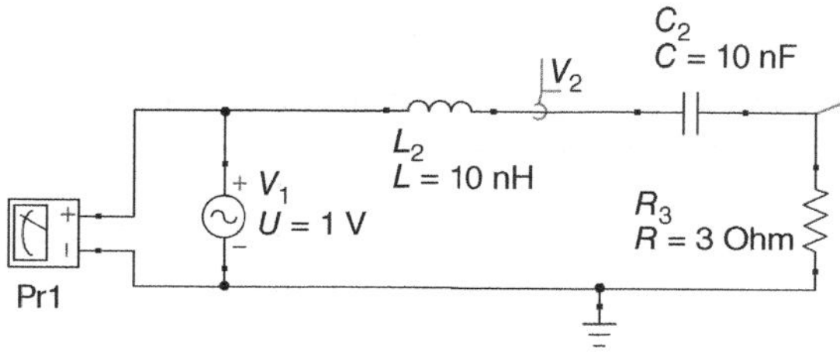

Figure 13.29 Series RLC simulation schematic with sinusoidal voltage source.

Based on Figure 13.28, the series RLC circuit simulation schematic in Figure 13.29 has a resistance value of 3 Ω. The voltage transfer function, $\mathbf{V}_{out}$ over $\mathbf{V}_{in}$ in Equation 13.114, corresponds to V_3 over V_1 in the simulation schematic in Figure 13.29, where V_1 is 1.

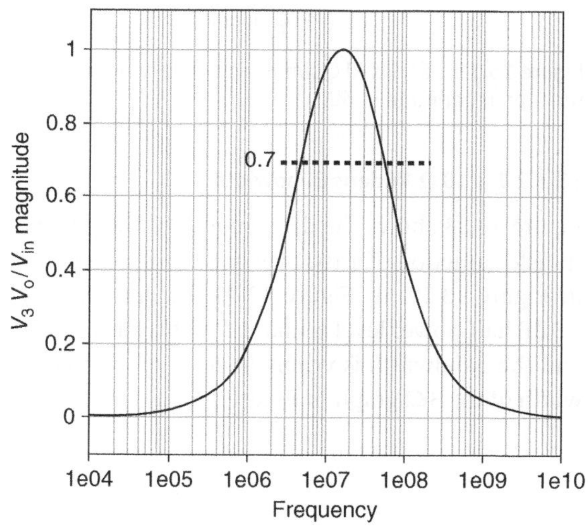

Figure 13.30 Series RLC simulation of voltage transfer function magnitude with 3 Ω.

Figure 13.30 shows the magnitude response of the voltage transfer function of the 3 Ω schematic in Figure 13.29. The approximate 0.707 level is marked in Figure 13.30 with a dashed line. Figure 1.30 is not in dB.

Figures 13.31 and 13.32 show the dB response of the voltage transfer function of the 3 Ω schematic in Figure 13.29. In Figure 13.32 the approximate −3 dB half-power level is marked with a dashed line.

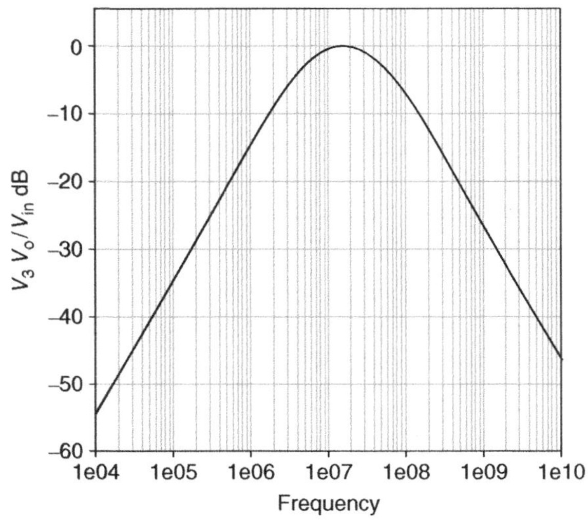

Figure 13.31 Series RLC simulation of voltage transfer function magnitude in dB.

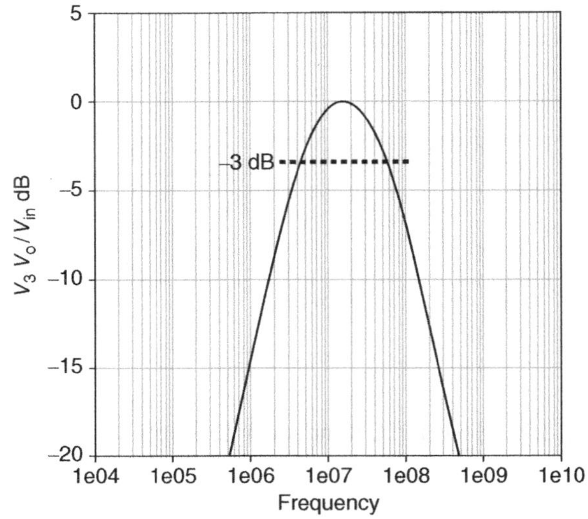

Figure 13.32 Series RLC simulation of voltage transfer function magnitude zoomed in.

13.3.2.1 10 Ω Widens Bandwidth

Figure 13.33 shows the dB response of the voltage transfer function of a 10 Ω version of the schematic in Figure 13.29. In Figure 13.33 the approximate −3 dB half-power level is marked with a dashed line and the bandwidth is visibly wider than the bandwidth of the 3 Ω graph in Figure 13.32.

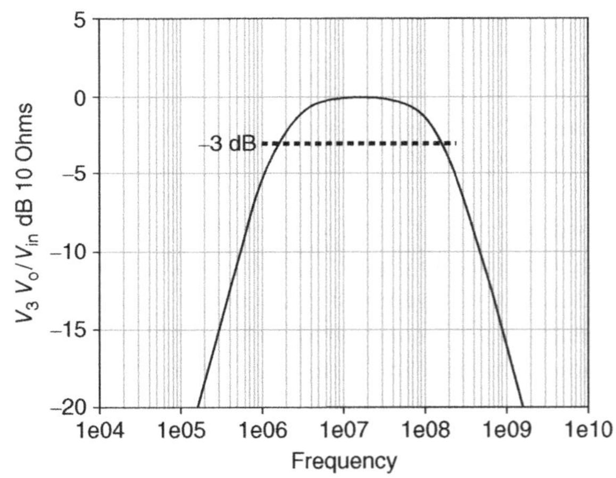

Figure 13.33 Series RLC 10 Ω simulation of voltage transfer function magnitude.

13.3.3 Parallel RLC Bandwidth

$$\frac{\boldsymbol{I}_{\text{out}}}{\boldsymbol{I}_{\text{in}}} = \frac{j\omega L}{j\omega L + j(j\omega^2 LC - j)R} \tag{13.121}$$

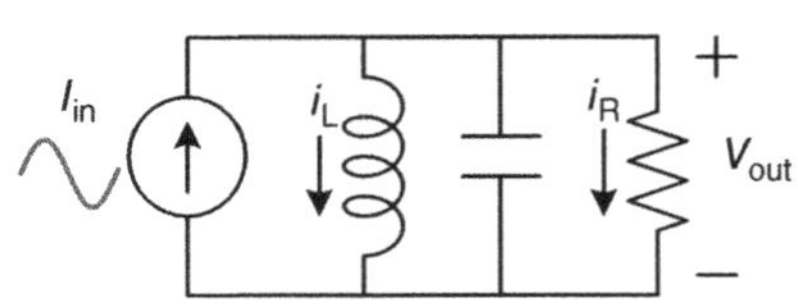

Figure 13.34 Parallel RLC circuit schematic driven by sinusoidal current.

Equation 13.121 is the parallel RLC current transfer function (from Equation 13.93), corresponding to Figure 13.34. Equation 13.122 sets the magnitude of Equation 13.121 equal to 0.707, corresponding to half-power.

$$\left|\frac{\boldsymbol{I}_{\text{out}}}{\boldsymbol{I}_{\text{in}}}\right| = \frac{j\omega L}{\sqrt{(R - \omega^2 RLC)^2 + (j\omega L)^2}} = \frac{1}{\sqrt{2}} = 0.707 \tag{13.122}$$

On either side of the parallel RLC transfer function resonance maximum, at ω equals ω_0, bandwidth is the frequency range between the two frequencies where the transfer function equals 0.707, as indicated in Equation 13.122.

$$\omega_1 = -\frac{1}{2RC} + \sqrt{\left(\frac{1}{2RC}\right)^2 + \frac{1}{LC}} \qquad (13.123)$$

$$\omega_2 = \frac{1}{2RC} + \sqrt{\left(\frac{1}{2RC}\right)^2 + \frac{1}{LC}} \qquad (13.124)$$

In Equations 13.123 and 13.124, ω_1 and ω_2 are the two cutoff frequencies where the parallel RLC current transfer function equals 0.707.

The bandwidth (BW), shown in Equation 13.125, lies between these two frequencies ω_1 and ω_2 in Equations 13.123 and 13.124.

$$\text{BW} = \omega_2 - \omega_1 \qquad (13.125)$$

Equation 13.126 calculates 1 over RC from Equations 13.123 and 13.124.

$$\omega_2 - \omega_1 = \left(\frac{1}{2RC} + \sqrt{\left(\frac{1}{2RC}\right)^2 + \frac{1}{LC}}\right) - \left(-\frac{1}{2RC} + \sqrt{\left(\frac{1}{2RC}\right)^2 + \frac{1}{LC}}\right) = \frac{1}{2RC} - -\frac{1}{2RC} = \frac{1}{RC} \qquad (13.126)$$

$$\text{BW} = \frac{1}{RC} \qquad (13.127)$$

For a parallel RLC circuit, larger resistance means less bandwidth. This is apparent from Equation 13.127, where BW, ω_2 minus ω_1, is proportional to one over the resistance R.

13.3.4 Parallel RLC Bandwidth Example

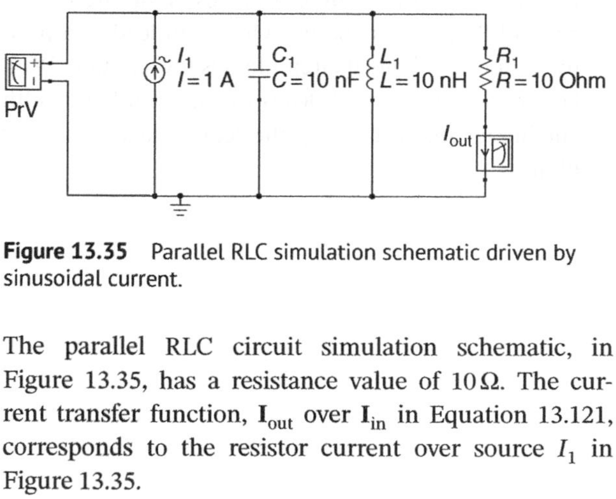

Figure 13.35 Parallel RLC simulation schematic driven by sinusoidal current.

The parallel RLC circuit simulation schematic, in Figure 13.35, has a resistance value of 10 Ω. The current transfer function, $\mathbf{I}_{out}$ over $\mathbf{I}_{in}$ in Equation 13.121, corresponds to the resistor current over source I_1 in Figure 13.35.

Figure 13.36 shows the magnitude response of the current transfer function of the 10 Ω schematic in Figure 13.35. The approximate 0.707 level is marked with a dashed line in Figure 13.36.

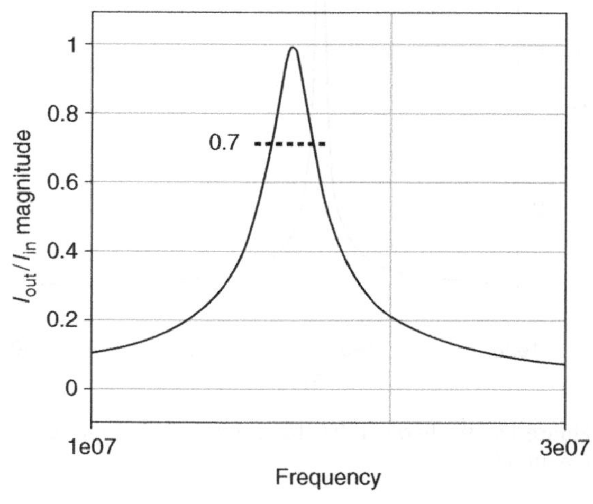

Figure 13.36 Parallel RLC 10 Ω simulation of current transfer function magnitude.

Figures 13.37 and 13.38 show the dB response of the current transfer function of the 10 Ω schematic in Figure 13.35. In Figure 13.38 the approximate −3 dB half-power level is marked with a dashed line.

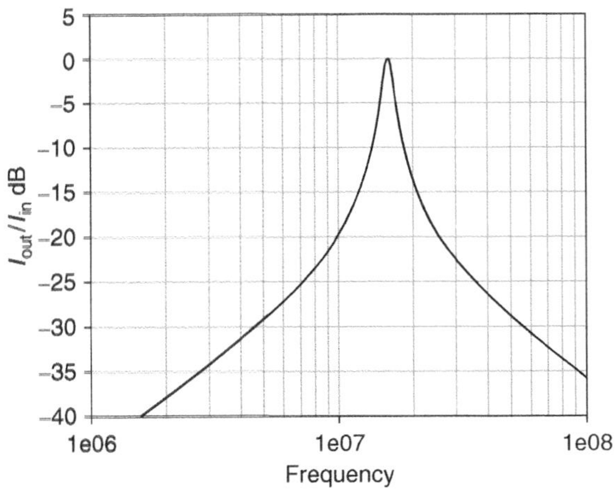

Figure 13.37 Parallel RLC 10 Ω simulation of current transfer function magnitude in dB.

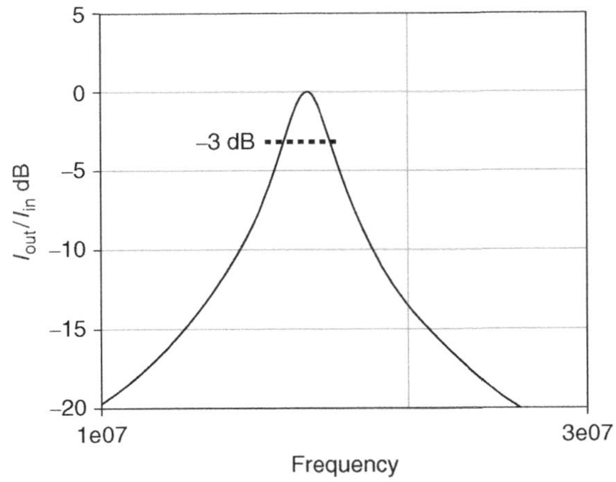

Figure 13.38 Parallel RLC 10 Ω simulation of current transfer function magnitude zoomed in.

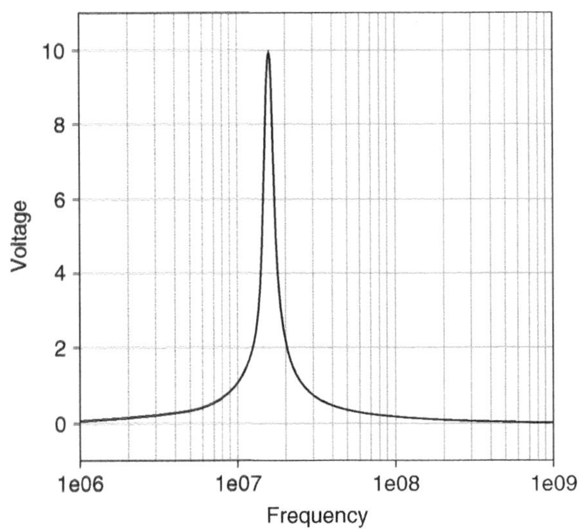

Figure 13.39 Parallel RLC 10 Ω simulation of voltage magnitude in dB.

Figure 13.39 shows the voltage of the parallel RLC circuit 10 Ω schematic in Figure 13.35. The peak value is 10 dB, a positive value. As a resonant circuit, the parallel RLC stores energy, which can lead to a positive value of voltage if circuit loss is low enough. With larger resistance, which leads to narrower bandwidth and higher quality factor Q, the peak voltage value goes higher.

13.3.4.1 3 Ω Widens Bandwidth

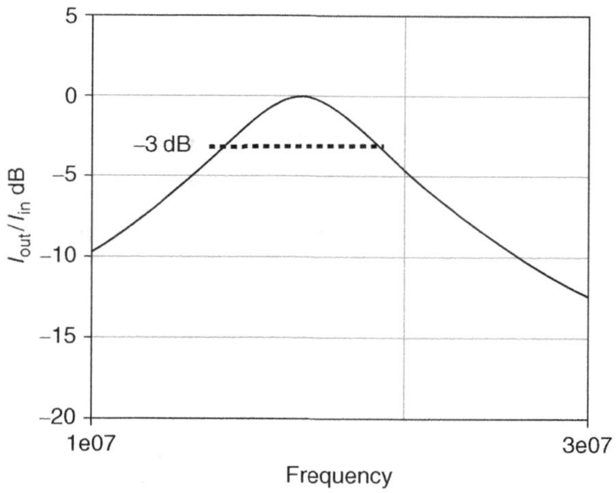

Figure 13.40 Parallel RLC 3 Ω simulation of current transfer function magnitude in dB.

Figure 13.40 shows the dB response of the current transfer function of a 3 Ω version of the schematic in Figure 13.35. In Figure 13.40 the approximate −3 dB half-power level is marked with a dashed line and the bandwidth is visibly wider than the 10 Ω graph in Figure 13.38.

13.3.5 Quality Factor *Q*

The quality factor *Q*, of a resonant circuit or electromagnetic cavity structure, is the amount of energy stored per cycle of sinusoidal signal, in comparison to the amount of energy lost or dissipated per cycle. Quality factor is related to bandwidth. In RLC circuits, both bandwidth and *Q* relate to energy dissipated in resistors.

$$Q = 2\pi \left(\frac{\text{energy stored}}{\text{energy dissipated per cycle}} \right) \tag{13.128}$$

As a ratio of two energies, *Q* is dimensionless.

For a sinusoidal steady-state RLC circuit analysis, Equation 13.128 defines quality factor *Q* as the energy stored, in the inductor and capacitor, divided by the energy dissipated in the resistor for a period of the input signal.

Equation 13.128 is one of two common definitions of quality factor *Q*. This verbal definition fits a wide variety of applications, from lumped element circuits to cavity filters to optical cavities. Equation 13.129 shows a second common definition of *Q* in terms of resonant frequency and bandwidth.

$$Q = \frac{\omega_0}{\Delta\omega} = \frac{\omega_0}{\text{BW}} \tag{13.129}$$

$$Q = \frac{\omega_0}{\Delta\omega} = \frac{\omega_0}{\omega_2 - \omega_1} \tag{13.130}$$

Equation 13.129 states that quality factor *Q* equals the resonant frequency ω_0 divided by the bandwidth $\Delta\omega$ or BW.
Substitute Equation 13.118, which states bandwidth equals $\omega_2 - \omega_1$, into Equation 13.129 to express bandwidth in terms of two cutoff frequencies as shown in Equation 13.130.

Equation 13.130, for quality factor *Q*, applies to both series and parallel RLC circuits. *Q* is always inversely proportional to bandwidth BW. Higher *Q* corresponds to narrower (smaller) bandwidth.

$$\omega_0 = \frac{1}{\sqrt{LC}} \tag{13.131}$$

$$\omega_0^2 = \frac{1}{LC} \tag{13.132}$$

In Equation 13.131, the resonant frequency ω_0, for both parallel and series RLC circuits, equals one over the square root of *LC*.
In Equation 13.132, the resonant frequency ω_0 squared, for both parallel and series RLC circuits, equals one over *LC*.

Series RLC

$$\mathrm{BW} = \frac{R}{L} \tag{13.133a}$$

Equation 13.133a is the bandwidth of a series RLC circuit.

Parallel RLC

$$\mathrm{BW} = \frac{1}{RC} \tag{13.133b}$$

Equation 13.133b is the bandwidth of a parallel RLC circuit.

Combine the above bandwidth Equations 13.133a and 13.133b, with the resonant frequency ω_0 Equation 13.131, to generate more expressions for series and parallel RLC Q.

13.3.6 Series RLC Quality Factor Q

13.3.6.1 Series RLC Equations for Q

$$Q = \frac{\omega_0 L}{R} \tag{13.134}$$

$$Q = \frac{1}{\omega_0 RC} \tag{13.135}$$

$$Q = \frac{1}{R}\sqrt{\frac{L}{C}} \tag{13.136}$$

The derivation, of Equation 13.134 for Q, is shown in Equation 13.137.

$$Q = \frac{\omega_0}{\mathrm{BW}} = \frac{\omega_0}{R/L} = \frac{\omega_0 L}{R} \tag{13.137}$$

For the BW in Equation 13.137, substitute in Equation 13.133a, bandwidth equal to resistance R over inductance L.

The derivation, of Equation 13.135 for Q, is shown in Equation 13.138.

In Equation 13.137, substitute in Equation 13.132 for resonant frequency ω_0 squared and Equation 13.133a bandwidth equal to R over L.

$$Q = \frac{\omega_0}{\mathrm{BW}} = \frac{\omega_0^2}{\omega_0^2}\frac{\omega_0}{\mathrm{BW}} = \frac{1/LC}{\omega_0^2}\frac{\omega_0}{R/L} = \frac{1/LC}{\omega_0}\frac{1}{R/L} = \frac{1}{\omega_0 LC}\frac{L}{R} = \frac{1}{\omega_0 RC} \tag{13.138}$$

The derivation, of Equation 13.136 for Q, is shown in Equation 13.139.

$$Q = \frac{\omega_0 L}{R} = \frac{1}{\sqrt{LC}}\frac{L}{R} = \frac{1}{R}\sqrt{\frac{L}{C}} \tag{13.139}$$

In Equation 13.134, substitute in Equation 13.131 for resonant frequency ω_0 (equal to one over the square root of LC).

13.3.6.2 Series RLC Q and Bandwidth Intuition

For a series RLC circuit, the bandwidth is proportional to the resistance. More resistance means more bandwidth.

In a series RLC circuit, shown in Figure 13.41, all the current goes through the resistor. For the same current, more resistance means more power lost through dissipation.

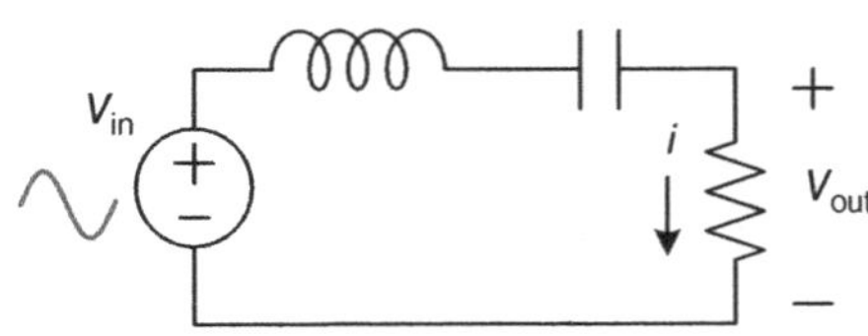

Figure 13.41 Series RLC circuit schematic driven by sinusoidal voltage.

Energy lost, per cycle of the sinusoidal signal, is the denominator of the quality factor expression – energy stored divided by energy lost. In a series RLC, more resistance means less Q. All the expressions for Q, Equations (13.134–13.136), are inversely proportional to R.

13.3.6.3 Series RLC Transfer Function in Terms of Q

$$\frac{\boldsymbol{V}_{\text{out}}}{\boldsymbol{V}_{\text{in}}} = \frac{1}{1 + j\left(\frac{\omega L}{R} - \frac{1}{\omega RC}\right)} \tag{13.140}$$

Equation 13.140 is the sinusoidal steady-state voltage transfer function for the series RLC circuit in Figure 13.41.

In the series RLC Equation 13.140 denominator, to express ωL over R and one over ωRC in terms of Q, start with Equations 13.134 and 13.135 for Q in terms of R, L, C, and ω_0, Equations 13.141 and 13.142.

$$Q = \frac{\omega_0 L}{R} \tag{13.141}$$

$$Q = \frac{1}{\omega_0 RC} \tag{13.142}$$

Multiply both sides of Equation 13.141 by ω over ω_0, to obtain Equation 13.143.

$$\frac{\omega}{\omega_0} Q = \frac{\omega}{\omega_0} \frac{\omega_0 L}{R} \tag{13.143}$$

Multiply both sides of Equation 13.142 by ω_0 over ω, to obtain Equation 13.144.

$$\frac{\omega_0}{\omega} Q = \frac{\omega_0}{\omega} \frac{1}{\omega_0 RC} \tag{13.144}$$

Simplify Equations 13.143 and 13.144 to obtain Equations 13.145 and 13.146.

$$\frac{\omega}{\omega_0} Q = \frac{\omega L}{R} \tag{13.145}$$

$$\frac{\omega_0}{\omega} Q = \frac{1}{\omega RC} \tag{13.146}$$

Substitute Equations 13.145 and 13.146 into the denominator of Equation 13.140 to express Equation 13.140 in terms of Q, generating Equations 13.147 and 13.148.

$$\frac{\boldsymbol{V}_{\text{out}}}{\boldsymbol{V}_{\text{in}}} = \frac{1}{1 + j\left(\frac{\omega}{\omega_0} Q - \frac{\omega_0}{\omega} Q\right)} \tag{13.147}$$

In Equation 13.147, factor the term Q outside of the brackets, to obtain Equation 13.148.

$$\frac{\boldsymbol{V}_{\text{out}}}{\boldsymbol{V}_{\text{in}}} = \frac{1}{1 + jQ\left(\frac{\omega}{\omega_0} - \frac{\omega_0}{\omega}\right)} = \frac{1}{1 + jQ(\omega/\omega_0 - \omega_0/\omega)} \tag{13.148}$$

Equation 13.148 is the transfer function of the series RLC circuit in Figure 13.41.

Notice that when expressed in terms of Q, ω, and ω_0, the right-hand side of the series RLC Equation 13.148 voltage transfer function is the same as the corresponding parallel RLC current transfer function in Equation 13.183.

13.3.6.4 Series RLC Transfer Function Magnitude in Terms of Q

$$\left|\frac{\boldsymbol{V}_{\text{out}}}{\boldsymbol{V}_{\text{in}}}\right| = \frac{1}{|1 + jQ(\omega/\omega_0 - \omega_0/\omega)|} \tag{13.149}$$

Apply Equation 13.61, for the magnitude of a ratio of complex functions equal to the ratio of the magnitudes, to Equation 13.148, to generate Equation 13.149.

Apply Equation 13.60, for magnitude equal to the square root of the sum of the squares of the real and imaginary terms, to Equation 13.149 get Equation 13.150.

$$\left|\frac{\mathbf{V}_{\text{out}}}{\mathbf{V}_{\text{in}}}\right| = \frac{1}{\sqrt{1+jQ(\omega/\omega_0-\omega_0/\omega)^2}} \quad (13.150)$$

Equation 13.150 is the magnitude, in terms of Q, of the series RLC voltage transfer function.

Notice that when expressed in terms of Q, ω, and ω_0, the right-hand side of this series RLC Equation 13.150 voltage transfer function magnitude is the same as the corresponding parallel RLC current transfer function magnitude in Equation 13.185.

13.3.6.5 Series RLC Transfer Function Phase in Terms of Q

$$\theta(\omega) = \angle H(\omega) = \tan^{-1}\frac{X(\omega)}{R(\omega)} \quad (13.151)$$

Equation 13.151 shows in the general formula for the phase angle of a complex function.

$$H(\omega) = \frac{\mathbf{V}_{\text{out}}}{\mathbf{V}_{\text{in}}} = \frac{1}{1+jQ(\omega/\omega_0-\omega_0/\omega)} = \frac{F(\omega)}{G(\omega)} \quad (13.152)$$

The series RLC transfer function, in Equation 13.148, is the quotient of two complex expressions $F(\omega)$ and $G(\omega)$, as shown in Equation 13.152.

$$F = 1 \quad (13.153)$$

$$G = 1 + jQ(\omega/\omega_0 - \omega_0/\omega) \quad (13.154)$$

In the case of this series RLC transfer function $H(j\omega)$ in Equation 13.148, the numerator is the purely real number 1, as shown in Equation 13.153.

Equation 13.155 shows how the phase angle of a quotient of complex functions is the phase angle of the numerator minus the phase angle of the denominator.

$$\angle H(j\omega) = \angle\frac{F(j\omega)}{G(j\omega)} = \frac{\angle\theta_{\text{F}}}{\angle\theta_{\text{G}}} = \angle(\theta_{\text{F}}) - \angle(\theta_{\text{G}}) \quad (13.155)$$

Into Equation 13.155, substitute the numerator and denominator from Equation 13.152, also shown in Equations 13.153 and 13.154, to obtain Equation 13.156.

$$\angle H(j\omega) = \frac{\angle 1}{\angle(1+jQ(\omega/\omega_0-\omega_0/\omega))} = \angle(1) - \angle(1+jQ(\omega/\omega_0-\omega_0/\omega)) \quad (13.156)$$

Apply the formula in Equation 13.151, for the phase of a complex function, separately to both terms in Equation 13.156, the numerator and the denominator, as shown in Equations 13.157 and 13.158.

*Numerator **F***

$$\angle(1) = \tan^{-1}[X/R] = \tan^{-1}[0/1] = \tan^{-1}(0) = 0 \quad (13.157)$$

The phase angle of the numerator is zero because the numerator is a purely real number.

*Denominator **G***

$$\angle(1+jQ(\omega/\omega_0-\omega_0/\omega)) = \tan^{-1}[Q(\omega/\omega_0-\omega_0/\omega)/(1)] \quad (13.158)$$

Substitute the results, in Equations 13.157 and 13.158, back into Equation 13.156, to obtain Equations 13.159 and 13.160.

$$\angle H(j\omega) = 0 - \tan^{-1}Q(\omega/\omega_0 - \omega_0/\omega) \quad (13.159)$$

$$\angle H(j\omega) = \angle\frac{\mathbf{V}_{\text{out}}}{\mathbf{V}_{\text{in}}} = -\tan^{-1}Q(\omega/\omega_0-\omega_0/\omega) = -\tan^{-1}Q\left(\frac{\omega}{\omega_0}-\frac{\omega_0}{\omega}\right) \quad (13.160)$$

Equation 13.160 is the sinusoidal steady-state phase response of the series RLC circuit in terms of quality factor Q.

The right-hand side of the series RLC Equation 13.160, for the phase response in terms of Q, is the same equation as the right-hand side of the parallel RLC circuit Equation 13.186, for phase response in terms of Q.

13.3.6.6 Express Series RLC Bandwidth Cutoff Frequencies in Terms of Q

This section expresses Equations 13.161a and 13.161b, for series RLC cutoff frequencies ω_1 and ω_2, in terms of Q and ω_0.

$$\omega_1 = -\frac{R}{2L} + \sqrt{\left(\frac{R}{2L}\right)^2 + \frac{1}{LC}} \quad (13.161a)$$

$$\omega_2 = \frac{R}{2L} + \sqrt{\left(\frac{R}{2L}\right)^2 + \frac{1}{LC}} \quad (13.161b)$$

In Equations 13.161a and 13.161b, ω_1 and ω_2 are the two cutoff frequencies where the series RLC voltage transfer function equals 0.707. The bandwidth (BW) lies between these two frequencies.

$$Q = \frac{\omega_0 L}{R} \quad (13.163)$$

$$\omega_0^2 = \frac{1}{LC} \quad (13.164)$$

Solve Equation 13.163 for R over $2L$ to obtain Equation 13.165.

$$\frac{R}{2L} = \frac{\omega_0}{2Q} \quad (13.165)$$

Equation 13.164 comes from earlier Equation 13.132 for the resonant frequency ω_0 squared equal to one over LC.

Substitute Equations 13.165 and 13.164, for series RLC Q and ω_0 squared, into Equations 13.161a and 13.161b for ω_1 and ω_2, to get Equations 13.166a and 13.166b.

$$\omega_1 = -\frac{\omega_0}{2Q} + \sqrt{\left(\frac{\omega_0}{2Q}\right)^2 + \omega_0^2} \quad (13.166a)$$

$$\omega_2 = \frac{\omega_0}{2Q} + \sqrt{\left(\frac{\omega_0}{2Q}\right)^2 + \omega_0^2} \quad (13.166b)$$

In Equations 13.166a and 13.166b, move the ω_0 squared term outside of the radical sign, as shown in Equations 13.167a and 13.167b.

$$\omega_1 = -\frac{\omega_0}{2Q} + \omega_0\sqrt{\left(\frac{1}{2Q}\right)^2 + 1} \quad (13.167a)$$

$$\omega_2 = \frac{\omega_0}{2Q} + \omega_0\sqrt{\left(\frac{1}{2Q}\right)^2 + 1} \quad (13.167b)$$

Equations 13.168a and 13.168b are another useful form of Equations 13.167a and 13.167b.

$$\omega_1 = -\frac{\omega_0}{2Q} + \omega_0\sqrt{1 + \frac{1}{4Q^2}} \quad (13.168a)$$

$$\omega_2 = \frac{\omega_0}{2Q} + \omega_0\sqrt{1 + \frac{1}{4Q^2}} \quad (13.168b)$$

13.3.7 Parallel RLC Quality Factor Q

13.3.7.1 Parallel RLC Equations for Q

$$Q = \omega_0 RC \quad (13.169)$$

$$Q = \frac{R}{\omega_0 L} \quad (13.170)$$

$$Q = R\sqrt{\frac{C}{L}} \quad (13.171)$$

The derivation of Equation 13.169, for Q, is shown in Equation 13.172.

For the BW in Equation 13.129, substitute in Equation 13.133b, bandwidth equals one over RC.

$$Q = \frac{\omega_0}{\mathrm{BW}} = \frac{\omega_0}{1/RC} = \omega_0 RC \tag{13.172}$$

The derivation of Equation 13.170, for Q, is shown in Equation 13.173.

In Equation 13.172, substitute in Equation 13.132, for resonant frequency ω_0 squared, and Equation 13.133b, bandwidth equal to 1 over RC, to obtain Equation 13.173.

$$Q = \frac{\omega_0}{\mathrm{BW}} = \frac{\omega_0^2}{\omega_0^2}\frac{\omega_0}{\mathrm{BW}} = \frac{1/LC}{\omega_0^2}\frac{\omega_0}{1/RC} = \frac{1/LC}{\omega_0}\frac{1}{1/RC} = \frac{R}{\omega_0 L} \tag{13.173}$$

The derivation of Equation 13.171, for Q, is shown in Equation 13.174.

$$Q = \omega_0 RC = \frac{1}{\sqrt{LC}}RC = R\sqrt{\frac{C}{L}} \tag{13.174}$$

In Equation 13.169, substitute in Equation 13.131 for resonant frequency ω_0 equal to one over the square root of LC.

13.3.7.2 Parallel RLC Q and Bandwidth Intuition

For a parallel RLC circuit, the bandwidth is inversely proportional to the resistance. More resistance means less bandwidth.

In a parallel RLC circuit, shown in Figure 13.42, current does not need to go through the resistor. More resistance means greater resistive impedance relative to the reactive impedances of the inductor and the capacitor. More resistance means more incentive for current to stay in the reactive elements and less loss.

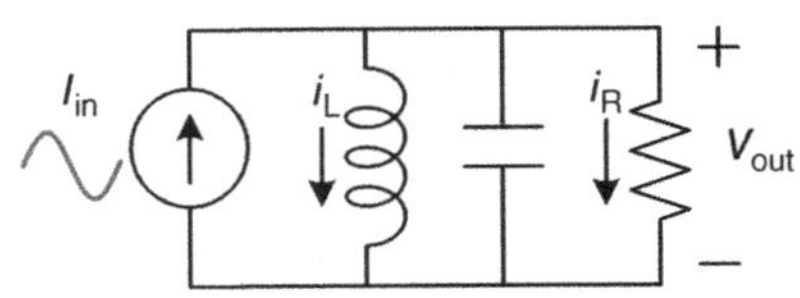

Figure 13.42 Parallel RLC circuit schematic driven by sinusoidal current.

Energy stored, per cycle of the sinusoidal signal, is the numerator of the quality factor expression – energy stored divided by energy lost. In a parallel RLC, more resistance means more Q. All the expressions for Q, Equations (13.169–13.171), are directly proportional to R.

13.3.7.3 Parallel RLC Transfer Function in Terms of Q

$$\frac{\mathbf{I}_{\text{out}}}{\mathbf{I}_{\text{in}}} = \frac{1}{1 + j\left(\omega RC - \frac{R}{\omega L}\right)} \tag{13.175}$$

Equation 13.175 is the sinusoidal steady-state current transfer function for the parallel RLC circuit in Figure 13.42.

This section transforms Equation 13.175, expressed in terms of R over ωL and ωRC into Equation 13.183 expressed in terms of Q and ω_0. Start by substituting two equations for Q, Equations 13.176 and 13.177, for Q in terms of R, L, C, and ω_0.

$$Q = \omega_0 RC \tag{13.176}$$

$$Q = \frac{R}{\omega_0 L} \tag{13.177}$$

Multiply both sides of Equation 13.176 by ω over ω_0, as shown in Equation 13.178.

$$\frac{\omega}{\omega_0}Q = \frac{\omega}{\omega_0}\omega_0 RC \tag{13.178}$$

Multiply both sides of Equation 13.177 by ω_0 over ω, as shown in Equation 13.179.

$$\frac{\omega_0}{\omega}Q = \frac{\omega_0}{\omega}\frac{R}{\omega_0 L} \tag{13.179}$$

Simplify Equations 13.178 and 13.179 to obtain Equations 13.180 and 13.181.

$$\frac{\omega}{\omega_0}Q = \omega RC \qquad (13.180)$$

$$\frac{\omega_0}{\omega}Q = \frac{R}{\omega L} \qquad (13.181)$$

Substitute Equations 13.180 and 13.181 into the denominator of Equation 13.175 to obtain Equations 13.182 and 13.183 in terms of Q.

$$\frac{\boldsymbol{I}_{\text{out}}}{\boldsymbol{I}_{\text{in}}} = \frac{1}{1 + j\left(\frac{\omega}{\omega_0}Q - \frac{\omega_0}{\omega}Q\right)} \qquad (13.182)$$

In Equation 13.182, factor the term Q outside of the brackets.

$$\frac{\boldsymbol{I}_{\text{out}}}{\boldsymbol{I}_{\text{in}}} = \frac{1}{1 + jQ\left(\frac{\omega}{\omega_0} - \frac{\omega_0}{\omega}\right)} = \frac{1}{1 + jQ(\omega/\omega_0 - \omega_0/\omega)} \qquad (13.183)$$

Equation 13.183 is the transfer function of the parallel RLC circuit in Figures 13.42 and 13.34.

Notice that when expressed in terms of Q, ω, and ω_0, the right-hand side of this parallel RLC Equation 13.183 current transfer function is the same as the corresponding series RLC voltage transfer function in Equation 13.148.

13.3.7.4 Parallel RLC Transfer Function Magnitude in Terms of Q

Apply Equation 13.61, for the magnitude of a ratio of complex functions equal to the ratio of the magnitudes, to Equation 13.183, to generate Equation 13.184.

$$\left|\frac{\boldsymbol{I}_{\text{out}}}{\boldsymbol{I}_{\text{in}}}\right| = \frac{1}{|1 + jQ(\omega/\omega_0 - \omega_0/\omega)|} \qquad (13.184)$$

Apply Equation 13.60, for magnitude equal to the square root of the sum of the squares of the real and imaginary terms, to get Equation 13.185.

$$\left|\frac{\boldsymbol{I}_{\text{out}}}{\boldsymbol{I}_{\text{in}}}\right| = \frac{1}{\sqrt{1 + jQ(\omega/\omega_0 - \omega_0/\omega)^2}} \qquad (13.185)$$

Equation 13.185 is the magnitude, in terms of Q, of the parallel RLC current transfer function.

Notice that when expressed in terms of Q, ω, and ω_0, the right-hand side of this parallel RLC Equation 13.185 transfer function magnitude is the same as the corresponding series RLC voltage transfer function in Equation 13.150.

13.3.7.5 Parallel RLC Transfer Function Phase in Terms of Q

Because the transfer function for parallel RLC, when expressed in terms of Q and ω and ω_0 in Equation 13.183, is identical to the series RLC transfer function in Equation 13.148, the phase of the parallel RLC transfer function, in Equation 13.186, is also the same as the phase for series RLC transfer function phase, shown in Equation 13.160.

$$\angle H(j\omega) = \angle\frac{\boldsymbol{V}_{\text{out}}}{\boldsymbol{V}_{\text{in}}} = -\tan^{-1}Q(\omega/\omega_0 - \omega_0/\omega) = -\tan^{-1}Q\left(\frac{\omega}{\omega_0} - \frac{\omega_0}{\omega}\right) \qquad (13.186)$$

Equation 13.186 is the sinusoidal steady-state phase response of the parallel RLC circuit in terms of quality factor Q.

13.3.7.6 Express Parallel RLC Bandwidth Cutoff Frequencies in Terms of Q

This section expresses Equations 13.187a and 13.187b, for cutoff frequencies ω_1 and ω_2, in terms of Q and ω_0.

$$\omega_1 = -\frac{1}{2RC} + \sqrt{\left(\frac{1}{2RC}\right)^2 + \frac{1}{LC}} \quad (13.187a)$$

$$\omega_2 = \frac{1}{2RC} + \sqrt{\left(\frac{1}{2RC}\right)^2 + \frac{1}{LC}} \quad (13.187b)$$

In Equations 13.187a and 13.187b, ω_1 and ω_2 are the two cutoff frequencies where the parallel RLC current transfer function equals 0.707. The bandwidth (BW) of the RLC circuit lies between these two frequencies.

$$Q = \omega_0 RC \quad (13.188)$$

Solve Equation 13.188 for 1 over $2RC$ to obtain Equation 13.190.

$$\frac{1}{2RC} = \frac{\omega_0}{2Q} \quad (13.190)$$

$$\omega_0^2 = \frac{1}{\mathrm{LC}} \quad (13.189)$$

Equation 13.189 comes from earlier Equation 13.132, for the resonant frequency ω_0 squared equal to one over LC.

Substitute Equations 13.190 and 13.189, for one over $2RC$ and ω_0 squared, into Equations 13.187a and 13.187b for ω_1 and ω_2, to get Equations 13.191a and 13.191b.

$$\omega_1 = -\frac{\omega_0}{2Q} + \sqrt{\left(\frac{\omega_0}{2Q}\right)^2 + \omega_0^2} \quad (13.191a) \qquad \omega_2 = \frac{\omega_0}{2Q} + \sqrt{\left(\frac{\omega_0}{2Q}\right)^2 + \omega_0^2} \quad (13.191b)$$

In Equations 13.191a and 13.191b, move the ω_0 squared term outside of the radical sign, as shown in Equations 13.192a and 13.192b.

$$\omega_1 = -\frac{\omega_0}{2Q} + \omega_0\sqrt{\left(\frac{1}{2Q}\right)^2 + 1} \quad (13.192a) \qquad \omega_2 = \frac{\omega_0}{2Q} + \omega_0\sqrt{\left(\frac{\omega_0}{2Q}\right)^2 + 1} \quad (13.192b)$$

Equations 13.193a and 13.193b are another useful form of Equations 13.192a and 13.192b.

$$\omega_1 = -\frac{\omega_0}{2Q} + \omega_0\sqrt{1 + \frac{1}{4Q^2}} \quad (13.193a) \qquad \omega_2 = \frac{\omega_0}{2Q} + \omega_0\sqrt{1 + \frac{1}{4Q^2}} \quad (13.193b)$$

14

Second-Order RLC Circuit Natural Response

14.1 Second-Order Natural Response Introduction

14.1.1 Second-Order Natural Response Overview

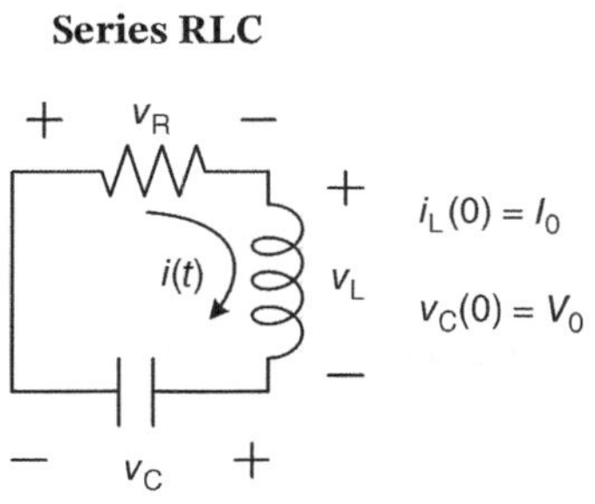

Figure 14.1a Series RLC circuit with two initial conditions.

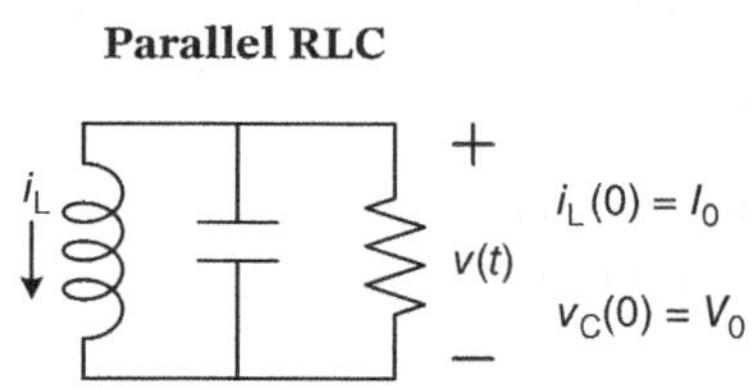

Figure 14.1b Parallel RLC circuit with two initial conditions.

Neither the series circuit in Figure 14.1a, nor the parallel RLC circuit in Figure 14.1b, has a current or voltage source driving it. The natural response, also called zero-input response (ZIR), comes from initial conditions on the circuit, such as initial inductor current $i_L(0) = I_0$ or an initial capacitor voltage $v_C(0) = V_0$. Both possible initial conditions are noted in Figures 14.1a and 14.1b. In circuit simulation, initial conditions can be specified, for a circuit element, like an inductor or capacitor, and used to simulate natural response.

In mathematical terms, the second-order RLC differential equations, series RLC in Equation 14.1a and parallel RLC in Equation 14.1b, are homogeneous linear equations with constant coefficients. The resistance R, the capacitance C, and the inductance L, which are constant with respect to time, are the constant coefficients. There are no current or voltage driving functions in the homogeneous differential equations in Equations 14.1a and 14.1b.

This chapter starts with analyses of two second-order RLC circuits, series and parallel, directly in terms of resistance R, inductance L, and capacitance C. RLC analyses are then repeated using two different damping variables, damping coefficient alpha α and then damping ratio zeta ζ. (α is also called damping constant or damping factor.)

14.1.2 Second-order RLC Analysis Outline

(1) Second-order RLC time domain circuit analysis often starts with Kirchhoff's current or voltage law to set up the differential equations.
(2) From the second-order differential equations, solve for the characteristic equations.

First and Second Order Circuits and Equations: Technical Background and Insights, First Edition. Robert O'Rourke.

(3) Generate the characteristic roots of the differential equations.
(4) The characteristic roots become part of the exponent of the general solution equation for current or voltage. Depending upon values of resistance R, inductance L, and capacitance C, there can be four response cases for the characteristic roots.
(5) Finally, solve for the general solution of the RLC differential equation, including the calculation of the constants based on initial conditions.

14.1.2.1 Set Up the Differential Equations from KCL or KVL

Series RLC differential equation

$$LC\frac{d^2i(t)}{dt^2} + RC\frac{di(t)}{dt} + i(t) = 0 \tag{14.1a}$$

Parallel RLC differential equation

$$LC\frac{d^2v(t)}{dt} + \frac{L}{R}\frac{dv(t)}{dt} + v(t) = 0 \tag{14.1b}$$

In Equation 14.1a, current $i(t)$ is the circuit variable and the current is the same for all the series circuit elements. In Equation 14.1b, voltage $v(t)$ is the circuit variable, and it is the same across all the parallel circuit elements. Other options for circuit variables include capacitor voltage in the series RLC and inductor current in the parallel RLC.

14.1.2.2 Solve for the Characteristic Equation

$$Ke^{st}(LCs^2 + RCs + 1) = 0 \tag{14.2a}$$

$$LCs^2 + RCs + 1 = 0 \tag{14.3a}$$

Equation 14.3a is the series RLC characteristic equation for Equation 14.1a, the differential equation for current.

$$Ke^{st}(LCs^2 + GL + 1) = 0 \tag{14.2b}$$

$$LCs^2 + GL + 1 = 0 \tag{14.3b}$$

Equation 14.3b is the parallel RLC characteristic equation for Equation 14.1b, the differential equation for voltage.

14.1.2.3 Generate the Characteristic Roots

Equations 14.4a and 14.4b show roots for series and parallel RLC. The characteristic roots, s_1 and s_2, are solutions to the characteristic Equations 14.3a and 14.3b. The characteristic roots are part of the exponent in the differential equations' general solution form Ke^{st} in Equations 14.5a and 14.5b.

Series RLC

$$s_1, s_2 = \frac{-R}{2L} \pm \sqrt{\left(\frac{R}{2L}\right)^2 - \frac{1}{LC}} \tag{14.4a}$$

$$i(t) = K_1e^{s_1t} + K_2e^{s_2t} \tag{14.5a}$$

Parallel RLC

$$s_1, s_2 = \frac{-1}{2RC} + \pm\sqrt{\left(\frac{1}{2RC}\right)^2 - \frac{1}{LC}} \tag{14.4b}$$

$$v(t) = K_1e^{s_1t} + K_2e^{s_2t} \tag{14.5b}$$

In the characteristic roots Equations 14.4a and 14.4b, the R, L, and C variables can become unwieldy. Other variables, like damping coefficient α (alpha) and damping ratio ζ (zeta), are often introduced to streamline and generalize the behavior descriptions to larger contexts. (α is also called damping constant or damping factor.)

$$s_{1,2} = -\alpha \pm \sqrt{\alpha^2 - \omega_0^2} \tag{14.6}$$

$$s_{1,2} = -\zeta\omega \pm \sqrt{\zeta^2 - 1} \tag{14.7}$$

Equations 14.6 and 14.7 show the roots in terms of damping factor alpha α, and damping ratio zeta ζ. Equations 14.8a and 14.8b show how α relates for R, L, and C for both series and parallel RLC circuits.

Damping Coefficient α

Series RLC

$$\alpha = \frac{R}{2L} \tag{14.8a}$$

$$\omega_0 = \frac{1}{\sqrt{LC}} \tag{14.9a}$$

Parallel RLC

$$\alpha = \frac{G}{2C} = \frac{1}{2RC} \tag{14.8b}$$

$$\omega_0 = \frac{1}{\sqrt{LC}} \tag{14.9b}$$

Equations 14.9a and 14.9b are the same ω_0 that frequency domain analysis uses. ω_0 is also called the undamped natural frequency, and it is angular or radian frequency. ω equals $2\pi f$.

14.1.2.4 Conditions for Four Cases of Characteristic Roots

For different values of R, L, and C, the characteristic roots can fall into four different cases, related to the mathematical form of the characteristic roots, including whether they are real, complex, or imaginary. The graphs of the responses in Table 14.1 are currents in series RLC.

Table 14.1 Simulated series RLC currents for the four damping cases in terms of alpha and omega-nought.

Overdamped Two real and unequal roots	$\alpha > \omega_0$ 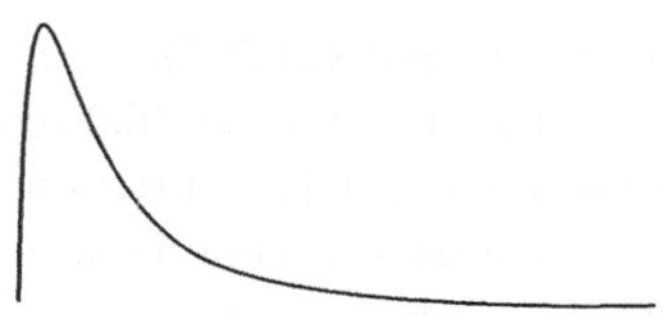
Critically damped Two real and equal roots	$\alpha = \omega_0$ 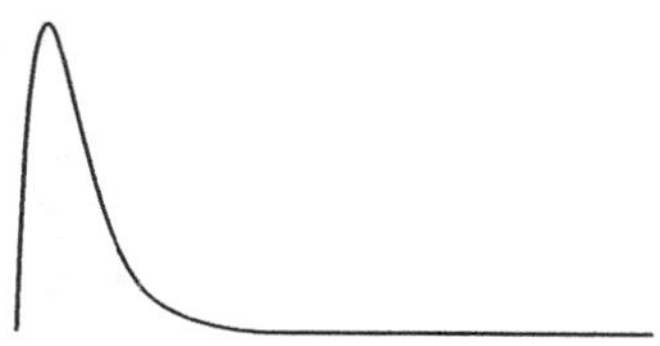
Underdamped Two complex conjugate roots	$\alpha < \omega_0$ 
Lossless oscillatory Two imaginary roots	$\alpha = 0$ 

14.1.2.5 General Solution Form

Series RLC

$$i(t) = K_1 e^{s_1 t} + K_2 e^{s_2 t} \quad (14.10a)$$

Parallel RLC

$$v(t) = K_1 e^{s_1 t} + K_2 e^{s_2 t} \quad (14.10b)$$

Equations 14.10a and 14.10b are the most general form of the second-order RLC differential equation solutions, before anything else about values or response behavior is known. The four different cases of the characteristic roots affect the final mathematical form and appearance of these solutions. The overdamped response is a decaying exponential function, while the underdamped response is a decaying sinusoid.

Equation 14.10a shows that loop current $i(t)$ has been chosen as the circuit variable in series RLC. Equation 14.10b shows that voltage $v(t)$ has been chosen and the circuit variable for parallel RLC.

14.2 Second-order Natural Response in Terms of *R*, *L*, and *C*

This section analyzes series and parallel RLC circuits and generates the characteristic roots in terms of R, L, and C. Following sections analyze in terms of the two damping variables, damping coefficient (also called damping constant) α and damping ratio ζ.

14.2.1 Second-order RLC Differential Equations

Natural response, also called **zero-input response (ZIR)**, depends only upon the initial conditions present in the circuit, e.g., an initial current I_0 in the inductor and an initial voltage V_0 across the capacitor. In natural response analysis, there is no input signal from an independent current or voltage source.

The first step in the RLC circuit analysis uses Kirchhoff's laws to generate homogeneous differential equations for the current and voltage natural response behaviors.

Series RLC

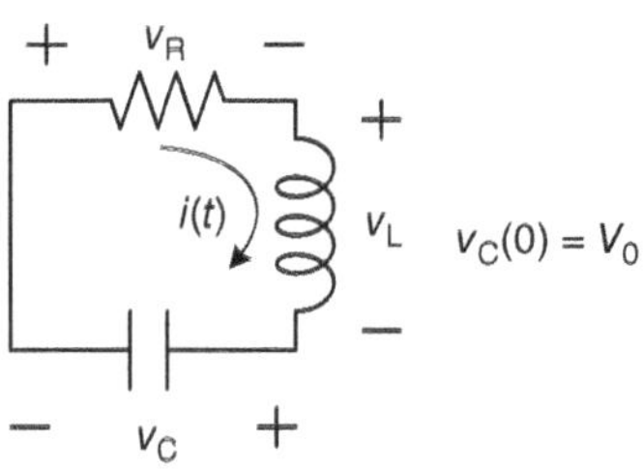

Figure 14.2a Series RLC circuit with one initial condition.

Parallel RLC

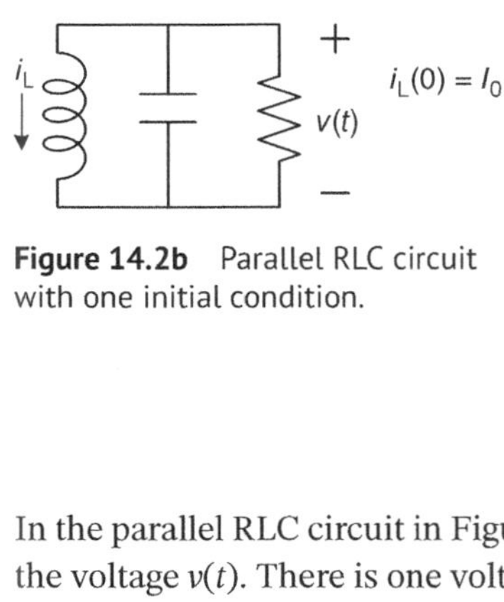

Figure 14.2b Parallel RLC circuit with one initial condition.

In the series RLC circuit in Figure 14.2a, we'll derive the loop current $i(t)$. There is one current in the entire loop; each circuit element has the same current, as indicated in Equation 14.11a.

$$i(t) = i_C(t) = i_R(t) = i_L(t) \quad (14.11a)$$

In the parallel RLC circuit in Figure 14.2b, we'll derive the voltage $v(t)$. There is one voltage in the circuit; the voltage across all three circuit elements is the same, as indicated in Equation 14.11b.

$$v_L(t) = v_C(t) = v_R(t) \quad (14.11b)$$

Write Kirchhoff's voltage law (KVL) around the series loop clockwise, starting with the inductor, as shown in Equation 14.12a.

$$v_L + v_C + v_R = 0 \tag{14.12a}$$

Recall the equations for the voltage across an inductor, in Equation 14.13a, the voltage across a capacitor, in Equation 14.14a, both expressed in terms of the current going through them.

$$v_L = L\frac{di_L(t)}{dt} \tag{14.13a}$$

$$v_C(t) = \frac{1}{C}\int_0^t i_C(\tau)\,d\tau + V_C(0) \tag{14.14a}$$

Rewrite the KVL Equation 14.12a, substituting in Equation 14.13a for inductor voltage and Equation 14.14a for capacitor voltage. Use Ohm's law for the resistor, to get Equation 14.15a.

$$L\frac{di(t)}{dt} + \frac{1}{C}\int_0^t i(\tau)\,d\tau + V_C(0) + i(t)R = 0 \tag{14.15a}$$

Notice how the constant of integration, $V_C(0)$ in Equations 14.14a and 14.15a, is a capacitor voltage that corresponds to the chosen series RLC initial condition indicated in Figure 14.2a.

Write Kirchhoff's current law (KCL) summing the currents going in or out of either node, as shown in Equation 14.12b.

$$I_C + I_R + I_L = 0 \tag{14.12b}$$

Recall the equations for the current through an inductor, in Equation 14.13b, and current through a capacitor, in Equation 14.14b, both expressed in terms of the voltage across them.

$$i_L(t) = \frac{1}{L}\int_0^t v_L(\tau)\,d\tau + I_L(0) \tag{14.13b}$$

$$i_C = C\frac{dv_C(t)}{dt} \tag{14.14b}$$

Rewrite the KCL Equation 14.12b, substituting in Equation 14.3b for inductor current and Equation 14.4b for capacitor current. Use Ohm's law for the resistor, to get Equation 14.15b.

$$C\frac{dv(t)}{dt} + \frac{v(t)}{R} + \frac{1}{L}\int_0^t v(\tau)\,d\tau + I_L(0) = 0 \tag{14.15b}$$

Notice how the constant of integration, $I_L(0)$ in Equations 14.13b and 14.15b, is an inductor current that corresponds to the chosen parallel RLC initial condition indicated in Figure 14.2b.

Series RLC | **Parallel RLC**

Rearrange the order of terms in Equations 14.15a and 14.15b to get Equations 14.16a and 14.16b.

$$L\frac{di(t)}{dt} + Ri(t) + \frac{1}{C}\int_0^t i(\tau)\,d\tau + V_C(0) = 0 \tag{14.16a}$$

$$C\frac{dv(t)}{dt} + \frac{v(t)}{R} + \frac{1}{L}\int_0^t v(\tau)\,d\tau + I_L(0) = 0 \tag{14.16b}$$

Take the derivative, with respect to time, of both sides of the series RLC Equation 14.16a. The constant $V_C(0)$ term in Equation 14.8a goes to zero in Equation 14.17a, which shows $i(t)$ as the circuit variable.

$$L\frac{d^2i(t)}{dt^2} + R\frac{di(t)}{dt} + \frac{1}{C}i(t) = 0 \tag{14.17a}$$

Multiply the entire Equation 14.17a by C to get Equation 14.18a.

Take the derivative, with respect to time, of the parallel RLC Equation 14.16b. The constant $I_L(0)$ term in Equation 14.16b goes to zero in Equation 14.17b, which shows $v(t)$ as the circuit variable.

$$C\frac{d^2v(t)}{dt} + \frac{dv(t)}{dt}\frac{1}{R} + \frac{1}{L}v(t) = 0 \tag{14.17b}$$

Multiply the entire Equation 14.17b by L to get Equation 14.18b.

Series RLC

$$LC\frac{d^2i(t)}{dt^2} + RC\frac{di(t)}{dt} + i(t) = 0 \quad (14.18a)$$

Equation 14.18a is the **series** RLC differential equation to be solved for **natural response** current.

Parallel RLC

$$LC\frac{d^2v(t)}{dt} + \frac{L}{R}\frac{dv(t)}{dt} + v(t) = 0 \quad (14.18b)$$

Equation 14.18b is the **parallel** RLC differential equation to be solved for **natural response** voltage.

Alternatively, divide the entire Equation 14.17a by LC to get Equation 14.19a. This makes the coefficient of the leading second derivative term equal to one.

Alternatively, divide the entire Equation 14.9b by LC to get Equation 14.19b. This makes the coefficient of the leading second derivative term equal to one.

Series RLC

$$\frac{di^2(t)}{dt^2} + \frac{R}{L}\frac{di(t)}{dt} + \frac{1}{LC}i(t) = 0 \quad (14.19a)$$

Equation 14.19a is the **series** RLC differential equation to be solved for **natural response** current.

Parallel RLC

$$\frac{d^2v(t)}{dt} + \frac{dv(t)}{dt}\frac{1}{RC} + \frac{1}{LC}v(t) = 0 \quad (14.19b)$$

Equation 14.19b is the **parallel** RLC differential equation to be solved for **natural response** voltage.

The next step, in the RLC circuit analysis, is solving the differential equations for the current or voltage that describe the circuit behavior. Keys to the solution are the exponential function Ke^{st} and its characteristic equation, which we use to calculate the roots, the s in the Ke^{st} exponential. The characteristic roots fall into four cases of response and provide insight to circuit behavior. After examining the four cases, overdamped, critically damped, underdamped, and oscillatory lossless, we need to solve for the constants K to get the complete solution to the RLC natural response current and voltage.

14.2.2 Second-order RLC Differential Equation Solutions

A common and standard approach to solving these differential equations, in Equations 14.18 and 14.19, is to make an educated guess about the general form of the solution to the differential equation. Notice in Equations 14.18 and 14.19 how there are derivative terms of the circuit variable function to be solved, $i(t)$ in the series RLC cases, in Equations 14.18a and 14.19a, and $v(t)$ in the parallel RLC cases, in Equations14.18b and 14.19b. Because the derivative of the exponential function e^{st} includes the original exponential function, we can imagine the exponential function e^{st} as a likely solution to the two RLC differential equations.

$$Ke^{st} \quad (14.20)$$

The exponential function in Equation 14.20 is the proposed differential equation solution function. K is a constant that is not yet known. The value of s is not yet known, but it may need to be a complex value. The notion of s as the complex frequency comes out of this mathematical derivation.

$$s = \sigma + j\omega \quad (14.21)$$

s is the complex frequency where ω is the radian frequency. The values of σ and ω, shown in Equation 14.21, are not yet known. The σ-and-ω notation is widely used to express s. σ (sigma) is the real part, related to damping, and ω (omega) is the imaginary component, related to sinusoidal response.

$$\frac{d(Ke^{st})}{dt} = Kse^{st} \quad (14.22)$$

The derivative of the exponential function e^{st} is the original exponential e^{st} function multiplied by s.

Proceed with the differential equation process by substituting the exponential function, from Equation 14.20, into each Equations 14.18a and 14.18b, in place of $i(t)$ and $v(t)$, respectively.

Series RLC

In Equation 14.18a, the series RLC differential equation, substitute Ke^{st} in place of $i(t)$.

$$LC\frac{d^2(Ke^{st})}{dt^2} + RC\frac{d(Ke^{st})}{dt} + Ke^{st} = 0 \quad (14.23a)$$

Parallel RLC

In Equation 14.18b, the parallel RLC differential equation, substitute Ke^{st} in place of $v(t)$.

$$LC\frac{d^2(Ke^{st})}{dt} + \frac{L}{R}\frac{d(Ke^{st})}{dt} + Ke^{st} = 0 \quad (14.23b)$$

Perform the differentiations in Equations 14.23a and 14.23b, to obtain Equations 14.24a and 14.24b.

$$LCs^2Ke^{st} + RCsKe^{st} + Ke^{st} = 0 \quad (14.24a)$$

$$LCs^2Ke^{st} + \frac{L}{R}sKe^{st} + Ke^{st} = 0 \quad (14.24b)$$

Factor out the Ke^{st} term from both Equations 14.24a and 14.24b.

Series RLC

$$Ke^{st}(LCs^2 + RCs + 1) = 0 \quad (14.25a)$$

Parallel RLC

$$Ke^{st}\left(LCs^2 + \frac{L}{R} + 1\right) = 0 \quad (14.25b)$$

These Equations 14.25a and 14.25b can be satisfied by either the exponential term, Ke^{st}, or the term in brackets being equal to zero. Since the exponential function Ke^{st} cannot be zero for finite t, Ke^{st} equals 0 is only a trivial solution. The solution to the RLC differential equation must be that the rest of the equation in parentheses in Equations 14.25a and 14.25b equals zero.

The exponential function Ke^{st} can be a solution to the differential equation if and only if Equation 14.26a is true.

$$LCs^2 + RCs + 1 = 0 \quad (14.26a)$$

The exponential function Ke^{st} can be a solution to the differential equation if and only if Equation 14.26b is true.

$$LCs^2 + \frac{L}{R} + 1 = 0 \quad (14.26b)$$

These Equations 14.26a and 14.26b are called the **characteristic equations** of the RLC differential equations for the circuits in Figure 14.3a,b. We'll solve these equations for s, the roots of the characteristic equations. The roots go into the exponential functions in the final differential equation solution.

14.2.3 Roots of the Second-order Characteristic Equations

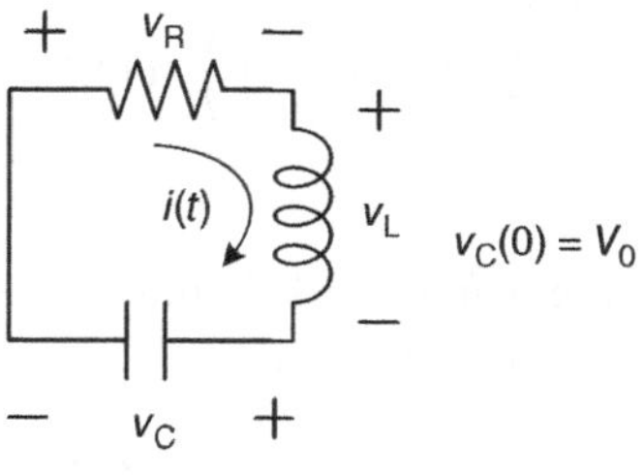

Figure 14.3a Series RLC circuit with one initial condition.

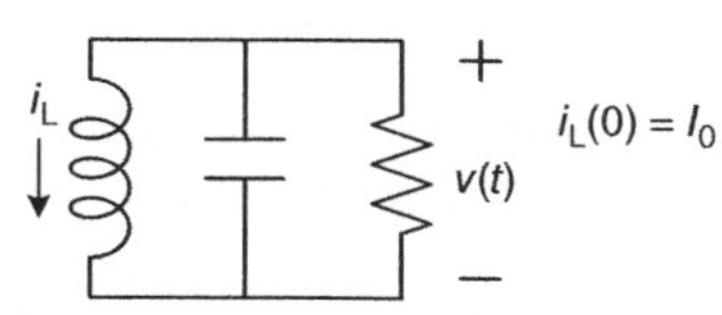

Figure 14.3b Parallel RLC circuit with one initial condition.

$$LCs^2 + RCs + 1 = 0 \qquad (14.27a)$$

Series RLC characteristic equation

$$LCs^2 + \frac{L}{R} + 1 = 0 \qquad (14.27b)$$

Parallel RLC characteristic equation

Substitute in G (conductance) for 1 over R in Equation 14.27b to obtain Equation 14.27c.

$$LCs^2 + GL + 1 = 0 \qquad (14.27c)$$

We'll test our proposed differential equation solution, the exponential function Ke^{st}. We'll use the characteristic equation to solve for s in terms of R, G, L, and C. After analyzing the behavior of two roots s_1 and s_2, we'll finally create a general solution using $K_1e^{s_1t}$ and $K_2e^{s_2t}$ and solving for the constants K_1 and K_2. Calculating the constants will utilize the two initial conditions, $v_C(0)$ in the series RLC and $i_L(0)$ in the parallel RLC.

To solve for the roots, s_1 and s_2, of these two characteristic Equations 14.27a,b,c apply the quadratic formula, in Equation 14.28, to Equations 14.27a, 14.27b, and 14.27c and solve for s.

$$\text{the quadratic formula : } ax^2 + bx + c = 0 \Rightarrow x = \frac{-b \pm \sqrt{b^2 - 4ac}}{2a} \qquad (14.28)$$

Series RLC

$$s_1, s_2 = \frac{-RC \pm \sqrt{(RC)^2 - 4LC}}{2LC} \qquad (14.29a)$$

Parallel RLC

$$s_1, s_2 = \frac{-GL \pm \sqrt{(GL)^2 - 4LC}}{2LC} \qquad (14.29b)$$

The s_1, s_2 notation, on the left side of Equations 14.29a and 14.29b, corresponds to the + and – notation before the radical (square root) sign. Each equation has two solutions, s_1 and s_2.

14.2.3.1 Changing the Form of the Parallel RLC Roots Equations with *R*

Different textbooks and treatments of RLC circuits use different notations, sometimes making it challenging to recognize and compare an equation.

This subsection shows the transformation, depicted by the right arrow, from Equation 14.30a to Equation 14.30b, for a parallel RLC circuit shown in Figures 14.3b and 14.4.

$$s_1, s_2 = \frac{-L/R \pm \sqrt{(L/R)^2 - 4LC}}{2LC} \qquad (14.30a)$$

$$\Longrightarrow$$

$$s_1, s_2 = \frac{-1}{2RC} + \left(\pm\sqrt{\left(\frac{1}{2RC}\right)^2 - \left(\frac{1}{LC}\right)}\right) \qquad (14.30b)$$

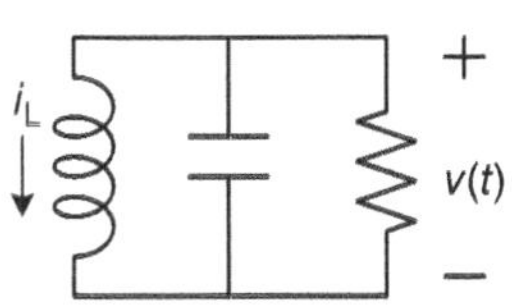

Figure 14.4 Parallel RLC circuit schematic.

Start the transformation with Equation 14.18b, the parallel RLC differential equation using $v(t)$ as the circuit variable, also shown as Equation 14.31.

$$LC\frac{d^2v(t)}{dt} + \frac{L}{R}\frac{dv(t)}{dt} + v(t) = 0 \qquad (14.31)$$

Equation 14.31 for parallel RLC was derived from Kirchhoff's current law (KCL).

$$LCs^2Ke^{st} + \frac{L}{R}sKe^{st} + Ke^{st} = 0 \tag{14.32}$$

Equation 14.32 has the presumed solution function, the exponential function Ke^{st}, substituted into the differential equation in place of the voltage function $v(t)$.

$$LCs^2 + \frac{L}{R} + 1 = 0 \tag{14.33}$$

Equation 14.33 is the characteristic equation for the parallel RLC circuit differential equation.

$$x = \frac{-b \pm \sqrt{b^2 - 4ac}}{2a} \tag{14.34}$$

Equation 14.34 is the quadratic equation, with which we can solve for s in the characteristic equation for the parallel RLC circuit. Substitute in the coefficients a equals LC, b equals L over R and C equals 1, to obtain Equation 14.35.

$$s_1, s_2 = \frac{-L/R \pm \sqrt{(L/R)^2 - 4LC}}{2LC} \tag{14.35}$$

Equation 14.35 shows the two roots equations, s_1 and s_2, that are the solutions to the parallel RLC differential equation characteristic equation.

From Equation 14.35, replace the slanted solidus division bar with the horizontal vinculum division bar to get Equation 14.36.

$$s_1, s_2 = \frac{-\frac{L}{R} \pm \sqrt{\left(-\frac{L}{R}\right)^2 - 4LC}}{2LC} \tag{14.36}$$

In Equation 14.36, distribute the denominator $2LC$ between the two terms in the numerator to obtain Equation 14.37a.

$$s_1, s_2 = \frac{-\frac{L}{R}}{2LC} + \frac{\pm\sqrt{\left(-\frac{L}{R}\right)^2 - 4LC}}{2LC} \tag{14.37a}$$

In Equation 14.37a, reconcile the denominator $2LC$ terms into the numerator terms with 1 over $2LC$ multipliers to generate Equation 14.38a.

$$s_1, s_2 = \frac{1}{2LC}\left(-\frac{L}{R}\right) + \left(\frac{1}{2LC}\right)\left(\pm\sqrt{\left(-\frac{L}{R}\right)^2 - 4LC}\right) \tag{14.38a}$$

In Equation 14.38a, express the 1 over $2LC$ as a squared term under the square root (radical) sign, as shown in Equation 14.39a.

$$s_1, s_2 = \frac{-1}{2RC} + \sqrt{\left(\frac{1}{2LC}\right)^2}\left(\pm\sqrt{\left(-\frac{L}{R}\right)^2 - 4LC}\right) \tag{14.39a}$$

In Equation 14.39a, distribute the square root of the quantity, 1-over-LC-quantity-squared, into the radical on the right to obtain Equation 14.40a.

$$s_1, s_2 = \frac{-1}{2RC} + \left(\pm\sqrt{\left(\frac{1}{2LC}\right)^2\left(-\frac{L}{R}\right)^2 - \left(\frac{1}{2LC}\right)^2 4LC}\right) \tag{14.40a}$$

Rationalize terms under the radical sign in Equation 14.40a to simplify and eliminate some terms, resulting in Equation 14.41a.

$$s_1, s_2 = \frac{-1}{2RC} + \left(\pm\sqrt{\left(\frac{1}{2RC}\right)^2 - \frac{1}{LC}}\right) \tag{14.41a}$$

Remove the unneeded brackets in Equation 14.41a to obtain Equation 14.42a, the equation for the roots of a parallel RLC natural response differential equation.

$$s_1, s_2 = \frac{-1}{2RC} + \pm\sqrt{\left(\frac{1}{2RC}\right)^2 - \frac{1}{LC}} \tag{14.42a}$$

$$s_1, s_2 = \frac{-1}{2RC} + \pm\left[\left(\frac{1}{2RC}\right)^2 - \frac{1}{LC}\right]^{\frac{1}{2}} \tag{14.43a}$$

Parallel RLC Roots Equations in Terms of Resistance R

Equation 14.43a is an alternative expression of Equation 14.42a, with the brackets taken to the one-half power instead of a radical sign in Equation 13.42.

14.2.3.2 Calculating Two Forms of RLC Roots Equations

Continuing from Equations 14.29a and 14.29b, shown here as Equations 14.36a and 14.36b, calculate the two notations of the roots equations for series and parallel RLC. Use conductance G in the parallel RLC equations here.

Series RLC **Parallel RLC**

$$s_1, s_2 = \frac{-RC \pm \sqrt{(RC)^2 - 4LC}}{2LC} \quad (14.36a) \qquad s_1, s_2 = \frac{-GL \pm \sqrt{(GL)^2 - 4LC}}{2LC} \quad (14.36b)$$

In Equations 14.36a and 14.36b, distribute the denominator $2LC$ between the two terms in the numerator, as shown in Equations 14.37b and 14.37c.

$$s_1, s_2 = \frac{-RC}{2LC} + \frac{\pm\sqrt{(RC)^2 - 4LC}}{2LC} \quad (14.37b) \qquad s_1, s_2 = \frac{-GL}{2LC} + \frac{\pm\sqrt{(GL)^2 - 4LC}}{2LC} \quad (14.37c)$$

In Equations 14.37b,c, reconcile the denominator $2LC$ terms into the numerator terms with 1 over $2LC$ multipliers, resulting in Equations 14.37b,c.

$$s_1, s_2 = \frac{-R}{2L} \pm \left(\frac{1}{2LC}\right)\sqrt{(RC)^2 - 4LC} \quad (14.38b) \qquad s_1, s_2 = \frac{-G}{2C} \pm \frac{1}{2LC}\sqrt{(GL)^2 - 4LC} \quad (14.38c)$$

In Equations 14.38b,c express the 1 over $2LC$ terms as square root terms, as shown in Equations 14.39b,c.

$$s_1, s_2 = \frac{-R}{2L} \pm \sqrt{\frac{1}{4L^2C^2}}\sqrt{(RC)^2 - 4LC} \quad (14.39b) \qquad s_1, s_2 = \frac{-G}{2C} \pm \sqrt{\frac{1}{4(LC)^2}}\sqrt{(GL)^2 - 4LC} \quad (14.39c)$$

In Equations 14.39b,c, distribute the square root of 1 over four L-squared C-squared into the radical to the right side to obtain Equations 14.40b,c.

$$s_1, s_2 = \frac{-R}{2L} \pm \sqrt{\frac{1}{4L^2C^2}(RC)^2 - \frac{1}{4L^2C^2}4LC} \quad (14.40b) \qquad s_1, s_2 = \frac{-G}{2C} \pm \sqrt{\frac{1}{4(LC)^2}(GL)^2 - \frac{1}{4(LC)^2}4LC} \quad (14.40c)$$

In Equations 14.40b,c, cancel terms, as shown in Equations 14.41b,c.

$$s_1, s_2 = \frac{-R}{2L} \pm \sqrt{\frac{R^2}{4L^2} - \frac{1}{4LC}} \quad (14.41b) \qquad s_1, s_2 = \frac{-G}{2C} \pm \sqrt{\frac{G^2}{4C^2} - \frac{1}{4LC}} \quad (14.41c)$$

Regroup terms in Equations 14.41b,c under a bracket under the radical sign in each equation, resulting in Equations 14.42b and 14.42c.

Series RLC

$$s_1, s_2 = \frac{-R}{2L} \pm \sqrt{\left(\frac{R}{2L}\right)^2 - \frac{1}{LC}} \tag{14.42b}$$

Parallel RLC

$$s_1, s_2 = \frac{-G}{2C} \pm \sqrt{\left(\frac{G}{2C}\right)^2 - \frac{1}{LC}} \tag{14.42c}$$

Series and Parallel RLC Roots Equations

Equations 14.42b,c show the general form of the characteristic roots of the series and parallel RLC differential equations. As these roots equations are applied to each of the four response cases, some roots will simplify in form.

14.2.4 Summary of RLC Differential Equations Solutions so Far

At this point in the solution process, for the two RLC circuits in Figures 14.5a,b, we know s_1 and s_2, but we do not know the values of the constants K_1 and K_2.

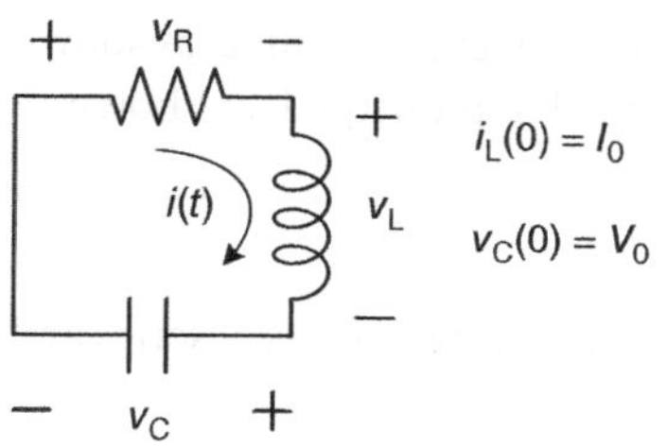

Figure 14.5a Series RLC circuit with two initial conditions.

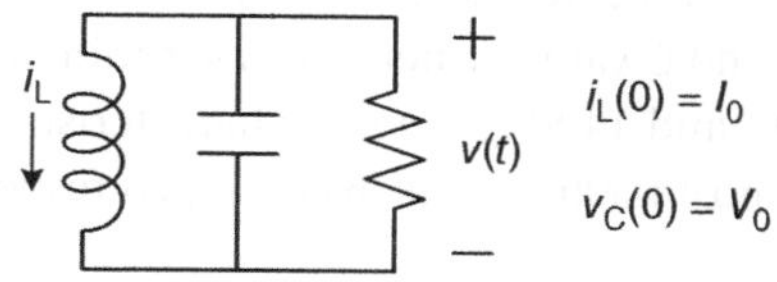

Figure 14.5b Parallel RLC circuit with two initial conditions.

Series RLC

$$i(t) = K_1 e^{s_1 t} + K_2 e^{s_2 t} \tag{14.43b}$$

Equation 14.43b, the general solution for the series RLC, is solved for $i(t)$, the current throughout the entire series loop.

Parallel RLC

$$v(t) = K_1 e^{s_1 t} + K_2 e^{s_2 t} \tag{14.43c}$$

Equation 14.43c, the general solution for the parallel RLC, is solved for $v(t)$, the voltage across all three circuit elements R, L, and C.

Series and Parallel RLC Roots Equations

$$s_1 = \frac{-RC + \sqrt{(RC)^2 - 4LC}}{2LC} \tag{14.44a}$$

$$s_1 = \frac{-GL + \sqrt{(GL)^2 - 4LC}}{2LC} \tag{14.44b}$$

$$s_2 = \frac{-RC - \sqrt{(RC)^2 - 4LC}}{2LC} \tag{14.45a}$$

$$s_2 = \frac{-GL - \sqrt{(GL)^2 - 4LC}}{2LC} \tag{14.45b}$$

Series RLC

$$s_1 = \frac{-R}{2L} + \sqrt{\left(\frac{R}{2L}\right)^2 - \frac{1}{LC}} \tag{14.46a}$$

$$s_2 = \frac{-R}{2L} - \sqrt{\left(\frac{R}{2L}\right)^2 - \frac{1}{LC}} \tag{14.47a}$$

Parallel RLC

$$s_1 = \frac{-G}{2C} + \sqrt{\left(\frac{G}{2C}\right)^2 - \frac{1}{LC}} \tag{14.46b}$$

$$s_2 = \frac{-G}{2C} - \sqrt{\left(\frac{G}{2C}\right)^2 - \frac{1}{LC}} \tag{14.47b}$$

Before solving for the constants, we'll look at four cases of circuit behavior reflected in different values of the roots s_1 and s_2.

Parallel RLC Roots Equations in Terms of Resistance R

$$s_1, s_2 = \frac{-1}{2RC} + \pm\sqrt{\left(\frac{1}{2RC}\right)^2 - \left(\frac{1}{LC}\right)} \tag{14.48}$$

$$s_1, s_2 = \frac{-1}{2RC} + \pm\left[\left(\frac{1}{2RC}\right)^2 - \left(\frac{1}{LC}\right)\right]^{\frac{1}{2}} \tag{14.49}$$

14.2.5 Four Cases of Characteristic Roots and RLC Response

The values of s_1, and s_2 can be complex valued if the quantity under the radical sign is negative, which leads to an imaginary term. Different values of R, L, and C generate different types of roots, e.g., real or complex valued, and there are different RLC response behaviors associated with four different cases of roots s_1 and s_2.

Case 1: Overdamped - roots real and unequal

In the case of R, L, and C values where the discriminant, the quantity under the radical sign, is positive, as shown in Equations 14.51a and 14.51b, the two characteristic roots s_1 and s_2 are real valued and unequal, as shown in Equations 14.50a and 14.50b. This is overdamped response.

Series RLC **Parallel RLC**

From Equations 14.42b and 14.48, Equations 14.50a and 14.50b are the general formulae for the roots of series and parallel RLC characteristic equations in terms of R, L, and C.

$$s_1, s_2 \ \frac{-R}{2L} \pm \sqrt{\left(\frac{R}{2L}\right)^2 - \frac{1}{LC}} \tag{14.50a}$$

$$\left(\frac{R}{2L}\right)^2 > \frac{1}{LC} \tag{14.51a}$$

$$s_1, s_2 = \frac{-1}{2RC} \pm \sqrt{\left(\frac{1}{2RC}\right)^2 - \frac{1}{LC}} \tag{14.50b}$$

$$\left(\frac{1}{2RC}\right)^2 > \frac{1}{LC} \tag{14.51b}$$

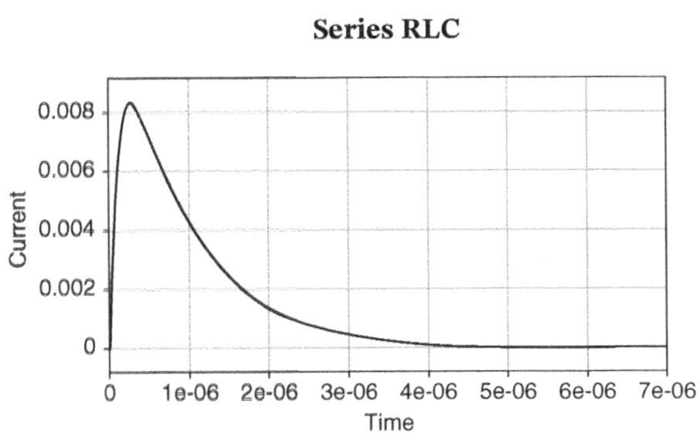

Figure 14.6a Overdamped series RLC simulated loop current.

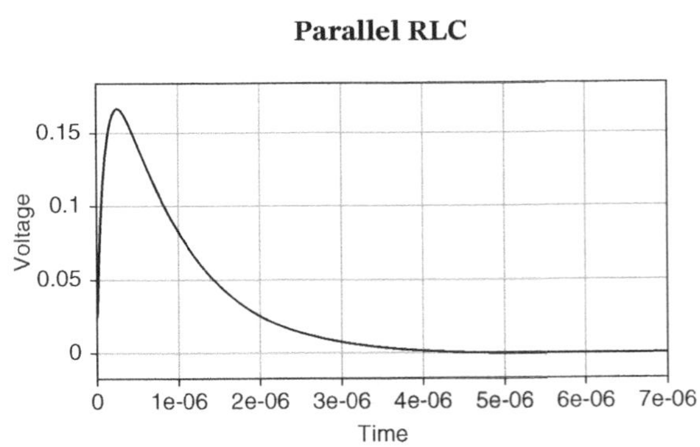

Figure 14.6b Overdamped parallel RLC simulated voltage.

Figure 14.6a shows overdamped series RLC loop current $i(t)$.

Figure 14.6b shows overdamped parallel RLC voltage $v(t)$.

In the overdamped case, the two characteristic roots s_1 and s_2 have no imaginary component, making both s_1 and s_2 real valued for both series and parallel RLC. Furthermore, because the difference in sign, before the square root quantity, plus for s_1 and minus for s_2, the two roots will be unequal. The general solution will be the sum of two decaying exponential functions.

$$i(t) = K_1 e^{s_1 t} + K_2 e^{s_2 t} \quad (14.52a) \qquad v(t) = K_1 e^{s_1 t} + K_2 e^{s_2 t} \quad (14.52b)$$

Equations 14.52a and 14.52b are the equations for the general solution of the series and parallel RLC circuits before the values for s_1, s_2, K_1, and K_2 are known. The overdamped case keeps all parts of the general solution equation form.

Case 2: Critically damped – roots real and equal

In the case of R, L, and C values where the discriminant, the quantity under the radical sign, is zero, characteristic roots s_1 and s_2 are real valued and equal. This is critically damped.

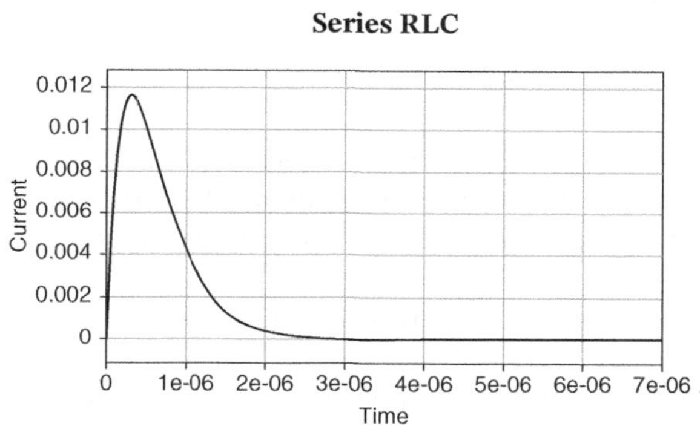

Figure 14.7a Critically damped series RLC simulated loop current.

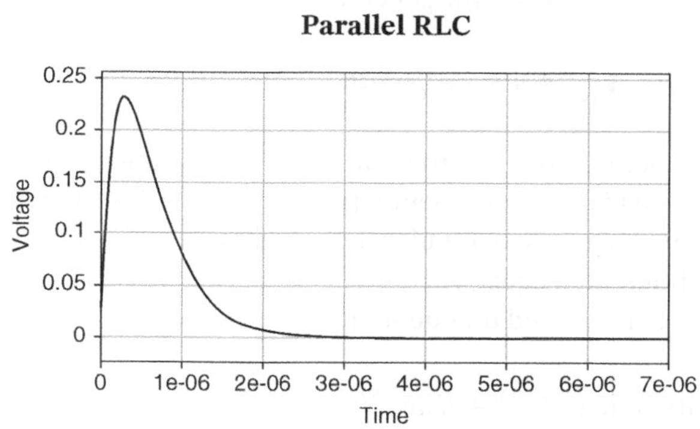

Figure 14.7b Critically damped parallel RLC simulated voltage.

Figure 14.7a shows critically damped series RLC loop current $i(t)$.

Figure 14.7b shows critically damped parallel RLC voltage $v(t)$.

In Figures 14.7a and 14.7b, for the critically damped case, the shape resembles the shape of the overdamped signals in Figures 14.6a and 14.6b. But notice how the critically damped signals have higher peak values and steeper descending slopes than the overdamped signals. The critically damped signals have less damping than the overdamped signals.

Series RLC

$$s_1, s_2 \; \frac{-R}{2L} \pm \sqrt{\left(\frac{R}{2L}\right)^2 - \frac{1}{LC}} \quad (14.53a)$$

Parallel RLC

$$s_1, s_2 = \frac{-1}{2RC} \pm \sqrt{\left(\frac{1}{2RC}\right)^2 - \frac{1}{LC}} \quad (14.53b)$$

Starting with Equations 14.53a and 14.53b, apply the conditions in Equations 14.54a and 14.54b, for critically damped series and parallel RLC circuits.

Series RLC

$$\left(\frac{R}{2L}\right)^2 = \frac{1}{LC} \quad (14.54a)$$

Parallel RLC

$$\left(\frac{1}{2RC}\right)^2 = \frac{1}{LC} \quad (14.54b)$$

In Equations 14.53a and 14.53b, the entire radical sign terms go to zero, resulting in Equations 14.55a and 14.55b for critically damped characteristic roots for series and parallel RLC circuits.

Two real and equal roots

Series RLC

$$s_1, s_2 = \frac{-R}{2L} \quad (14.55a)$$

Parallel RLC

$$s_1, s_2 = \frac{-1}{2RC} \quad (14.55b)$$

$$s_1, s_2 = \frac{-GL}{2LC} = \frac{-G}{2C} \quad (14.55c)$$

Series RLC

$$i(t) = K_1 e^{s_1 t} + K_2 e^{s_2 t} \quad (14.56a)$$

Parallel RLC

$$v(t) = K_1 e^{s_1 t} + K_2 e^{s_2 t} \quad (14.56b)$$

Substitute the critically damped roots, in Equations 14.55a and 14.55b, into Equations 14.56a and 14.56b, the formulae for the general solution of the RLC differential equations, to obtain Equations 14.57a and 14.57b.

$$i(t) = K_1 e^{(-R/2L)t} + K_2 e^{(-R/2L)t} \quad (14.57a) \qquad v(t) = K_1 e^{(-1/2RC)t} + K_2 e^{(-1/2RC)t} \quad (14.57b)$$

Because the two roots s_1 and s_2, in Equations 14.55a and 14.55b, are the same, Equations 14.57a and 14.57b, have identical exponent terms and K_1 and K_2 can be consolidated. This means that Equations 14.57a and 14.57b are each only one independent solution. Two unique solutions are needed for a general solution to each RLC differential equation.

For a second unique solution function, te^{st} will work. This appears in Equations 14.58a and 14.58b.

$$i(t) = K_1 e^{(-R/2L)t} + K_2 te^{(-R/2L)t} \quad (14.58a) \qquad v(t) = K_1 e^{(-1/2RC)t} + K_2 te^{(-1/2RC)t} \quad (14.58b)$$

The two identical exponential terms can be factored out from Equations 14.58a and 14.58b, grouping the new t term with the constant K_2, as shown in Equations 14.59a and 14.59b. Notice the factor of t multiplied by K_2 in Equations 14.59a,b.

Series RLC

$$i(t) = (K_1 + K_2 t)e^{(-R/2L)t} \quad (14.59a)$$

Parallel RLC

$$v(t) = (K_1 + K_2 t)e^{(-R/2L)t} \quad (14.59b)$$

For critically damped RLC circuits, both series and parallel, the quantity under the radical sign in the characteristic roots equation (Equations 14.55a and 14.55b) is zero. The two characteristic roots s_1 and s_2 have no imaginary component. This makes both s_1 and s_2 real valued for both series and parallel RLC circuits. Furthermore, because the quantity under the square root is zero, it is no longer added in s_1 and subtracted in s_2, giving the two characteristic roots s_1 and s_2 the same value.

Because s_1 and s_2 have the same value, in this case Equations 14.59a and 14.59b do not provide two independent constants K_1 and K_2. Therefore, in order to calculate two arbitrary constants, another solution to the differential equation, in addition to Ke^{st}, is needed.

Critically damped is the boundary between overdamped and underdamped natural responses.

Case 3: Underdamped – complex conjugate roots

In the case of R, L, and C values where the discriminant, the quantity under the radical sign, is negative, the characteristic roots s_1 and s_2 are complex and conjugates of one another. This is underdamped and the signals are decaying oscillatory signals.

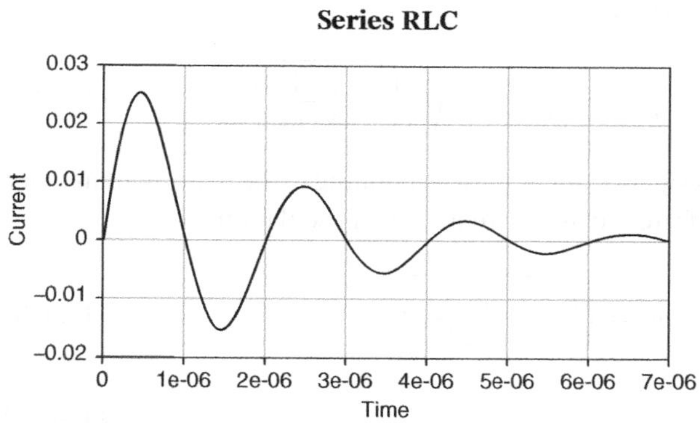

Figure 14.8a Underdamped series RLC simulated loop current.

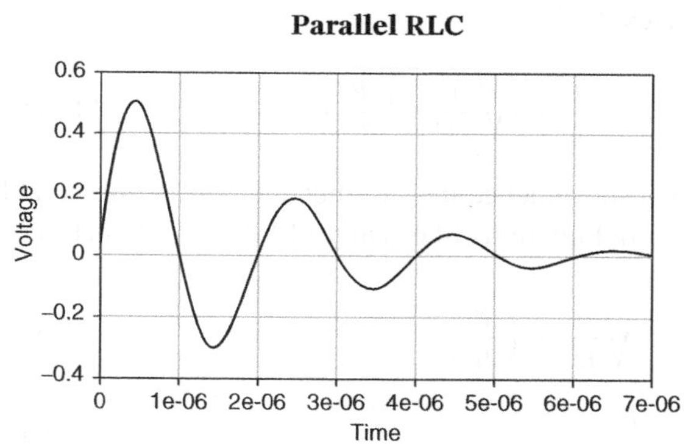

Figure 14.8b Underdamped parallel RLC simulated voltage.

Figure 14.8a shows underdamped series RLC loop current $i(t)$.

Figure 14.8b shows underdamped parallel RLC voltage $v(t)$.

In Figures 14.8a and 14.8b, for the underdamped cases, the signals have higher peak values than the corresponding critically damped signals, due to less damping.

Series RLC

$$\left(\frac{R}{2L}\right)^2 < \frac{1}{LC} \tag{14.60a}$$

Parallel RLC

$$\left(\frac{1}{2RC}\right)^2 < \frac{1}{LC} \tag{14.60b}$$

Equations 14.60a and 14.60b show the conditions for the underdamped case, which lead to the discriminants, the quantities under the radical signs, being imaginary valued in Equations 14.61a and 14.61b.

$$s_1, s_2 = \frac{-R}{2L} \pm \sqrt{\left(\frac{R}{2L}\right)^2 - \frac{1}{LC}} \tag{14.61a}$$

$$s_1, s_2 = \frac{-1}{2RC} + \pm\sqrt{\left(\frac{1}{2RC}\right)^2 - \frac{1}{LC}} \tag{14.61b}$$

Knowing that the discriminants in Equations 14.61a and 14.61b are imaginary valued, begin rearranging Equations 14.61a and 14.61b by multiplying the discriminant by 1 as shown in Equations 14.62a and 14.62b. Minus one squared equals 1.

$$s_1, s_2 = \frac{-R}{2L} \pm \sqrt{(-1)^2\left(\left(\frac{R}{2L}\right)^2 - \frac{1}{LC}\right)} \tag{14.62a}$$

$$s_1, s_2 = \frac{-1}{2RC} + \pm\sqrt{(-1)^2\left(\left(\frac{1}{2RC}\right)^2 - \frac{1}{LC}\right)} \tag{14.62b}$$

In Equations 14.62a and 14.62b, factor one −1 term outside of the radical sign and distribute the other minus one out as shown in Equations 14.63a and 14.63b.

$$s_1, s_2 = \frac{-R}{2L} \pm \sqrt{-1}\sqrt{\frac{1}{LC} - \left(\frac{R}{2L}\right)^2} \tag{14.63a}$$

$$s_1, s_2 = \frac{-1}{2RC} + \pm\sqrt{-1}\sqrt{\frac{1}{LC} - \left(\frac{1}{2RC}\right)^2} \tag{14.63b}$$

In Equations 14.63a and 14.63b, substitute j for the square root of minus one terms, as shown in Equations 14.64a and 14.64b. Series and parallel underdamped RLC have complex conjugate roots. In Equations 14.64a and 14.64b, there are both real and imaginary components.

Series RLC

$$s_1, s_2 = \frac{-R}{2L} \pm j\sqrt{\frac{1}{LC} - \left(\frac{R}{2L}\right)^2} \quad (14.64a)$$

Parallel RLC

$$s_1, s_2 = \frac{-1}{2RC} + \pm j\sqrt{\frac{1}{LC} - \left(\frac{1}{2RC}\right)^2} \quad (14.64b)$$

For convenience and succinct notation, define variables sigma σ and ω_d, corresponding to the real and imaginary parts of Equations 14.64a and 14.64b. ω_d is sometimes referred to as the frequency of oscillation.

$$\omega_d = \sqrt{\frac{1}{LC} - \left(\frac{R}{2L}\right)^2} \quad (14.65a) \qquad \omega_d = \sqrt{\frac{1}{LC} - \left(\frac{1}{2RC}\right)^2} \quad (14.65b)$$

$$\sigma = \frac{R}{2L} \quad (14.66a) \qquad \sigma = \frac{1}{2RC} \quad (14.66b)$$

$$i(t) = K_1 e^{s_1 t} + K_2 e^{s_2 t} \quad (14.67a) \qquad v(t) = K_1 e^{s_1 t} + K_2 e^{s_2 t} \quad (14.67b)$$

In the general format for the solutions of the series RLC current and the parallel RLC voltage, in Equations 14.67a and 14.67b, substitute in the variables sigma σ and $j\omega_d$, from Equations 14.65a, 14.65b, 14.66a, and 14.66b to obtain Equations 14.68a and 14.68b.

Series RLC

$$i(t) = K_1 e^{(-\sigma + j\omega_d)t} + K_2 e^{(-\sigma - j\omega_d)t} \quad (14.68a)$$

Parallel RLC

$$v(t) = K_1 e^{(-\sigma + j\omega_d)t} + K_2 e^{(-\sigma - j\omega_d)t} \quad (14.68b)$$

Rearrange the signs in the exponents of Equations 14.68a and 14.68b so that the entire exponent quantity has a negative sign, as shown in Equations 14.69a and 14.69b. This will be convenient for a later step.

$$i(t) = K_1 e^{-(\sigma - j\omega_d)t} + K_2 e^{-(\sigma + j\omega_d)t} \quad (14.69a) \qquad v(t) = K_1 e^{-(\sigma - j\omega_d)t} + K_2 e^{-(\sigma + j\omega_d)t} \quad (14.69b)$$

The exponents of Equations 14.69a and 14.69b show a real part σ, corresponding to exponential decay, and an imaginary part ω_d, corresponding to frequency. The response is an oscillation decaying in amplitude over time.

If damping goes to zero, that is, resistance R equals zero in series RLC or conductance G equals zero in parallel RLC, then ω_d equals the resonant frequency ω_0.

Case 4: Oscillatory lossless – two imaginary roots

At the extreme of underdamped is an ideal case, oscillatory lossless, where the damping goes away completely. This results in oscillatory signals going on indefinitely.

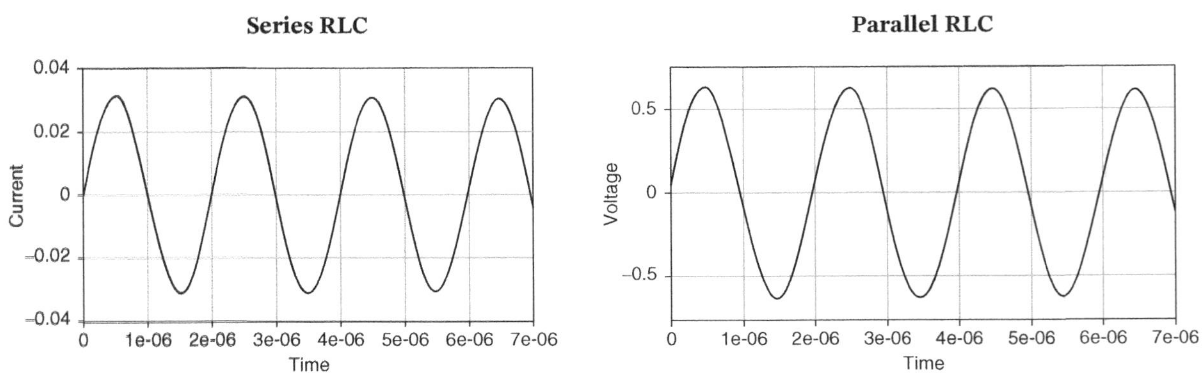

Figure 14.9a Lossless oscillatory series RLC simulated current.

Figure 14.9b Lossless oscillatory parallel RLC simulated voltage.

Figure 14.9a shows oscillatory series RLC loop current $i(t)$.

Figure 14.9b shows oscillatory parallel RLC voltage $v(t)$.

In Figures 14.9a and 14.9b for the oscillatory lossless ideal case, the signals have higher peak values than the underdamped series current and parallel voltage signals.

In practice this ideal lossless case is not achievable for an undriven circuit. Even for the circuit simulations, used to make these plots in Figures 14.9a and 14.9b, there is still a small amount damping.

$$R = 0 \qquad (14.70a) \qquad G = 0 \qquad (14.70b)$$

Equations 14.70a and 14.70b show the conditions for the undamped lossless case, which lead to the sinusoidal response, zero resistance for series RLC, and zero conductance for parallel RLC.

Series RLC **Parallel RLC**

$$s_1, s_2 = \frac{-R}{2L} \pm j\sqrt{\frac{1}{LC} - \left(\frac{R}{2L}\right)^2} \qquad (14.71a) \qquad s_1, s_2 = \frac{-1}{2RC} + \pm\sqrt{\left(\frac{1}{2RC}\right)^2 - \frac{1}{LC}} \qquad (14.71b)$$

Apply the lossless conditions, in Equations 14.70a and 14.70b, to Equations 14.71a and 14.71b, the general formulae for the roots of series and parallel RLC characteristic equations. The result is two imaginary roots, in Equations 14.72a and 14.72b, which lead to undamped sinusoidal response signals.

$$s_1, s_2 = \frac{\pm\sqrt{-4LC}}{2LC} = \frac{\pm j}{\sqrt{LC}} = \pm j\omega_0 \qquad (14.72a) \qquad s_1, s_2 = \frac{\pm\sqrt{-4LC}}{2LC} = \frac{\pm j}{\sqrt{LC}} = \pm j\omega_0 \qquad (14.72b)$$

When the resistance R drops out of the general formulae, for the roots of series and parallel RLC characteristic equations, shown in Equations 14.72a and 14.72b, the result is the resonant frequency ω_0 shown in Equations 14.73a and 14.73b.

$$s_1, s_2 = \pm j\omega_0 \qquad (14.73a) \qquad s_1, s_2 = \pm j\omega_0 \qquad (14.73b)$$

$$i(t) = K_1 e^{s_1 t} + K_2 e^{s_2 t} \qquad (14.74a) \qquad v(t) = K_1 e^{s_1 t} + K_2 e^{s_2 t} \qquad (14.74b)$$

In the general format for the solutions of the series RLC current and the parallel RLC voltage, in Equations 14.74a and 14.74b, substitute in the roots from Equations 14.73a and 14.73b, to obtain Equations 14.75a and 14.75b.

$$i(t) = K_1 e^{j\omega_0 t} + K_2 e^{-j\omega_0 t} \qquad (14.75a) \qquad v(t) = K_1 e^{j\omega_0 t} + K_2 e^{-j\omega_0 t} \qquad (14.75b)$$

Equations 14.75a and 14.75b correspond to sinusoidal behavior, using Euler's law. This is addressed in a later section where the constants K_1 and K_2 are calculated.

14.2.6 Series RLC Critical Damping Resistance Calculation Example

Consider a series RLC circuit, shown in Figure 14.10, with inductance $L = 10\,\mu\text{H}$ and capacitance $C = 10\,\text{nF}$. The capacitor has an initial voltage value of 1 V.

Different values of the resistor R_1 can make this circuit overdamped, underdamped, or critically damped. This example explores critically damped.

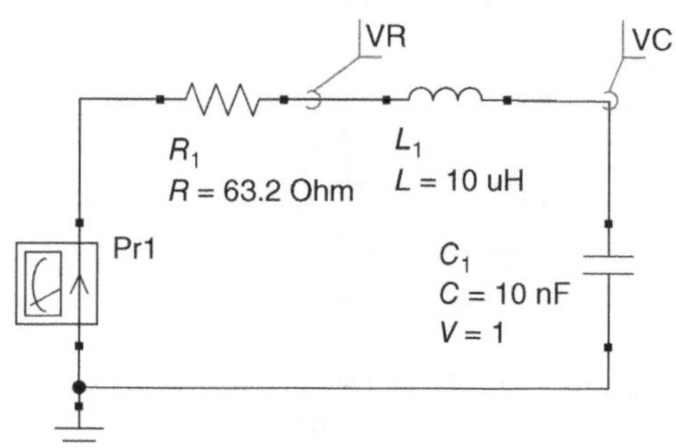

Figure 14.10 Series RLC simulation schematic with initial capacitor voltage.

14.2.6.1 Calculate the Critical Damping Resistance

$$\left(\frac{R}{2L}\right)^2 = \frac{1}{LC} \quad (14.76)$$

In the series RLC characteristic roots equations, Equation 14.76 is the equation for the critical damping condition. Solve Equation 14.76 for R as follows.

$$\frac{R^2}{4L^2} = \frac{1}{LC} \quad (14.77)$$

In Equation 14.77, multiply both sides by $4L$-squared to obtain Equation 14.78.

$$R^2 = \frac{4L^2}{LC} \quad (14.78)$$

In Equation 14.78, simplify L terms to obtain Equation 14.79.

$$R^2 = \frac{4L}{C} \quad (14.79)$$

Take the square root of both sides of Equation 14.79.

$$R_{\text{CRIT-ser}} = \sqrt{\frac{4L}{C}} \quad (14.80)$$

In Equation 14.80, substitute in values L equals 10e−6 and C equals 10e−9, as shown in Equation 14.80.

$$R_{\text{CRIT-ser}} = \sqrt{\frac{40e-6}{10e-9}} = \sqrt{4000} = 63.245 \quad (14.81)$$

For resistor R_1 in the schematic in Figure 14.10, a resistance value of 63.2 yields a critically damped natural response.

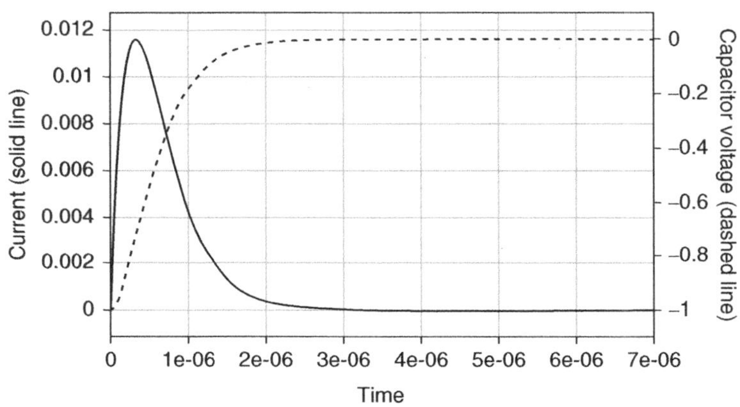

Figure 14.11 Series RLC critically damped simulated current and capacitor voltage.

Figure 14.11 shows the critically damped loop current with the solid line. The capacitor voltage, with the dashed line and labeled on the right-hand vertical axis, moves from an initial condition of minus one volt relative to ground, to zero.

By Ohm's law, the shape of the current response tells us the shape of the voltage across the series resistor.

14.2.7 Parallel RLC Critical Damping Resistance Calculation Example

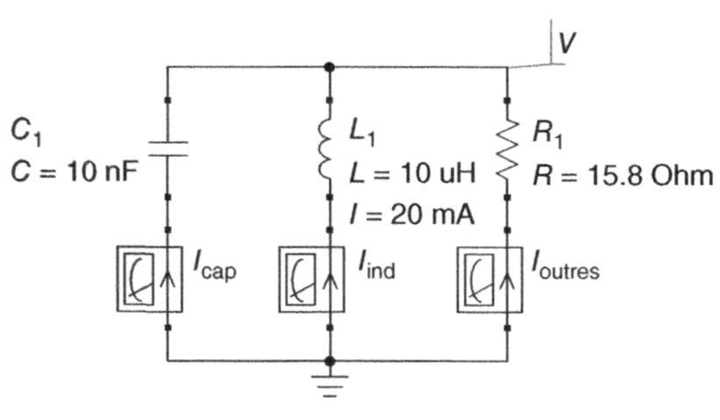

Figure 14.12 Parallel RLC simulation schematic with initial inductor current.

Consider a parallel RLC circuit, shown in Figure 14.12, with inductance L of 10 µH and capacitance C of 10 nF. The inductor has an initial current value of 20 mA.

Different values of the resistor R_1 can make this circuit overdamped, underdamped, or critically damped. This example explores critically damped.

14.2.7.1 Calculate Critical Damping for a Parallel RLC Example

$$\left(\frac{1}{2RC}\right)^2 = \frac{1}{LC} \quad (14.82)$$

In the parallel RLC characteristic roots equations, Equation 14.82 states the critical damping condition. Solve Equation 14.82 for R as follows.

$$\frac{1}{4R^2C^2} = \frac{1}{LC} \quad (14.83)$$

Multiply both sides of Equation 14.83 by R squared.

$$\frac{1}{4C^2} = \frac{R^2}{LC} \quad (14.84)$$

Cancel a capacitance term in each denominator of Equation 14.84 (by multiplying both sides of Equation 14.84 by C).

$$\frac{1}{4C} = \frac{R^2}{L} \quad (14.85)$$

Multiply both sides of Equation 14.85 by L.

$$\frac{L}{4C} = R^2 \quad (14.86)$$

Take the square root of both sides of Equation 14.86.

$$R_{\text{CRIT-par}} = \frac{1}{2}\sqrt{\frac{L}{C}} \quad (14.87)$$

In Equation 14.87, substitute in values L equals 10e−6 and C equals 10e−9, as shown in Equation 14.88.

$$R_{\text{CRIT-par}} = \frac{1}{2}\sqrt{\frac{10\text{e}-6}{10\text{e}-9}} = \frac{1}{2}\sqrt{1000} = 15.8\,\Omega \quad (14.88)$$

For resistor R_1, a resistance value of 15.8 yields a critically damped natural response in the parallel RLC circuit in Figure 14.12.

Figure 14.13 shows the critically damped voltage with the solid line. The dashed line is the inductor current, labeled on the right-hand vertical axis, decaying from the initial 20 mA initial condition down to zero.

By Ohm's law, the shape of the current response tells us the shape of the resistor voltage.

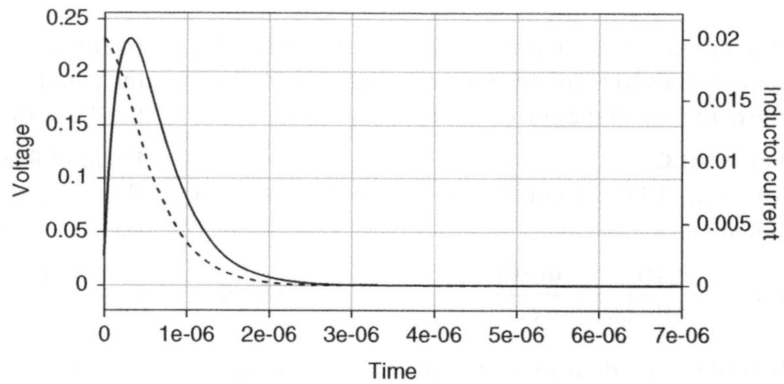

Figure 14.13 Parallel RLC critically damped simulated voltage and inductor current.

14.2.8 Calculating Constants for the RLC General Solutions

After looking at the roots of the characteristic equations, and how they determine circuit response, this subsection returns to the mathematical steps to derive general solutions for the natural response differential equations for series and parallel RLC circuits, shown in Figures 14.14a and 14.4b. Once the characteristic roots, s_1 and s_2, are known, the final step to a general solution, for the RLC differential equation, is to calculate the arbitrary constants, K_1 and K_2, in Equations 14.89a and 14.89b.

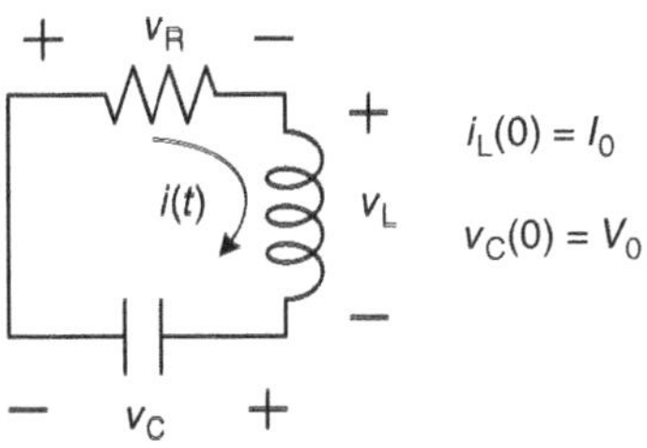

Figure 14.14a Series RLC circuit with two initial conditions.

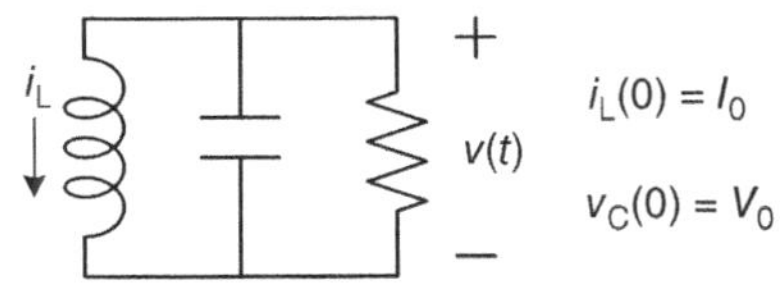

Figure 14.14b Parallel RLC circuit with two initial conditions.

Series RLC

$$i(t) = K_1 e^{s_1 t} + K_2 e^{s_2 t} \tag{14.89a}$$

Parallel RLC

$$v(t) = K_1 e^{s_1 t} + K_2 e^{s_2 t} \tag{14.89b}$$

The mathematical approach, to calculating the unknown constants, K_1 and K_2 in Equations 14.89a and 14.89b, uses the initial condition of the chosen circuit variable, and the initial condition of the derivative of the circuit variable. The initial condition is the value at time t equals 0.

$$i(0) = i(t)|_{t=0} = I_0 \tag{14.90a}$$

$$v(0) = v(t)|_{t=0} = V_0 \tag{14.90b}$$

$$i'(0) = \left.\frac{di(t)}{dt}\right|_{t=0} = I_0' \tag{14.91a}$$

$$v'(0) = \left.\frac{dv(t)}{dt}\right|_{t=0} = V_0' \tag{14.91b}$$

In Equation 14.90a, current is the circuit variable $i(t)$, for the series RLC circuit, at time 0. Equation 14.91a is the derivative of the current at time t equals 0.

In Equation 14.90b, voltage is the state variable $v(t)$, for the parallel RLC circuit, at time 0. Equation 14.92b is the derivative of the voltage at time t equals 0.

Series RLC

The series RLC current at time t equals 0 is shown in Equation 14.92a.

$$i'(t_0) = \left.\frac{di_L(t)}{dt}\right|_{t=0} = \frac{di_L(0)}{dt} \tag{14.92a}$$

Parallel RLC

The voltage at time t equals 0 is shown in Equation 14.92b.

$$v'(t_0) = \left.\frac{dv_C(t)}{dt}\right|_{t=0} = \frac{dv_C(0)}{dt} \tag{14.92b}$$

Substitute the derivative of current at time t equals 0, into the inductor voltage equation, as shown in Equation 14.93a.

$$v_L(0) = L\frac{di_L(0)}{dt} \tag{14.93a}$$

Equation 14.93a provides a value for initial inductor voltage in terms of initial inductor current. This can be used for calculating arbitrary constants using initial conditions.

Plug the derivative of voltage at time t equals 0 into the capacitor current equation, as shown in Equation 14.93b.

$$i_C(0) = C\frac{dv_C(0)}{dt} \tag{14.93b}$$

Equation 14.93b provides a value for initial capacitor current in terms of initial capacitor voltage. This can be used for calculating arbitrary constants using initial conditions.

14.2.8.1 Calculation of K_1 and K_2 for Overdamped RLC

This subsection calculates the arbitrary constants K_1 and K_2 for the overdamped RLC circuits in Figures 14.15a and 14.15b. The initial condition for the series RLC will be the capacitor voltage, at time equals 0, $v_C(0)$ equals V_0. The initial condition for the parallel RLC will be the inductor current, at time t equals 0, $i_L(0) = I_0$.

Start calculating the arbitrary constants K_1 and K_2, for the RLC series and parallel differential equation solutions, with general solution form Equations 14.52a and 14.52b from Case 1, rewritten here as Equations 14.94a and 14.94b.

Series RLC

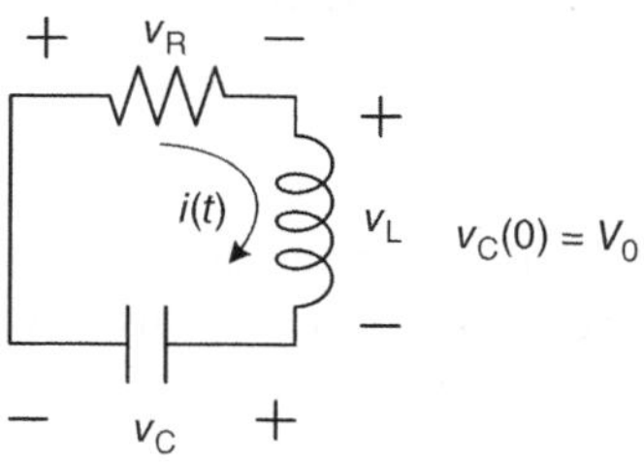

Figure 14.15a Series RLC circuit with one initial condition.

Parallel RLC

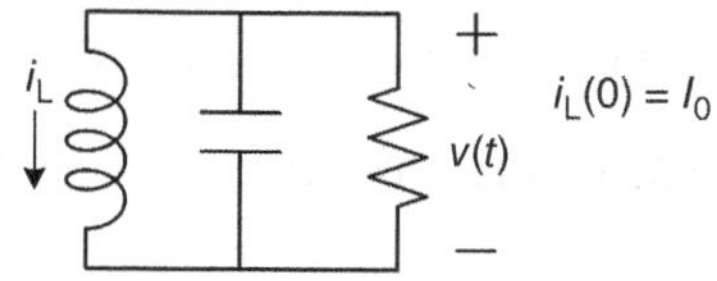

Figure 14.15b Parallel RLC circuit with one initial condition.

Series RLC

$$i(t) = K_1 e^{s_1 t} + K_2 e^{s_2 t} \tag{14.94a}$$

Parallel RLC

$$v(t) = K_1 e^{s_1 t} + K_2 e^{s_2 t} \tag{14.94b}$$

Equations 14.95a and 14.95b are the equations for the series and parallel RLC circuit characteristic roots s_1 and s_2, in terms of R, L, and C.

Series RLC

$$s_1, s_2 \; \frac{-R}{2L} \pm \sqrt{\left(\frac{R}{2L}\right)^2 - \frac{1}{LC}} \tag{14.95a}$$

Parallel RLC

$$s_1, s_2 = \frac{-1}{2RC} \pm \sqrt{\left(\frac{1}{2RC}\right)^2 - \frac{1}{LC}} \tag{14.95b}$$

To start using initial conditions of the circuit equations to solve for the arbitrary constants K_1 and K_2, evaluate both Equations 14.94a and 14.94b at time t equals 0.

In a series RLC circuit with an initial condition V_0 on the capacitor, there is no initial current at time t equals 0. As a result, the exponential terms from Equation 14.94a go to 1, resulting in Equations 14.96a and 14.97a. The $i(0)$ term on the left side of Equation 14.96a goes to zero in Equation 14.97a.

In a parallel RLC circuit with an initial condition I_0 in the inductor, there is no initial voltage at time t equals 0. As a result, the exponential terms from Equation 14.94b go to 1, resulting in Equations 14.96b and 14.97b. The $v(0)$ term on the left side of Equation 14.96a goes to zero in Equation 14.97a.

Series RLC

$$i(0) = i(t)|_{t=0} = 0 = K_1 + K_2 \tag{14.96a}$$

$$K_1 + K_2 = 0 \tag{14.97a}$$

Parallel RLC

$$v(0) = v(t)|_{t=0} = 0 = K_1 + K_2 \tag{14.96b}$$

$$K_1 + K_2 = 0 \tag{14.97b}$$

Equations 14.97a and 14.97b are the first set of two sets of equations needed to calculate the two arbitrary constants K_1 and K_2.

For a second set of equations for calculating the arbitrary constants of the overdamped series and parallel RLC natural response, look at the initial conditions of the derivatives of the circuit variables.

Next, differentiate the general form Equations 14.94a and 14.94b with respect to time to obtain Equations 14.99a and 14.99b.

Series RLC

$$i'(t) = s_1K_1e^{s_1t} + s_2K_2e^{s_2t} \tag{14.98a}$$

$$i'(0) = i'(t)\big|_{t=0} = s_1K_1 + s_2K_2 \tag{14.99a}$$

Parallel RLC

$$v'(t) = s_1K_1e^{s_1t} + s_2K_2e^{s_2t} \tag{14.98b}$$

$$v'(0) = v'(t)\big|_{t=0} = s_1K_1 + s_2K_2 \tag{14.99b}$$

Next, evaluate the left side of Equations 14.99a and 14.99b, for values of $i'(0)$ and $v'(0)$ to finish the second set of equations for calculating the arbitrary constants K_1 and K_2.

14.2.8.2 Overdamped Series RLC $i'(0)$ Substitution for K_1 and K_2 Calculation

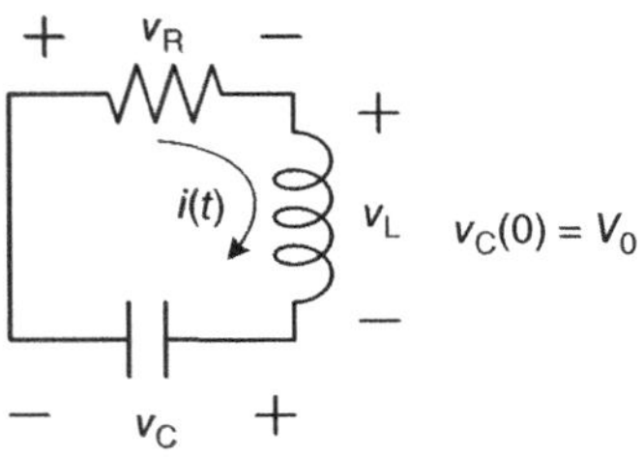

Figure 14.16 Series RLC circuit with one initial condition.

In a series RLC circuit with an initial condition voltage V_0 on the capacitor, as shown in Figure 14.16, there is no initial resistor current at time t equals 0. As a result, the initial capacitor voltage means $-V_0$ across the inductor.

$$v_L(t) = L\frac{di}{dt} \tag{14.100}$$

The branch relationship of an ideal inductor, in Equation 14.100, states that the voltage across an inductor equals the time derivative of the current through the inductor times the inductance L.

Divide Equation 14.100 by L and apply to the series RLC Equation 14.99a at time t equals 0 to get Equation 14.101. v_L equals $-V_0$, from the initial capacitor voltage.

$$\left.\frac{di(t)}{dt}\right|_{t=0} = \frac{-v(0)}{L} = \frac{-V_0}{L} \tag{14.101}$$

Equation 14.101 provides a value for the derivative of the loop current in the series RLC circuit at time t equals 0. Substitute Equations 14.101 into 14.99a for $i'(0)$.

$$\frac{-V_0}{L} = s_1K_1 + s_2K_2 \tag{14.102}$$

For series RLC, Equation 14.102 provides the second equation needed to calculate the two arbitrary constants K_1 and K_2.

Rearrange Equation 14.97a to solve for K_2 equal to minus K_1, as shown in Equation 14.103.

$$K_2 = -K_1 \tag{14.103}$$

Substitute Equation 14.103 for K_2 into Equation 14.102 to eliminate the K_2 term, obtaining Equation 14.104.

$$\frac{-V_0}{L} = s_1K_1 + s_2(-K_1) \tag{14.104}$$

In Equation 14.104, gather the two K_1 terms, as shown in Equation 14.105.

$$\frac{-V_0}{L} = K_1(s_1 - s_2) \tag{14.105}$$

Solve Equation 14.105 for K_1, leading to Equation 14.106 for both K_1 and K_2.

$$K_1 = -K_2 = -\frac{V_0}{L(s_1 - s_2)} \tag{14.106}$$

Substitute Equation 14.106 values for K_1 and K_2 into Equation 14.94a to get a general solution to the series RLC overdamped current $i(t)$ in Equation 14.107.

$$i(t) = \left(-\frac{V_0}{L(s_1 - s_2)}\right)e^{s_1t} + \left(\frac{V_0}{L(s_1 - s_2)}\right)e^{s_2t} \tag{14.107}$$

In Equation 14.107 gather the exponential terms, noticing the minus signs, to obtain Equation 14.108.

$$i(t) = \left(e^{s_2t} - e^{s_1t}\right)\left(\frac{V_0}{L(s_1 - s_2)}\right) \tag{14.108}$$

For series RLC with one initial condition capacitor voltage V_0, Equation 14.108 is the solution for current, including values for the two constants K_1 and K_2.

14.2.8.3 Overdamped Parallel RLC i′(0) Substitution for K_1 and K_2 Calculation

In a parallel RLC circuit, with an initial condition current I_0 in the inductor, as shown in Figure 14.17, there is no initial voltage at time t equals 0. As a result, right at time t equals 0, all the inductor current goes into the capacitor. By KCL, the capacitor current equals minus the inductor current I_0.

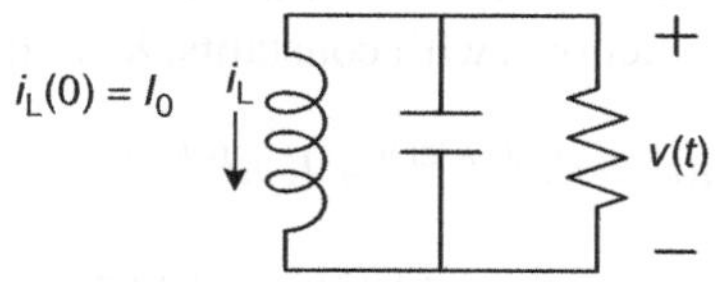

Figure 14.17 Parallel RLC circuit with two initial conditions.

$$i_C(t) = C\frac{dv}{dt} \tag{14.109}$$

The branch relationship of an ideal capacitor, in Equation 14.109, states that the current through a capacitor equals the time derivative of the voltage across the capacitor times the capacitance C.

$$\left.\frac{dv_C(t)}{dt}\right|_{t=0} = \frac{-i_L(0)}{C} = \frac{-I_0}{C} \tag{14.110}$$

Equation 14.110 provides a value for the derivative of the voltage in the parallel RLC circuit at time t equals 0. Substitute Equation 14.110 into Equation 14.99b for $v'(0)$.

$$\frac{-I_0}{C} = s_1K_1 + s_2K_2 \tag{14.111}$$

Divide Equation 14.109 by C and apply to the parallel RLC Equation 14.99b at time t equals 0 to get Equation 14.111. i_C equals $-I_0$, from the initial inductor current.

For parallel RLC, Equation 14.111 provides the second equation needed to calculate the two arbitrary constants K_1 and K_2.

$$K_2 = -K_1 \tag{14.112}$$

Equation 14.103 shows K_2 equal to minus K_1, as shown in Equation 14.112.

$$\frac{-I_0}{C} = s_1K_1 + s_2(-K_1) \tag{14.113}$$

Substitute Equation 14.112 for K_2 into Equation 14.111 to eliminate the K_2 term, as shown in Equation 14.113.

$$\frac{-I_0}{C} = K_1(s_1 - s_2) \tag{14.114}$$

In Equation 14.113, gather the two K_1 terms to obtain Equation 14.114.

Solve Equation 14.114 for K_1, as shown in Equation 14.115.

$$K_1 = -K_2 = -\frac{I_0}{C(s_1 - s_2)} \tag{14.115}$$

Substitute Equation 14.115 into Equation 14.94b to get a general solution to the parallel RLC overdamped current $i(t)$ in Equations 14.116 and 14.117.

$$v(t) = \left(-\frac{I_0}{C(s_1 - s_2)}\right)e^{s_1t} + \left(\frac{I_0}{C(s_1 - s_2)}\right)e^{s_2t} \tag{14.116}$$

In Equation 14.116 gather the exponential terms, noticing the minus signs, to obtain Equation 14.117.

$$v(t) = \left(e^{s_2t} - e^{s_1t}\right)\left(\frac{I_0}{C(s_1 - s_2)}\right) \tag{14.117}$$

For parallel RLC, with one initial condition inductor current, Equation 14.117 is the solution for the overdamped natural response voltage, including values for the two constants K_1 and K_2.

14.2.8.4 The Function tKe^{s_2t} for Critically Damped RLC

For critically damped RLC, we use a general form of the solution with a function te^{st} in the second term, as shown in Equations 14.118a and 14.118b. Otherwise, using the $K_1e^{s_1t}$ plus $K_2e^{s_2t}$ general solution, the K_1 and K_2 terms aren't unique because for critically damped RLC, s_1 and s_2 are equal. We need to be able to determine two unique functions, with constants, K_1 and K_2, to finish the general solution for critically damped RLC.

$$i(t) = K_1 e^{(-R/2L)t} + K_2 t e^{(-R/2L)t} \quad (14.118a) \qquad v(t) = K_1 e^{(-1/2RC)t} + K_2 t e^{(-1/2RC)t} \quad (14.118b)$$

Equations 14.118a and 14.118b reduce to Equations 14.119a and 14.119b.

$$i(t) = (K_1 + K_2 t)e^{(-R/2L)t} \quad (14.119a) \qquad v(t) = (K_1 + K_2 t)e^{(-1/2RC)t} \quad (14.119b)$$

To completely solve the series and parallel RLC differential equations, we need two solutions to the equation, two different functions that both solve the RLC differential equations. For the other cases, like overdamped and underdamped, the characteristic roots are different from one another and the general form, with $K_1 e^{s_1 t}$ plus $K_2 e^{s_2 t}$ provides two unique solutions. For critically damped, where s_1 and s_2 are the same, we only get one solution; we need another function that solves the RLC equations. The function te^{st} suffices.

14.2.8.5 Calculation of K_1 and K_2 for Critically Damped RLC

To use initial conditions to solve for the arbitrary constants K_1 and K_2, evaluate both Equations 14.119a and 14.119b for time t equals 0, as shown in Equations 14.120a and 14.120b. The t in the K_2 term makes the entire K_2 term equal to zero, resulting in Equations 14.121a and 14.121b where K_1 equals 0.

Series RLC | **Parallel RLC**

$$i(0) = i(t)|_{t=0} = 0 = K_1 \quad (14.120a) \qquad v(0) = v(t)|_{t=0} = 0 = K_1 \quad (14.120b)$$

$$K_1 = 0 \quad (14.121a) \qquad K_1 = 0 \quad (14.121b)$$

Equations 14.121a and 14.121b are the first set of two sets of equations needed to calculate the two arbitrary constants K_1 and K_2 for series and parallel RLC.

For the derivation of the second set of simultaneous equations, needed to calculate the two arbitrary constants K_1 and K_2 for critically damped RLC, partition the analyses between series and parallel. First there is a derivation of the second series equation and then another analysis for the second parallel equation.

14.2.8.6 Calculation of K_1 and K_2 for Critically Damped Series RLC

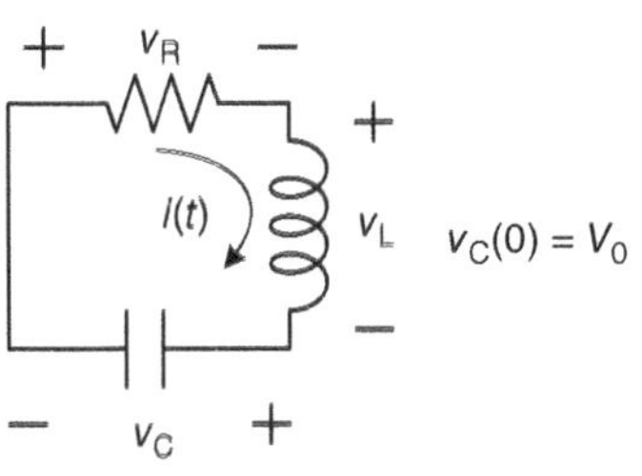

Figure 14.18 Series RLC circuit with one initial condition.

Two simultaneous equations are needed to calculate the two arbitrary constants, K_1 and K_2, for the series RLC equation solution for current $i(t)$, corresponding to the circuit shown in Figure 14.18. The previous section calculated the first equation K_1 equals 0. This section calculates the second simultaneous equation which will get used with the first equation to solve for K_1 and K_2.

Differentiate the general form Equation 14.118a, shown here as Equation 14.122, where s is not designated s_1 or s_2 because s_1 and s_2 are the same for critically damped.

$$i(t) = K_1 e^{st} + K_2 t e^{st} \quad (14.122)$$

Differentiate Equation 14.122, using the chain rule in Equation 14.123 for differentiating the K_2 term in Equation 14.122, to obtain Equation 14.124.

$$f(x) = g(x)h(x) = g'(x)h(x) + g(x)h'(x) \quad (14.123)$$

$$i'(t) = sK_1 e^{st} + K_2 e^{st} + sK_2 t e^{st} \quad (14.124)$$

In Equation 14.124, gather $K_2 e^{st}$ terms to obtain Equation 14.125.

$$i'(t) = sK_1e^{st} + K_2e^{st}(1 + st) \quad (14.125)$$

Evaluate Equation 14.125 at time t equals 0. The t term goes to zero and the e^{st} term goes to 1, as shown in Equation 14.126.

$$i'(0) = i'(t)\big|_{t=0} = sK_1 + K_2 \quad (14.126)$$

Simplify Equation 14.126 to obtain Equation 14.127.

$$i'(0) = sK_1 + K_2 \quad (14.127)$$

Equation 14.127 has an $i'(0)$ term. The next step is to find a circuit quantity for that $i'(0)$ term.

In the case of the series RLC, with an initial condition V_0 on the capacitor, there will be no initial current $i(0)$. Because there is no current in the circuit at time t equals zero, there is no voltage across the resistor. By Kirchhoff's voltage law (KVL) that means that the positive V_0 across the capacitor is also $-V_0$ across the inductor.

The branch relationship of an ideal inductor, in Equation 14.128, states that the voltage across an inductor equals the time derivative of the current through the inductor times the inductance L.

$$v_L(t) = L\frac{di}{dt} \quad (14.128)$$

Divide Equation 14.128 by L and apply to the series RLC Equation 14.92a at time t equals 0 to get Equation 14.129. v_L equals $-V_0$, from the initial capacitor voltage.

$$\left.\frac{di(t)}{dt}\right|_{t=0} = \frac{-v(0)}{L} = \frac{-V_0}{L} \quad (14.129)$$

Equation 14.129 provides a value for the derivative of the loop current in the series RLC circuit at time t equals 0. Substitute Equation 14.129 into Equation 14.127 for $i'(0)$ to get Equation 14.130.

$$\frac{-V_0}{L} = sK_1 + K_2 \quad (14.130)$$

Equation 14.130 is the second simultaneous equation for calculating the values of K_1 and K_2 for the series RLC current solution.

From Equation 14.121a, substitute in K_1 equals 0 into Equation 14.130 to get the value for K_2 in Equation 14.131. This substitution represents the simultaneous solution of the two equations to calculate K_1 and K_2.

$$K_2 = \frac{-V_0}{L} \quad (14.131)$$

Substitute Equation 14.131 for K_2 into Equation 14.119a with K_1 equal to 0, to get Equation 14.132.

$$i(t) = \frac{-V_0}{L}te^{st} \quad (14.132)$$

Equation 14.132 is the critically damped series RLC current response.

14.2.8.7 Calculation of K_1 and K_2 for Critically Damped Parallel RLC

Two simultaneous equations are needed to calculate the two arbitrary constants, K_1 and K_2, for the parallel RLC equation solution for voltage $v(t)$, corresponding to the circuit shown in Figure 14.19. A previous section calculated the first equation K_1 equals 0 in Equation 14.121b. This section calculates the second simultaneous equation which will get used with the first equation to solve for K_1 and K_2.

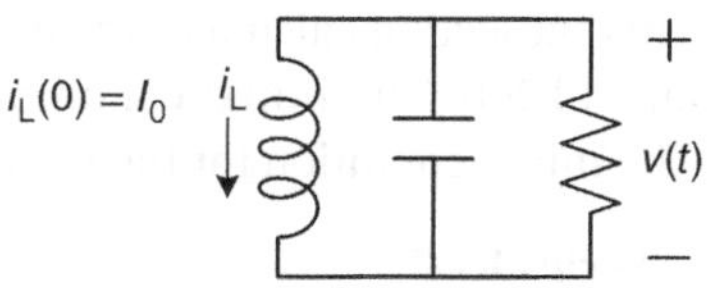

Figure 14.19 Parallel RLC circuit with two initial conditions.

Differentiate the general form Equation 14.118b, shown here as Equation 14.133, where s is not designated s_1 or s_2 because s_1 and s_2 are the same for critically damped.

$$v(t) = K_1e^{st} + K_2te^{st} \quad (14.133)$$

Differentiate Equation 14.133, using the chain rule in Equation 14.134 for differentiating the K_2 term in Equation 14.133, to obtain Equation 14.135.

$$f(x) = g(x)h(x) = g'(x)h(x) + g(x)h'(x) \quad (14.134)$$

$$v'(t) = sK_1e^{st} + K_2e^{st} + sK_2te^{st} \quad (14.135)$$

In Equation 14.135, gather K_2e^{st} terms to obtain Equation 14.136.

$$v'(t) = sK_1e^{st} + K_2e^{st}(1 + st) \quad (14.136)$$

Evaluate Equation 14.136 at time t equals 0. The t term goes to zero and the e^{st} term goes to 1, as shown in Equation 14.137.

$$v'(0) = v'(t)|_{t=0} = sK_1 + K_2 \quad (14.137)$$

Simplify Equation 14.137 to obtain Equation 14.138.

$$v'(0) = sK_1 + K_2 \quad (14.138)$$

Equation 14.138 has an $v'(0)$ term. The next step is find a circuit quantity for that $v'(0)$ term.

In the case of the parallel RLC circuit with an initial condition current I_0 in the inductor, there will be no initial voltage $v(0)$. Because there is no voltage across the circuit at time t equals zero, there is no current in the resistor. By Kirchhoff's current law (KCL) that means that the positive I_0 in the inductor is also $-I_0$ in the capacitor.

The branch relationship of an ideal capacitor, in Equation 14.139, states that the current through an ideal capacitor equals the time derivative of the voltage across the capacitor times the capacitance C.

$$i_C(t) = C\frac{dv}{dt} \quad (14.139)$$

Divide Equation 14.139 by C and apply to the parallel RLC Equation 14.91b at time t equals 0 to get Equation 14.140. i_C equals $-I_0$, from the initial inductor current.

$$\left.\frac{dv_C(t)}{dt}\right|_{t=0} = \frac{-i_L(0)}{C} = \frac{-I_0}{C} \quad (14.140)$$

Equation 14.140 provides a value for the derivative of the voltage in the parallel RLC circuit at time t equals 0. Substitute Equation 14.140 into Equation 14.138 for $v'(0)$ to obtain Equation 14.141.

$$\frac{-I_0}{C} = s_1K_1 + s_2K_2 \quad (14.141)$$

Equation 14.141 is the second simultaneous equation for calculating the values of K_1 and K_2 for the parallel RLC current solution.

$$K_2 = \frac{-I_0}{C} \quad (14.142)$$

From Equation 14.121b substitute in K_1 equals 0 into Equation 14.141 to get the value for K_2 in Equation 14.142. This substitution represents the simultaneous solution of the two equations to calculate K_1 and K_2.

$$i(t) = \frac{-I_0}{C}te^{st} \quad (14.143)$$

Equation 14.143 is the critically damped parallel RLC voltage natural response.

14.2.8.8 Calculation of K_1 and K_2 for Underdamped RLC

This subsection calculates the arbitrary constants K_1 and K_2 for the underdamped RLC circuits in Figures 14.20a and 14.20b. The initial condition for the series RLC is a capacitor voltage, at time equals 0, $v_C(0)$ equals V_0. The initial condition for the parallel RLC is an inductor current, at time t equals 0, $i_L(0) = I_0$.

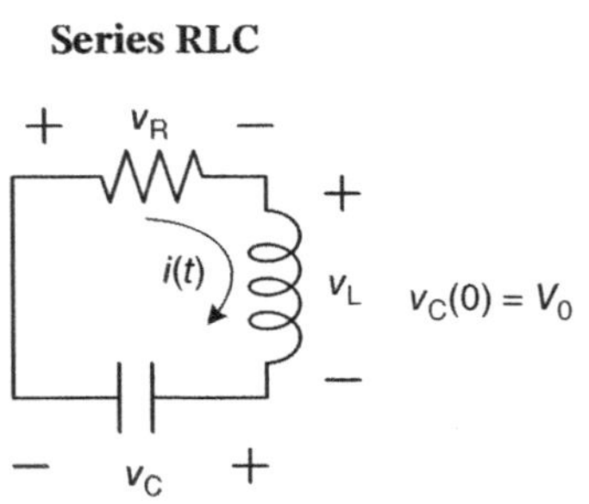

Figure 14.20a Series RLC circuit with one initial condition.

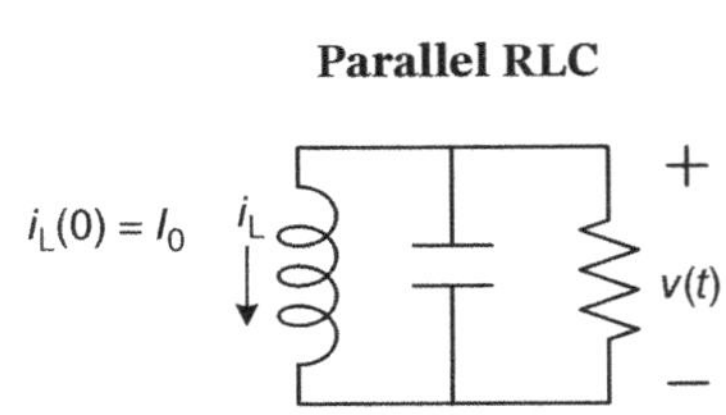

Figure 14.20b Parallel RLC circuit with one initial condition.

Start with the general forms for the RLC series and parallel differential equation solutions, Equations 14.69a and 14.69b from underdamped Case 3, repeated here as Equations 14.144a and 14.144b.

Series RLC

$$i(t) = K_1 e^{-(\sigma - j\omega_d)t} + K_2 e^{-(\sigma + j\omega_d)t} \quad (14.144a)$$

Parallel RLC

$$v(t) = K_1 e^{-(\sigma - j\omega_d)t} + K_2 e^{-(\sigma + j\omega_d)t} \quad (14.144b)$$

To start using initial conditions of the circuit equations to solve for the arbitrary constants K_1 and K_2, evaluate both Equations 14.144a and 14.144b for time t equals 0.

In a series RLC circuit with an initial condition V_0 on the capacitor, there is no initial current at time t equals 0. As a result, the exponential terms from Equation 14.144a go to 1, resulting in Equations 14.145a and 14.146a. The $i(0)$ term on the left side of Equation 14.145a goes to zero in Equation 14.146a.

In a parallel RLC circuit with an initial condition I_0 in the inductor, there is no initial voltage at time t equals 0. As a result, the exponential terms from Equation 14.144b go to 1, resulting in Equations 14.145b and 14.146b. The $v(0)$ term on the left side of Equation 14.145a goes to zero in Equation 14.146a.

Series RLC

$$i(0) = i(t)|_{t=0} = 0 = K_1 + K_2 \quad (14.145a)$$

Parallel RLC

$$v(0) = v(t)|_{t=0} = 0 = K_1 + K_2 \quad (14.145b)$$

$$K_1 + K_2 = 0 \quad (14.146a) \qquad K_1 + K_2 = 0 \quad (14.146b)$$

Equations 14.146a and 14.146b are the first set of simultaneous equations for finding the arbitrary constants, K_1 and K_2, of the series and parallel RLC.

For a second set of equations, for calculating the arbitrary constants of the underdamped series and parallel RLC natural response, look at the initial conditions of the derivatives of the circuit variables $i(t)$ and $v(t)$ as follows.

Differentiate the general form solution Equations 14.144a and 14.144b with respect to time to obtain Equations 14.147a and 14.147b.

Series RLC

$$i'(t) = -(\sigma - j\omega_d)K_1 e^{-(\sigma - j\omega_d)t} - (\sigma + j\omega_d)K_2 e^{-(\sigma + j\omega_d)t} \quad (14.147a)$$

Parallel RLC

$$v'(t) = -(\sigma - j\omega_d)K_1 e^{-(\sigma - j\omega_d)t} - (\sigma + j\omega_d)K_2 e^{-(\sigma + j\omega_d)t} \quad (14.147b)$$

Evaluate Equations 14.147a and 14.147b, at time t equals 0, as shown in Equations 14.148a and 14.148b. The exponential terms go to 1 when t equals 0.

Series RLC

$$i'(0) = -(\sigma - j\omega_d)K_1 - (\sigma + j\omega_d)K_2 \quad (14.148a)$$

Parallel RLC

$$v'(0) = -(\sigma - j\omega_d)K_1 - (\sigma + j\omega_d)K_2 \quad (14.148b)$$

Evaluate the left side of Equations 14.148a and 14.148b, for values of $i'(0)$ and $v'(0)$ to finish the second set of equations for calculating the arbitrary constants K_1 and K_2.

Equation 14.149a states that the voltage across an inductor equals the time derivative of the current through the inductor times the inductance L.

$$v_L(t) = L\frac{di}{dt} \quad (14.149a)$$

Equation 14.149b states that the current through a capacitor equals the time derivative of the voltage across the capacitor times the capacitance C.

$$i_C(t) = C\frac{dv}{dt} \quad (14.149b)$$

In a series RLC circuit with an initial condition voltage V_0 on the capacitor, there is no initial resistor current at time t equals 0. As a result, the initial capacitor voltage means $-V_0$ across the inductor.

Divide Equation 14.149a by L and apply to the series RLC Equation 14.91a at time t equals 0, to get Equation 14.150a. v_L equals $-V_0$, from the initial capacitor voltage.

In a parallel RLC circuit with an initial condition current I_0 in the inductor, there is no initial voltage at time t equals 0. As a result, right at time t equals 0, all the inductor current goes into the capacitor. By KCL, the capacitor current equals minus the inductor current I_0. Divide Equation 14.149b by C and apply to the parallel RLC Equation 14.91b at time t equals 0 to get Equation 14.150b. i_C equals $-I_0$, from the initial inductor current.

Series RLC

$$\left.\frac{di(t)}{dt}\right|_{t=0} = \frac{-v(0)}{L} = \frac{-V_0}{L} \quad (14.150a)$$

Equation 14.150a provides a value for the derivative of the loop current in the series RLC circuit at time t equals 0.

$$\frac{di}{dt} = \frac{-V_0}{L} \quad (14.151a)$$

Substitute Equation 14.151a into Equation 14.148a for $i'(0)$ to get Equation 14.152a.

$$\frac{-V_0}{L} = -(\sigma - j\omega_d)K_1 - (\sigma + j\omega_d)K_2 \quad (14.152a)$$

For series RLC, Equation 14.152a provides the second equation needed to calculate the two arbitrary constants K_1 and K_2.

$$K_1 = -K_2 \quad (14.153a)$$

Parallel RLC

$$\left.\frac{dv_C(t)}{dt}\right|_{t=0} = \frac{-i_L(0)}{C} = \frac{-I_0}{C} \quad (14.150b)$$

Equation 14.150b provides a value for the derivative of the voltage in the parallel RLC circuit at time t equals 0.

$$\frac{dv}{dt} = \frac{-I_0}{C} \quad (14.151b)$$

Substitute Equation 14.151b into Equation 14.148b for $v'(0)$ to get Equation 14.152b.

$$\frac{-I_0}{C} = -(\sigma - j\omega_d)K_1 - (\sigma + j\omega_d)K_2 \quad (14.152b)$$

For parallel RLC, Equation 14.152b provides the second equation needed to calculate the two arbitrary constants K_1 and K_2.

$$K_1 = -K_2 \quad (14.153b)$$

Substitute Equations 14.153a and 14.153b, the first set of simultaneous equations for finding the arbitrary constants K_1 and K_2, into Equations 14.152a and 14.152b to obtain Equations 14.154a and 14.154b. This is the simultaneous solution of the two equations.

Series RLC

$$\frac{-V_0}{L} = (\sigma - j\omega_d)K_2 - (\sigma + j\omega_d)K_2 \quad (14.154a)$$

Parallel RLC

$$\frac{-I_0}{C} = (\sigma - j\omega_d)K_2 - (\sigma + j\omega_d)K_2 \quad (14.154b)$$

In Equations 14.154a and 14.154b, gather the K_2 terms to obtain Equations 14.155a and 14.155b.

$$\frac{-V_0}{L} = K_2[(\sigma - j\omega_d) - (\sigma + j\omega_d)] \quad (14.155a)$$

$$\frac{-I_0}{C} = K_2[(\sigma - j\omega_d) - (\sigma + j\omega_d)] \quad (14.155b)$$

Series RLC **Parallel RLC**

In Equations 14.155a and 14.155b, remove the curved brackets to obtain Equations 14.156a and 14.156b.

$$\frac{-V_0}{L} = K_2[\sigma - j\omega_d - \sigma - j\omega_d] \quad (14.156a)$$

$$\frac{-I_0}{C} = K_2[\sigma - j\omega_d - \sigma - j\omega_d] \quad (14.156b)$$

Simplify terms in Equations 14.156a and 14.156b, resulting in Equations 14.157a and 14.157b.

$$\frac{-V_0}{L} = -K_2 2j\omega_d \quad (14.157a) \qquad \frac{-I_0}{C} = -K_2 2j\omega_d \quad (14.157b)$$

In Equations 14.157a and 14.157b, solve for K_2, as shown in Equations 14.158a and 14.158b.

$$K_2 = \frac{V_0}{2j\omega_d L} \quad (14.158a) \qquad K_2 = \frac{I_0}{2j\omega_d C} = 2j\omega_d \quad (14.158b)$$

Solve Equations 14.158a and 14.158b for K_1 using Equations 14.146a and 14.146b, which say that K_1 plus K_2 equals 0, as shown in Equations 14.159a and 14.159b.

$$K_1 = -\frac{V_0}{2j\omega_d L} \quad (14.159a) \qquad K_1 = -\frac{I_0}{2j\omega_d C} \quad (14.159b)$$

Substitute Equations 14.158a, 14.159a, 14.158b, and 14.159b into Equations 14.160a and 14.160b, to obtain Equations 14.161a and 14.161b.

$$i(t) = K_1 e^{-(\sigma - j\omega_d)t} + K_2 e^{-(\sigma + j\omega_d)t} \quad (14.160a) \qquad v(t) = K_1 e^{-(\sigma - j\omega_d)t} + K_2 e^{-(\sigma + j\omega_d)t} \quad (14.160b)$$

Series RLC **Parallel RLC**

$$i(t) = -\frac{V_0}{2j\omega_d L} e^{-(\sigma - j\omega_d)t} + \frac{V_0}{2j\omega_d L} e^{-(\sigma + j\omega_d)t} \quad (14.161a) \qquad v(t) = -\frac{I_0}{2j\omega_d C} e^{-(\sigma - j\omega_d)t} + \frac{I_0}{2j\omega_d C} e^{-(\sigma + j\omega_d)t} \quad (14.161b)$$

In the second terms of Equations 14.161a and 14.161b, multiply twice by minus 1 and gather the exponential term into brackets, as shown in Equations 14.162a and 14.162b.

Series RLC **Parallel RLC**

$$i(t) = -\frac{V_0}{2j\omega_d L} \left[e^{-(\sigma - j\omega_d)t} - e^{-(\sigma + j\omega_d)t}\right] \quad (14.162a) \qquad v(t) = -\frac{I_0}{2j\omega_d C} \left[e^{-(\sigma - j\omega_d)t} - e^{-(\sigma + j\omega_d)t}\right] \quad (14.162b)$$

In Equations 14.162a and 14.162b, partition the exponential term and the denominator term, as shown in Equations 14.163a and 14.163b.

Series RLC **Parallel RLC**

$$i(t) = -\frac{V_0}{\omega_d L} e^{-\sigma t} \frac{\left(e^{j\omega_d t} - e^{-j\omega_d t}\right)}{2j} \quad (14.163a) \qquad v(t) = -\frac{I_0}{\omega_d C} e^{-\sigma t} \frac{\left(e^{j\omega_d t} - e^{-j\omega_d t}\right)}{2j} \quad (14.163b)$$

$$e^{j\theta} - e^{-j\theta} = 2j \sin\theta \quad (14.164)$$

Apply Equation 14.164 to Equation 14.163a to replace the exponential terms, resulting in Equations 14.165a and 14.165b.

Series RLC **Parallel RLC**

$$i(t) = -\frac{V_0}{\omega_d L} e^{-\sigma t} \sin\omega_d t \quad (14.165a) \qquad v(t) = -\frac{I_0}{\omega_d C} e^{-\sigma t} \sin\omega_d t \quad (14.165b)$$

Equations 14.165a and 14.165b are the **underdamped** RLC series current response and parallel voltage responses.

For reference, Equations 14.166a, 14.166b, 14.167b, and 14.167b are provided here to relate to circuit quantities R, L, and C.

$$\sigma = \frac{R}{2L} \quad (14.166a)$$

$$\sigma = \frac{1}{2RC} \quad (14.166b)$$

$$\omega_d = \sqrt{\frac{1}{LC} - \left(\frac{R}{2L}\right)^2} \quad (14.167a)$$

$$\omega_d = \sqrt{\frac{1}{LC} - \left(\frac{1}{2RC}\right)^2} \quad (14.167b)$$

14.2.8.9 Calculation of K_1 and K_2 for Oscillatory Lossless RLC Natural Response

The oscillatory lossless natural response of series and parallel RLC is an ideal case that can be derived from the underdamped response case.

Start with Equations 14.163a and 14.163b from the previous underdamped response section, restated here as Equations 14.168a and 14.168b.

Series RLC

$$i(t) = -\frac{V_0}{\omega_d L} e^{-\sigma t} \frac{\left(e^{j\omega_d t} - e^{-j\omega_d t}\right)}{2j} \quad (14.168a)$$

Parallel RLC

$$v(t) = -\frac{I_0}{\omega_d C} e^{-\sigma t} \frac{\left(e^{j\omega_d t} - e^{-j\omega_d t}\right)}{2j} \quad (14.168b)$$

Using the conductance version of the characteristic roots Equation 14.169, generate ω_d in terms of conductance G, as shown in Equation 14.170b. ω_d is the discriminant of Equation 14.169.

$$s_1, s_2 = \frac{-G}{2C} \pm \sqrt{\left(\frac{G}{2C}\right)^2 - \frac{1}{LC}} \quad (14.169)$$

Series RLC

$$\omega_d = \sqrt{\frac{1}{LC} - \left(\frac{R}{2L}\right)^2} \quad (14.170a)$$

Parallel RLC

$$\omega_d = \sqrt{\left(\frac{G}{2C}\right)^2 - \frac{1}{LC}} \quad (14.170b)$$

In Equations 14.170a and 14.170b, when both resistance R and conductance G go to zero, ω_d become ω_0 as shown in Equation 14.171.

$$\omega_0 = \frac{1}{\sqrt{LC}} \quad (14.171)$$

In Equations 14.172a and 14.172b, sigma σ, also referred to as alpha α in other contexts, goes to zero in the ideal oscillatory lossless case.

$$\sigma = \frac{R}{2L} \quad (14.172a)$$

$$\alpha = \frac{G}{2C} = \frac{1}{2RC} \quad (14.172b)$$

With resistance R and conductance G equal to zero, Equations 14.168a and 14.168b reduce to Equations 14.173a and 14.173b.

$$i(t) = -\frac{V_0}{\omega_0 L} \frac{\left(e^{j\omega_0 t} - e^{-j\omega_0 t}\right)}{2j} \quad (14.173a)$$

$$v(t) = -\frac{I_0}{\omega_0 C} \frac{\left(e^{j\omega_0 t} - e^{-j\omega_0 t}\right)}{2j} \quad (14.173b)$$

Apply Equation 14.174 to Equations 14.173a and 14.173b to obtain Equations 14.175a and 14.175b.

$$e^{j\theta} - e^{-j\theta} = 2j \sin\theta \quad (14.174)$$

Series RLC

$$i(t) = -\frac{V_0}{\omega_0 L}\sin \omega_0 t \qquad (14.175a)$$

Parallel RLC

$$v(t) = -\frac{I_0}{\omega_0 C}e^{-\sigma t}\sin \omega_0 t \qquad (14.175b)$$

Equations 14.175a and 14.175b are the **lossless oscillatory** RLC series current response and parallel voltage response.

14.2.9 Series RLC ZIR Differential Equations – $v_C(t)$ and $i(t)$

This section derives series RLC differential equations, corresponding to the circuit shown in Figure 14.21, in terms of both the loop current $i(t)$ and the capacitor voltage $v_c(t)$, with derivations adjacent to one another for comparison.

Note that in series RLC i_C equals i_L equals i_R.

Write Kirchhoff's voltage law (KVL) around the series loop clockwise, starting with the inductor.

$$v_L + v_C + v_R = 0 \qquad (14.176)$$

The voltage across an inductor equals the inductance L times the time derivative of the current through the inductor, as shown in Equation 14.177.

$$v_L = L\frac{di_L(t)}{dt} \qquad (14.177)$$

Series RLC

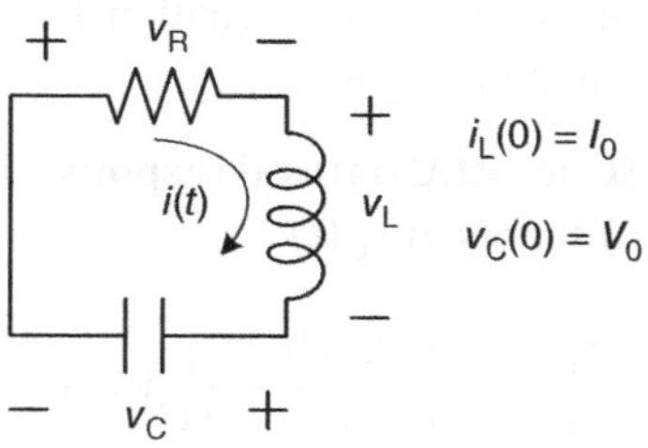

Figure 14.21 Series RLC circuit with two initial conditions.

The current through a capacitor equals the capacitance C times the time derivative of the capacitor voltage, as shown in Equation 14.178.

$$i_C = C\frac{dv_C(t)}{dt} \qquad (14.178)$$

$v_L + v_C + v_R = 0$ KVL

$v_L = L\frac{di_L(t)}{dt}$ $\qquad v = iR$

Substitute Equation 14.177 into KVL Equation 14.176 for the inductor voltage term and apply Ohm's law for the resistor voltage term, resulting in Equation 14.179.

$$L\frac{di_L(t)}{dt} + v_C(t) + Ri(t) = 0 \qquad (14.179)$$

From the KVL Equation 14.179 we can derive differential equations for the series RLC natural response in two different ways, one way in terms of the capacitor voltage $v_C(t)$ and the other way in terms of the loop current $i(t)$.

Series RLC in terms of $v_C(t)$

$$i_C = C\frac{dv_C(t)}{dt} \qquad (14.180a)$$

Substitute Equation 14.180a into Equation 14.179 for the inductor and resistor current terms, as shown in Equation 14.81a.

$$LC\frac{d^2v_C(t)}{dt^2} + v_C(t) + RC\frac{dv_C(t)}{dt} + = 0 \qquad (14.181a)$$

Series RLC in terms of $i(t)$

$$v_C(t) = \frac{1}{C}\int_0^t i_C(\tau)\,d\tau + V_C(0) \qquad (14.180b)$$

Substitute Equation 14.180b into Equation 14.179 for the $v_C(t)$ capacitor voltage term, as shown in Equation 14.181b.

$$L\frac{di(t)}{dt} + \frac{1}{C}\int_0^t i(\tau)\,d\tau + V_C(0) + i(t)R = 0 \qquad (14.181b)$$

Series RLC in terms of $v_C(t)$

In Equation 14.181a, rearrange terms to obtain Equation 14.182.

$$LC\frac{d^2v(t)}{dt^2} + RC\frac{dv_C(t)}{dt} + v_C(t) = 0 \tag{14.182}$$

Divide the entire Equation 14.182 by LC, resulting in Equation 14.185.

Series RLC natural response differential equation in terms of $v_C(t)$

$$\frac{d^2v_C(t)}{dt^2} + \frac{R}{L}\frac{dv_C(t)}{dt} + \frac{1}{LC}v_C(t) = 0 \tag{14.185}$$

Series RLC in terms of $i_L(t)$

In Equation 14.181b, take the time derivative d/dt of the whole equation. The constant term $V_C(0)$ goes to zero, resulting in Equation 14.183.

$$L\frac{d^2i(t)}{dt^2} + \frac{1}{C}i(t) + R\frac{di(t)}{dt} = 0 \tag{14.183}$$

In Equation 14.183, rearrange terms to obtain Equation 14.184.

$$L\frac{d^2i(t)}{dt^2} + R\frac{di(t)}{dt} + \frac{1}{C}i(t) = 0 \tag{14.184}$$

Divide the entire Equation 14.184 by L, resulting in Equation 14.186.

Series RLC natural response differential equation in terms of $i_L(t)$

$$\frac{d^2i(t)}{dt^2} + \frac{R}{L}\frac{di(t)}{dt} + \frac{1}{LC}i(t) = 0 \tag{14.186}$$

Notice how the coefficients in Equations 14.185 and 14.186 match.

14.2.10 Parallel RLC ZIR Differential Equations – $v(t)$ and $i_L(t)$

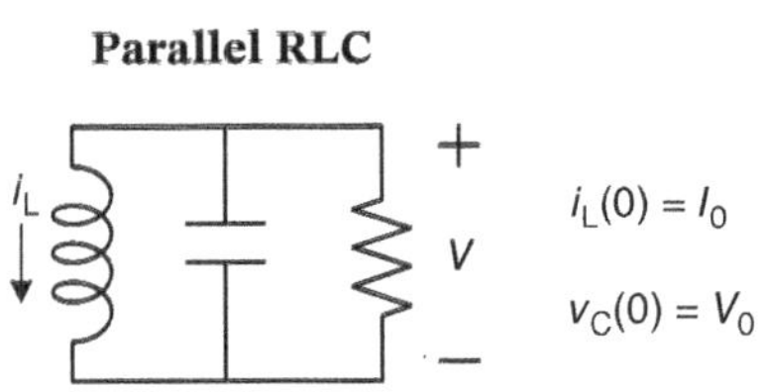

Figure 14.22 Parallel RLC circuit with two initial conditions.

This section derives parallel RLC differential equations, corresponding to the circuit shown in Figure 14.22, in terms of both the inductor current $i_L(t)$ and the voltage $v(t)$, with derivations adjacent to one another for comparison.

Note that v_C equals v_L equals v_R.

Write Kirchhoff's current law (KCL) summing the currents going in or out of either node, as shown in Equation 14.187.

$$i_C + i_R + i_L = 0 \tag{14.187}$$

$$i_L(t) = \frac{1}{L}\int_0^t v_L(\tau)\,d\tau + V_L(0) \tag{14.188}$$

Equation 14.188 gives the current through an inductor in terms of the voltage across the inductor.

$$i_C = C\frac{dv_C(t)}{dt} \tag{14.189}$$

Equation 14.189 states that the current through a capacitor equals the capacitance C times the time derivative of the capacitor voltage $v(t)$.

$$v_L = L\frac{di_L(t)}{dt} \tag{14.190}$$

Equation 14.190 shows the branch relationship for an ideal inductor; the voltage across an inductor equals the inductance L times the time derivative of the current through the inductor.

$$C\frac{dv(t)}{dt} + \frac{v(t)}{R} + i_L(t) = 0 \quad (14.191)$$

Substitute Equation 14.189 into Equation 14.187 for the capacitor term and apply Ohm's law for the resistor voltage term, to obtain Equation 14.191.

Parallel RLC in terms of $v(t)$

Substitute Equation 14.188 into Equation 14.191 for the inductor current term, as shown in Equation 14.192a.

$$C\frac{dv(t)}{dt} + \frac{v(t)}{R} = \frac{1}{L}\int_0^t v(\tau)\,d\tau + V(0) = 0 \quad (14.192a)$$

Take the time derivative d/dt of the whole Equation 14.192a, to obtain Equation 14.193. The constant term $V_C(0) = V_0$ goes to zero.

Parallel RLC in terms of $v(t)$

$$C\frac{d^2v(t)}{dt^2} + \frac{dv(t)}{Rdt} + \frac{1}{L}v(t) = 0 \quad (14.193)$$

Multiply the entire Equation 14.193 by L, resulting in Equation 14.195.

$$LC\frac{d^2v(t)}{dt^2} + \frac{L}{R}\frac{dv(t)}{dt} + v(t) = 0 \quad (14.195)$$

Parallel RLC natural response differential equation in terms of $v(t)$

Parallel RLC in terms of $i_L(t)$

Substitute Equation 14.190 into Equation 14.191 for the voltage in terms of inductor current, as shown in Equation 14.192b.

$$C\frac{d\left(L\frac{di_L(t)}{dt}\right)}{dt} + \frac{L\frac{di_L(t)}{dt}}{R} + i_L(t) = 0 \quad (14.192b)$$

Rearrange terms in Equation 14.192b to obtain Equation 14.194.

Parallel RLC in terms of $i_L(t)$

$$LC\frac{d^2i_L(t)}{dt^2} + \frac{L}{R}\frac{di_L(t)}{dt^2} + i_L(t) = 0 \quad (14.194)$$

Parallel RLC natural response differential equation in terms of $i_L(t)$

14.3 Second-order Damping Variables α and ω_0

14.3.1 Second-order Damping and Frequency Variables Overview

To work with second-order characteristics, such as damping and frequency, it is useful to introduce variables that will apply similarly to different electrical topologies, series and parallel RLC, as well as to non-electrical systems including mechanical systems.

This section shows series and parallel RLC analysis, and the four cases of natural response, in terms of two different damping variables, first damping coefficient α (alpha – also called damping constant) and then damping ratio ζ (zeta).

$$\frac{d^2v_C}{dt^2} + 2\alpha\frac{dv_C}{dt} + \omega_0^2 v_C = 0 \quad (14.196a)$$

$$s^2 + 2\alpha s + \omega_0^2 = 0 \quad (14.197a)$$

$$s_1, s_2 = -\alpha \pm \sqrt{\alpha^2 - \omega_0^2} \quad (14.198a)$$

$$\frac{d^2y(t)}{dt^2} + 2\zeta\omega_0\frac{dy(t)}{dt} + \omega_0^2 y(t) = 0 \quad (14.196b)$$

$$s^2 + 2\zeta\omega_0 s + \omega_0^2 = 0 \quad (14.197b)$$

$$s_1, s_2 = \omega_0(-\zeta \pm \sqrt{\zeta^2 - 1}) \quad (14.198b)$$

Equations 14.196a – 14.198b show second-order differential equations, characteristic equations, and characteristic roots in terms of two different damping variables, damping coefficient α, and damping ratio ζ.

This resonant frequency ω_0 is the same ω_0 that we see in frequency domain analysis. ω_0 is also called the undamped natural frequency, and it is angular or radian frequency.

14.3.2 Second-order Differential Equations with Variables α and ω_0

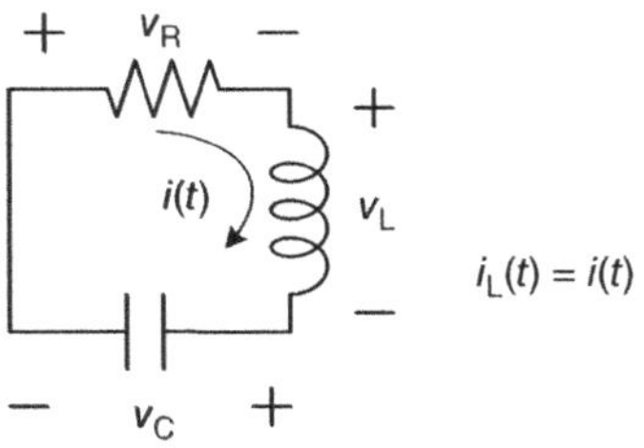

Figure 14.23a Series RLC circuit with one initial condition.

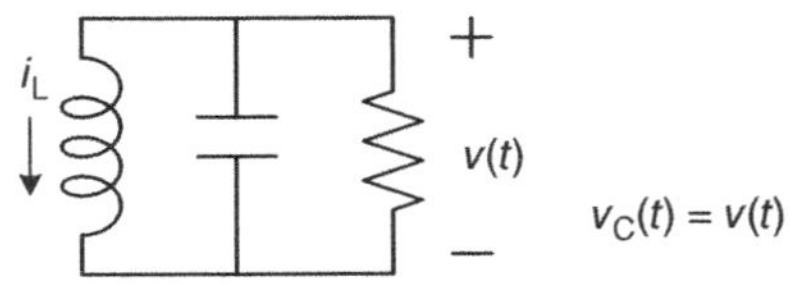

Figure 14.23b Parallel RLC circuit with one initial condition.

Start with the series and parallel RLC differential equations in terms of R, L, and C based on Figures 14.23a and 14.23b. This form, shown in Equations 14.199a, 14.199b, 14.200a, and 14.200b, has LC as the coefficient of the second derivative term.

Series RLC

$$LC\frac{d^2v_C(t)}{dt^2} + RC\frac{dv_C(t)}{dt} + v_C(t) = 0 \qquad (14.199a)$$

Series RLC natural response differential equation in terms of $v_C(t)$

$$LC\frac{d^2i(t)}{dt^2} + RC\frac{di(t)}{dt} + i(t) = 0 \qquad (14.200a)$$

Series RLC natural response differential equation in terms of $i(t)$

Parallel RLC

$$LC\frac{d^2i_L(t)}{dt^2} + \frac{L}{R}\frac{di_L(t)}{dt^2} + i_L(t) = 0 \qquad (14.199b)$$

Parallel RLC natural response differential equation in terms of $i_L(t)$

$$LC\frac{d^2v(t)}{dt^2} + \frac{L}{R}\frac{dv(t)}{dt} + v(t) = 0 \qquad (14.200b)$$

Parallel RLC natural response differential equation in terms of $v(t)$

Divide all the series and parallel RLC differential Equations 14.199a, 14.199b, 14.200a, and 14.200b, by LC to make the coefficient of the second derivative term equal to 1, as shown in Equations 14.201a, 14.201b, 14.202a, and 14.202b.

Series RLC

$$\frac{d^2v_C(t)}{dt^2} + \frac{R}{L}\frac{dv_C(t)}{dt} + \frac{1}{LC}v_C(t) = 0 \qquad (14.201a)$$

Series RLC natural response differential equation in terms of $v_C(t)$

$$\frac{di^2(t)}{dt^2} + \frac{R}{L}\frac{di(t)}{dt} + \frac{1}{LC}i(t) = 0 \qquad (14.202a)$$

Series RLC natural response differential equation in terms of $i_L(t)$

Parallel RLC

$$\frac{d^2i_L(t)}{dt^2} + \frac{1}{RC}\frac{di_L(t)}{dt^2} + \frac{1}{LC}i_L(t) = 0 \qquad (14.201b)$$

Parallel RLC natural response differential equation in terms of $i_L(t)$

$$\frac{dv^2(t)}{dt^2} + \frac{1}{RC}\frac{dv(t)}{dt} + \frac{1}{LC}v(t) = 0 \qquad (14.202b)$$

Parallel RLC natural response differential equation in terms of $v_C(t)$

Notice how the current in Equation 14.202a and the voltage in Equation 14.202b do not have subscripts. The reason is that in a series RLC, in Equation 14.202a, the current is the same current in all components; there is only one current in the circuit. There is no need to subscript the current for any particular component in the circuit. Similarly, in parallel RLC, in Equation 14.202b, there is only one voltage for the entire circuit; there is no need to identify the voltage for any specific component in the circuit.

One advantage of the damping and frequency variables, over expressing RLC response in terms of R, L, and C, is that there is a common set of differential equation coefficients. The common coefficients apply to series and parallel RLC topologies, as well as to different choices of circuit variable currents or voltages.

Introduce variables α and ω_0 to the second-order time domain response analysis equations, as shown in Equation 14.203a, for series, and Equation 14.203b for parallel.

Series RLC **Parallel RLC**

Damping Coefficient α

$$\alpha = \frac{R}{2L} \tag{14.203a}$$

$$\alpha = \frac{G}{2C} = \frac{1}{2RC} \tag{14.203b}$$

α, in Equations 14.203a and 14.203b, is the damping coefficient or damping constant. α (alpha) is also called the exponential damping coefficient, neper frequency, or exponential attenuation rate. These quantities α also appear in the frequency domain analyses.

In characteristic roots Equations 14.204a and 14.204b, the damping coefficient α is marked by a dashed line rectangular box.

Series RLC **Parallel RLC**

$$s_1, s_2 = \boxed{\frac{-R}{2L}} \pm \sqrt{\left(\frac{R}{2L}\right)^2 - \frac{1}{LC}} \tag{14.204a}$$

$$s_1, s_2 = \boxed{\frac{-G}{2C}} \pm \sqrt{\left(\frac{G}{2C}\right)^2 - \frac{1}{LC}} \tag{14.204b}$$

In Equations 14.204a and 14.204b, the characteristic roots equations in terms of R, L, and C, the α (alpha) variable corresponds to the purely real term of the roots. The discriminant, the quantity under the radical sign, can also contribute to damping, depending upon its value, as we saw in the four cases.

Series RLC **Parallel RLC**

Resonant Frequency ω_0

$$\omega_0 = \frac{1}{\sqrt{LC}} \tag{14.205a}$$

$$\omega_0 = \frac{1}{\sqrt{LC}} \tag{14.205b}$$

ω_0 is the resonant frequency, the same equation for both series and parallel RLC, as indicated by identical Equations 14.205a and 14.205b. This is the same ω_0 that frequency domain analysis uses. ω_0 is also called the undamped natural frequency, and it is angular or radian frequency.

14.3.3 Alpha and Omega Variables for 2nd-order Characteristic Roots

$$\frac{d^2i(t)}{dt^2} + \frac{R}{L}\frac{di(t)}{dt} + \frac{1}{LC}i(t) = 0 \tag{14.206a}$$

$$\frac{d^2v(t)}{dt^2} + \frac{1}{RC}\frac{dv(t)}{dt} + \frac{1}{LC}v(t) = 0 \tag{14.206b}$$

Start the RLC analyses with Equation 14.206a, the series RLC differential equation for current and Equation 14.206b, the parallel RLC differential equation for voltage.

From Equations 14.203a and 14.203b, and Equations 14.205a and 14.205b, substitute variables α and ω_0 into Equations 14.206a and 14.206b, obtaining Equations 14.207a and 14.207b in terms of α and ω_0.

Series RLC

$$\frac{d^2 i(t)}{dt^2} + 2\alpha \frac{di(t)}{dt} + \omega_0^2 i(t) = 0 \tag{14.207a}$$

Parallel RLC

$$\frac{d^2 v(t)}{dt^2} + 2\alpha \frac{dv(t)}{dt} + \omega_0^2 v(t) = 0 \tag{14.207b}$$

Both series and parallel RLC differential equations in Equations 14.207a and 14.207b have the same coefficients in this α and ω format.

The α and ω variable substitution works the same for the other set of differential equations, Equations 14.201a and 14.201b, using capacitor voltage as the state variable in the series RLC and inductor current as the state variable in the parallel RLC.

Continuing the differential equation solution process, substitute the presumed exponential solution Ke^{st} into Equations 14.207a and 14.207b. Because the differential equations for the series circuit in Equation 14.207a and the parallel circuit Equation 14.207b have the same coefficients, we can combine series and parallel into one resulting Equation 14.208.

$$\frac{d^2 (Ke^{st})}{dt^2} + 2\alpha \frac{d(Ke^{st})}{dt} + \omega_0^2 (Ke^{st}) = 0 \tag{14.208}$$

$$s = \sigma + j\omega \tag{14.209}$$

In Equation 14.208, the variables $i(t)$ for the series circuit and $v(t)$ for the parallel circuit don't appear. We can treat the two differential equation versions, series and parallel, as one version for several steps of this solution process.

In Equation 14.208, perform the differentiations on Ke^{st}, as shown in Equation 14.210.

$$s^2 Ke^{st} + 2\alpha s Ke^{st} + \omega_0^2 Ke^{st} = 0 \tag{14.210}$$

In Equation 14.210, factor out the characteristic polynomial from Ke^{st}, to obtain Equation 14.211.

$$\left(s^2 + 2\alpha s + \omega_0^2\right) Ke^{st} = 0 \tag{14.211}$$

In Equation 14.211, the product means that two solutions are possible to make the equation equal to zero. But the solution Ke^{st} equals zero is a trivial solution. Consider the case where the term in the brackets equals zero.

$$s^2 + 2\alpha s + \omega_0^2 = 0 \tag{14.212}$$

Equation 14.212 is the characteristic polynomial, in terms of α and ω, for both series and parallel undriven RLC circuits.

$$ax^2 + bx + c = 0 \Rightarrow x = \frac{-b \pm \sqrt{b^2 - 4ac}}{2a} \tag{14.213}$$

Use the quadratic formula Equation 14.213 to solve Equation 14.212 for the two natural frequencies s_1 and s_2 in Equations 14.214a and 14.214b.

$$s_1 = -\alpha + \sqrt{\alpha^2 - \omega_0^2} \tag{14.214a}$$

$$s_2 = -\alpha - \sqrt{\alpha^2 - \omega_0^2} \tag{14.214b}$$

Equations 14.214a and 14.214b are the solutions to the RLC quadratic characteristic frequency Equation 14.212. Equations 14.214a and 14.214b are the RLC natural frequencies in terms of α and ω.

These characteristic roots s_1 and s_2, in terms of α and ω, apply to both series and parallel RLC circuits. Substitute Equations 14.214a and 14.214b into the general solution Equations 14.215a and 14.215b, to obtain Equations 14.216a and 14.216b.

Series RLC

$$i(t) = K_1 e^{s_1 t} + K_2 e^{s_2 t} \quad (14.215a)$$

Equation 14.215a, the general solution for the series RLC, is solved for $i(t)$, the current throughout the entire series loop.

Parallel RLC

$$v(t) = K_1 e^{s_1 t} + K_2 e^{s_2 t} \quad (14.215b)$$

Equation 14.215b, the general solution for the parallel RLC, is solved for $v(t)$, the voltage across all three circuit elements R, L, and C.

The exponent st must be dimensionless, which implies that s must have units of $1/t$ or frequency.

Series $$i(t) = K_1 e^{\left(-\alpha+\sqrt{\alpha^2-\omega_0^2}\right)t} + K_2 e^{\left(-\alpha-\sqrt{\alpha^2-\omega_0^2}\right)t} \quad (14.216a)$$

Parallel $$v(t) = K_1 e^{\left(-\alpha+\sqrt{\alpha^2-\omega_0^2}\right)t} + K_2 e^{\left(-\alpha-\sqrt{\alpha^2-\omega_0^2}\right)t} \quad (14.216b)$$

14.3.4 Four Cases of Characteristic Roots in Terms of α

Variations in the values of R, L, and C lead to various values of alpha α and omega-nought ω_0. Because of the subtraction under the radical, variations in value can lead to four distinct cases or classes of roots.

This subsection provides natural response graphs, of series RLC current versus time, and the complex plane locations are plotted for the four cases or classes of roots. The alpha α term, in the characteristic roots, gets plotted along the horizontal sigma σ real axis of the complex plane.

Here is a summary of the four cases of characteristic root values. The graphs and equations correspond to current $i(t)$ in series RLC. Root locations on graphs are approximate and unquantified.

Case 1: Overdamped

$$\alpha > \omega_0 \quad (14.217)$$ Two real and unequal roots

$$\left.\begin{matrix} s_1 \\ s_2 \end{matrix}\right\} = -\alpha \pm \sqrt{\alpha^2 - \omega_0^2} \quad (14.218)$$

$\alpha^2 - \omega_0^2 > 0$

Series	Parallel
$(RC)^2 - 4LC > 0$	$(GL)^2 - 4LC > 0$

The two roots s_1 and s_2 are real, unequal, and negative.

$$i(t) = k_1 e^{s_1 t} + k_2 e^{s_2 t} \quad (14.219)$$

$$i(t) = k_1 e^{\left(-\alpha+\sqrt{\alpha^2-\omega_0^2}\right)t} + k_2 e^{\left(-\alpha-\sqrt{\alpha^2-\omega_0^2}\right)t} \quad (14.220)$$

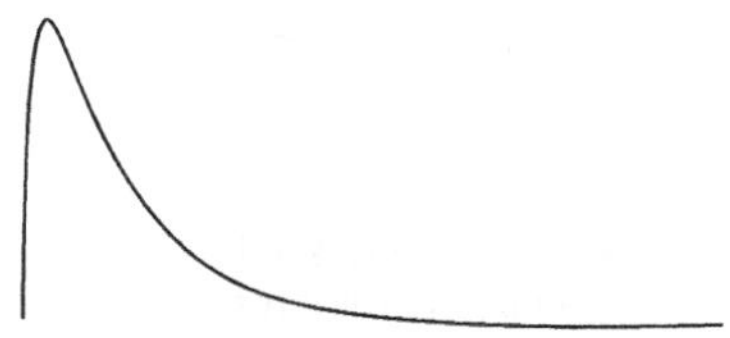

Figure 14.24 Series RLC overdamped natural response current.

Plotted in Figure 14.24, the natural response is the sum of two damped exponentials in Equation 14.220.

$$s_1, s_2 = -\alpha \pm \sqrt{\alpha^2 - \omega_0^2} \quad (14.221)$$

In the overdamped case α^2 is greater than $\omega_0{}^2$. Because the quantity under the radical sign is positive, and the two characteristic roots s_1 and s_2 are real valued, there is no imaginary component, and the two roots lie on the negative real axis of the complex plane, as shown in Figure 14.25.

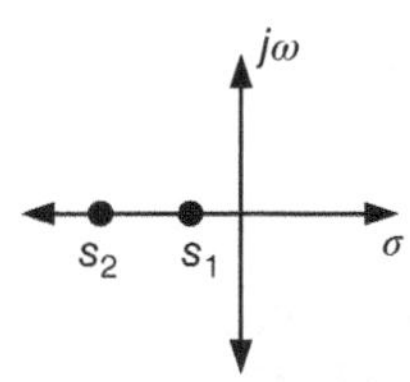

Figure 14.25 Series RLC overdamped root locations in the complex plane.

Case 2: Critically damped Two real, equal, and negative roots

The two roots s_1 and s_2 are real, equal, and negative.

$$\alpha = \omega_0 \tag{14.222}$$

Series	Parallel
$(RC)^2 - 4LC = 0$	$(GL)^2 - 4LC = 0$

$$s_1 = s_2 = -\alpha \tag{14.223}$$

The response is a decaying exponential in Equation 14.224 and plotted in Figure 14.26.

$$i_L(t) = (k_1 + k_2 t)e^{-\alpha t} \tag{14.224}$$

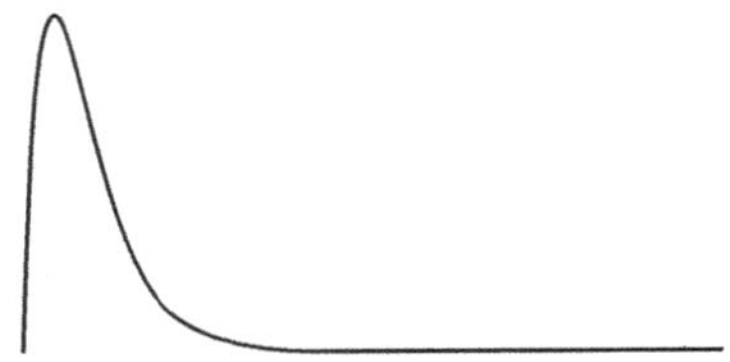

Figure 14.26 Series RLC critically damped natural response current.

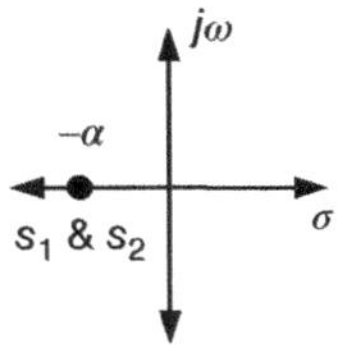

Figure 14.27 Series RLC critically damped root locations in the complex plane.

In the critically damped case α^2 equals $\omega_0{}^2$ and the quantity under the radical sign in Equation 14.216a equals zero. The two characteristic roots s_1 and s_2 are both $-\alpha$. There is no imaginary component to the two roots, and they lie on the negative real axis of the complex plane, as shown in Figure 14.27.

Case 3: Underdamped

$$\alpha < \omega_0 \tag{14.225}$$

Two complex, unequal roots

$$\left.\begin{matrix} s_1 \\ s_2 \end{matrix}\right\} = -\alpha \pm \sqrt{\alpha^2 - \omega_0^2} \tag{14.226}$$

Series	Parallel
$(RC)^2 - 4LC < 0$	$(GL)^2 - 4LC < 0$

Figure 14.28 Series RLC underdamped natural response current.

$$\left.\begin{matrix} s_1 \\ s_2 \end{matrix}\right\} = -\alpha \pm \sqrt{-1\left(\omega_0^2 - \alpha^2\right)} \tag{14.227}$$

Because ω_0 is larger than α, as shown in Equation 14.225, create a new variable ω_d, in Equation 14.229, damped resonant frequency, so that the quantity under the radical sign in Equation 14.226 can be positive.

$$\left.\begin{matrix} s_1 \\ s_2 \end{matrix}\right\} = -\alpha \pm j\sqrt{\omega_0^2 - \alpha^2} \tag{14.228}$$

$$\omega_d^2 \equiv \omega_0^2 - \alpha^2 \tag{14.229}$$

$$\left.\begin{matrix} s_1 \\ s_2 \end{matrix}\right\} = \left\{\begin{matrix} -\alpha + j\omega_d \\ -\alpha - j\omega_d \end{matrix}\right. \tag{14.230}$$

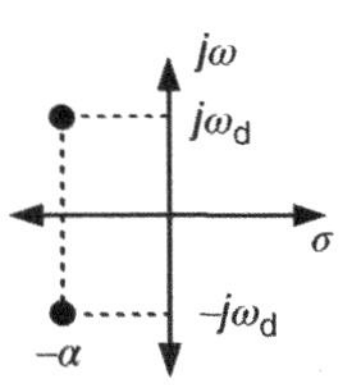

Figure 14.29 Series RLC underdamped root locations in the complex plane.

The two roots s_1 and s_2, in Equation 14.228, are complex, unequal, and complex conjugates. In the complex plane, the two roots lie in the negative real half plane. In the vertical imaginary dimension, the terms plus and minus $j\omega_d$ lie above and below the real axis, as shown in Figure 14.29.

$$i_L(t) = K_1 e^{(-\alpha + j\omega_d)t} + K_2 e^{(-\alpha - j\omega_d)t} \quad (14.231)$$

$$i_L(t) = ke^{-\alpha t}\cos(\omega_d t + \theta) \quad (14.232)$$

The two roots, s_1 and s_2, are expressed in terms of ω_d in Equation 14.230 and in the exponent of Equation 14.231.

The response is a decaying oscillatory exponential. K_1 and K_2, in Equation 14.231, can be combined into one coefficient k in Equation 14.232 and plotted in Figure 14.28.

Case 4: Lossless oscillatory

$\alpha = 0$ Two imaginary and unequal roots

The lossless oscillatory Case 4 is an ideal limit of the underdamped Case 3. In Case 4 the damping goes to zero (lossless) and the signal becomes a pure sinusoid (oscillatory) rather than the damped sinusoid in the underdamped Case 3.

The two roots s_1 and s_2 are imaginary and unequal.

$$s_1 = j\omega_0 \quad (14.233)$$

$$s_2 = -j\omega_0 \quad (14.234)$$

Series $R = 0$ Parallel $G = 0$

For the ideal lossless oscillatory case, the α term in the exponent of Equation 14.231 for Case 3, goes to zero and the ω_d term becomes ω_0, as shown in Equation 14.235.

$$i(t) = K_1 e^{j\omega_0 t} + K_2 e^{-j\omega_0 t} \quad (14.235)$$

$$\cos\theta = \frac{e^{j\theta} + e^{-j\theta}}{2} \quad (14.236)$$

$$i_L(t) = k\cos(\omega_0 t + \theta) \quad (14.237)$$

When α equals zero, then ω_d equals ω_0. That is, when the damping goes to zero, the response is lossless oscillatory.

The Case 4 response is an oscillation in Equation 14.237 and Figure 14.30.

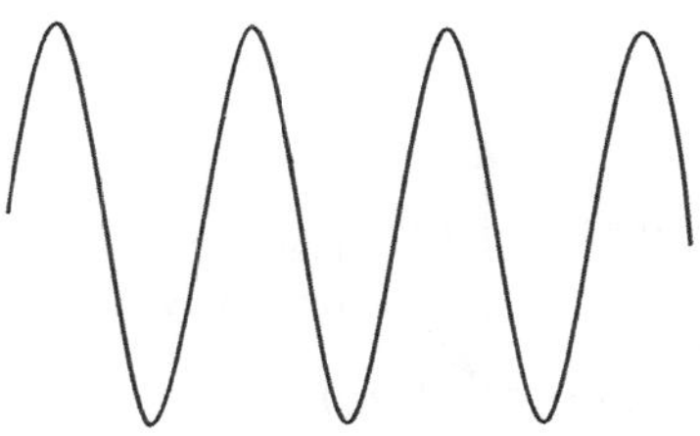

Figure 14.30 Series RLC lossless oscillatory natural response current.

In the ideal case where there is no damping at all, the two roots s_1 and s_2 lie on the vertical imaginary axis with no damping in the horizontal real dimension as shown in Figure 14.31. The current becomes a pure sinusoid.

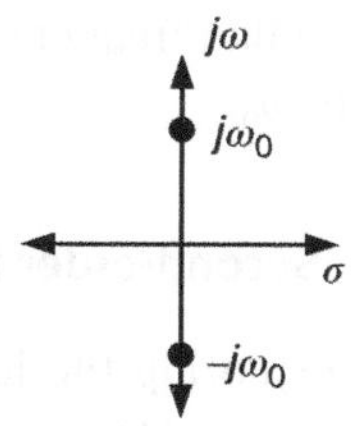

Figure 14.31 Series RLC lossless oscillatory root locations in the complex plane.

14.3.5 Q and ω_d

Expressing the second-order RLC −3 dB cutoff frequencies, ω_1 and ω_2, in terms of Q, connects the RLC circuit time domain response with the frequency domain response.

$$s_1, s_2 = -\alpha \pm \sqrt{\alpha^2 - \omega_0^2} \quad (14.238)$$

From Equations 12.214a and 12.214b, Equation 14.238 provides the characteristic roots for series and parallel RLC in terms of α and ω_0.

For cutoff frequencies, ω_1 and ω_2 in terms of Q, we have Equations 14.239a and 14.239b for series and Equations 14.240a and 14.240b for parallel. Combine these equations, with minus/plus sign notation, in Equation 14.241.

Series RLC

$$\omega_1 = -\frac{\omega_0}{2Q} + \omega_0\sqrt{1 + \frac{1}{4Q^2}} \quad (14.239a)$$

$$\omega_2 = \frac{\omega_0}{2Q} + \omega_0\sqrt{1 + \frac{1}{4Q^2}} \quad (14.239b)$$

Parallel RLC

$$\omega_1 = -\frac{\omega_0}{2Q} + \omega_0\sqrt{1 + \frac{1}{4Q^2}} \quad (14.240a)$$

$$\omega_2 = \frac{\omega_0}{2Q} + \omega_0\sqrt{1 + \frac{1}{4Q^2}} \quad (14.240b)$$

$$\omega_1, \omega_2 = \mp\frac{\omega_0}{2Q} + \omega_0\sqrt{1 + \frac{1}{4Q^2}} \quad (14.241)$$

$$\alpha = \frac{\omega_0}{2Q} \quad (14.242)$$

$$\omega_d^2 \equiv \omega_0^2 - \alpha^2 \quad (14.243)$$

$$\omega_d = \sqrt{\omega_0^2 - \alpha^2} \quad (14.244)$$

$$\omega_d = \omega_0\sqrt{1 - \frac{1}{4Q^2}} \quad (14.245)$$

Comparing Equation 14.238 and Equation 14.241, we see that α equals ω_0 over $2Q$, as shown in Equation 14.242.

In Equations 14.243 and 14.244, ω_d is defined in terms of ω_0 and α.

Comparing Equation 14.241 to Equation 14.238 and Equation 14.244, we express ω_d in terms of Q in Equation 14.245.

Equation 14.245 expresses ω_d, the damped radian frequency, in terms of quality factor Q. Resistance does not appear in this equation, which applies to both series and parallel RLC circuits.

In the frequency domain, we know that higher Q means narrower bandwidth and less loss, less energy dissipated in an RLC resistor. From Equation 14.245 we see that higher Q means less damping, lower values of damped radian frequency ω_d, in the time domain response of the RLC circuit. In the limit, as Q goes to infinity, ω_d goes to ω_0.

14.3.6 Second-order Root Locus - α and ω_0

Further exploring the locations of the characteristic roots, the natural frequencies of RLC natural response, in the complex sigma σ and j-omega $j\omega$ plane, examine the whole path of possible root locations, the root locus, for all values of damping.

The complex plane in Figure 14.32 shows a semicircle corresponding to the range of underdamped characteristic roots s_1 and s_2.

The coordinate pairs in Figure 14.32, $(-\alpha, \omega_d)$ and $(-\alpha, -\omega_d)$, correspond to underdamped characteristic root locations.

In Figure 14.32, the coordinate pairs $(0, \omega_0)$ and $(0, -\omega_0)$ correspond to the s_1 and s_2 root locations where damping goes to zero and underdamped becomes lossless oscillatory response.

The duplicate characteristic roots pairs, at $(\omega_0, 0)$ in Figure 14.32, correspond to critically damped, where the oscillatory behavior stops completely.

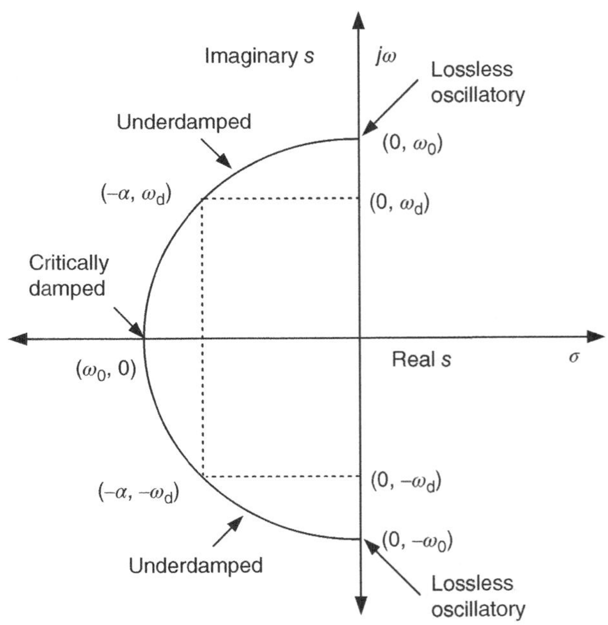

Figure 14.32 RLC natural response root locus locations in the complex plane.

Figure 14.33 is the same complex plane as Figure 14.32, partially cut away and focused on the negative real axis to clarify the overdamped region.

$$s_1 = -\alpha + \sqrt{\alpha^2 - \omega_0^2} \quad (14.246a)$$

$$s_2 = -\alpha - \sqrt{\alpha^2 - \omega_0^2} \quad (14.246b)$$

$$\left(-\alpha + \sqrt{\alpha^2 - \omega_0^2}, 0\right) \quad (14.247a)$$

$$\left(-\alpha - \sqrt{\alpha^2 - \omega_0^2}, 0\right) \quad (14.247b)$$

The coordinate in Equation 14.247a corresponds to overdamped characteristic root s_1 in Equation 14.246a. The coordinate in Equation 14.247b corresponds to overdamped characteristic root s_2 in Equation 14.246b.

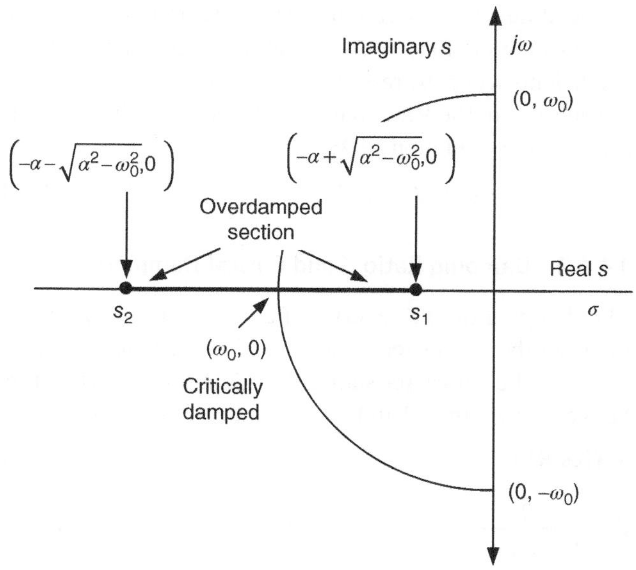

Figure 14.33 RLC natural response overdamped and critically damped root locations.

In Figure 14.33, on the horizontal axis, moving away from the critically damped coordinate location in both directions, going past s_1 and s_2, damping goes to infinity.

14.4 Second-order Damping Ratio ζ - Zeta

14.4.1 Introduction to RLC Damping Ratio ζ

The damping ratio ζ (zeta) is a ratio between the circuit resistance and the resistance R_{crit} at critical damping, as shown in Equations 14.248a and 14.248b.

Series RLC

$$\zeta = \frac{R}{R_{\text{CRIT-Series}}} \quad (14.248a)$$

Parallel RLC

$$\zeta_{\text{par}} = \frac{R_{\text{CRIT-parallel}}}{R} \quad (14.248b)$$

$$\frac{d^2 v_c}{dt^2} + \frac{R}{L}\frac{dv_c}{dt} + \frac{1}{LC}v_c = 0 \quad (14.249a) \qquad \frac{d^2 i_L}{dt^2} + \frac{1}{RC}\frac{di_L}{dt} + \frac{1}{LC}i_L = 0 \quad (14.249b)$$

$$\frac{d^2 y(t)}{dt^2} + 2\zeta\omega_0\frac{dy(t)}{dt} + \omega_0^2 y(t) = 0 \quad (14.250)$$

$$s^2 + 2\zeta\omega_0 s + \omega_0^2 = 0 \quad (14.251)$$

The differential equations for series and parallel RLC, shown in Equations 14.249a and 14.249b, share common coefficients when expressed using ζ (zeta).

The roots of the RLC characteristic equation in terms of ζ are shown in Equation 14.252.

$$s_1, s_2 = \omega_0(-\zeta \pm \sqrt{\zeta^2 - 1}) \tag{14.252}$$

> The four natural response cases with zeta are:
> $\zeta > 1$ overdamped
> $\zeta = 1$ critically damped
> $\zeta < 1$ underdamped
> $\zeta = 0$ lossless oscillatory

14.4.2 Damping Ratio ζ and Critical Damping

The damping ratio ζ (zeta), (different from the damping factor alpha a from the previous sections), is a ratio between the circuit resistance and the resistance R_{crit} at critical damping, as shown in Equations 14.253a and 14.253b. The critical resistance R_{crit}, Equation 14.254a for series and Equation 14.254b for parallel, is the boundary between overdamped and underdamped RLC response.

Series RLC

$$\zeta = \frac{R}{R_{\text{CRIT-Series}}} \tag{14.253a}$$

$$R_{\text{CRIT-ser}} = 2\sqrt{\frac{L}{C}} \tag{14.254a}$$

$$\zeta_{\text{ser}} = \frac{R}{2\sqrt{\frac{L}{C}}} = \frac{R}{2}\sqrt{\frac{C}{L}} \tag{14.255a}$$

$$\omega_0 = \frac{1}{\sqrt{LC}} \tag{14.256}$$

Parallel RLC

$$\zeta_{\text{par}} = \frac{R_{\text{CRIT-parallel}}}{R} \tag{14.253b}$$

$$R_{\text{CRIT-par}} = \frac{1}{2}\sqrt{\frac{L}{C}} \tag{14.254b}$$

$$\zeta_{\text{par}} = \frac{1}{2R}\sqrt{\frac{L}{C}} \tag{14.255b}$$

Equations 14.255a and 14.255b define ζ in terms of R, L, and C.

The zeta ζ variable relates directly to the four damping cases, with the ω factor included in ζ. By comparison, the alpha α variable compares to ω_0, the resonant frequency in Equation 14.256, to determine the four natural response cases.

The quantity $2\zeta\omega_0$ is the coefficient of the first derivative, in both the series and parallel RLC differential equations in terms of ζ, in Equation 14.250 in the previous introduction section.

To derive the quantity $2\zeta\omega_0$, introduce Equation 14.256, for ω_0, into Equations 14.255a and 14.255b to generate Equations 14.257a and 14.257b.

Series RLC

$$2\zeta_{ser}\omega_0 = 2\frac{R}{2}\sqrt{\frac{C}{L}}\left(\frac{1}{\sqrt{LC}}\right) \tag{14.257a}$$

Parallel RLC

$$2\zeta_{par}\omega_0 = 2\frac{1}{2R}\sqrt{\frac{L}{C}}\left(\frac{1}{\sqrt{LC}}\right) \tag{14.257b}$$

In Equations 14.257a and 14.257b, cancel the "2" terms, to obtain Equations 14.258a and 14.258b.

$$2\zeta_{ser}\omega_0 = R\sqrt{\frac{C}{L}}\left(\frac{1}{\sqrt{LC}}\right) = \frac{R}{L} \tag{14.258a}$$

$$2\zeta_{par}\omega_0 = \frac{1}{R}\sqrt{\frac{L}{C}}\left(\frac{1}{\sqrt{LC}}\right) = \frac{1}{RC} \tag{14.258b}$$

Simplify Equations 14.258a and 14.258b, resulting in Equations 14.259a and 14.259b.

$$2\zeta_{ser}\omega_0 = \frac{R}{L} \quad (14.259a) \qquad 2\zeta_{par}\omega_0 = \frac{1}{RC} \quad (14.259b)$$

The right-hand sides of Equations 14.259a and 14.259b also match the coefficients in Equations 14.249a and 14.249b in the previous introduction section.

In series RLC, more resistance means more damping. In parallel RLC, less resistance means more damping. The damping ratio zeta ζ is naturally different for series and parallel RLC. In both definitions of ζ, series in Equation 12.259a, and parallel in Equation 14.259b, zero damping makes ζ equal to zero.

The ratio zeta ζ can also be expressed in terms of the damping factor α from previous sections, as shown in Equation 14.260.

$$\zeta = \frac{\alpha}{\omega_0} \quad (14.260)$$

14.4.3 RLC Differential Equation Analysis in Terms of ζ

For RLC differential equation analysis, a general notation can be used and applied to both series and parallel RLC including those in Figures 14.34a,b. The function $y(t)$, shown in Equation 14.261,* can refer to either $i(t)$ or $v(t)$ in an RLC circuit, where the "a" terms, a_2, a_1, and a_0, are constant coefficients. (This analysis approach can also apply more generally than RLC circuits.)

$$a_2\frac{d^2y(t)}{dt^2} + a_1\frac{dy(t)}{dt} + a_0y(t) = 0 \quad (14.261)$$

Divide Equation 14.261 by a_2 to obtain Equation 14.262,* where the coefficient of the leading second derivative term equals 1.

$$\frac{d^2y(t)}{dt^2} + \frac{a_1}{a_2}\frac{dy(t)}{dt} + \frac{a_0}{a_2}y(t) = 0 \quad (14.262)$$

Define new variables, in Equations 14.263a and 14.263b,* in terms of these "a" coefficients.

$$2\zeta\omega_0 = \frac{a_1}{a_2} \quad (14.263a) \qquad \omega_0^2 = \frac{a_0}{a_2} \quad (14.263b)$$

ζ (zeta) in Equation 14.263a is the damping ratio. ω_0 in Equation 14.263b is called the undamped natural frequency. ω_0 is also the resonant frequency. In some textbooks and discussions, undamped natural frequency might be called ω_n.

Equating Equation 14.263a to coefficients in the series and parallel RLC differential Equations 14.264a and 14.264b, solve for ζ (zeta) in Equations 14.267a and 14.267b as follows.

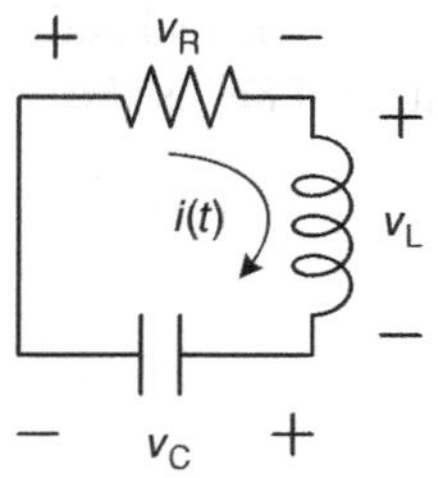

Figure 14.34a Series RLC undriven circuit schematic.

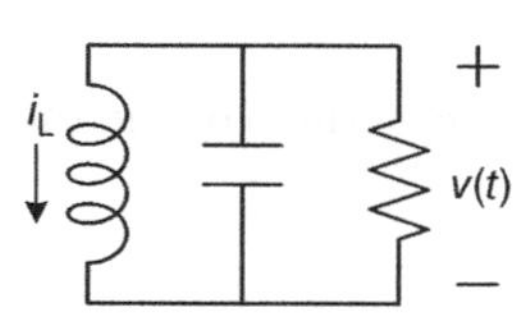

Figure 14.34b Parallel RLC undriven circuit schematic.

---

* The coefficient notation a_2, a_2, a_0, and the function notation $y(t)$, in Equations 14.261, 14.262, and 14.263, appear in "The Analysis and Design of Linear Circuits" fourth edition, by Roland E. Thomas and Albert J. Rosa, 2004, John Wiley & Sons, Inc., on page 326 in step response Section 7.7.

Series RLC

$$\frac{d^2i(t)}{dt^2} + \frac{R}{L}\frac{di(t)}{dt} + \frac{1}{LC}i(t) = 0 \tag{14.264a}$$

$$\omega_0^2 = \frac{a_0}{a_2} = \frac{1}{LC} \tag{14.265}$$

$$\omega_0 = \frac{1}{\sqrt{LC}} \tag{14.266}$$

$$2\zeta_{\text{ser}}\omega_0 = \frac{a_1}{a_2} = \frac{R}{L} \tag{14.267a}$$

Parallel RLC

$$\frac{d^2i_L}{dt^2} + \frac{1}{RC}\frac{di_L}{dt} + \frac{1}{LC}i_L = 0 \tag{14.264b}$$

In Equation 14.263b, the resonant frequency ω_0 squared equals 1 over LC for both series and parallel RLC, as shown in Equation 14.265.

The resonant frequency ω_0 equals 1 over the square root of LC for both series and parallel RLC, as shown in Equation 14.266.

$$2\zeta_{\text{par}}\omega_0 = \frac{a_1}{a_2} = \frac{1}{RC} \tag{14.267b}$$

Solve Equations 14.267a and 14.267b for ζ, as shown in Equations 14.268a and 14.268b.

$$\zeta_{\text{ser}} = \frac{R}{2L\omega_0} = \frac{R\sqrt{LC}}{2L\omega_0} \tag{14.268a}$$

$$\zeta_{\text{par}} = \frac{1}{2RC\omega_0} = \frac{\sqrt{LC}}{2RC} \tag{14.268b}$$

Substitute Equation 14.266 for ω_0 in Equations 14.268a and 14.268b, resulting in expressions for ζ in terms of R, L, and C, for both series and parallel RLC, in Equations 14.269a and 14.269b.

$$\zeta_{\text{ser}} = \frac{R}{2}\sqrt{\frac{C}{L}} \tag{14.269a}$$

$$\zeta_{\text{par}} = \frac{1}{2R}\sqrt{\frac{L}{C}} \tag{14.269b}$$

Substituting Equations 14.267a and 14.267b into Equation 14.262 results in the general form of the second-order differential equation in terms of ζ and ω_0 in Equation 14.270.

$$\frac{d^2y(t)}{dt^2} + 2\zeta\omega_0\frac{dy(t)}{dt} + \omega_0^2y(t) = 0 \tag{14.270}$$

Following the standard analysis approach, insert the exponential function Ke^{st}, as a trial solution, into Equation 14.270 to obtain Equation 14.271, which applies to both series and parallel RLC.

$$\frac{d^2(Ke^{st})}{dt^2} + 2\zeta\omega_0\frac{d(Ke^{st})}{dt} + \omega_0^2(Ke^{st}) = 0 \tag{14.271}$$

$$s = \sigma + j\omega \tag{14.272}$$

As in previous analyses, s in Equation 14.272 is the complex frequency variable where sigma σ is the real part. Perform the differentiations on Equation 14.271 to obtain Equation 14.273. The time derivative of Ke^{st} is sKe^{st}.

$$s^2Ke^{st} + 2\zeta\omega_0 sKe^{st} + \omega_0^2Ke^{st} = 0 \tag{14.273}$$

In Equation 14.273, factor out the Ke^{st} term to obtain the characteristic Equation 14.274.

$$Ke^{st}\left(s^2 + 2\zeta\omega_0 s + \omega_0^2\right) = 0 \tag{14.274}$$

$$s_1, s_2 = \omega_0(-\zeta \pm \sqrt{\zeta^2 - 1}) \tag{14.275}$$

$$s_1 = -\zeta\omega_0 + \omega_0\sqrt{\zeta^2 - 1} \tag{14.276a}$$

$$s_2 = -\zeta\omega_0 - \omega_0\sqrt{\zeta^2 - 1} \tag{14.276b}$$

The characteristic roots of Equation 14.274 are shown in Equations 14.275, 14.276a, and 14.276b.

Series RLC

$$i(t) = K_1 e^{s_1 t} + K_2 e^{s_2 t} \tag{14.277a}$$

Equation 14.277a shows the general solution for the series RLC current $i(t)$ in terms of characteristic frequencies s_1 and s_2 and constants K_1 and K_2.

Parallel RLC

$$v(t) = K_1 e^{s_1 t} + K_2 e^{s_2 t} \tag{14.277b}$$

Equation 14.277b shows the general solution for the parallel RLC voltage $v(t)$ in terms of characteristic frequencies s_1 and s_2 and constants K_1 and K_2.

The exponent st must be dimensionless, which implies that s must have units of $1/t$ or frequency.

$$\textbf{Series} \quad i(t) = K_1 e^{(-\zeta\omega_0 + \omega_0\sqrt{\zeta^2-1})t} + K_2 e^{(-\zeta\omega_0 - \omega_0\sqrt{\zeta^2-1})t} \tag{14.278a}$$

$$\textbf{Parallel} \quad v(t) = K_1 e^{(-\zeta\omega_0 + \omega_0\sqrt{\zeta^2-1})t} + K_2 e^{(-\zeta\omega_0 - \omega_0\sqrt{\zeta^2-1})t} \tag{14.278b}$$

14.4.4 Four Cases of Characteristic Roots in Terms of ζ

Variations in the values of R, L, and C lead to various values of alpha ζ and omega-nought ω_0. Because of the subtraction under the radical sign, variations in value can lead to four distinct cases or classes of roots.

Here is a summary of the four cases of characteristic root values and the related RLC natural response behaviors. The graphs and equations correspond to current $i(t)$ in series RLC.

Case 1: Overdamped

$$\zeta > 1 \tag{14.279}$$

Two real and unequal roots

$$\left.\begin{matrix} s_1 \\ s_2 \end{matrix}\right\} = -\zeta\omega_0 \pm \omega_0\sqrt{\zeta^2 - 1} \tag{14.280}$$

The two roots, s_1 and s_2 in Equation 14.280, are real, unequal, and negative.

$$i(t) = K_1 e^{(-\zeta\omega_0 + \omega_0\sqrt{\zeta^2-1})t} + K_2 e^{(-\zeta\omega_0 - \omega_0\sqrt{\zeta^2-1})t} \tag{14.281}$$

The overdamped series RLC natural response, in terms of ζ, is the sum of two damped exponentials in Equation 14.281 and plotted in Figure 14.35.

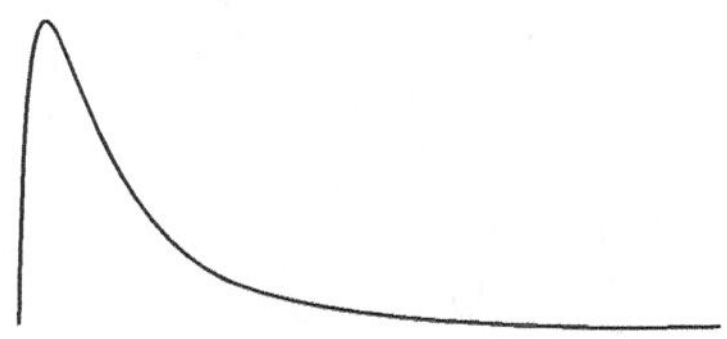

Figure 14.35 Series RLC overdamped natural response current.

Case 2: Critically damped

$$\zeta = 1 \tag{14.282}$$

Two real and equal roots

$$s_1 = s_2 = -\zeta\omega_0 \tag{14.283}$$

The two roots s_1 and s_2 are real, equal, and negative.

$$i(t) = K_1 e^{(-\zeta\omega_0)t} + tK_2 e^{(-\zeta\omega_0)t} \tag{14.284}$$

$$i(t) = (K_1 + tK_2)e^{(-\zeta\omega_0)t} \tag{14.285}$$

The critically damped series RLC natural response, in terms of ζ, is a decaying exponential in Equation 14.285 and plotted in Figure 14.36.

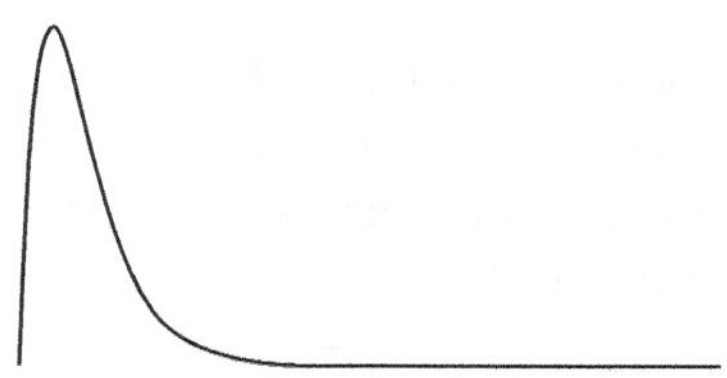

Figure 14.36 Series RLC critically damped natural response current.

Case 3: Underdamped

$$\zeta < 1 \quad (14.286)$$ Two complex, unequal roots

$$s_1 = -\zeta\omega_0 + \omega_0\sqrt{\zeta^2 - 1} = -\sigma + j\omega_{\mathrm{d}} \quad (14.287a)$$

$$s_2 = -\zeta\omega_0 - \omega_0\sqrt{\zeta^2 - 1} = -\sigma - j\omega_{\mathrm{d}} \quad (14.287b)$$

Figure 14.37 Series RLC underdamped natural response current.

The underdamped natural response is a decaying oscillatory exponential. The general solution form is shown in Equations 14.290 and 14.291, and the response is plotted in Figure 14.37.

Because ζ is less than 1, create a new variable ω_{d}, damped resonant frequency, so that the quantity under the radical sign can be positive.

$$\omega_d = \omega_0\sqrt{1 - \zeta^2} \quad (14.288)$$

$$\sigma = \zeta\omega_0 \quad (14.289)$$

$$i(t) = K_1 e^{(-\zeta\omega_0 + \omega_0\sqrt{\zeta^2 - 1})t} + K_2 e^{(-\zeta\omega_0 - \omega_0\sqrt{\zeta^2 - 1})t} \quad (14.290)$$

$$i(t) = K_1 e^{-(\sigma - j\omega_{\mathrm{d}})t} + K_2 e^{-(\sigma + j\omega_{\mathrm{d}})t} \quad (14.291)$$

$$i(t) = e^{-\sigma t}\left(K_1 e^{j\omega_{\mathrm{d}}t} + K_2 e^{-j\omega_{\mathrm{d}}t}\right) \quad (14.292)$$

Case 4: Lossless oscillatory

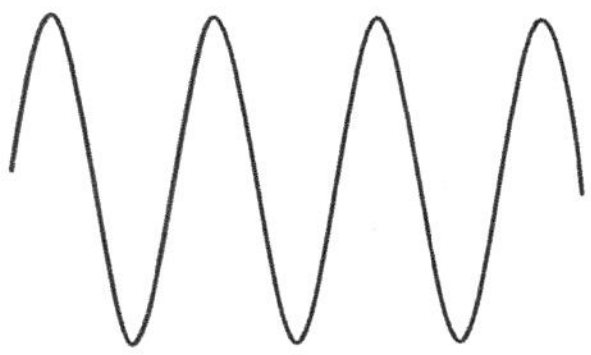

Figure 14.38 Series RLC lossless oscillatory natural response current.

$$\zeta = 0 \quad (14.293)$$ Two imaginary and unequal roots

The two roots s_1 and s_2 are imaginary and unequal.

$$s_1 = j\omega_0 \quad (14.294)$$

$$s_2 = -j\omega_0 \quad (14.295)$$

$$i(t) = K_1 e^{j\omega_0 t} + K_2 e^{-j\omega_0 t} \quad (14.296)$$

$$\cos\theta = \frac{e^{j\theta} + e^{-j\theta}}{2} \quad (14.297)$$

$$i_L(t) = k\cos(\omega_0 t + \theta) \quad (14.298)$$

The response is an oscillation in Equation 14.298 and Figure 14.38.

When ζ equals zero, then ω_{d} equals ω_0. That is, when the damping goes to zero, the response is lossless oscillatory.

14.4.5 Second-order Root Locus with ζ

Explore the locations of the characteristic roots in terms of ζ, in the complex sigma σ and j-omega $j\omega$ plane.

Examine the path of root locations, the root locus, for all values of damping, from zero damping on the vertical axes to infinite damping along the horizontal axis moving away from the critically damped location, shown in Figure 14.39.

$$\omega_d = \omega_0\sqrt{1 - \zeta^2} \quad (14.299)$$

$$\sigma = \zeta\omega_0 \quad (14.300)$$

The variable ω_{d}, in Equation 14.299, is expressed in terms of ζ (zeta) here. In previous discussion of damping, in terms of α alpha, ω_{d} is defined differently, but ω_{d} is functionally the same variable, whether defined in terms of α or ζ.

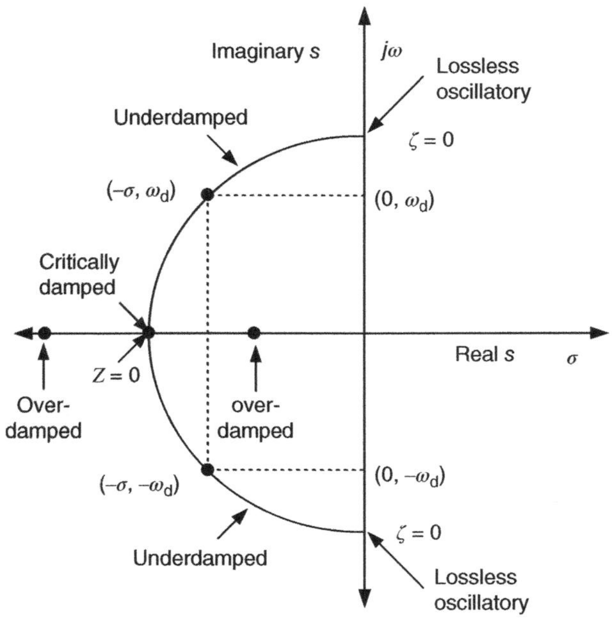

Figure 14.39 RLC natural response root locus locations in the complex plane.

The coordinate pairs in Figure 14.39, $(-\sigma,\ \omega_d)$ and $(-\sigma,\ -\omega_d)$, correspond to underdamped characteristic roots.

The underdamped section of the characteristic roots locus goes from the two ζ equals 0 points, at plus and minus ω_0 on the vertical axis, around the quarter circles to the critically damped location on the horizontal real axis.

At the locations where ζ equals 0, on the vertical $j\omega$ axis in Figure 14.39, damping goes to zero and underdamped becomes lossless oscillatory response.

15

Second-Order RLC Step and Complete Response

15.1 RLC Step Response Intuitive Overview

Based on the behaviors of capacitors and inductors, in response to step inputs of voltage or current, we can intuitively describe the expected step response of series and parallel RLC circuits.

15.1.1 Series RLC Branch Relationship Intuition

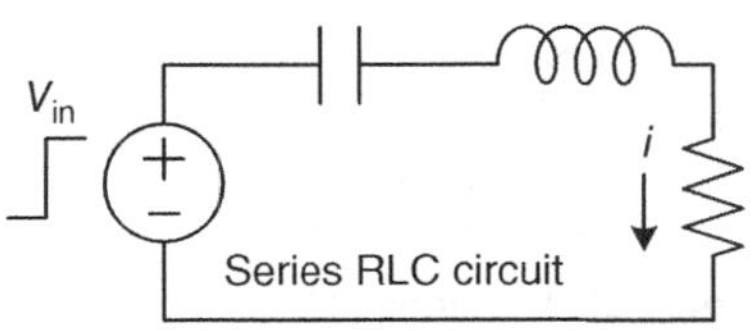

Figure 15.1 RLC circuit driven by a voltage source step function with magnitude V_{in}.

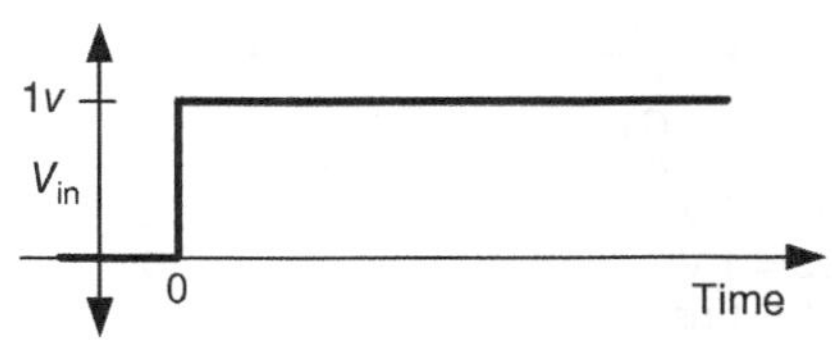

Figure 15.2 Step function graph versus time with DC amplitude of V_{in}.

When a 1 V step function is applied instantaneously to this series RLC circuit (Figure 15.1), what voltage appears across the capacitor, the inductor, and the resistor? Is there an instantaneous current?

An instantaneous step function (Figure 15.2) can be thought of as having an infinitely short rise time. A finite voltage or current step gets applied to the circuit in an infinitesimally small amount of time, ideally in zero time.

Because the voltage across a capacitor cannot change instantaneously, at the instant of the step function, there will be no instantaneous voltage across the capacitor.

$$i = C\frac{dv}{dt} \qquad (15.1)$$

The current through a capacitor equals the capacitance multiplied by the derivative, with respect to time, of the voltage across the capacitor, as shown in Equation 15.1.

$$\Delta i \neq C\frac{\Delta v}{dt} \qquad (15.2)$$

A finite instantaneous voltage cannot be applied across an ideal capacitor. A finite-sized delta voltage, over an infinitesimally small period of time dt, applied across a capacitor implies infinite current. The inequality in Equation 15.2 indicates this impossibility.

First and Second Order Circuits and Equations: Technical Background and Insights, First Edition. Robert O'Rourke.

Because the current in an inductor cannot change instantaneously, at the instant of the voltage step function, there will be no instantaneous current through the inductor.

$$v = L\frac{di}{dt} \quad (15.3)$$

The voltage across an inductor equals the inductance multiplied by the derivative, with respect to time, of the current through the inductor.

$$v \neq L\frac{\Delta i}{dt} \quad (15.4)$$

A finite instantaneous current cannot flow through an ideal inductor. A finite-sized current step through an inductor, in an infinitesimally small period of time dt, implies infinite voltage across the inductor. The inequality in Equation 15.4 indicates this impossibility.

Because there is no instantaneous current through the inductor, there is no instantaneous loop current when the voltage step is applied. Because there is no loop current, there is no instantaneous voltage across the resistor. Because there can be no instantaneous voltage across the capacitor, when the voltage step function is applied to the RLC circuit, all the voltage is initially dropped across the inductor. Underdamped step response plots in the following example reflect this initial inductor voltage.

15.1.2 Series RLC Step Response Simulation Example

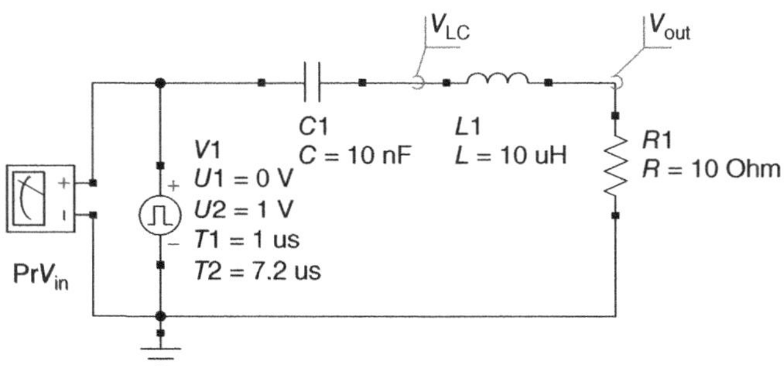

Figure 15.3 Series RLC simulation schematic driven by step voltage.

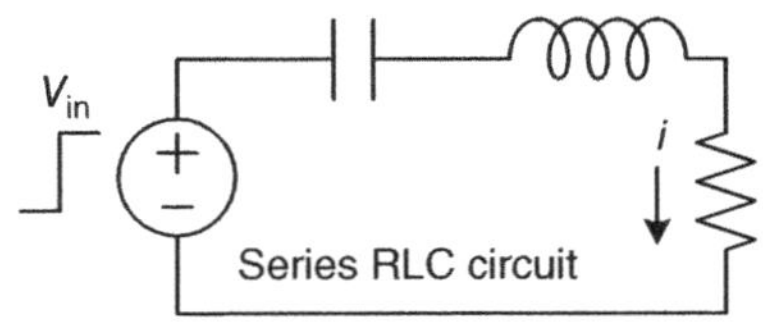

Figure 15.4 Series RLC circuit schematic driven by step voltage.

Figure 15.3 shows a circuit simulation schematic for the series RLC circuit schematic, in Figure 15.4, with a step function input voltage. The circuit values result in underdamped behavior. The critical resistance for this circuit is 63.2 Ω.

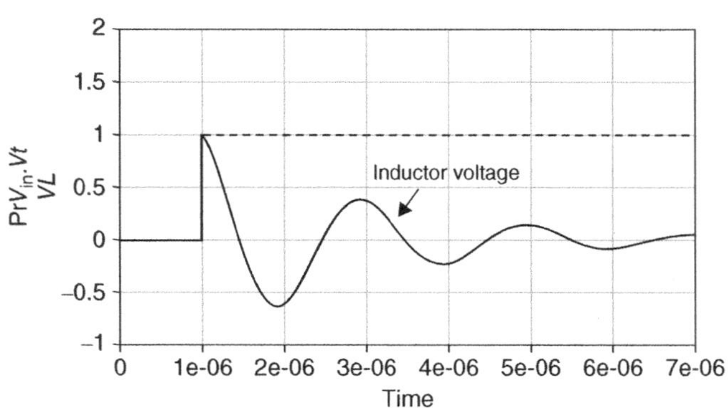

Figure 15.5 Series RLC simulated inductor and input voltages.

Figure 15.5 shows the series RLC input voltage step function where the 1-V value is visible with a dashed line. The solid line is the voltage across the inductor.

At the instant of the voltage step, no voltage gets applied across the capacitor nor the resistor. All the stepped source voltage gets applied across the inductor, going up to the 1-V value of the input voltage source before starting underdamped decaying oscillation.

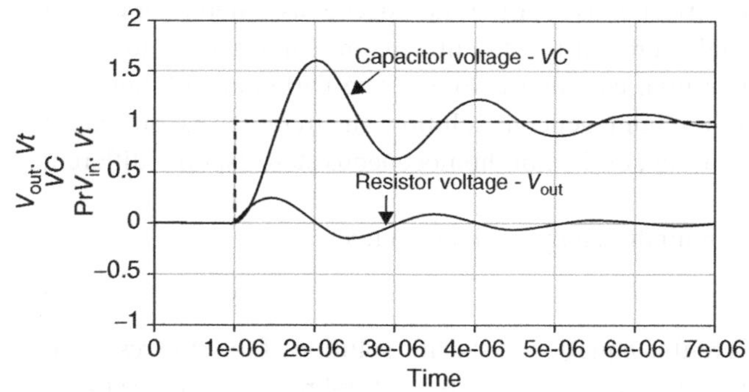

Figure 15.6 Series RLC simulated capacitor, resistor, and input voltages.

Figure 15.6 shows the capacitor voltage and the resistor voltage. They both start at zero. There can be no resistor voltage without capacitor voltage. Notice how the capacitor voltage decaying oscillation is out of phase with the resistor voltage. The circuit is configured and labeled to have the resistor voltage be the output voltage V_{out}.

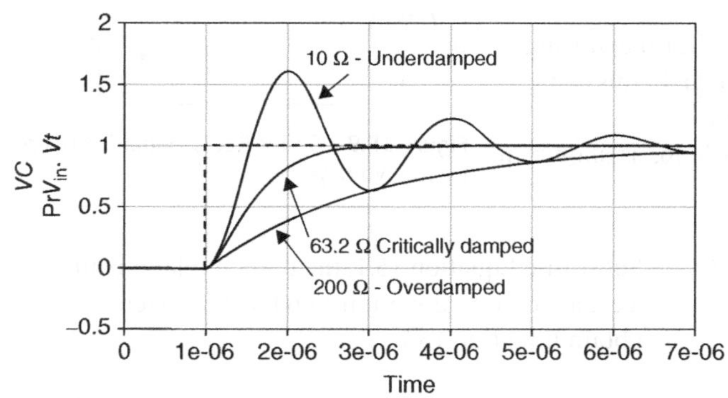

Figure 15.7 Series RLC simulated capacitor voltages for various dampings.

Figure 15.7 shows the capacitor voltage for resistance swept over three values, 10 Ω, which is underdamped, 63.2 Ω, which is critically damped, and 200 Ω, which is overdamped. Notice how the overdamped capacitor voltage line represents part of an envelope for the underdamped decaying oscillation.

15.2 RLC Step Response Detailed Analyses

15.2.1 Step Response Detailed Description

Step response is a forced response, driven by a signal source, in comparison to natural response, which comes from initial conditions already present in the circuit. Step response is one form of zero-state response (ZSR), whose definition dictates that there are no initial conditions present in the circuit.

Mathematically and rigorously, for all previous initial conditions to decay away to zero state, before the application of the step function, this means that an infinite amount of time must precede the step input. Thus, the step function in Figure 15.8 is marked as extending out to infinity in negative time before zero.

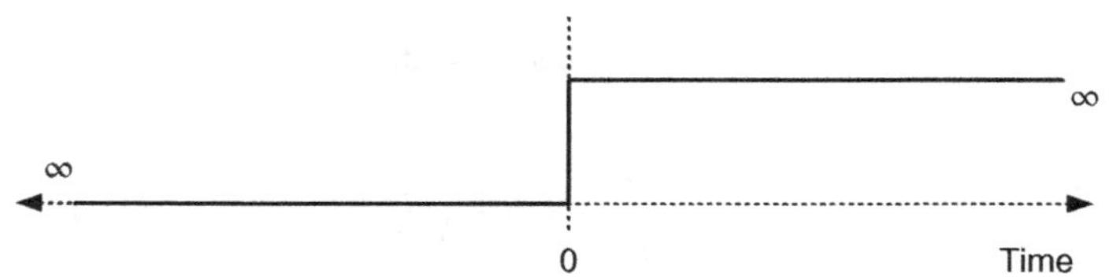

Figure 15.8 Step function graph.

In mathematical terms, RLC circuit step response corresponds to a non-homogeneous differential equation, which has a function term corresponding to the signal source in the circuit. A common approach to calculating the general solution to a non-homogeneous differential equation is to add one particular solution, of the non-homogeneous solution, to the general solution of the homogeneous solution. In circuit terms, one can find the step response of a circuit by adding one specific solution, of the non-homogeneous step response differential equation, to the natural response of that circuit.

Equation 15.5 shows the general solution v_{out}, to a circuit driven by a voltage source.

$$v_{\mathrm{out}} = v_{\mathrm{h}} + v_{\mathrm{p}} \tag{15.5}$$

v_{p} = Particular solution to the non-homogeneous differential equation (a specific solution of the step response)
v_{h} = Solution to the homogeneous differential equation (general solution to the natural response of the circuit)

15.2.2 RLC Step Response Homogeneous Differential Equation Solution

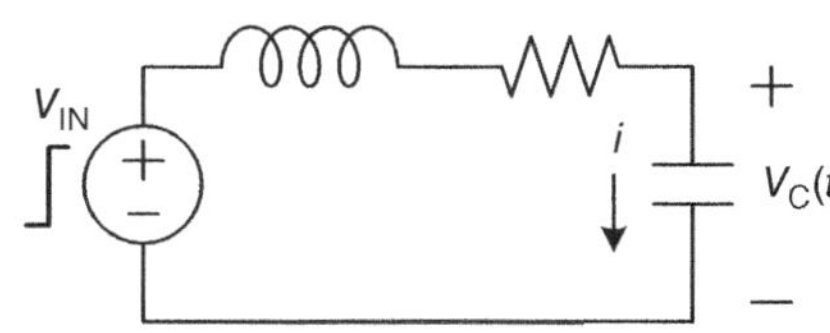

Figure 15.9 Series RLC schematic driven by step voltage.

For the series RLC schematic in Figure 15.9, consider the time starting at t equals 0, when the source voltage has just been applied. In this context, V_{IN} appears in the differential equation as a constant.

Use KVL to sum the voltages around the circuit loop, as shown in Equation 15.6.

$$L\frac{di(t)}{dt} + Ri(t) + v_{\mathrm{C}}(t) = V_{\mathrm{IN}} \tag{15.6}$$

Substitute Equation 15.1, the branch relationship for a capacitor, into Equation 15.6 for the current $i(t)$, to obtain Equation 15.7.

$$LC\frac{d^2v_{\mathrm{C}}(t)}{dt^2} + RC\frac{dv_{\mathrm{C}}(t)}{dt} + v_{\mathrm{C}}(t) = V_{\mathrm{IN}} \tag{15.7}$$

Divide the whole Equation 15.7 by LC to make the coefficient of the second derivative term 1, as shown in Equation 15.8.

$$\frac{d^2v_{\mathrm{C}}(t)}{dt^2} + \frac{R}{L}\frac{dv_{\mathrm{C}}(t)}{dt} + \frac{1}{LC}v_{\mathrm{C}}(t) = \frac{V_{\mathrm{IN}}}{LC} \tag{15.8}$$

We need a general solution to the homogeneous differential equation, with the source removed, as shown in Equation 15.9, and one particular solution to the nonhomogeneous Equation 15.8.

$$\frac{d^2v_{\mathrm{C}}(t)}{dt^2} + \frac{R}{L}\frac{dv_{\mathrm{C}}(t)}{dt} + \frac{1}{LC}v_{\mathrm{C}}(t) = 0 \tag{15.9}$$

The coefficients of the terms in Equation 15.9, the homogeneous differential equation for the natural response, match coefficients in previously solved series *RLC* differential equations with $i(t)$ as the circuit variable. In Equations 15.8 and 15.9, $v_{\mathrm{C}}(t)$ is the state variable. We can use all the following previous results, in Equations 15.10–15.15, including the characteristic equations and their roots, the natural frequencies.

Damping coefficient α | **Damping ratio ζ**

Series RLC

$$\alpha = \frac{R}{2L} \tag{15.10a}$$

$$\zeta = \frac{R}{R_{CRIT-Series}} = \frac{R}{2}\sqrt{\frac{C}{L}} \tag{15.10b}$$

Damping coefficient α

Resonant Frequency ω_0

$$\omega_0 = \frac{1}{\sqrt{LC}} \tag{15.11}$$

Damping ratio ζ

$$2\zeta_{ser}\omega_0 = R\sqrt{\frac{C}{L}}\left(\frac{1}{\sqrt{LC}}\right) = \frac{R}{L} \tag{15.12}$$

$$\frac{d^2v_C}{dt^2} + 2\alpha\frac{dv_C}{dt} + \omega_0^2 v_C = 0 \tag{15.13a}$$

$$\frac{d^2v_C}{dt^2} + 2\zeta\omega_0\frac{dv_C(t)}{dt} + \omega_0^2 v_C(t) = 0 \tag{15.13b}$$

$$s^2 + 2\alpha s + \omega_0^2 = 0 \tag{15.14a}$$

$$s^2 + 2\zeta\omega_0 s + \omega_0^2 = 0 \tag{15.14b}$$

$$s_1, s_2 = -\alpha \pm \sqrt{\alpha^2 - \omega_0^2} \tag{15.15a}$$

$$s_1, s_2 = \omega_0\left(-\zeta \pm \sqrt{\zeta^2 - 1}\right) \tag{15.15b}$$

The solutions for each of the four cases, already developed for natural response (ZIR), also apply here for step response.

15.2.3 Particular Solution to the Non-Homogeneous Differential Equation

In addition to the solution of the natural response Equation 15.9, the step response general solution also requires one particular solution to the non-homogeneous Equation 15.8.

Looking at time increasing to infinity, after all transients settle, the series RLC circuit in Figure 15.9 looks like a DC circuit driven by V_{IN}. To DC and very low frequencies, the inductor looks like a short circuit and the capacitor looks like an open circuit. The result is all the input signal voltage V_{IN} drops across the capacitor, and this provides one particular solution to Equation 15.8.

$$v_C(\infty) = V_{IN} \tag{15.16}$$

Equation 15.16 shows how the steady-state capacitor voltage $v_C(t)$, of the series RLC step response circuit, settles to the value of the input voltage V_{IN} after an infinite amount of time has passed after the step.

$$v_C(t) = v_h + v_p \tag{15.17}$$

The general solution to the series RLC step response, Equation 15.17, includes the natural response v_h, the solution to the homogeneous differential equation. The natural response solution depends upon which of the four cases the circuit falls into.

$$v_C(t) = v_h(t) + v_{IN} \tag{15.18}$$

The general solution, Equation 15.18, also includes the particular solution v_P, which is v_{IN} for this circuit.

15.3 Parallel RLC Intuitive Step Response Example

Figure 15.10 shows a circuit simulation schematic for the parallel RLC circuit schematic in Figure 15.11 with a step function input current.

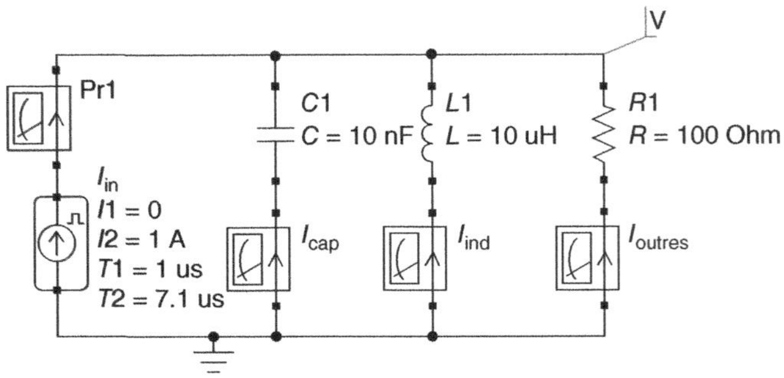

Figure 15.10 Parallel RLC simulation schematic driven by step current.

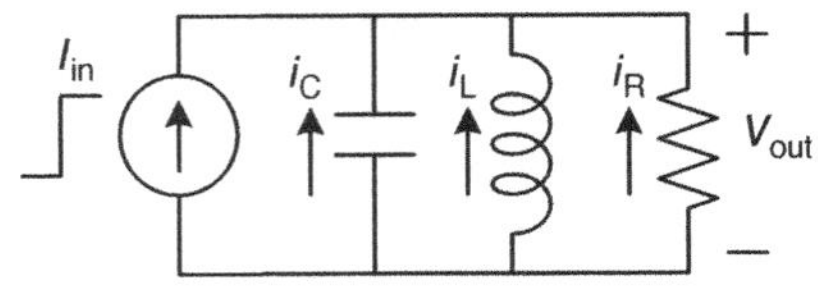

Figure 15.11 Parallel RLC circuit schematic driven by step current.

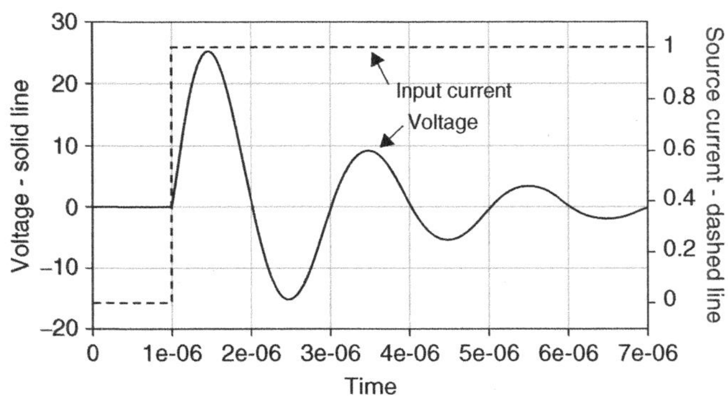

Figure 15.12 Parallel RLC simulated voltage and input current.

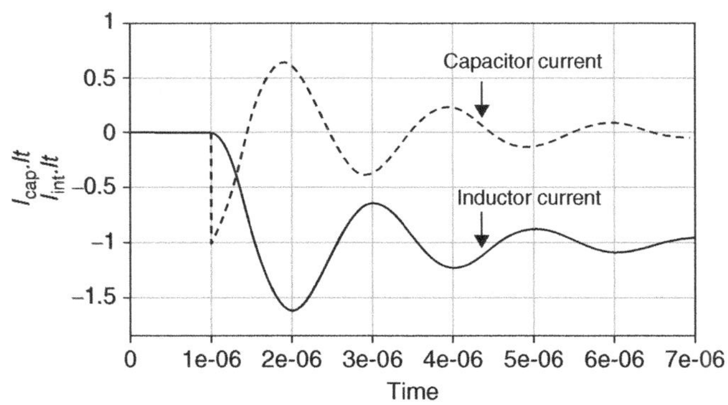

Figure 15.13 Parallel RLC simulated inductor and capacitor currents.

The 100 Ω resistor in Figure 15.10 results in underdamped behavior. The critical resistance for this circuit is 15.8 Ω.

Figure 15.12 shows the parallel RLC input current step function which is visible with a dashed line and the scale on the right-hand vertical axis. The solid line is the voltage across all the circuit elements. The voltage builds up gradually after the step is applied. No instantaneous voltage is dropped across the three elements because one element is the capacitor.

Figure 15.13 shows the parallel RLC inductor current, with the solid line, and the capacitor current with the dashed line. Notice how there is an instantaneous current into the capacitor, as soon as the step input current is applied. The current starts at minus one amp; initially all the source current goes into the capacitor.

Notice also in Figure 15.13 how there is no instantaneous change to the inductor current as soon as the step input current is applied. The inductor current, the voltage, and the resistor current all build gradually. We know that the resistor current will look like the voltage, scaled by the resistance value to current.

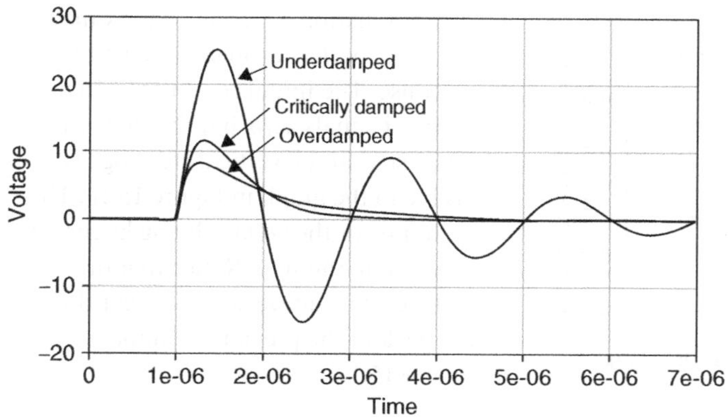

Figure 15.14 Parallel RLC simulated inductor currents for various dampings.

Figure 15.14 shows the parallel RLC inductor current for three different values of resistance. 100 Ω corresponds to the underdamped tallest signal. 15.8 Ω corresponds to critical damping and 10 Ω was used for overdamped.

15.4 Complete RLC Time Domain Response

15.4.1 RLC Complete Response is ZSR Plus ZIR

Step response is a zero-state response (ZSR) which does not consider any initial conditions or stored energy in the circuit. Natural response is a zero-input response (ZIR), which considers only initial conditions like capacitor voltage and inductor current. ZIR does not consider any input signals driving the circuit.

The compete response of a circuit considers both ZSR and ZIR. A circuit can have stored energy and an input signal.

Complete response equals ZSR plus ZIR.

Complete response equals step response plus natural response.

15.4.2 RLC Complete Response Simulation Example

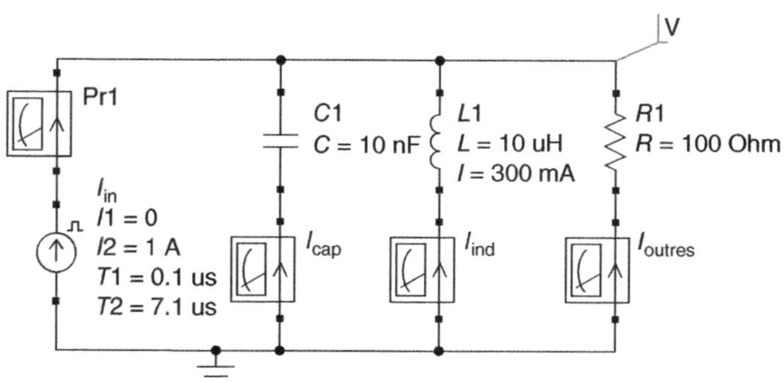

Figure 15.15 Parallel RLC simulation schematic with initial condition inductor current.

To simulate a complete response, this circuit in Figure 15.15 uses a 1 A input current step and an inductor initial condition current of 300 mA. The initial inductor current is added to the schematic from the previous parallel RLC example in Figure 15.10. The start of the step also comes sooner than in the previous example. T1 equals 0.1 μs here.

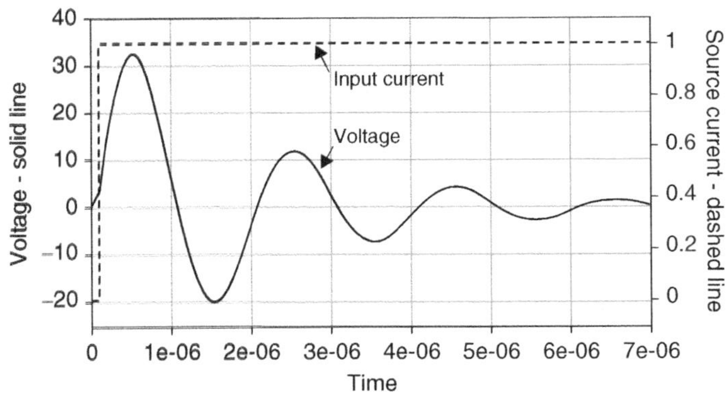

Figure 15.16 Parallel RLC simulated complete response voltage and input current.

This simulation has some approximation; it is not a perfect expression of complete response. The initial condition is in place at time *t* equals zero, but the step is at 0.1 μs so that the vertical step edge is visible in the graph in Figure 15.16. This accounts for the voltage being larger than zero before the step. Notice that the voltage peak, above 30 V, is larger than the peak in the previous example in Figure 15.12.

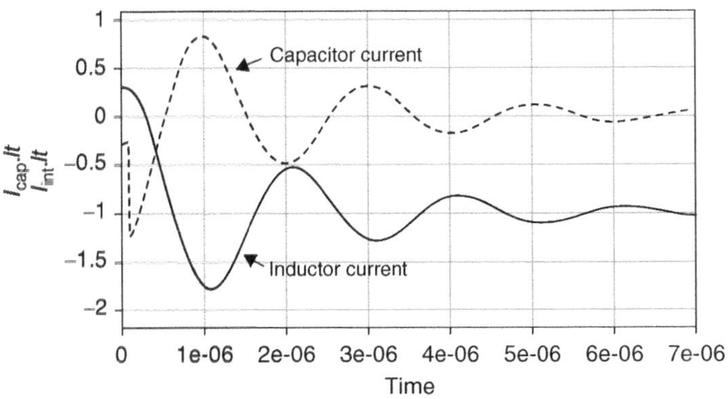

Figure 15.17 Parallel RLC simulated complete response capacitor and inductor currents.

In Figure 15.17, the inductor current starts at 300 mA, due to the initial condition, as does the capacitor current. The input current step, at 0.1 μs, still instantaneously pulls down the capacitor current by 1 A. The capacitor and inductor current peak values, in the plus and minus directions, are greater than the corresponding peak values in the step response example with no initial conditions.

Part 4

Technical Background Topics

16

Complex Numbers, Exponentials, and Phasors

16.1 Imaginary and Complex Numbers

16.1.1 Imaginary Operators and Complex Numbers

Imaginary numbers are the square roots of negative numbers. An imaginary number is a real number times the imaginary unit i, also called the imaginary operator.

Mathematics, physics, and other non-electrical engineering disciplines use the letter i as the complex operator. Since electrical engineering uses i for current, we use j for the complex operator in electrical engineering.

$j = \sqrt{-1}$ (16.1a) Equations 16.1a and 16.1b say that the square root of negative 1 is the imaginary operator. $i = \sqrt{-1}$ (16.1b)

$j^2 = -1$ (16.2a) Equations 16.2a and 16.2b say that the square of the imaginary operator is minus 1. $i^2 = -1$ (16.2b)

$\frac{1}{j} = -j$ (16.3a) Equations 16.3a and 16.3b are useful to keep in mind for reconciling imaginary operators in the denominator of a quantity like impedance or admittance. $\frac{1}{i} = -i$ (16.3b)

A complex number z, shown in Figure 16.1, is the sum of a real number x and an imaginary number jy.

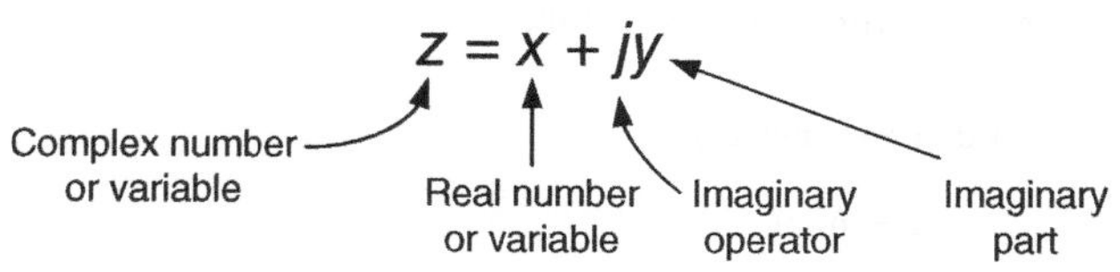

Figure 16.1 Impedance equation and real and imaginary parts.

$\text{Re}(z) = x$ (16.4) Equation 16.4 says that the real part of the complex number z equals $x + jy$ is the real quantity x.

$\text{Im}(z) = y$ (16.5) Equation 16.5 says that the imaginary part of the complex quantity z equals $x + jy$ is the real quantity y.

$jy = z = 0 + jy$ (16.6) Equation 16.6 says that jy is a purely imaginary complex number z with the real part x equal to 0.

First and Second Order Circuits and Equations: Technical Background and Insights, First Edition. Robert O'Rourke.

Impedance is an example of a complex quantity in electrical engineering. Signals, admittance, and transfer functions can all be complex quantities, either numbers or functions.

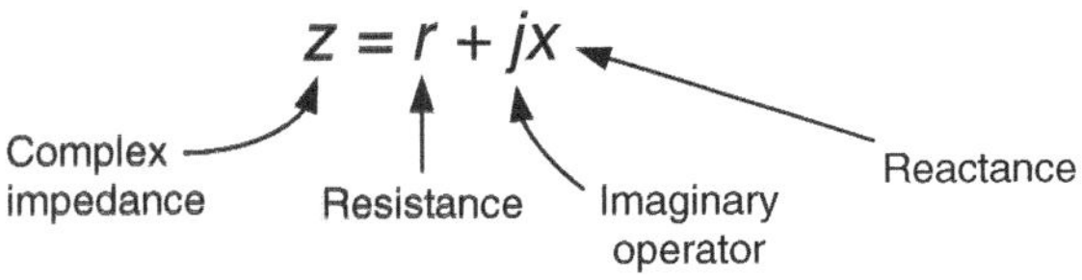

Figure 16.2 Impedance equation with resistance and reactance.

In Figure 16.2, z represents impedance. In Figure 16.1 z represented a complex number in general. Another common general notation for complex numbers is $c = a + ib$.

$$Y = \frac{1}{Z} = G + jB \tag{16.7}$$

Admittance Y, in Equation 16.7 and Figure 16.3, is a complex quantity and the inverse of impedance. Conductance G is the inverse of resistance and susceptance B is the inverse of admittance Y (Figure 16.3).

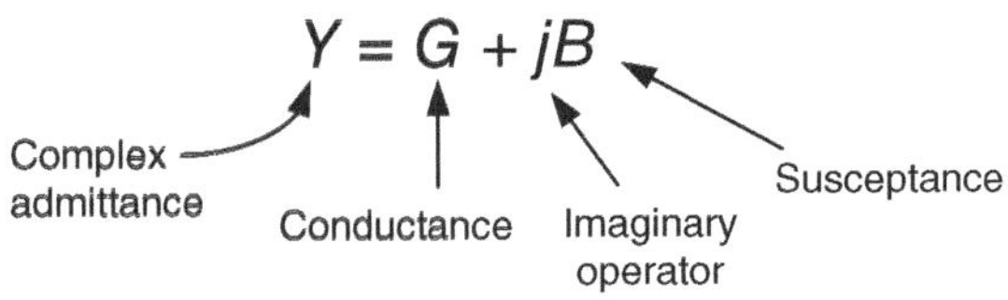

Figure 16.3 Admittance equation with conductance and susceptance.

Different discussions of complex numbers and impedance might refer to the imaginary part of a complex number, $z = x + iy$, without clearly distinguishing between y and iy.

The imaginary part of z is y and y is a real number. The quantity iy, the imaginary part and the imaginary unit together, is an imaginary number.

Complex number terminology is important in electrical engineering to distinguish a purely imaginary quantity, like the impedance of an ideal inductor, $j\omega L$, from the reactance of the inductor ωL. Similarly, a purely imaginary admittance can be equal to jB, where **B** is the susceptance, which is the imaginary part of the admittance.

16.1.2 The Complex Plane and Polar Notation

Complex numbers can be plotted in a complex plane where the vertical axis plots the imaginary portion of the complex number, and the horizontal axis plots the real portion of the complex number.

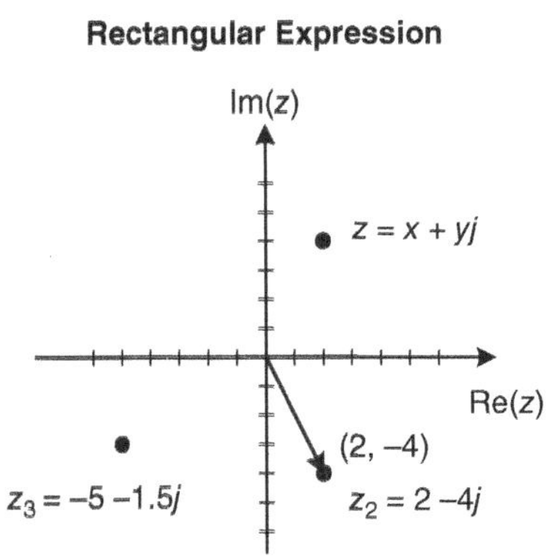

Figure 16.4a Complex plane with rectangular expression of complex number.

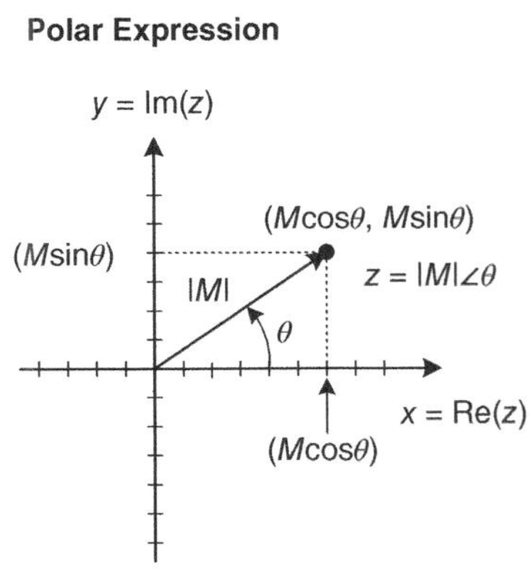

Figure 16.4b Complex plane with polar expression of complex number.

The rectangular expression of a complex number, in terms of the x (real) portion and the y (imaginary) portion, is shown in Figure 16.4a. The $\mathbf{z_2}$ coordinate point in the complex plane is also shown as an ordered pair (2, −4).

The polar expression of a complex number looks at the magnitude M, the distance from the origin of the complex plane, and the phase angle θ, the angle in the counterclockwise (anticlockwise) direction from the positive x axis, as shown in Figure 16.4b and Equation 16.9a.

In Figure 16.4b, the complex number with magnitude M can be represented in the complex plane in terms of length and angle by the ordered pair ($M\cos\theta$, $M\sin\theta$).

16.1.2.1 Alternative Expressions for the Complex number z

Rectangular: $z = x + jy$ (16.8)

Polar or Steinmetz: $z = M\angle\theta$ (16.9a) $z = r\angle\theta$ (16.9b)

M and r are two interchangeable common notations for the magnitude of a complex number, where r corresponds to radius of a circle as utilized for the unit circle in Figure 16.5.

$$\theta = \tan^{-1}\left[\frac{y}{x}\right] \quad (16.10)$$

Arctangent is also denoted by $\tan^{-1}$. The angle of a complex number is the arctangent of the real part divided by the imaginary part, as shown in Equation 16.10 and Figure 16.7.

$$M = |z| = |x + jy| = \sqrt{x^2 + y^2} \quad (16.11)$$

The magnitude of M is shown in Equation 16.11.

In trigonometric terms, a complex number can be represented by Equation 16.12a, where $\cos\theta$ is the real part, $\sin\theta$ is the imaginary part, and M or r is the magnitude. An ordered pair ($M\cos\theta$, $M\sin\theta$) is shown in Figure 16.4b to describe the complex number z.

$$z = M(\cos\theta + j\sin) \quad (16.12a) \qquad z = r(\cos\theta + j\sin) \quad (16.12b)$$

16.2 Exponentials, Complex Numbers, and Trigonometry

16.2.1 The Unit Circle and Trigonometry

Overlaying a unit circle on the complex plane, shown in Figure 16.5, facilitates understanding how complex numbers relate to exponential and trigonometric functions.

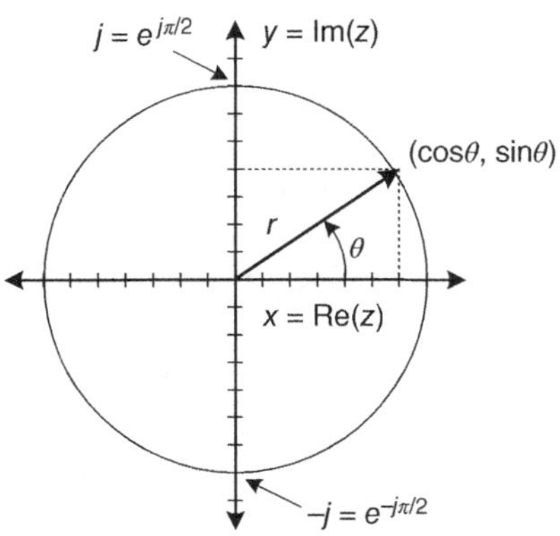

Figure 16.5 Complex plane with unit circuit and labeled with j.

The unit circle has a radius of 1, as shown in Equation 16.13, the equation for a circle.

$$r^2 = x^2 + y^2 = 1 \quad (16.13)$$

The unit circle can be expressed in terms of cosine and sine, as shown in Equation 16.14.

$$1 = \cos^2\theta + \sin^2\theta \quad (16.14)$$

Euler's formula, in Equation 16.15, relates the cosine and to the exponential function.

$$e^{j\theta} = \cos\theta + j\sin\theta \quad (16.15)$$

In exponential form, the complex number z can be expressed by M times e to the j-theta power, as shown in Equation 16.16.

$$z = M\angle\theta = Me^{j\theta} \tag{16.16}$$

A 360° circle corresponds to 2π radians, and 90° is π over 2 radians, as shown in Equation 16.17.

$$\frac{\pi}{2}\text{ radians} = 90\text{ degrees} \tag{16.17}$$

Equation 16.18 derives the equation j-equals-e-to-the-$j\pi$-over-2 radians, shown at the top of Figure 16.5.

$$e^{j\pi/2} = \cos(\pi/2) + j\sin(\pi/2) = 0 + j\sin(90) = j \tag{16.18}$$

The unit circle, with a radius equal to 1, illustrates trigonometry formulae in Equations 16.19 and 16.20, and Figure 16.6.

$$\sin\theta = \frac{\text{opposite}}{\text{hypotenuse}} = \text{opposite} \tag{16.19}$$

$$\cos\theta = \frac{\text{adjacent}}{\text{hypotenuse}} = \text{adjacent} \tag{16.20}$$

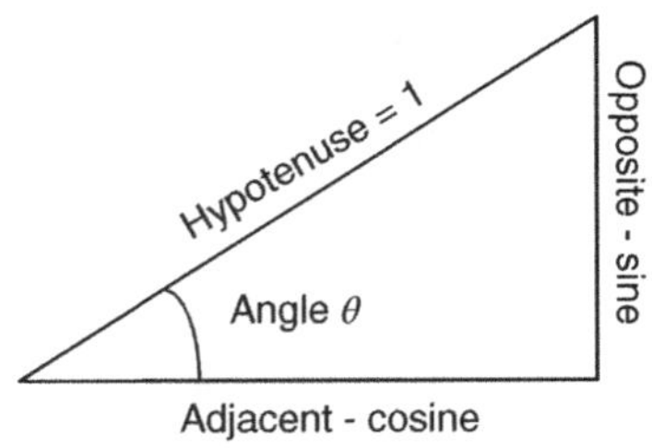

Figure 16.6 Sine and cosine in terms of sides of a triangle and unit circle.

As indicated in both Figures 16.4b and 16.5, in the complex plane, sinθ is the vertical imaginary component and cosθ is the horizontal real component of a complex number or function.

In terms of a unit circle, the tangent of the angle θ is the length of the tangent to the unit circle from the angle θ, as shown in Figure 16.7.

Be aware that this picture in Figure 16.7 shows theta in the positive x and positive y quadrant of the complex plane. Special consideration is needed for calculating arctangent in all four quadrants. Look up the *a*tan2 function for more information and insight.

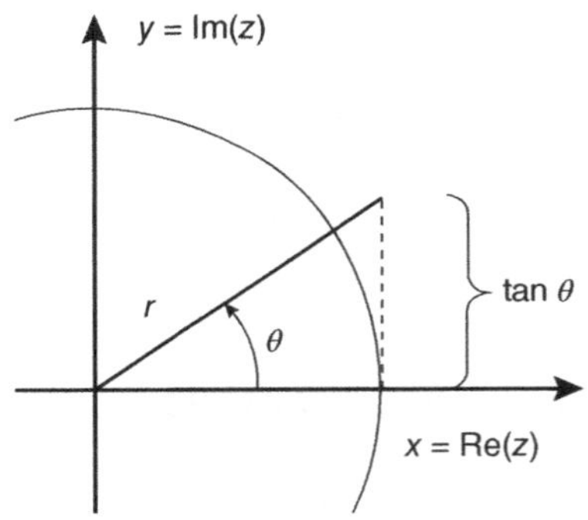

Figure 16.7 Tangent in terms of unit circuit on the complex plane.

16.2.2 The Complex Conjugate

The conjugate of a complex number is formed by negating the imaginary part, as shown in Equations 16.21b and 16.22b and Figure 16.8. Complex conjugate is denoted by a bar over the number or an asterisk-looking symbol.

Figure 16.8 shows complex number z equals 8 plus j5 and its conjugate, 8 minus j5, in the complex plane. Negating the imaginary part of the complex number reflects the vector below the horizontal axis.

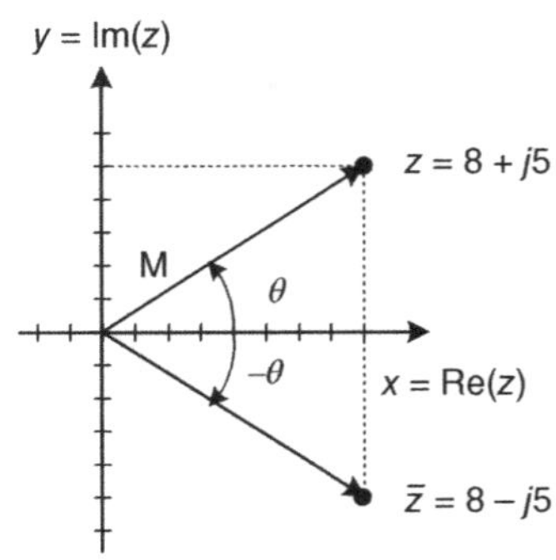

Figure 16.8 Complex plane showing two complex conjugate complex numbers.

$$z = x + jy \quad (16.21a)$$

$$z^* = (x + jy)^* = x - jy \quad (16.21b)$$

The conjugate of the complex number z, in Equations 16.21a and 16.22a, is z^* in Equations 16.21b and 16.22b.

$$z = M\angle\theta \qquad z = Me^{j\theta} \qquad z = M(\cos\theta + j\sin\theta) \quad (16.22a)$$

$$\bar{z} = z^* = M\angle\theta \qquad z^* = Me^{-j\theta} \qquad z^* = M(\cos\theta - j\sin\theta) \quad (16.22b)$$

Equation 16.23a derives the complex number magnitude Equation 16.23b using the complex conjugate z^*.

$$|z| = \sqrt{zz^*} = \sqrt{(x + jy)(x - jy)} = \sqrt{x^2 + jxy - jxy + y^2} = \sqrt{x^2 + y^2} \quad (16.23a)$$

$$|z| = M = \sqrt{zz^*} = \sqrt{x^2 + y^2} \quad (16.23b)$$

Notice how Equation 16.23b gets the same result for the length of the complex number vector as a calculation of that length using the Pythagorean theorem, as shown in Equations 16.24a and 16.24b.

$$M^2 = x^2 + y^2 \quad (16.24a) \qquad |z| = M = \sqrt{x^2 + y^2} \quad (16.24b)$$

For the quotient of two complex numbers, $A + jB$ and $C + jD$, in Equations 16.25a and 16.25b shows the complex conjugate.

$$z = \frac{A + jB}{C + jD} \quad (16.25a) \qquad z^* = \overline{\frac{A + jB}{C + jD}} = \frac{A - jB}{C - jD} \quad (16.25b)$$

16.2.3 Complex Numbers and Functions

Electrical circuit network functions are complex, either in the variable s (which includes the time domain element) or in ω (omega) which corresponds to sinusoidal steady-state analysis (no transients). Functions of complex numbers follow the same principles and rules that we study for complex numbers.

16.2.3.1 Complex Functions

$$Z(\omega) = R(\omega) + jX(\omega) \quad (16.26)$$

In Equation 16.26, $R(\omega)$ is the real part of the complex function and $X(\omega)$ is the imaginary part of the complex function. If Equation 16.26 were a complex impedance, as a function of angular frequency ω, then R would be the resistance and X would be reactance.

16.2.3.2 Magnitude and Phase Calculations

Summaries of complex number math include the calculation of the magnitude of a complex number and the calculation of the angle. In the steady-state frequency domain analysis of electrical circuits we calculate the magnitude and phase of circuit responses such as current or voltage. The (ω) notation, in Equations 16.26 and 16.27b, indicates a function of ω.

The magnitude M of a complex number z is the square root of the sum of the square of the real part R and the square of the imaginary part X, as shown in Equation 16.27a.

$$M = \sqrt{R^2 + X^2} \tag{16.27a}$$

The magnitude of the complex impedance function Z is the square root of the sum of the square of the (resistive) real part $R(\omega)$ and the square of the (reactive) part $X(\omega)$, as shown in Equation 16.27b.

$$|Z(\omega)| = \sqrt{|R(\omega)|^2 + |X(\omega)|^2} \tag{16.27b}$$

The angle theta of the complex number z is the arctangent of the quantity x over y, as shown in Equation 16.28a.

$$\angle Z = \theta = \tan^{-1}\left[\frac{X}{R}\right] \tag{16.28a}$$

The phase of the complex impedance function $Z(\omega)$ is the arctangent of the real part $R(\omega)$ over the imaginary part $X(\omega)$, as shown in Equation 16.28b.

$$\angle Z(\omega) = \phi(\omega) = \tan^{-1}\frac{X(\omega)}{R(\omega)} \tag{16.28b}$$

16.3 Phasors and Sinusoidal Steady State

This section explores phasors and sinusoidal steady state in electric circuit theory. Sinusoidal steady state allows the creation and use of phasors. Phasors allow the analysis of reactive circuits, described by complex functions, using algebra as we do with resistive circuit analysis.

16.3.1 Phasors Allow Algebraic Analysis of Reactive Circuits

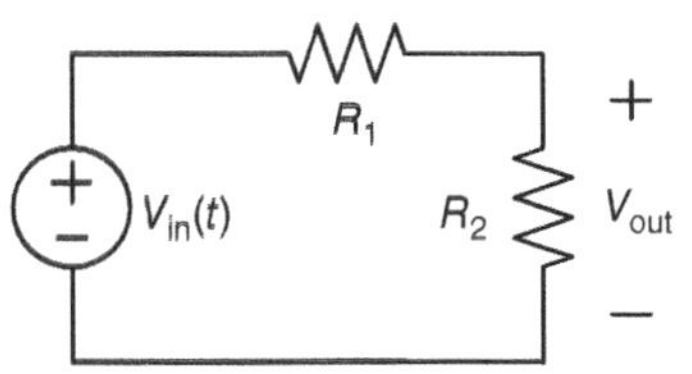

Figure 16.9a Resistive voltage divider circuit.

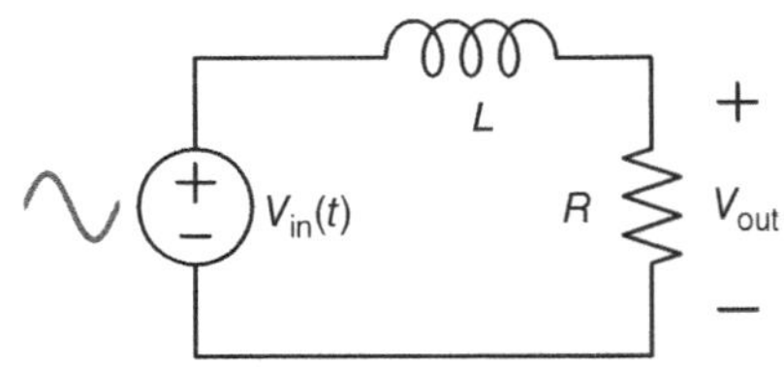

Figure 16.9b RL voltage divider circuit in terms of phasors.

$$\frac{v_{\text{out}}}{v_{\text{in}}} = \frac{R_2}{R_1 + R_2} \tag{16.29a}$$

$$\frac{\mathbf{V}_{\text{out}}}{\mathbf{V}_{\text{in}}} = \frac{R}{j\omega L + R} = H(j\omega) \tag{16.29b}$$

The analysis of the voltage divider circuit, shown in Figure 16.9a, using Kirchhoff's laws and the voltage divider formula, uses algebraic equations including resistance, voltage, and currents, as shown in Equation 16.29a.

Using phasors, indicated by the bold-face type for $\mathbf{V}_{\text{out}}$ over $\mathbf{V}_{\text{in}}$ in Equation 16.29b, the schematic in Figure 16.1b can also be analyzed algebraically using frequency domain impedances.

16.3.2 Phasor Derivation from Sine or Cosine

Phasors relate time domain sinusoidal functions to complex numbers and functions through Euler's law.

In Equation 16.30, a pure cosine function $A_m\cos(\omega t+\theta)$, can be the real part of the cosine plus an imaginary sine term.

$$A_m \cos(\omega t + \phi) = \text{Re}[A_m \cos(\omega t + \phi) + jA_m \sin(\omega t + \phi)] \tag{16.30}$$

Apply Euler's law, in Equation 16.31, to the right side of Equation 16.30, to obtain Equation 16.32.

$$e^{j\omega t+\theta} = \cos(\omega t + \theta) + j\sin(\omega t + \theta) \tag{16.31}$$

$$A_{\rm m}\cos(\omega t + \phi) = \text{Re}(A_{\rm m}e^{j(\omega t+\phi)}) \tag{16.32}$$

In Equation 16.32, apply the property of exponents, in Equation 16.33, that the exponent of a sum equals the product of exponents, to obtain Equation 16.34.

$$A_{\rm m}e^{j(\omega t+\theta)} = A_{\rm m}e^{j\theta}e^{j\omega t} \tag{16.33}$$

$$A_{\rm m}\cos(\omega t + \phi) = \text{Re}(A_{\rm m}e^{j\phi}e^{j\omega t}) \tag{16.34}$$

The $e^{j\phi}$ term in Equation 16.34 is not time varying, and the time-varying $e^{j\omega t}$ term is sinusoidal steady state. The phasor, on the right-hand side of Equation 16.34, is created by isolating the $e^{j\phi}$ term from the $e^{j\omega t}$ term as shown on the right side of Equation 16.35.

$$A_{\rm m}\cos(\omega t + \phi) \Leftrightarrow A_{\rm m}e^{j\phi} \tag{16.35}$$

The beauty and utility of phasors, as a transform from time-varying functions to complex numbers, is the ability to analyze sinusoidal steady-state functions conveniently. The phasor can be converted back to the time-varying exponential or sinusoidal function at the end of the analysis.

The transform notion in electronics is a big and important topic. For further information on transforms to sinusoidal steady-state phasors or to the Laplace *s*-domain, see "The Analysis and Design of Linear Circuits" by Roland E. Thomas and Albert J. Rosa, from John Wiley & Sons, Inc.

16.3.3 Sinusoidal Steady-State Frequency Domain

When electrical signals, consisting of pure sinusoids (cosine and sine functions), extend in time from minus infinity to plus infinity, we call that sinusoidal steady state, as illustrated in Figure 16.10.

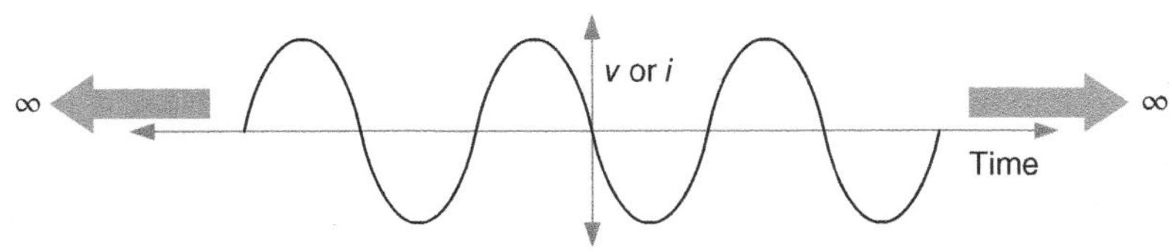

Figure 16.10 Graph of sine wave marked as extending to infinity.

Frequency domain sinusoidal steady state means that there are no transient signal components to take into account. Sinusoidal steady affords simplicity in analysis.

For contrast, this is an example of a sinusoidal signal with a transient signal component arising from the signal start up transition at $t = 0$. Laplace transforms are well suited to handling complex signals like this, which mix steady state sinusoidal with transient time domain (Figure 16.11).

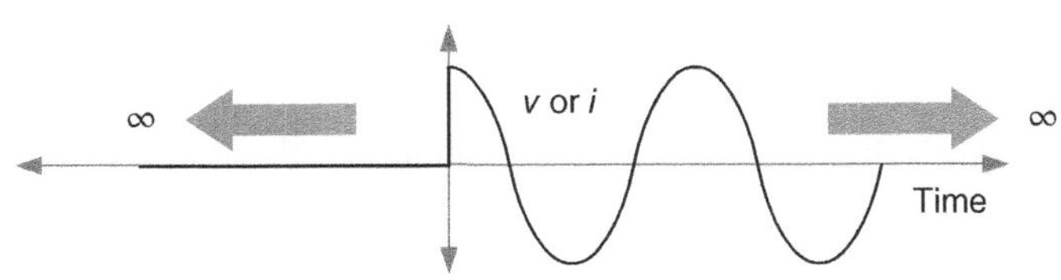

Figure 16.11 Stepped sine wave.

16.3.4 Rotating Phasor $e^{j\omega t}$

$$A_m e^{j\phi} \quad (16.36)$$

The phasor in Equation 16.36 is related to the exponential function in Equation 16.37.

$$A_m e^{j\theta} e^{j\omega t} = A_m \cos(\omega t + \phi) + jA_m \sin(\omega t + \phi) \quad (16.37)$$

$$A_m e^{j\omega t} = A_m \cos(\omega t) + jA_m \sin(\omega t) \quad (16.38)$$

Equation 16.38 shows the time-varying portion of Equation 16.37. The exponential on the left-hand side of Equation 16.38 can be viewed as phasor rotating, as a function of time, counterclockwise in the complex plane. A_m denotes the magnitude of the rotating phasor, shown in Figure 16.12.

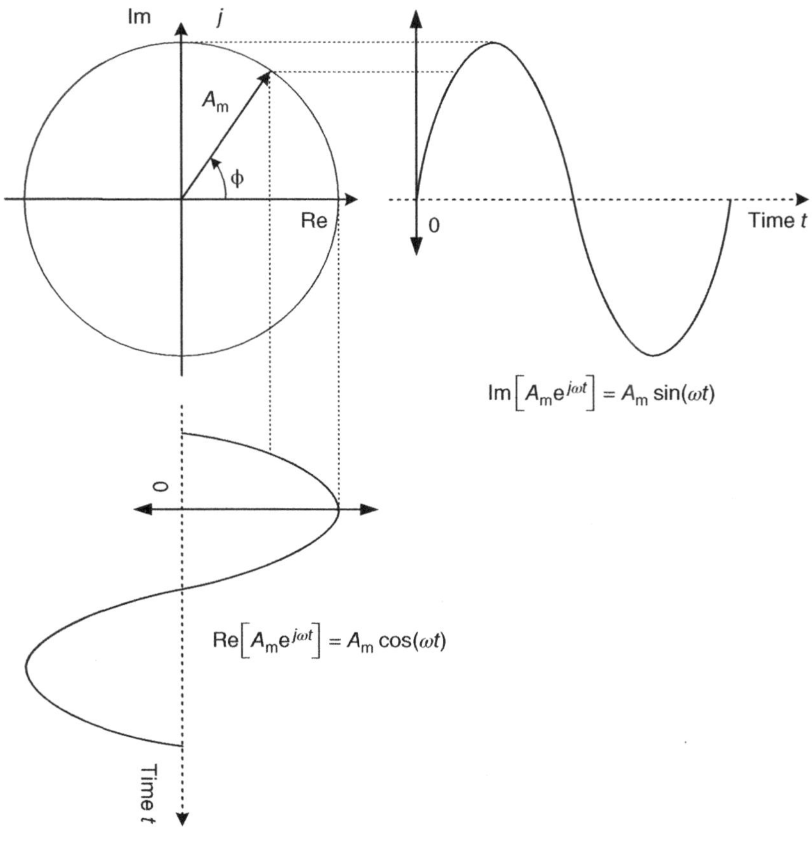

Figure 16.12 Projections of sine on the imaginary axis and cosine on the real axis.

In Figure 16.12, the projection of the phasor rotation circle onto the imaginary axis is the time domain sine function in Equation 16.39.

$$\text{Im}[A_m e^{j\omega t}] = A_m \sin(\omega t) \quad (16.39)$$

The projection of the rotating phasor onto the real axis is the cosine function in Equation 16.40.

$$\text{Re}[A_m e^{j\omega t}] = A_m \cos(\omega t) \quad (16.40)$$

Notice that the two are offset in phase as indicated by the time 0. Both sine and cosine are real functions. The imaginary operator *j* does not appear in the projected sine wave function.

16.3.5 Sinusoidal Phase and Phasor Diagrams

The relationship between sine and cosine functions, shown in Equation 16.41, indicates a phase shift, between sine and cosine, of 90° or pi over 2 radians, illustrated in Figure 16.13.

$$\sin\left(\theta + \frac{\pi}{2}\right) = cos\theta \tag{16.41}$$

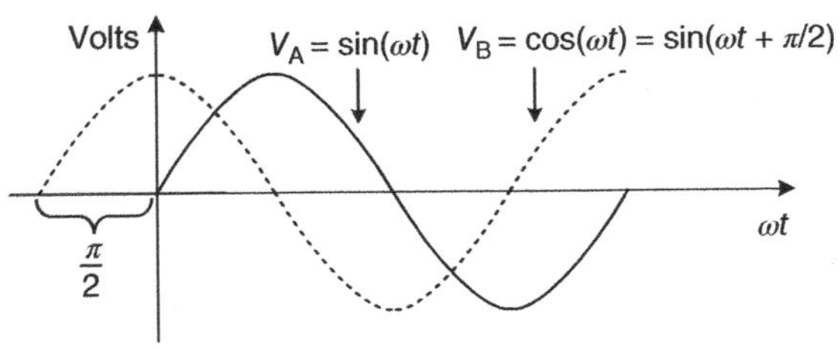

Figure 16.13 Graph of sine and cosine versus time showing 90°phase shift.

In Figure 16.13, the dashed line for V_B cosine equals 1 at time t equals 0. The V_A sine function equals 0 at time *t* equals 0. The V_B cosine function appears to start before the sine function because the cosine zero value is to the left of the sine zero value. The V_B cosine function leads the V_A sine function by 90° or pi over 2 radians.

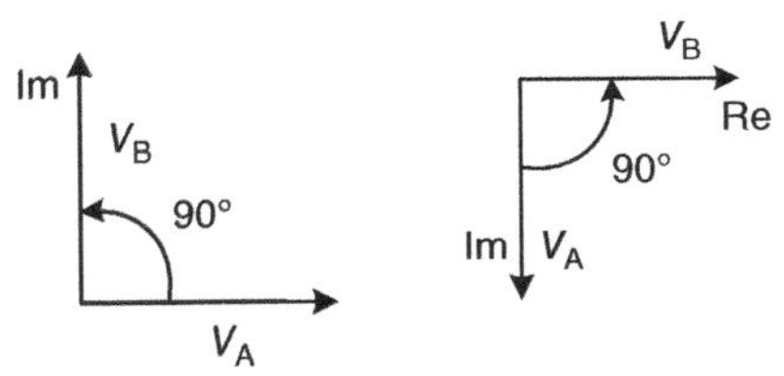

Figure 16.14 Two graphs of example phasor V_B leading phasor V_A by 90°.

Phase lead and lag can be expressed with phasor diagrams in the complex plane as shown in Figure 16.14. The V_B phasor, representing the cosine function from Figure 16.13, leads the V_A phasor representing the sine function. The V_A phasor lags the V_B phasor. Figure 16.14 shows two different possible orientations of the phasors, depending upon the specific phase.

16.3.6 Phasor Properties Summary

Phasors are complex numbers that represent a time-varying sinusoidal signal.

Phasors simplify sinusoidally time-varying signals and facilitate circuit analysis, by removing the time-varying portion of the sinusoidal signal.

Phasors can be used to analyze linear sinusoidal steady-state circuits, where all the signals in the circuit have the same frequency.

Phasors do not contain frequency information.

Phasors can be added vectorially, in a head-to-tail fashion. But phasors themselves are not vectors in the context of three-dimensional space.

Phasor addition adds the magnitudes together and the phases together. The polar form of complex numbers, magnitude and phase, is convenient for this.

Phasors obey Kirchhoff's current law (KCL) and Kirchhoff's voltage law (KVL). When writing KCL and KVL equations, the magnitudes of each phasor are added together and the phases of each phasor are added together.

Impedances and admittances are complex numbers, with a magnitude and a phase. Impedances can be used with Ohm's law to for circuit analysis in the phasor notation.

Current phasors and voltage phasors, plotted together on the same complex plane in a phasor diagram, share a common angle but different magnitude scales, one for current magnitude and another for voltage magnitude.

Phasor rotation counterclockwise of one revolution represents 1 period (2π radians) of the signal. In the complex plane, counterclockwise represents the direction of positive phase rotation.

Linear sinusoidal functions, in the sinusoidal steady state, can be added subtracted, multiplied, divide, integrated, and differentiated, without changing the frequency ω.

Index

Page numbers in italics refer to figures. Page numbers in bold refer to tables.

First and Second Order Circuits and Equations: Technical Background and Insights, First Edition. Robert O'Rourke.

d

Printed and bound by CPI Group (UK) Ltd, Croydon, CR0 4YY
17/09/2026
14964260-0001